Phoma: Diversity, Taxonomy, Bioactivities, and Nanotechnology

Mahendra Rai
Beata Zimowska • György János Kövics
Editors

Phoma: Diversity, Taxonomy, Bioactivities, and Nanotechnology

Springer

Editors
Mahendra Rai
Department of Biotechnology
SGB Amravati University
Amravati, Maharashtra, India

Visiting Professor
Nicolaus Copernicus University
Torun, Poland

György János Kövics
Plant Protection Institute
University of Debrecen
Debrecen, Hungary

Beata Zimowska
Department of Plant Protection
Institute of Plant Pathology
and Mycology
University of Life Sciences in Lublin
Lublin, Lubelskie, Poland

ISBN 978-3-030-81217-1 ISBN 978-3-030-81218-8 (eBook)
https://doi.org/10.1007/978-3-030-81218-8

This Springer imprint is published by the registered company Springer Nature Switzerland AG
The registered company address is: Gewerbestrasse 11, 6330 Cham, Switzerland

"Inter Silvas Fungorum Quaerere Verum" (In English: "I was seeking the Truth about the fungi)

Dr. G.H. Boerema

A great Phoma researcher of 20th Century

Dedicated to Gerhard H. Boerema (1925–2008) who made remarkable contributions to our understanding of the taxonomy of phomoid (Phoma-like) fungi

Foreword

Phoma is an important group of fungi in terms of their impacts on plants, and it is a genus of fungi that is dear to my heart. My very first taxonomic papers were on *Phoma* and other pycnidial fungi (Coelomycetes). I have many fond memories of the long and very late evenings spent studying the features of these fungi and the excitement of figuring out how they developed their structures. I cut my first "scientific teeth" on these fungi so to speak. *Phoma* is a unique genus in containing many very similar species structurally. It is easy to recognize the genus but getting to species is difficult. *Phoma* may be considered one of the most challenging groups of fungi because of the identification issues. Molecular systematics is now used to identify species in many groups of fungi, including *Phoma*-like organisms. However, these taxonomic treatments generally do not cover the groups comprehensively, including ecology, pathology and applications. And few books focused on specific taxonomic groups of fungi address comprehensive aspects of the fungi. But, with this book on genus *Phoma*, the international group of editors have done just that. This book covers essential areas regarding *Phoma*. Chapter 1 provides an overview of the genus, while the remaining chapters cover areas of taxonomy (Chap. 2), diversity, pathology and host ranges (Chaps. 3–9), secondary metabolites and bioactivities (Chaps. 10–16) and evolutionary biology and applications in nanotechnology (Chaps. 17 and 18).

This book is an important book for anyone that is interested in fungi and their impacts and applications. It will help readers to understand the importance of *Phoma* and related species. Anyone planning research in the Coelomycetes and *Phoma*-like fungi should have a copy for reference and perspective. I will maintain a copy of this book in the lab for reference by me and my students.

Department of Plant Biology
Rutgers University,
New Brunswick, NJ, USA

James White

Preface

Phoma is widely distributed fungus belonging to division *Ascomycota*, class *Dothidiomycetes*, order *Pleosporales* and family *Didymellaceae*. It occurs in soil, plants (as parasites, endophytes and saprophytes), humans, animals and even in water and air. The fungus is well known as plant pathogen causing several diseases of plants. It is often considered as one of the largest genus with more than 3000 intrageneric taxa. The taxonomy of this genus has long been a matter of dispute.

There are some species of *Phoma* that inhabits as endophytes in medicinal plants. These *Phoma* species have already demonstrated beneficial applications in medical and agriculture field. Some *Phoma* species including *Phoma macrostoma* and *P. herbarum* are potentially used as herbicides and also for management of plant diseases, which can be used in eco-friendly and sustainable agriculture. These species have exhibited antimicrobial activities particularly antibiotic properties. In the era of antimicrobial resistance, there is a greater need of development of antimicrobial agents and in this context, a wide range of species of *Phoma* is very useful. Moreover, some species have the potential for growth promotion of the plants.

This book has been divided into four distinct parts: Part I introduces the genus *Phoma,* provides an overview of the genus and critically discusses the disputed taxonomy; Part II discusses diseases caused by different *Phoma* spp. in plants and human, diversity and host ranges, and also the strategies for its management; Part III incorporates secondary metabolites secreted by various *Phoma* species and also discusses their role as bioherbicides, biocontrol of plant diseases and potent plant growth promoters; Part IV-the last part includes an interesting application of *Phoma* spp. for synthesis of silver nanoparticles; in addition, a chapter is devoted to the fossil records of *Phoma*. Overall, the book has covered almost all the important areas concerning the genus *Phoma*.

This book will be useful for the postgraduate students and researchers working in the areas of fungal biology, plant biology, chemical biology, microbiology, plant pathology, biotechnology, and industries involved in production of biocontrol agents and their formulations.

We take the opportunity to thank all the contributors for their generous cooperation and efforts in offering up-to-date chapters. Further, we express our sincere thanks to the publisher and the authors of the chapters, whose research work has been cited in this book. We are also thankful to Carolyn Spence, Senior Publishing Editor, Springer New York, and her team for their cooperation and patience in the publication of this book Finally, MR is thankful to the Polish National Agency for Academic Exchange (NAWA) for financial support (Project No. PPN/ULM/2019/1/00117/A/DRAFT/00001) to visit the Department of Microbiology, Nicolaus Copernicus University, Torun, Poland.

Amravati, India — Mahendra Rai
Lublin, Poland — Beata Zimowska
Debrecen, Hungary — György János Kövics

Contents

About the Editors

Mahendra Rai is a Senior Professor and UGC-Basic Science Research Faculty Fellow at the Department of Biotechnology, Sant Gadge Baba Amravati University at Amravati, Maharashtra, India. He was a Visiting Scientist at the following institutions and universities: Department of Bioenergetics, University of Geneva, Switzerland and at the Department of Plant Protection of Debrecen University, Debrecen, Hungary. Institute of Chemistry, University of Campinas, Brazil under Indo-Brazil Research Programme (DST-CNPq collaboration). Professor in Nicolaus Copernicus University, Torun, Poland, State University of Campinas, Brazil, Ostrava, Czech Republic and in National University of Rosario, Argentina. His areas of expertise include microbial biotechnology and nanobiotechnology. Moreover, he studied different aspects of *Phoma* extensively. His present research interest includes application of nanobiotechnology in medicine and agriculture with particular reference to the development of nanoantimicrobials. He has published more than 400 research papers in national and international journals. In addition, he has edited/authored more than 60 books and 6 patents.

Beata Zimowska is Associate Professor, at the Department of Plant Pathology, Subdepartment Phytopathology and Mycology, Faculty of Horticulture and LandScape, University of Life Sciences in Lublin, Poland. She has published more than 60 research papers, 30 popular articles in Polish and abroad. She is a member of several scientific societies. She has won several awards: grade I team from HM Rector the Academy of Agriculture in Lublin—2000, grade II team from HM Rector the University of Life Sciences in Lublin—2009, individual grade I from HM Rector the University of Life Sciences for scientific activity in the years 2010–2014, individual grade II from HM Rector the University of Life Sciences for scientific activity in the years 2014–2017. She was co-organizer and member of Scientific Board 1st Eurasian Mycological Congress 3–5.07.2016, Manisa, Turkey and XIII Congress of Ecology and Environment with International

Participation 12–15.09.2017, Edirne, Turkey. Since 2014 she cooperates with Bulgarian Institute of Genetics in Sofia, since 2015 with Department of Agriculture, University of Naples Federico II, Portici and Institute for Sustainable Plant Protection (IPSP)/National Research Council (CNR) Portici, Italy, since 2015 with Department of Biology Trakya University, Edirne, Turkey. Presently, her research activities are aimed at: *Phoma* sensu lato group of fungi, endophytic fungi, fungal symbiont and their relationships with gall midges, mutualistic interactions between the gall causing species of dipterans from Asphondylia genus (Diptera: Cecidomyiidae) and on fungi developing in the inner wall of galls, aetiology and epidemiology of infectious diseases of MAPs (medicinal and aromatic plants), with special regard to new and emerging infectious diseases (EIDs).

György János Kövics is professor emeritus of plant pathology at the Institute of Plant Protection, University of Debrecen, Hungary. After some years of research experience, his university career has been started in 1989 and during the last three decades he taught plant pathology. He has published more than 310 scientific publications, including 28 books and book chapters. He started the Scientific Trans-Tisza National Conference of Plant Protection in 1996, which includes International Plant Protection Symposia (IPPS). He was visiting professor at the University of Reading, UK (1993), Wageningen University, the Netherlands (1994), Mysore University and Amravati University, India (2004). Prof. Kövics serves as a referee for about 15 international journals and is a member of editorial boards of numerous national and international journals. He was awarded"*Pro re rustica promovenda*" medal for the development of agriculture (1997), "*Pro Universitate et Scientia*" medal of the World Council of Hungarian Professors (2005), "*György Linhart medal*" recognition of plant pathologists (2006) and *Eminent Plant Doctor* of Hungarian Plant Doctors' Chamber (2014). Prof. Kövics is engaged in mycotaxonomy and research on the biology of recently emerged and spreading plant pathogens.

Contributors

Ghulam Abbas Department of Biological Sciences and Chemistry, College of Arts and Sciences, University of Nizwa, Nizwa, Sultanate of Oman

Krishnendu Acharya Centre of Advanced Study, Department of Botany, University of Calcutta, Kolkata, India

Iftikhar Ali Shandong Key Laboratory of TCM Quality Control Technology, Shandong Analysis and Test Center, Jinan, P.R. China

Department of Chemistry, Karakoram International University, Gilgit, Pakistan

Maroof Ali College of life Sciences, Anhui Normal University, Wuhu, China

Omar Atik Department of Plant Protection, Aleppo Center, General Commission for Scientific Agricultural Research, Aleppo, Syria

Graciela Dolores Avila-Quezada Facultad de Ciencias Agrotecnológicas, Universidad Autónoma de Chihuahua, Chihuahua, México

Karen L. Bailey AAFC Saskatoon Research and Development Centre, Saskatoon, SK, Canada

Meghma Bera Centre of Advanced Study, Department of Botany, University of Calcutta, Kolkata, India

Department of Botany, Vidyanagar College, Vidyanagar, West Bengal, India

Subir Bera Centre of Advanced Study, Department of Botany, University of Calcutta, Kolkata, India

Santa Olga Maria Cacciola Department of Agriculture, Food and Environment (Di3A), University of Catania, Catania, Italy

Ravishankar Chauhan National Centre for Natural Resources (NCNR), Pandit Ravishankar Shukla University, Raipur, India

Tássia C. Confortin Department of Chemical Engineering, Federal University of Santa Maria, Santa Maria, Brazil

Elizbit Department Materials Engineering, National University of Sciences and Technology (NUST), Islamabad, Pakistan

Mohsen Mohamed Elsharkawy Agricultural Botany Department, Faculty of Agriculture, Kafrelsheikh University, Kafr Elsheikh, Egypt

Aniket Gade Nanobiotechnology Laboratory, Department of Biotechnology, Sant Gadge Baba Amravati University Amravati, Amravati, Maharashtra, India

Ivan R. Green Department of Chemistry and Polymer Science, University of Stellenbosch, Stellenbosch, South Africa

Md Motaher Hossain Department of Plant Pathology, Bangabandhu Sheikh Mujibur Rahman Agricultural University, Gazipur, Bangladesh

Michelle Hubbard Agriculture and Agri-Food Canada (AAFC) Swift Current Research and Development Centre, Swift Current, SK, Canada

Hidayat Hussain Department of Bioorganic Chemistry, Leibniz Institute of Plant Biochemistry, Halle (Saale), Germany

Russell K. Hynes AAFC Saskatoon Research and Development Centre, Saskatoon, SK, Canada

László Irinyi Molecular Medical Mycology, Medicine, Westmead Clinical School, The Westmead Institute for Medical Research, University of Sydney, Westmead, Australia

Mahasin Ali Khan Department of Botany, Sidho-Kanho-Birsha University, Purulia, India

Vladimir S. Kiyan Research Platform of Agricultural Biotechnology, Saken Seifullin Kazakh Agrotechnical University, Nur-Sultan, Kazakhstan

György J. Kövics Plant Protection Institute, University of Debrecen, Debrecen, Hungary

Elena V. Kukhar Research Platform of Agricultural Biotechnology, Saken Seifullin Kazakh Agrotechnical University, Nur-Sultan, Kazakhstan

Luciana Luft Department of Chemical Engineering, Federal University of Santa Maria, Santa Maria, Brazil

Pramod Kumar Mahish Department of Biotechnology, Government Digvijay (Autonomous) P.G. College, Rajnandgaon, India

Nilufar Z. Mamadalieva Institute of the Chemistry of Plant Substances of the Academy Sciences of Uzbekistan, Tashkent, Uzbekistan

Marcio A. Mazutti Department of Chemical Engineering, Federal University of Santa Maria (UFSM), Santa Maria, Brazil

Reza Mostowfizadeh-Ghalamfarsa Department of Plant Protection, School of Agriculture, Shiraz University, Shiraz, Iran

José R. C. Neto Laboratory of Agroindustrial Processes Engineering (LAPE), Federal University of Santa Maria (UFSM), Cachoeira do Sul, RS, Brazil

Mahendra Rai Department of Biotechnology, SGB Amravati University, Amravati, Maharashtra, India

Department of Microbiology, Nicolaus Copernicus University, Torun, Poland

Fatemeh Salmaninezhad Department of Plant Protection, School of Agriculture, Shiraz University, Shiraz, Iran

Maicon S. N. dos Santos Laboratory of Agroindustrial Processes Engineering (LAPE), Federal University of Santa Maria (UFSM), Cachoeira do Sul, Brazil

Sudhir Shende Nanobiotechnology Laboratory, Department of Biotechnology, Sant Gadge Baba Amravati University Amravati, Amravati, Maharashtra, India

Shweta Singh Department of Biotechnology, Government Digvijay (Autonomous) P.G. College, Rajnandgaon, India

Farjana Sultana College of Agricultural Sciences, International University of Business Agriculture and Technology, Dhaka, Bangladesh

Izelmar Todero Department of Chemical Engineering, Federal University of Santa Maria, Santa Maria, Brazil

Marcus V. Tres Laboratory of Agroindustrial Processes Engineering (LAPE), Federal University of Santa Maria (UFSM), Cachoeira do Sul, Brazil

Giovani L. Zabot Laboratory of Agroindustrial Processes Engineering (LAPE), Federal University of Santa Maria (UFSM), Cachoeira do Sul, Brazil

Gohar Zaman Department of Chemistry, Quaid-i-Azam University, Islamabad, Pakistan

Beata Zimowska Department of Plant Protection, Institute of Plant Pathology and Mycology, University of Life Sciences in –Lublin, Lublin, Poland

Part I

General, Identity and Taxonomy

The Genus *Phoma*: What We Know and What We Need to Know?

1

Mahendra Rai, Beata Zimowska, and György János Kövics

1.1 Introduction

In the current taxonomy, the genus *Phoma* has been classified as the phylum *Ascomycota*, class *Dothideomycetes*, order *Pleosporales* and family *Didymellaceae* (Hou et al. 2020). The genus *Phoma* occurs ubiquitously as phytopathogens, saprophytes, endophytes and aquatic, air and also human and animal pathogens (Rai 1989; Aveskamp et al. 2008; de Gruyter et al. 2009; Kövics et al. 2013; Kukhar et al. 2020). Majority of *Phoma* species parasitize on plants (mainly leaves and stems) including medicinal and aromatic plants which is responsible for huge loss (Zimowska 2007a, b, 2008, 2012). It has been reported globally on different hosts/substrates. Interestingly, Bera et al. (2018) reported *Phomites* (morphologically comparable to typical *Phoma* sp.) from the cuticular surface of fossil leaf fragments recovered from Siwalik sediments (middle Miocene to early Pleistocene) of eastern Himalaya.

The history of genus *Phoma* dates back to 1880 when the first, very simple, based mainly on morphology of asexual spores and host-orientation, taxonomical description of fungi imperfecti has been introduced by Italian mycologist Pier Andrea Saccardo (1880, 1884). The generic name was later emended by Boerema and Bollen in 1975. In the Saccardoan system, the genus name *Phoma* referred to filamentous fungi which were capable of forming pycnidial bodies with hyaline, aseptate conidia that could occur on plant stems. Fungi with the same morphological features, but associated with leaf spots, were classified in the genus *Phyllosticta* (Van der Aa and Vanev 2002; Boerema et al. 2004). This very simple, and in the end, very confusing taxonomical system was valid until the fourth decade of the twentieth century. Wollenweber and Hochapfel (1936) were the first who pointed to the advantages of using growth characteristics on artificial media in *Phoma* identification. Dennis (1946) passed a remark on the use of basic morphology, physiology and biochemical tests in vitro. In a study on *Hyphomycetes*, Hughes (1953) proposed conidiogenesis as an important taxonomic criterion, a feature that was later used for all anamorphic fungi including species of *Phoma* (Sutton 1964). Moreover, new criteria for morphological differentiation were incorporated to differentiate *Phoma* species from other

M. Rai
Department of Biotechnology, SGB Amravati University, Amravati, Maharashtra, India
e-mail: mahendra.rai@v.umk.pl

B. Zimowska (✉)
Department of Plant Protection, Institute of Plant Pathology and Mycology, University of Life Sciences in Lublin, Lublin, Lubelskie, Poland
e-mail: beata.zimowska@up.lublin.pl

G. J. Kövics
Plant Protection Institute, Debrecen University, Debrecen, Hungary

M. Rai et al. (eds.), *Phoma: Diversity, Taxonomy, Bioactivities, and Nanotechnology*,
https://doi.org/10.1007/978-3-030-81218-8_1

Coelomycetes such as *Ascochyta* (Boerema and Bollen 1975), *Phyllosticta* (Van der Aa et al. 1990), *Coniothyrium* and *Paraconiothyrium* (Verkley et al. 2004).

The most remarkable progress in the taxonomy of *Phoma* has been done by Dutch researchers during 1992–2003. New criteria for morphological differentiation were incorporated, to distinguish *Phoma* species from other *Coelomycetes*, which resulted in presenting genus concept including numerous species description using also in vitro studies. Intensive work carried out by Dutch mycologists who had studied the morphology of those fungi in constant conditions of artificial cultures since 1992 led to their division into nine sections (Boerema 1997). The results of 40 years of taxonomic research based on the properties of *Phoma* species were presented in articles concerning their identification, and as a consequence, the number of species was reduced to 223 (Boerema et al. 2004). The highly phylogenetic heterogeneity of the species in the *Phoma* sections negated the existing division of the genus into sections and caused the necessity to reclassify (De Gruyter et al. 2009, 2010, 2012; Aveskamp et al. 2010). That conventional taxonomical system based on morphological features in in vitro conditions is still valid but insufficient because the absence of outstanding differences in conidial characters and moreover micromorphological features depend on strains, place of origin and host plants. Consequently, this results in discrepancies and misidentifications. Increasingly, in order to achieve the correct identification of *Phoma sensu lato*, secondary metabolites, the protein profile, chemotaxonomy and nucleotide sequences using modern molecular techniques are being examined (Aveskamp et al. 2008, 2010; Rai et al. 2014; Keirnan et al. 2021). Although the internal transcribed spacers (ITS) have been allied as the standard DNA barcode for fungi (Schoch et al. 2012), it turned out to not be sufficient to distinguish all *Phoma* species. Internal transcribed spacer regions 1 and 2 and 5.8S nrDNA (ITS), partial large subunit 28S nrDNA region (LSU) and β-tubulin gene (TUB) (Woudenberg et al. 2009; Aveskamp et al. 2009a, 2010) in combination with morphological characteristics have been routinely used to resolve phylogenetic relationships in *Phoma*. RNA polymerase II gene (RPB2) has also been used for several species (Chilvers et al. 2009), while actin (ACT) (Aveskamp et al. 2009b; de Gruyter et al. 2012; Zimowska et al. 2018), chitin synthase 1 (CHS-1) (de Gruyter et al. 2012) region and the translation elongation factor gene (TEF1) (Irinyi et al. 2009) have been applied for *Boeremia exigua* (Desm.) Aveskamp, Gruyter & Verkley, *Boeremia strasseri* (Moesz) Gruyter et Verkley and *Leptosphaeria doliolum* (Pers.) Ces. & De Not. complexes and intraspecific taxa due to their high variations below the species level. Results of phylogenetic studies yielded, *Didymellaceae* family, have been defined.

The *Didymellaceae* was established to accommodate *Phoma*, *Ascochyta*, *Didymella* and other allied *Phoma*-like genera (de Gruyter et al. 2009; Aveskamp et al. 2010; Keirnan et al. 2021). To date, more than 5.400 species from 36 genera have been recorded, including recently established genera such as *Dimorphoma* and *Macroascochyta* (Hou et al. 2020). Species of *Didymellaceae* are cosmopolitan and occupy a broad range of environments. Many species are plant pathogens that cause leaf and stem lesions, often with a broad host range (Aveskamp et al. 2009a, b, 2010; Chen et al. 2015b). This taxon comprises terrestrial species which are well-known phytopathogens, including quarantine species that cause significant losses in commercial crops. Only in some cases, the pathogenic nature of *Phoma* is regarded as helpful, by means of biocontrol agent of weeds and plant pathogens. These fungi produce structurally diverse secondary metabolites including the highly selective as well as less specific. Within *Phoma* spp. there are also endophytes, considered as a rich source of biologically active metabolites. Furthermore, *Phoma* complex comprises species and varieties that are recognized as saprophytes and pathogens of humans and other vertebrates such as cattle and fish. *Phoma* species not only occur on land but are also frequently found in the marine environment from which a number of facultative marine species have been described which yielded new natural products.

Multilocus phylogenetics and a polyphasic approach to classify species have helped to revise taxa and refine systematic relationships in the *Didymellaceae* (Aveskamp et al. 2009a, b, 2010; de Gruyter et al. 2009, 2012; Chen et al. 2015a; Hou et al. 2020; Keirnan et al. 2021). In fact, the easier access to DNA sequencing and online databases is ongoingly supporting the distinction of novel species; however, a large number of unknown *Didymellaceae* taxons still exist in nature, especially from previously ignored ecosystems and diverse substrates. That's why to better understand their taxonomical status, global community-wide efforts to investigate species diversity of *Didymellaceae* are needed, by use of comprehensive morphological, phylogenetic and chemotaxonomy tools.

1.2 *Phoma* as a Rich Source of Secondary Metabolites with a Wide Range of Bioactivities

Different species of *Phoma* are the storehouse of secondary metabolites that possess a wide range of bioactivities (Rai et al. 2009, 2018). Endophytic *Phoma* species inhabiting plants/animals are capable to produce bioactive compounds (Hussain et al. 2015; Nalli et al. 2019; Li et al. 2020; Rai et al. 2020).

Deliberating the importance of bioactivities, *Phoma* can be used as antimicrobial, anti-inflammatory, bio-herbicidal, cytotoxic, antidiabetic, anti-Alzheimer, antiangiogenic and antioxidant agent. *Phoma* species in general and endophytic species in particular secretes several secondary metabolites which have demonstrated antibacterial (Arora et al. 2016; Huang et al. 2017; Chen et al. 2019), antifungal (Aoyagi et al. 2007; Hussain et al. 2014) and antiviral (Liu et al. 2019; Peng et al. 2020) activities. Chen et al. (2019) isolated five new polyketides from *Phoma* species SYSU-SK-7 living as an endophyte in a marine plant *Kandelia candel* mangrove of China. This endophyte demonstrated significant antimicrobial activity against *Pseudomonas aeruginosa* Gram-positive methicillin-resistant *Staphylococcus aureus* and *Candida albicans*. The isolated compounds also showed α-glucosidase inhibitory activity (Fig. 1.1).

More recently, Peng et al. (2020) isolated ergocytochalasin A from *Phoma multirostrata* inhabiting as an endophyte in *Parasenecio albus* occurring in China. Ergocytochalasin A represents a novel kind of merocytochalasan and has garnered attention in current pharmacological research. This compound showed significant activity against influenza A virus (H1N1), human dengue virus type 3 (DV3), and respiratory syncytial virus (RSV).

α-Pyrones contain a varied unit of fungal polyketide with a wide range of antimicrobial activities (Bräse et al. 2009). In this context, in a study, Sang et al. (2017) reported four a-pyrone derivatives phomones (Fig. 1.2) isolated from the endophytic fungus *Phoma* sp. YN02-P-3, which demonstrated cytotoxic activities against human prostatic carcinoma PC-3, human leukaemia HL-60, and human leukaemia HL-60 cell lines.

The bioherbicidal potential of the *Phoma* species has been much exploited; for example, *P. macrostoma* has been extensively worked out against broadleaf weeds such as dandelion (*Taraxacum officinale*), clover (*Trifolium repens* (L.)) and chickweed (*Stellaria media* (L.) Vill.) (Graupner et al. 2003; Bailey and Derby 2010, 2012; Bailey et al. 2011). More importantly, this fungus is safer to agricultural crops. *P. macrostoma* strain 94-44B was isolated from Canada thistle (*Cirsium arvense* (L.) Scop.) growing in Saskatchewan with symptoms of chlorotic leaves. The characteristic feature of this fungus is that it enters into the vascular system of the target weeds. The bioactivity of this fungus is due to secretion of macrocidins which damages the photosystem; as a result chlorosis takes place. The fungus acts in a natural way showing less toxicity. Professor Karen Bailey and her group have been studying *P. macrostoma* as a bioherbicide for the last two decades. The fungus has been registered and approved by the Canadian Pest Management Regulatory Agency (PMRA) as a bioherbicide. Unfortunately, there are various challenges in commercialization that hinders its applicability owing to the higher cost involved in

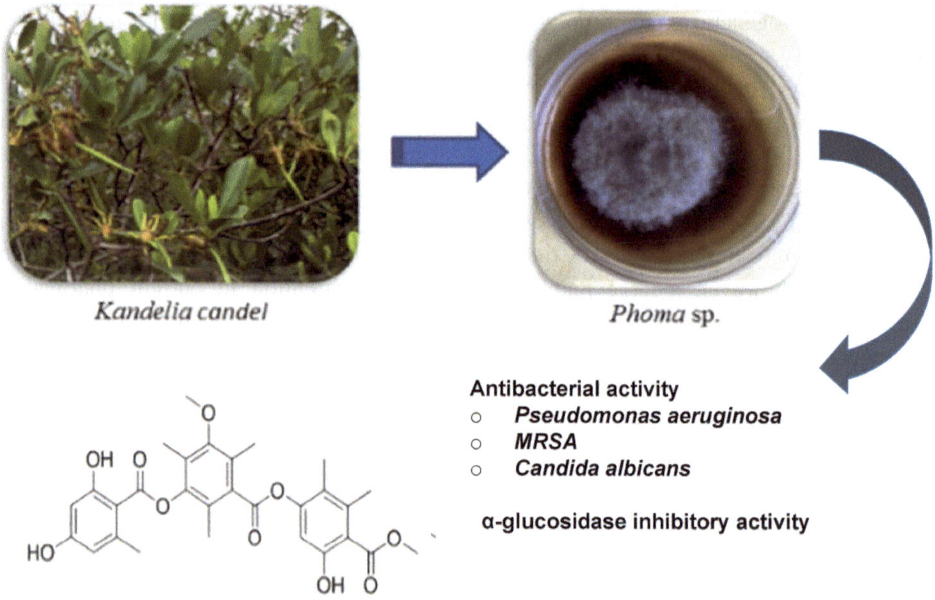

Fig. 1.1 An endophytic *Phoma* sp. SYSU-SK-7 inhabiting *Kandelia candel* mangrove plant. The isolated bioactive polyketides showed strong antibacterial and anti-α-glucosidase activity. Adapted with permission from Chen et al. (2019)

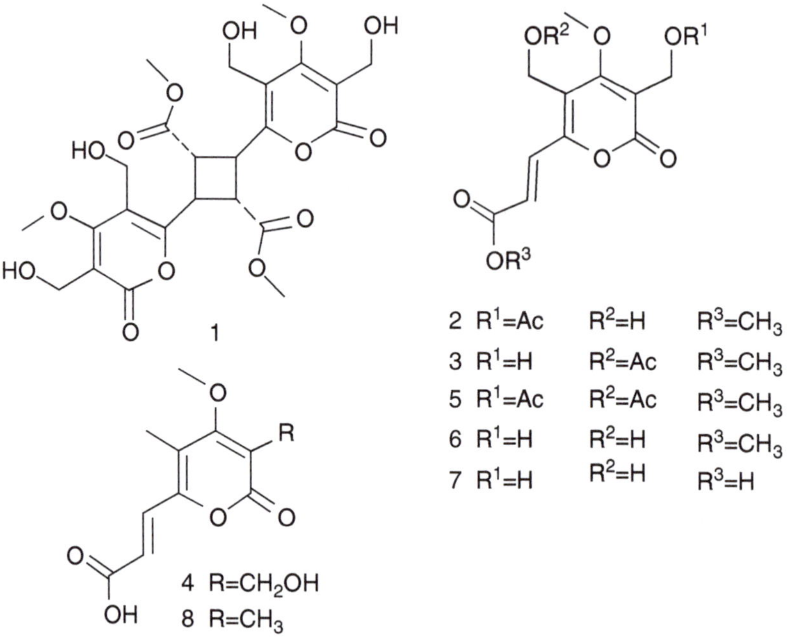

Fig. 1.2 *α*-Pyrone derivatives phomones demonstrated cytotoxic activity. Reproduced with permission from Sang et al. (2017)

the process and less production of bioactive macrocidins.

Recently, Chaves Neto et al. (2020) isolated *Phoma dimorpha* from *Senna obtusifolia* and applied fermented broth on *S. obtusifolia* as this weed deteriorates the grasslands of Brazil. The authors evaluated the effect of fermented broth at pre-emergence, detached leaves and post-

Fig. 1.3 Phytotoxic effects on leaves of *Senna obtusifolia* 3, 6 and 9 days after the application (DAP) of *Phoma dimorpha* fermented broth (T1, control treatment without the fermented broth; T2, treatment with the fermented broth). Reproduced from Chaves Neto et al. (2020) under creative common licence

emergence period. During the pre-emergence period, 100% inhibition was recorded; the necrosis and death of detached leaves were reported on the ninth day (Fig. 1.3), while in the post-emergence stage, the moderate efficacy was recorded.

1.3 *Phoma* as Agents for Biocontrol of Diseases and Growth Promotion of Plants

There are various non-pathogenic species of *Phoma* that can be used for the management of plant diseases (Hossain et al. 2008). This approach is a novel step towards eco-friendly and sustainable agriculture. These *Phoma* species are reported to suppress the soil-borne as well as air-borne diseases caused by phytopathogens. Among the *Phoma*, some species that are responsible for the control of diseases include *P. betae* which could suppress the growth of *Fusarium culmorum* in wheat (Diamond and Cooke 2003). Similarly, in another study, *P. glomerata* remarkably inhibited the growth of *F. graminearum* and *F. culmorum* in vitro and reduced the level of Fusarium head blight (FHB) infection in a detached wheat spikelet which provides evidence that *Phoma* species can be used as a biocontrol agent. *Phoma* can also be applied to suppress Fusarium wilt caused in different plants (Comby et al. 2017). In this context, a significant contribution has been made by Egyptian researchers (Elsharkawy and El-Khateeb 2019; Elsharkawy et al. 2020). They found that *Phoma* sp. if mixed in soil can control soil-borne pathogens and induce resistance in plants.

The *Phoma* species have already demonstrated antibacterial (Hossain et al. 2008; Sultana et al. 2009), antifungal (Sullivan and White 2000) and antiviral efficacy by inducing the resistance in host plants (Elsharkawy 2019).

1.4 *Phoma* as Plant Growth Promoter

There has been increasing attention of the researchers on endophytes from the plants for their bioactivities including growth promotion. In this context, there are reports of growth-promoting *Phoma* species from the plants (Shivanna et al. 2005; Hamayun et al. 2010; Zhou et al. 2014; Kedar et al., 2014), for example, Shivanna et al. (2005) reported remarkable growth enhancement in cucumber after treatment with *Phoma* species. In another experiment,

Kedar et al. (2014) isolated two endophytic *Phoma* species inhabiting *Tinospora cordifolia* and *Calotropis procera* and evaluated them on maize for growth promotion. The authors reported significant growth promotion of treated plants in terms of shoot and root length and biomass (fresh and dry) as compared to untreated control plants of maize. They also found that the inoculated plants were healthier as compared to noninoculated which suggests that *Phoma* species are excellent growth promoters and induce disease resistance. The maize plants were also grown on Murashige and Skoog medium and treated with extract of two *Phoma* species, which demonstrated proliferation in treated roots as compared to untreated plants. The proliferation of roots provides a better opportunity for the absorption of mineral nutrients. In addition, the chlorophyll a fluorescence of treated and untreated maize plants was analysed by using Handy Pea, and it was found that the performance index of inoculated plants was higher as compared to uninoculated ones. The plant growth promotion by different *Phoma* species is related to phytohormones, nutrient uptake or production of volatile compounds that are responsible for protection from disease. However, the exact mechanism of action needs to be elucidated.

1.5 *Phoma* Species as Novel Nanofactories

The increasing interest in the biogenic synthesis of nanoparticles is now the main platform for synthesis of eco-friendly and economically viable bionanoparticles which has given birth to a new branch of myconanotechnology (Rai et al. 2009). In this context, many fungi have been evaluated for the synthesis of nanoparticles. Among these, *Phoma* species also play a key role in synthesis of nanoparticles. In most cases, silver nanoparticles have been synthesized extracellularly. In a study, Birla et al. (2009) synthesized silver nanoparticles from *P. glomerata* and reported profound efficacy against pathogenic bacteria. Gade et al. (2013) screened 18 species of *Phoma* for the synthesis of silver nanoparticles and proposed three-step mechanism that includes activation, nucleation and reduction. The silver nanoparticles are capped by proteins. In another study, Rai et al. (2015) also used three species of *Phoma* including *P. capsulatum*, *P. putaminum* and *P. citri* for the synthesis of silver nanoparticles which exhibited significant antimicrobial activity against human pathogenic bacteria and fungi. Sheikhloo and Salouti (2012) synthesized gold nanoparticles using *P. macrostoma*. The authors reported the intracellular synthesis of gold nanoparticles. There is much scope to utilize the *Phoma* species for the synthesis of different nanoparticles.

1.6 Conclusions

To sum up, the genus *Phoma* is a complex genus with an enigmatic taxonomy. The trend in taxonomy was initially based on 'host-specificity' as a criterion for determining the identity of the species which was later changed to comparative cultural and morphological studies on different media. However, in the mid-1990s after the discovery of polymerase chain reaction, there has been a paradigm shift in taxonomy from the use of morphological to molecular markers which has made tremendous changes in nomenclature and evolutionary relationship of *Phoma* species. The species of *Phoma* have a huge potential of secreting different bioactive compounds which can be used in the biomedical and agricultural fields. On one hand, the bioactive compounds have already demonstrated antimicrobial, anti-inflammatory, cytotoxic and antidiabetic activities to enable them to use against human diseases, while on the other hand, the potential of these bioactive metabolites can be harnessed for the development of sustainable agriculture by developing bioherbicides, plant growth promoters and biocontrol agents. There is a pressing need to screen more and more *Phoma* species in general and endophytic species in particular to tap their potential for the development of chemical-free, bio-based agriculture. Another area of research that has attracted the attention of researchers is the synthesis of nanoparticles and their potential applications in different fields. So far much emphasis has been made on synthesis of silver nanoparticles by *Phoma* spp.; the fabrication of nanoparticles such as zinc, titania and silica among others may open up new vistas in the new and emerging field of myconanotechnology.

Acknowledgement MR is thankful to the Polish National Agency for Academic Exchange (NAWA) for financial support (Project No. PPN/ULM/2019/1/00117/A/DRAFT/00001) to visit the Department of Microbiology, Nicolaus Copernicus University, Toruń, Poland.

References

Aoyagi A, Yano T, Kozuma S, Takastu T (2007) Pleofungins, Novel Inositol Phosphorylceramide Synthase Inhibitors, from *Phoma* sp. SANK 13899. J Antibiot 60:143-152.

Arora P, Wani ZA, Nalli Y, Ali A, Hassan SR (2016) Antimicrobial potential of Thiodiketopiperazine derivatives produced by *Phoma* sp., an endophytes of *Glycyrrhiza glabra* Linn. Microb Ecol 72:802-812.

Aveskamp MM, de Gruyter J, Crous PW (2008) Biology and recent developments in the systematics of *Phoma*, a complex genus of major quarantine significance. Fungal Divers 31: 1–18.

Aveskamp MM, Verkley GJM, de Gruyter J, Murace MA, Perelló A, Woudenberg JHC, Groenewald JZ, Crous PW (2009a) DNA phylogeny reveals polyphyly of *Phoma* section *Peyronellaea* and multiple taxonomic novelties. Mycologia 101: 363–382. doi: https://doi.org/10.3852/08-199

Aveskamp MM, Woudenberg JHC, de Gruyter J, Turco E, Groenewald JZ, Crous PW (2009b) Development of taxon-specific sequence characterized amplified region (SCAR) markers based on actin sequences and DNA amplification fingerprinting (DAF): a case study in the *Phoma exigua* species complex. Mol Plant Pathol 10: 403–414. https://doi.org/10.1111/j.1364-3703.2009.00540.x

Aveskamp MM, de Gruyter J, Woudenberg JHC, Verkley GJM, Crous PW (2010) Highlights of the *Didymellaceae*: A polyphasic approach to characterise *Phoma* and related pleosporalean genera. Stud Mycol 65: 1–60.

Bailey KL, Derby J, inventors (2010) Her Majesty the Queen in right of Canada, as represented by the Minister of Agriculture and Agri-Food Canada (Saskatoon, CA). Continuation in Part. , assignee. Fungal isolates and biological control compositions for the control of weeds. United States Patent # 7,772,155. August 10, 2010.

Bailey KL, Derby J, inventors (2012) Her Majesty the Queen in right of Canada, as represented by the Minister of Agriculture and Agri-Food Canada (Saskatoon, CA), assignee. Fungal isolates and biological control compositions for the control of weeds. United States Patent # 8,211,830. July 3, 2012.

Bailey KL, Pitt WM, Falk S, Derby J (2011) The effects of *Phoma macrostoma* on nontarget plant and target weed species. Biol Control 58: 379–386. doi: https://doi.org/10.1016/j.biocontrol.2011.06.001

Bera, M., Khan, M.A., Bera, S. (2018). Two new species of *Phomites Fritel* from the phyllosphere of Siwalik. *J. Mycopathol. Res.* 56(1), 11–14.

Birla, S. S., Tiwari, V. V., Gade, A. K., Ingle, A. P., Yadav, A. P and Rai, M. (2009). Fabrication of silver nanoparticles by *Phoma glomerata* and its combined effect against *Escherichia coli*, *Pseudomonas aeruginosa* and *Staphylococcus aureus*. Letters in Applied Microbiology, 48: 173-179.

Boerema GH (1997) Contributions towards a monograph of *Phoma* (Coelomycetes) – V. Subdivision of the genus in sections. Mycotaxon 64: 321–333.

Boerema, G.H. and Bollen, G.J. (1975). Conidiogenesis and conidial septation as differentiating criteria between *Phoma* and *Ascochyta*. Persoonia 8: 111-144.

Boerema GH, de Gruyter J, Noordeloos ME, Hamers MEC (2004) Phoma *identification manual. Differentiation of specific and infraspecific taxa in culture*. CABI Publishing, United Kingdom.

Bräse S, Encinas A, Keck J, Nising CF (2009). Chemistry and biology of mycotoxins and related fungal metabolites. Chem Rev. 109(9):3903-90. doi: https://doi.org/10.1021/cr050001f. PMID: 19534495

Chaves Neto José Roberto, Mazutti, Marcio Antonio, Zabot, Giovani Leone, & Tres, Marcus Vinícius. (2020). Bioherbicidal action of *Phoma dimorpha* fermented broth on seeds and plants of *Senna obtusifolia* 1. *Pesquisa Agropecuária Tropical*, *50*, e56894. Epub April 03, 2020.doi: https://doi.org/10.1590/1983-40632020v5056894

Chen Q, Jiang JR, Zhang GZ, et al. (2015a) Resolving the *Phoma* enigma. Stud Mycol 82: 137–217.

Chen Q, Zhang K, Zhang GZ, et al. (2015b). A polyphasic approach to characterise two novel species of *Phoma* (*Didymellaceae*) from China.Phytotaxa197: 267–281.

Chen Y, Yang W, Zou G, Chen S, Pang J and She Z (2019). Bioactive polyketides from the mangrove endophytic fungi *Phoma* sp. SYSU-SK-7. Fitoterapia.139:104369. DOI: https://doi.org/10.1016/j.fitote.2019.104369.

Chilvers MI, Rogers JD, Dugan FM, Stewart JE, Chen WD, Peever L (2009) *Didymella pisi sp. nov.*, the teleomorph of *Ascochyta pisi*. Mycological Research 113: 391–400. doi: https://doi.org/10.1016/j.mycres.2008.11.017

Comby M, Gacoin M, Robineau M, Rabenoelina F, Ptas S, Dupont J, Profizi C, Baillieul F (2017) Screening of wheat endophytes as biological control agents against *Fusarium* head blight using two different in vitro tests. Microbiological Research 202: 11-20.

de Gruyter J, Aveskamp MM, Woudenberg JHC, Verkley GJM, Groenewald JZ, Crous PW (2009) Molecular phylogeny of *Phoma* and allied anamorph genera:

towards a reclassification of the *Phoma* complex. Mycological Research 113: 508–519.
de Gruyter J, Woudenberg JHC, Aveskamp MM, Verkley GJM, Groenewald JZ, Crous PW (2010) Systematic reappraisal of species in *Phoma* section *Paraphoma*, *Pyrenochaeta* and *Pleurophoma*. Mycologia 102: 1066–1081.
de Gruyter J, Woudenberg JHC, Aveskamp MM, Verkley GJM, Groenewald JZ, Crous PW (2012) Redisposition of *Phoma*-like anamorphs in *Pleosporales*. Stud Mycol 75: 1–36.
Dennis RWG (1946) Notes on some British fungi ascribed to *Phoma* and related genera. Transactions of the British Mycological Society 29: 11-42.
Diamond H, Cooke BM (2003) Preliminary studies on biological control of the *Fusarium* ear blight complex of wheat. Crop Protection 22:99-107.
Elsharkawy MM (2019). Induced systemic resistance against Cucumber mosaic virus by *Phoma* sp. GS8-2 stimulates transcription of pathogenesis-related genes in Arabidopsis. Pest Manag Sci. 75(3):859-866. doi: https://doi.org/10.1002/ps.5193. Epub 2018 Oct 15.
Elsharkawy MM, El-Khateeb NMM (2019) Antifungal activity and resistance induction against *Sclerotium cepivorum* by plant growth-promoting fungi in onion plants. Egyptian Journal of Biological Pest Control 29:68.
Elsharkawy MM, Suga H, Shimizu M (2020) Systemic resistance induced by *Phoma* sp. GS8-3 and nanosilica against *Cucumber mosaic virus*. Environ Sci Pollut Res 27: 19029–19037.
Gade A, Gaikwad S, Duran N, Rai M. (2013). Screening of different species of *Phoma* for the synthesis of silver nanoparticles. Biotechnol Appl Biochem. 60(5):482-93. doi: https://doi.org/10.1002/bab.1141.
Graupner PR, Carr A, Clancy E, Gilbert J, Bailey KL, Derby JA, Gerwick BC (2003) The macrocidins: novel cyclic tetramic acids with herbicidal activity produced by *Phoma macrostoma*. J Nat Prod 66:1558–61. https://doi.org/10.1021/np030193e
Hamayun M, Khan SA, Khan AL et al. (2010) Growth promotion of cucumber by pure cultures of gibberellin-producing *Phoma* sp. GAH7. World J Microbiol Biotechnol 26:889–894.
Hossain MM, Sultana F, Kubota M, Koyama H, Hyakumachi M (2008) Systemic resistance to bacterial leaf speck pathogen in *Arabidopsis thaliana* induced by the culture filtrate of a plant growth-promoting fungus (PGPF) *Phoma* sp. GS8-1. Journal of General Plant Pathology 74 (3), pp. 213-221
Hou LW, Groenewald Z, Pfenning LH, Yarden O, Crous PW, Cai L (2020) The phoma-like dilemma. Stud Mycol 96: 309-396.
Huang S, Xu J, Fenqi Li, Zhou D, Xu L, Chunyuan Li (2017) Identification and antifungal activity of metabolites from the mangrove fungus *Phoma* sp. Chem Nat Compd 53:1-4.
Hughes SJ (1953) Conidiophores, conidia and classification. Canadian Journal of Botany 31:577-599.
Hussain H, Kock I, Harrasi AA, Rawahi AA, Abbas G, Green IR, Shah A, Badshah A, Saleem M, Draeger S, Schulz B, Krohn K (2014) Antimicrobial chemical constituents from endophytic fungus *Phoma* sp. Asian Pac J Trop Med 7:699-702.
Hussain H, John M, Al-Harrasi A, Shah A, Hassan Z, Abbas G, Rana UA, Green IR, Schulz B, Krohn, K (2015) Phytochemical investigation and antimicrobial activity of an endophytic fungus *Phoma* sp. J King Saud Univ Sci 27:92-95
Irinyi L, Kövics GJ, Sándor E (2009) Taxonomical re-evaluation of *Phoma*-like soybean pathogenic fungi. Mycological Research 113: 249–260. doi: https://doi.org/10.1016/j.mycres.2008.11.003
Kedar, A., Rathod, D.P., Yadav, A., Agarkar, G., Rai, M.K. (2014). Endophytic *Phoma* sp. isolated from medicinal plants promote the growth of *Zea mays*. *Nusantara Bioscience*. 6:132-139
Keirnan EC, Tan YP, Laurence MH, Mertin AA, Liew EC, Summerell BA, Shivas RG (2021) Cryptic diversity found in *Didymellaceae* from Australian native legumes. MycoKeys, 78, 1. doi: https://doi.org/10.3897/mycokeys.78.60063
Kövics, G.J., Sándor, E., Rai, M. and Irinyi, L. (2013). *Phoma*-like fungi on soybeans. *Critical Reviews in Microbiology* 40(1):49-62. doi: https://doi.org/10.3109/1040841X.2012.755948
Kukhar E, Smagulova A, Kiyan V (2020) Biological properties of Phoma macrostoma related to non-dermatophyte onychomycosis. Medical Mycology Case Reports 27:55–58.
Li HT, Liu T, Yang R, Xie F, Yang Z, Yang Y, Zhou H, Ding ZT (2020) Phomretones A–F, C 12 polyketides from the co-cultivation of *Phoma* sp YUD17001 and *Armillaria* sp. RSC Adv 10:18384-18389
Liu SS, Jiang JX, Huang R, Wang YT, Jiang BG, Zheng KH, Wu SH (2019) A new antiviral 14 Nordrimane Sesquiterpenoid from an endophytic fungus *Phoma* sp. Phytochem Lett 29:75-79.
Nalli Y, Arora P, Khan S, Malik F, Riyaz-Ul-Hassan S, Gupta V, Ali A (2019) Isolation, structural modification of macrophin from endophytic fungus *Phoma macrostoma* and their cytotoxic potential. Med Chem Res 28:260-266
Peng X, Duan F, He Y, Gao Y, Chen J, Chang J, Ruan H (2020) Ergocytochalasin A, a polycyclic merocytochalasan from an endophytic fungus *Phoma multirostrata* XJ-2-1. Org Biomol Chem 18:4056-4062.
Rai M. (1989) *Phoma sorghina* infection in human being. Mycopathologia (Den Haag) 105:167–170.
Rai, M., Yadav, A., Bridge, P. and Gade, A. (2009). Myconanotechnology: a new and emerging science. In *Applied Mycology*, CAB international. 258-267
Rai M., Tiwari VV Irinyi L, Kövics GJ (2014) Advances in taxonomy of genus *Phoma*: Polyphyletic nature and role of phenotypic traits and molecular systematics. Indian J Microbiol 54: 123-128.
Rai M, Ingle AP, Gade AK, Duarte MC, Duran N (2015). Three *Phoma* species. synthesised novel silver nanoparticles that possess excellent antimicrobial

efficacy. IET Nanobiotechnol. 2015;9(5):280-7. doi: https://doi.org/10.1049/iet-nbt.2014.0068.

Rai M, Gade A, Zimowska B, Ingle AP, Ingle P (2018). Marine-derived *Phoma*-the gold mine of bioactive compounds, *Appl Microbiol Biotechnol,* 102:9053–9066; doi: https://doi.org/10.1007/s00253-018-9329-2

Rai, M., Gade, A., Zimowska, B., Ingle, A.P., Ingle, P. (2020). Harnessing the potential of novel bioactive compounds produced by endophytic *Phoma* spp. – biomedical and agricultural applications. *Acta Sci. Pol. Hortorum Cultus*, 19(6): 31–45. DOI: https://doi.org/10.24326/asphc.2020.6.3

Saccardo PA (1880) *Conspectus generum fungorum Italiae inferiorum nempe ad Sphaeropsideas, Melanconieas et Hyphomyceteas pertinentium systemate sporologico dispositorum.* Michelia 2: 1-38.

Saccardo PA (1884) *Sylloge Sphaeropsidearum et Melanconiearum omnium hucusque cognitorum.* Sylloge Fungorum 3: 1-860.

Sang XN, Chen SF, Tang MX, Wang HF, An X, Lu XJ, Zhao D, Wang YB, Bai J, Hua HM, Chen G, Pei YH (2017). α-Pyrone derivatives with cytotoxic activities, from the endophytic fungus *Phoma* sp. YN02-P-3. Bioorg Med Chem Lett. 27(16): 3723-3725. doi: https://doi.org/10.1016/j.bmcl.2017.06.079. Epub 2017 Jun 29.

Schoch CL, Seifert KA, Huhndorf S, Robert V, Spouge JL, Levesque CA, Chen W (2012) Nuclear ribosomal internal transcribed spacer (ITS) region as a universal DNA barcode marker for Fungi. Proceedings of the National Academy of Sciences of the United States of America 9(16): 6241–6246. doi: https://doi.org/10.1073/pnas.1117018109

Sheikhloo Z. and Salouti M (2012). Intracellular Biosynthesis of Gold Nanoparticles by Fungus *Phoma macrostoma*, Synthesis and Reactivity in Inorganic, Metal-Organic, and Nano-Metal Chemistry, 42:1, 65-67

Shivanna MB, Meera MS, Kubota M, Hyakumachi M (2005) Promotion of growth and yield in cucumber by zoysiagrass rhizosphere fungi. Microbes Environ 20:34–40.

Sullivan RF, White JR JF (2000) *Phoma glomerata* as a mycoparasite of powdery mildew. Applied and Environmental Microbiology 66:425–427.

Sultana F, Hossain MM, Kubota M, Hyakumachi M (2009) Induction of systemic resistance in *Arabidopsis thaliana* in response to culture filtrate from a plant growth-promoting fungus, *Phoma* sp. GS8-3. Plant Biology 11 (1):97-104.

Sutton BC (1964) *Phoma* and related genera. Trans Brit Mycol Soc 47: 497- 509.

van der Aa HA, and Vanev S (2002) A Revision of the Species Described in *Phyllosticta.* Centraalbureau voor Schimmelcultures, Utrecht, The Netherlands.

van der Aa HA, Noordeloos ME, de Gruyter J (1990) Species concepts in some larger genera of the *Coelomycetes*. Stud Mycol 32: 3-19.

Verkley GJM, da Silva M, Wicklow DT, Crous PW (2004) *Paraconiothyrium*, a new genus to accommodate the mycoparasite *Coniothyrium minitans*, anamorphs of *Paraphaeosphaeria*, and four new species. Studies in Mycology 50: 323-335.

Wollenweber HW, Hochapfel H (1936) Beiträge zur Kenntnis parasitärer und saprophytotischer Pilze. I. *Phomopsis*, *Dendrophoma*, *Phoma* und *Ascochyta* und ihre Beziehung zur Fruchtfäule. Z Parasitenk 8: 561-605.

Woudenberg JHC, Aveskamp MM, de Gruyter J, Spiers AG, Crous PW (2009) Multiple *Didymella* teleomorphs are linked to the *Phoma clematidina* morphotype. Persoonia 22: 56–62. doi: https://doi.org/10.3767/003158509X427808

Zhou Z, Zhang C, Zhou W, Li W, Chu L, Yan J, et al. (2014) Diversity and plant growth-promoting ability of endophytic fungi from the five flower plant species collected from Yunnan, Southwest China. Journal of Plant Interactions. 9(1):585-591.

Zimowska B (2007a) New *Phoma* species on *Leonurus cardiaca*. Acta Mycol 42: 119-123

Zimowska B (2007b) Fungi colonizing and damaging different parts of peppermint (*Mentha piperita* L.) cultivated in south-eastern Poland. Herba Pol 53: 97-105.

Zimowska B (2008) Biodiversity of fungi colonizing and damaging selected parts of motherwort (*Leonurus cardiaca* L.). Herba Pol 54: 30-40.

Zimowska B (2012) Pathogenicity and ultrastructural studies of the mode of penetration by *Phoma strasseri* in peppermint stems and rhizomes. Pol J Microbiol 61: 273-279.

Zimowska B, Król ED, Furmańczyk A, Abramczyk B, Okoń S (2018) Molecular characterization of *Boeremia strasseri* the causal agent of black stems and rhizomes rot of peppermint. J. Plant Pathol 100: 13–24, doi: https://doi.org/10.1007/s42161-018-0003-4

Taxonomical Evaluation of *Phoma*: History of Classification, Current Status and Future Directions

2

Beata Zimowska

2.1 Introduction

The genus *Phoma* has always been considered by mycologists to be one of the most problematic and challenging fungal genera regarding identity. Previously the taxonomy was based on host with combination of micromorphological characters, mainly on shape and size of pycnidia (asexual fruiting bodies) and pycnidiospores (asexual spores) (Saccardo 1880, 1884; Aveskamp et al. 2010). This approach has resulted in the fact that the systematics of the genus never has been fully understood (Aveskamp et al. 2008, 2010). Progress in molecular phylogenetic analyses has greatly contributed to a substantial revision of *Phoma*-like genera and turned out to be helpful in accurate and fast identification of taxa which was crucial. Besides several harmless saprobic species, *Phoma* has also been considered to be an important fungal plant pathogenic genus occurring on economically important crops (Mendes-Pereira et al. 2003). Several *Phoma* species are also of quarantine importance, posing serious problems to organizations that are involved in plant health quarantine regulation (Somai et al. 2002; Koch and Utkhede 2004; Balmas et al. 2005; Licciardella et al. 2006; Cullen et al. 2007; EPPO Global Database 2019). Identification of strains found on possible infected samples of plant material is frequently carried out under extreme time constraints. Because morphological studies are time-consuming, are expensive and require highly skilled employees, the chance of successfully identifying strains to species level in such laboratories is often low. Therefore, besides the necessity of reclassification of *Phoma* species within the sections, there was a prompt need for the development of fast, reliable molecular methods for the detection of quarantine actionable *Phoma* species.

2.2 History of Classification and Morphological Concept of *Phoma*

Phoma is a polyphyletic genus of fungi belonging to the phylum *Ascomycota*, class *Dothideomycetes*, order *Pleosporales* and family *Didymellaceae* (Fig. 2.1).

In the first taxonomical description of fungi imperfecti introduced by Italian mycologist Pier Andrea Saccardo in 1880 (Saccardo 1884), which was based mainly on morphology of conidia, the genus *Phoma* was capable to form pycnidial conidiomata with aseptate, hyaline conidia (asexual spores). Besides, the Saccardoan system pointed at substrate out, as

B. Zimowska (✉)
Department of Plant Protection, Institute of Plant Pathology and Mycology, University of Life Sciences in Lublin, Lublin, Poland
e-mail: beata.zimowska@up.lublin.pl

M. Rai et al. (eds.), *Phoma: Diversity, Taxonomy, Bioactivities, and Nanotechnology*,
https://doi.org/10.1007/978-3-030-81218-8_2

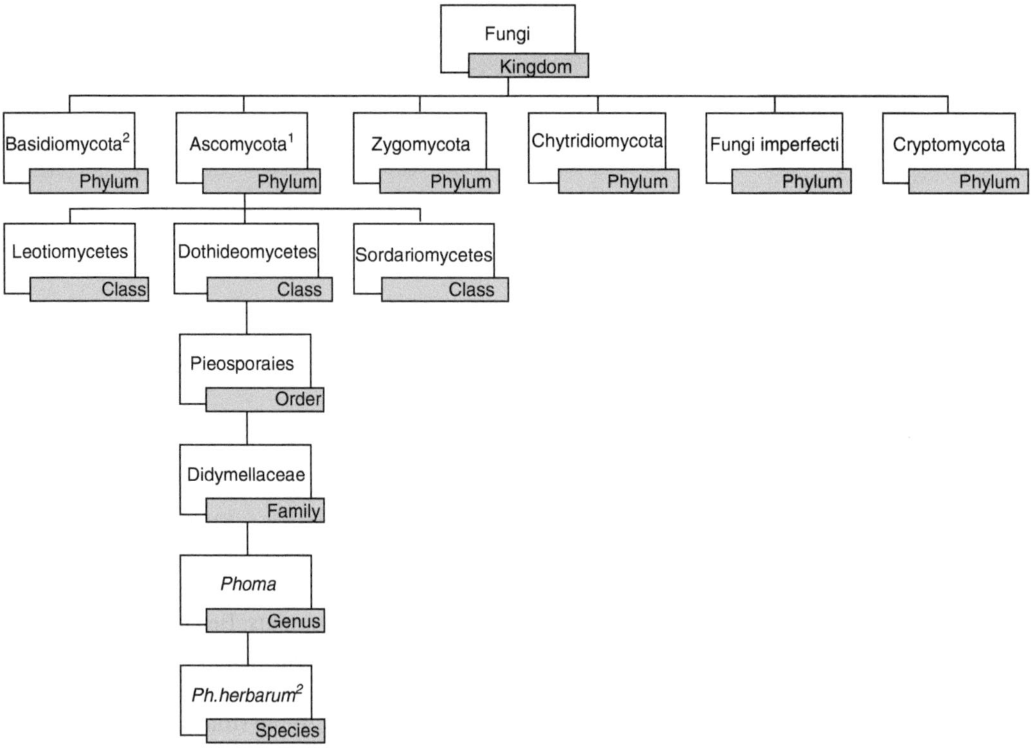

Fig. 2.1 *Phoma* species scheme based on current classification data utilizing morphological and molecular characterization data. (1) *Basidiomycota* and *Ascomycota* are more closely related to one another than to other phyla. (2) Use of molecular-based phylogenetic analyses has restricted the *Phoma* genus to *P. herbarum* sp. within family *Didymellaceae* (Banett et al. 2018)

the relevant criterion for identification. Thus, considering the presence or absence of conidial septum, as well as the plant's organ where pycnidia were formed, the closely related genera of fungi were distinguished. Amongst the many scientists that attempted to order the complexity in genus *Phoma*, Wollenweber and Hochapfel (1936) were the first who indicated the advantages of using growth characteristics on artificial substrates in *Phoma* taxonomy. Dennis (1946) implemented the use of basic morphology, physiology and biochemical tests in vitro. In a study on hyphomycetes, Hughes (1953) introduced conidiogenesis as an important taxonomic criterion, a feature that was later applied to all conidial fungi including species of *Phoma* (Sutton 1964). It turned out that the taxonomical system introduced by Saccardo based only on features created by fungi in vivo is too simple and completely not practical, since in pycnidia developing both on leaves and stems, there were found multi-celled conidia as well as one-cell. Consequently, the usage of this system resulted in confusion, due to many species similar to *Phoma* having been described by the use of different names which were treated as synonyms in several morphological closely related genera (Boerema and Dorenbosh 1973; Boerema 1976). To get the confusing system right, according to the rules of the International Code of Botanical Nomenclature, the designed type species have been proposed for morphologically similar genera. *P. herbarum* Westendorp nom. Cons. has been the type species of genus *Phoma* Saccardo (Boerema 1964). Morphology, synonymy and ecology of this type species were later extensively described by Boerema (1970) and Morgan-Jones (1988).

The use of modern techniques in taxonomic studies and especially the use of electron microscopy to examine the process of conidiogenesis become the important criterion in differentiation of similar genera like *Phoma* and *Ascochyta* (Brewer and Boerema 1965; Boerema and Bollen 1975). These studies enabled us to find out the structure and type of conidia-forming cell, which is a significant taxonomic feature in genus *Phoma*. The key achievement at this stage of research was showing that in the case of *Phoma*, conidia are formed in the process of enteroblastic phialidial ontogeny. The characteristic of this process was formation of conidia on monophialidic, doliiform to flask-shaped conidiogenous cells, termed as phialide. The development of the first conidium begins with the thickening on the apex of these cells, where after the first conidium breaks off, a characteristic collarette is formed, which distinguishes the phialide from the annelide, a conidiogenous cell, which is created by fungi from *Ascochyta* genus. The collarette surrounded the conidia-forming point of the phialide, from which successive conidia were formed. The studies pointed to the formation of the phialidial mucous substance during the enteroblastic ontogeny. That substance was formed at the final stage of the conidial wall differentiation to its final structure (Brewer and Boerema 1965; Zimowska 2011).

The genus *Phoma* initially comprised more than 3000 species (Monte et al. 1991). Such a large number of taxa described within *Phoma* were related to the use of nomenclature mainly based on the characteristics of the host plant and marginalization of micromorphological properties (Aveskamp et al. 2008). The first key for identification of species from the genus *Phoma* has been introduced by Sutton in 1980. The key included 28 most commonly occurring species (Sutton 1980). In the 1980s, the studies concerning species from genus *Phoma* were also carried out in the United States. Many species have been described, including the new one, *P. americana* (Morgan-Jones and White 1983). The greatest effort made on *Phoma* systematics was by Boerema and co-workers from Dutch Plant Protection Service (Plantenziektenkundige Dienst, PD). They started very extensive studies on *Phoma* taxonomy. The origin of the host stopped to be the main criterion for differentiation within species. The main presumption in Boeremean classification was identification of taxon according to stable morphological features observed both in vivo and in vitro, in cultures developing in standard condition; as a consequence the number of species was reduced to 223 (Boerema and Bollen 1975; Boerema 1976, 1997; van der Aa et al. 1990). The next step in *Phoma* taxonomy was the classification of the genus into nine sections (van der Aa et al. 1990; Boerema 1997). The *Phoma* sections *Phoma*, *Heterospora*, *Macrospora*, *Paraphoma*, *Peyronellaea*, *Phyllostictoides*, *Pilosa*, *Plenodomus* and *Sclerophomella* (Table 2.1) were recognized based on the characteristics of the type species of each section. The sections included both specific pathogens of (cultivated) plants and plurivorous, saprophytic or opportunistic species. Standardized descriptions of the *Phoma* species per section followed in a series of "Contributions towards a monograph of *Phoma*", published in *Persoonia* in the period 1992–2003 (de Gruyter and Noordeloos 1992; Boerema 1993, 2003; de Gruyter et al. 1993, 1998, 2002; Boerema et al. 1994, 1996, 1997; Boerema and de Gruyter 1998, 1999; van der Aa et al. 2000; de Gruyter 2002; de Gruyter and Boerema 2002) including various new taxa. These contributions gave the base to "*Phoma* Identification Manual" (Boerema et al. 2004). The manual provided keys and descriptions of more than 200 specific and infraspecific *Phoma* taxa.

Although the key was helpful for identification of species, there was still uncertainty if this division into sections can be considered natural from an evolutionary point of view. In this classification system, most sections have been created based on morphological features, like the conidial and pycnidial morphology (Figs. 2.1 and 2.2), at the presence of enlarged, thick-walled chlamydospores within hyphae or at hyphal tips (Figs. 2.3 and 2.4).

Unfortunately, several features that were linked to a certain section sometimes also seemed to occur in species that were placed in other taxo-

Table 2.1 The nine sections of the genus *Phoma*, with their type species and associated teleomorph genera

Section	Type species	Associated teleomorph genus
Heterospora	*Phoma heteromorphospora*	–
Macrospora	*P. zeae-maydis*	*Didymella*
Paraphoma	*P. radicina*	–
Peyronellaea	*P. glomerata*	–
Phoma	*P. herbarum*	*Didymella*
Phyllostictoides	*P. exigua* var. *exigua*	*Didymella*
Pilosa	*P. betae*	*Pleospora*
Plenodomus	*P. lingam*	*Leptosphaeria*
Sclerophomella	*P. complanata*	*Didymella*

Reproduced from Aveskamp et al. (2008) under Creative Commons Attribution 4.0 International licence

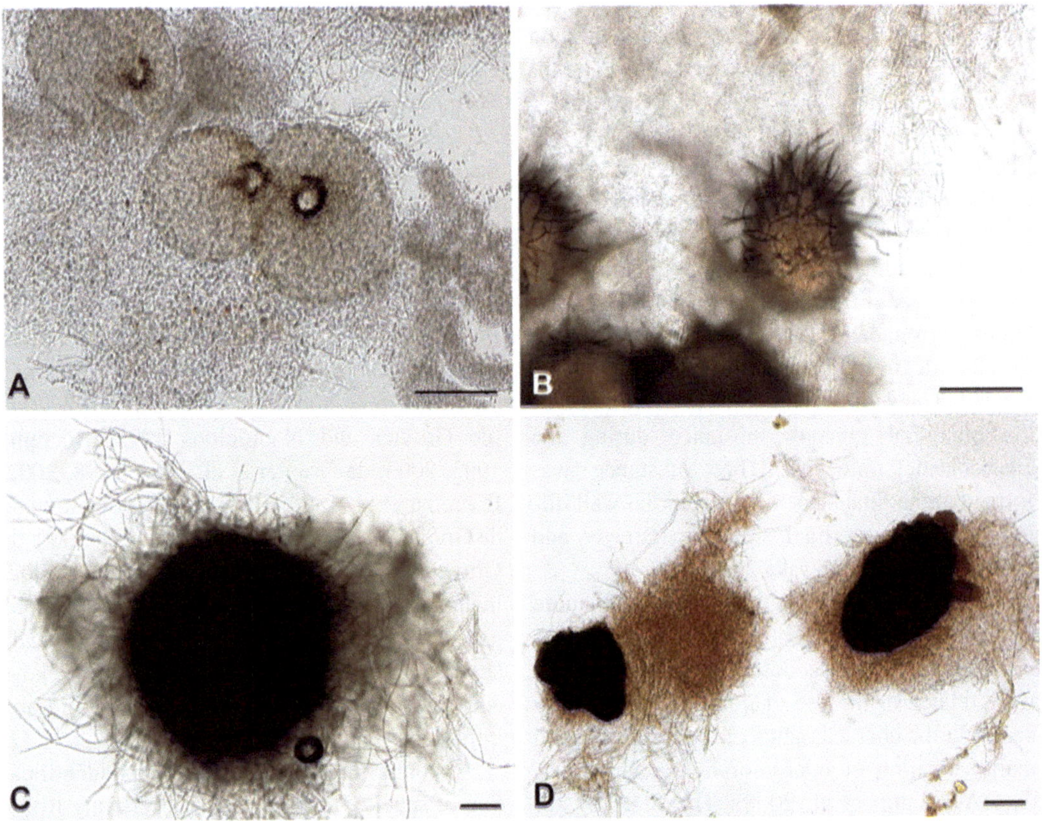

Fig. 2.2 (**a–d**) Pycnidial types. (**a**) Regular glabrous as in *P. herbarum*. (**b**) Setose as in *P. carteri*. (**c**) Pilose as in *P. betae*. (**d**) Pycnosclerotia-like pycnidia as in *P. incompta*. Scale bars = 100 μm. Reproduced from Aveskamp et al. (2008) under Creative Commons Attribution 4.0 International licence

nomic groups. For example, *Phoma lingam* mentioned the slightly pilose species, *P. anserina* and *P. leonuri*, that were not accommodated in either section *Pilosa* or *Paraphoma*. According to the appearance of their conidia, these two species have been included in the sections *Phoma* and *Plenodomus*, respectively. This inconclusive approach has resulted in multiple overlaps between the separate sections. Furthermore, the sections *Phoma* and *Phyllostictoides* were also considered to be artificial. Instead of incorporating species with a shared feature, both sections

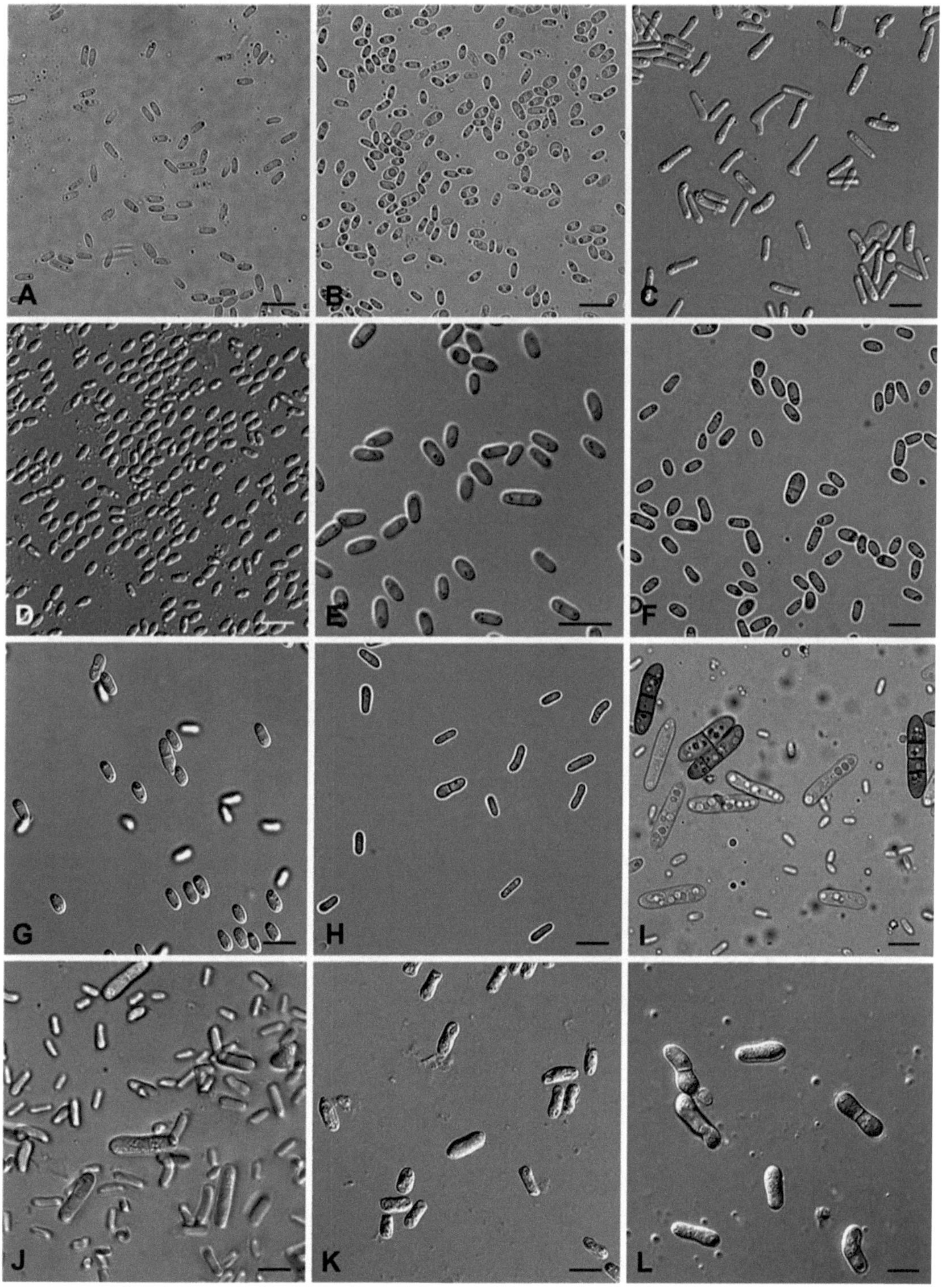

Fig. 2.3 (**a–l**) Conidial morphology in *Phoma*. (**a–d**) Conidia hyaline, aseptate small-sized as in (**a**) *P. herbarum*, (**b**) *P. multirostrata*, (**c**) *P. astragali* and (**d**) *P. eupyrena*. (**e–h**) Conidia hyaline, occasionally septate, small-sized as in (**e**) *P. exigua*, (**f**) *P. cucurbitacearum*, (**g**) *P. macrostoma* and (**h**) *P. polemonii*. (**i, j**) Conidia generally hyaline small-sized, but always large septate and often pigmented as in (**i**) *P. actaeae* and (**j**) *P. schneiderae*. (**k, l**). Conidia hyaline, occasionally septate, but all relatively large as in (**k**) *P. rabiei* and (**l**) *P. xanthina*. Scale bars = 10 μm. Reproduced from Aveskamp et al. (2008) under Creative Commons Attribution 4.0 International licence

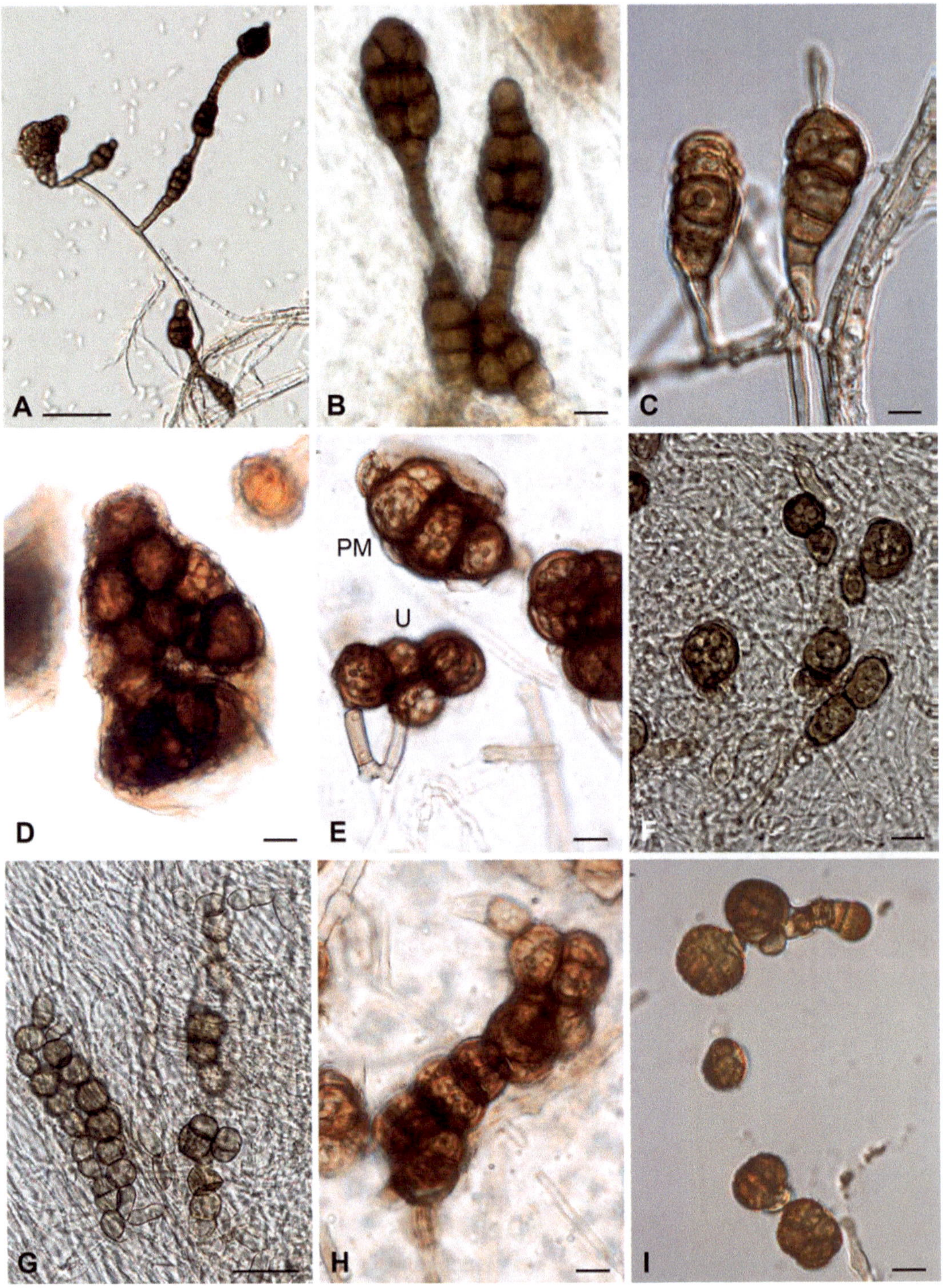

Fig. 2.4 (**a–i**) Chlamydospore morphology in *Phoma*. (**a**) Chain of chlamydospores of *P. glomerata*. (**b**) Alternaroid chlamydospores of *P. glomerata*. (**c**) Alternaroid chlamydospores of *P. jolyana*. (**d**) Botryoid chlamydospore as in *P. sorghina*. (**e**) Unicellular (u) and pseudosclerotoid multicellular chlamydospores (pm) in *P. chrysanthemicola*. (**f**) Unicellular chlamydospores of *P. pinodella*. (**g**) Chain of unicellular chlamydospores in *P. clematidina*. (**h**) Chain of unicellular chlamydospores in *P. chrysanthemicola*. (**i**) Epicoccoid chlamydospores in *P. epicoccina*. Scale bars: A = 100 μm, G = 50 μm, B–F and H–I = 10 μm. Reproduced from Aveskamp et al. (2008) under Creative Commons Attribution 4.0 International licence

seemed to be a dump for species that lack the presence of good sectional features. According to Boerema et al. (2004), section *Phoma* included species "that have much in common with *P. herbarum*". The delimitation of *Phoma* into sections was ambiguous, because the main character of a *Phoma* section could also be shared by some species classified in other sections (Boerema et al. 2004; Punithalingam 2004). In addition, morphological features of certain *Phoma* sections can be shared by related Coelomycetes including *Ascochyta*, *Asteromella*, *Microsphaeropsis*, *Phomopsis*, *Phyllosticta*, *Pleurophoma*, *Pyrenochaeta* and *Stagonospora*. Especially the delimitation of the genera *Pyrenochaeta, Pleurophoma* and *Phoma* section *Paraphoma* was still unclear (Grodona et al. 1997). It was therefore strongly recommended that the system of *Phoma* classification should have been improved by evaluating their phylogenetic placement in order to establish monophyletic groups (Torres et al. 2005).

To conclude, the morphological species concept of *Phoma* was highly polyphyletic. The classification of the genus within sections was useful for identification based on morphological features that are clearly visible in cultures grown in vitro, but this classification was rather artificial because morphological characters may be shared amongst sections. Moreover, delimitation with related Coelomycetes can be controversial, and there are multiple teleomorphs associated with *Phoma*, including *Didymella*, *Leptosphaeria*, *Mycosphaerella* and *Pleospora*. Even these teleomorphs are shared with morphologically similar coelomycete genera.

2.3 Molecular Phylogeny of *Phoma*: Towards a Reclassification of the *Phoma* Complex

The highly phylogenetic heterogeneity of the species in the *Phoma* sections negated the existing division of the genus into sections and caused the necessity to reclassify (De Gruyter et al. 2009, 2010, 2012; Aveskamp et al. 2010). Introducing molecular studies turned out to be also useful to determine the phylogeny of *Phoma* species with an unknown teleomorph and provide an answer where teleomorphs associated with *Phoma* species are disputed, e.g. *Didymella* or *Mycosphaerella* (Punithalingam 2004). Finally, the phylogeny of *Phoma* spp. with *Epicoccum*, *Phaeomoniella*, *Phialophora*, *Sclerotium* and *Stagonosporopsis* synanamorphs (Boerema and Bollen 1975; Sutton 1977; Boerema 1993, 1997; Boerema et al. 1994, 1997; Crous and Gams 2000) still left required to be clarified. Results of a phylogenetic study including the type species of all nine *Phoma* sections and allied coelomycetous genera demonstrated that all nine sections grouped in the *Pleosporales* (de Gruyter et al. 2009). The type species of the sections *Macrospora*, *Peyronellaea*, *Phoma*, *Phyllostictoides* and *Sclerophomella* resided in *Didymellaceae* (de Gruyter et al. 2009, 2012). However, the four other sections, namely, *Heterospora*, *Paraphoma*, *Pilosa and Plenodomus*, clustered in several distinct clades outside *Didymellaceae* and were thus excluded from *Phoma* (de Gruyter et al. 2009; Aveskamp et al. 2010).

Although the internal transcribed spacers (ITS) have been selected as the standard DNA barcode for fungi (Schoch et al. 2012), it turned out to not be sufficient to distinguish all *Phoma* species. Internal transcribed spacer regions 1 and 2 and 5.8S nrDNA (ITS), partial large subunit 28S nrDNA region (LSU) and β-tubulin gene (TUB) (Woudenberg et al. 2009; Aveskamp et al. 2009a, 2010) in combination with morphological characteristics have been routinely used to resolve phylogenetic relationships in *Phoma*. RNA polymerase II gene (RPB2) has also been used for several species (Chilvers et al. 2009), while actin (ACT) (Aveskamp et al. 2009b; de Gruyter et al. 2012; Zimowska et al. 2018), chitin synthase 1 (CHS-1) (de Gruyter et al. 2012) region and the translation elongation factor gene (TEF1) (Irinyi et al. 2009) have been applied for *Boeremia exigua* (Desm.) Aveskamp, Gruyter & Verkley, *Boeremia strasseri* (Moesz) Gruyter et Verkley and *L. doliolum* (Pers.) Ces. & De Not. complexes and intraspecific taxa due to their high variations below species level. In the studies carried out by de Gruyter et al. (2012), the status of the genus *Phoma* and related genera and teleomorph associations were examined by means of DNA phylogenetic analyses, using sequence data of the 18S (SSU) and the 28S (LSU) nrDNA regions of 15 representative strains of the type species of the *Phoma* sections (Boerema et al. 2004), and 3 reference strains of *Phoma* were compared with those of 93 strains representing 39 anamorph and 11 teleomorph genera. The outcomes from these studies were compared with related molecular data, as well as morphological and ecological information. Additionally, the anamorph-teleomorph relations within this complex have been discussed, and a new family *Didymellaceae* in the *Pleosporales* has been proposed. Currently, *Didymellaceae* is considered as one of the largest families in the fungal kingdom, which belongs to the largest *Dothideomycetes* order, *Pleosporales* (*Pleosporomycetidae*, *Dothideomycetes*, *Pezizomycotina*, *Ascomycota*). This family includes extremely large number of species (Fig. 2.5). To date, more than 5400 species belonging to at least 36 genera have been recorded, including recently established genera such as *Neoascochyta* and *Paraboeremia* and historical genera such as *Ascochyta*, *Didymella* and *Phoma* that remain highly problematic for the identification of new isolates (De Gruyter et al. 2009; Chen et al. 2015, 2017; Hou et al. 2020). Molecular characteristics have become increasingly essential in the identification of species in *Didymellaceae*. Based on sequence data obtained from 324 strains, Aveskamp et al. (2010) divided the family into 18 distinct clusters and indicated that some important genera, such as *Ascochyta*, *Didymella* and *Phoma*, were highly polyphyletic. At the same time, *Phoma*-like species that are located outside of this family were revealed to have affiliations to more than 25 families after phylogenetic analyses (De Gruyter et al. 2009, 2010, 2013). In a further study, Chen et al. (2015) provided an updated clarification towards the understanding of the taxonomy and evolution of *Didymellaceae* by including rpb2 as the most effective locus for the identification of *Didymellaceae* at species level, and it has been proposed as a secondary DNA marker for the family, which resulted in embedding 17 genera in *Didymellaceae* (Chen et al. 2015). Several additional new genera, namely, *Briansuttonomyces*, *Cumuliphoma*, *Ectophoma*, *Juxtiphoma*, *Neomicrosphaeropsis*, *Pseudoascochyta*, *Remotididymella*, *Similiphoma* and *Vacuiphoma*, and a large number of new species have recently been introduced and accepted in the family (Ariyawansa et al. 2015; Chen et al. 2015; Thambugala et al. 2016; Wijayawardene et al. 2016; Valenzuela-Lopez et al. 2018). Chen et al. (2017) in their further studies described 32 new taxa and 2 new combinations in 9 genera, mostly based on specimens collected from Asia. In total 19 genera have been recognized in the family, with *Heracleicola* being reduced to synonymy under *Ascochyta*.

The current status of *Didymellaceae* taxonomy has been updated by Hou et al. (2020). They examined isolates originating from seven continents; most isolates came from Europe, South Africa and America and 92 countries. Their studies resulted in 36 genera of which 7 have been described as the new one, including 40 new species and 21 new combinations (Fig. 2.5). Moreover, six epitypifications and six neotypifications were provided, and four basionyms, namely, *Epicoccum mezzettii* (Goidànich 1937), *Epi. oryzae* (Ito and Iwadare 1934), *Epi. purpurascens* and *Toruloidea tobaica* (Von Szilvinyi 1936), were renatured according to morphological and phylogenetic analysis of the ex-type or reference strains. Although the genera *Didysimulans*, *Endophoma*, *Platychora* and *Vacuiphoma* were revealed to belong to *Didymellaceae*, their taxonomic status still abides unresolved.

Didymellaceae and outgroup sequences. The RAxML bootstrap support values (MLBS) above 50% and Bayesian posterior probabilities (BPP) above 0.80 are given at the nodes (BPP/MLBS). Some of the basal branches were shortened to facilitate layout (the fraction in round parentheses refers to the presented length compared to the actual length of the branch). The scale bar represents the expected number of changes per site. Genera are delimited in coloured boxes, with the

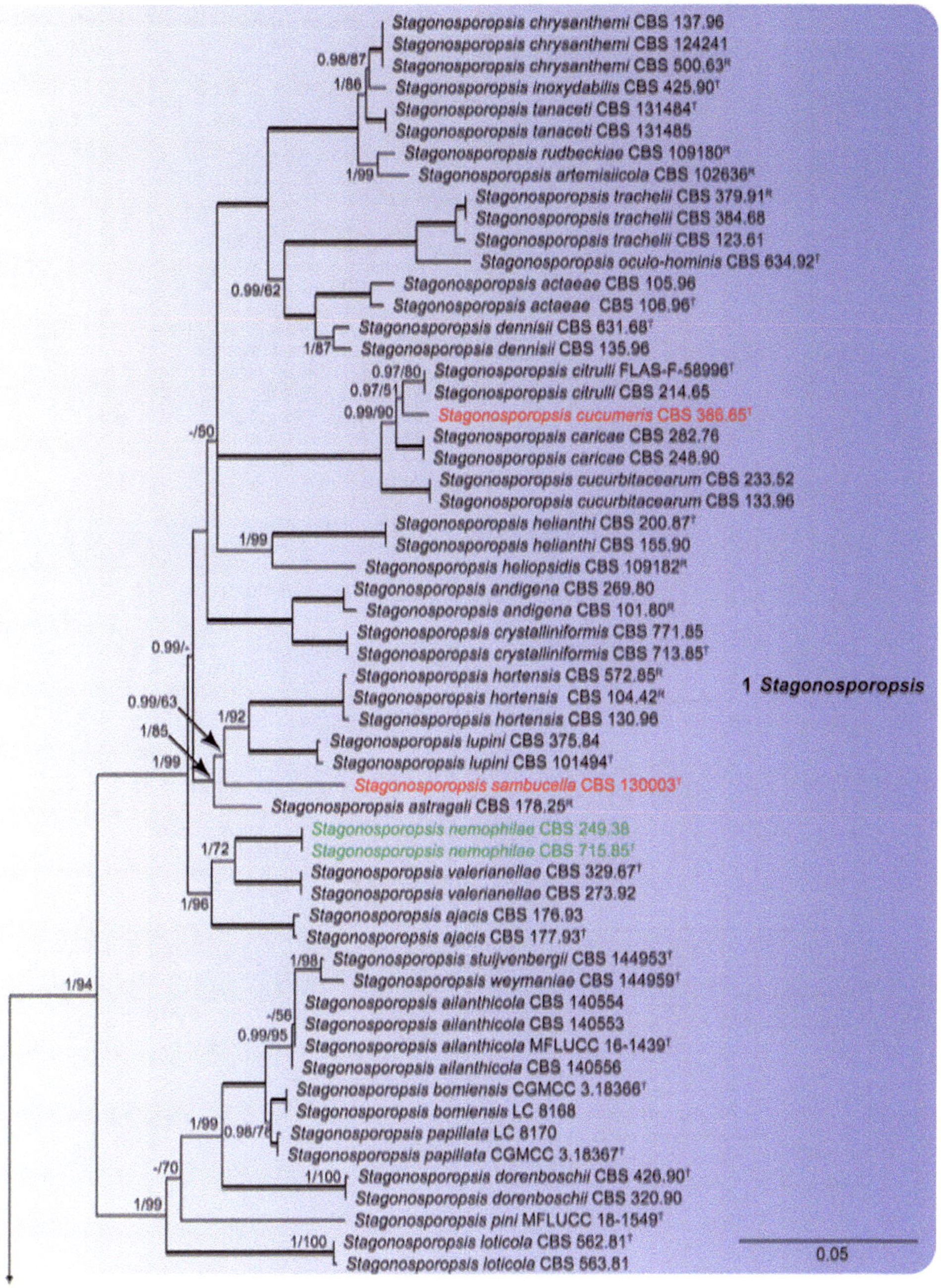

Fig. 2.5 Phylogenetic tree inferred from a maximum likelihood analysis based on a concatenated alignment of LSU, ITS, rpb2 and tub2 sequences of 597 strains representing

genus name indicated to the right. Strains with special status are indicated with a superscript letter after the accession number (R, representative; T, ex-type). The new species are printed in red font and new combinations in green font.

The tree is rooted to *C. palmarum* culture CBS 400.71, *Neocucurbitaria aquatica* culture CBS 297.74 and *Pleiochaeta setosa* cultures CBS 496.63 and CBS 118.25 (Hou et al. 2020).

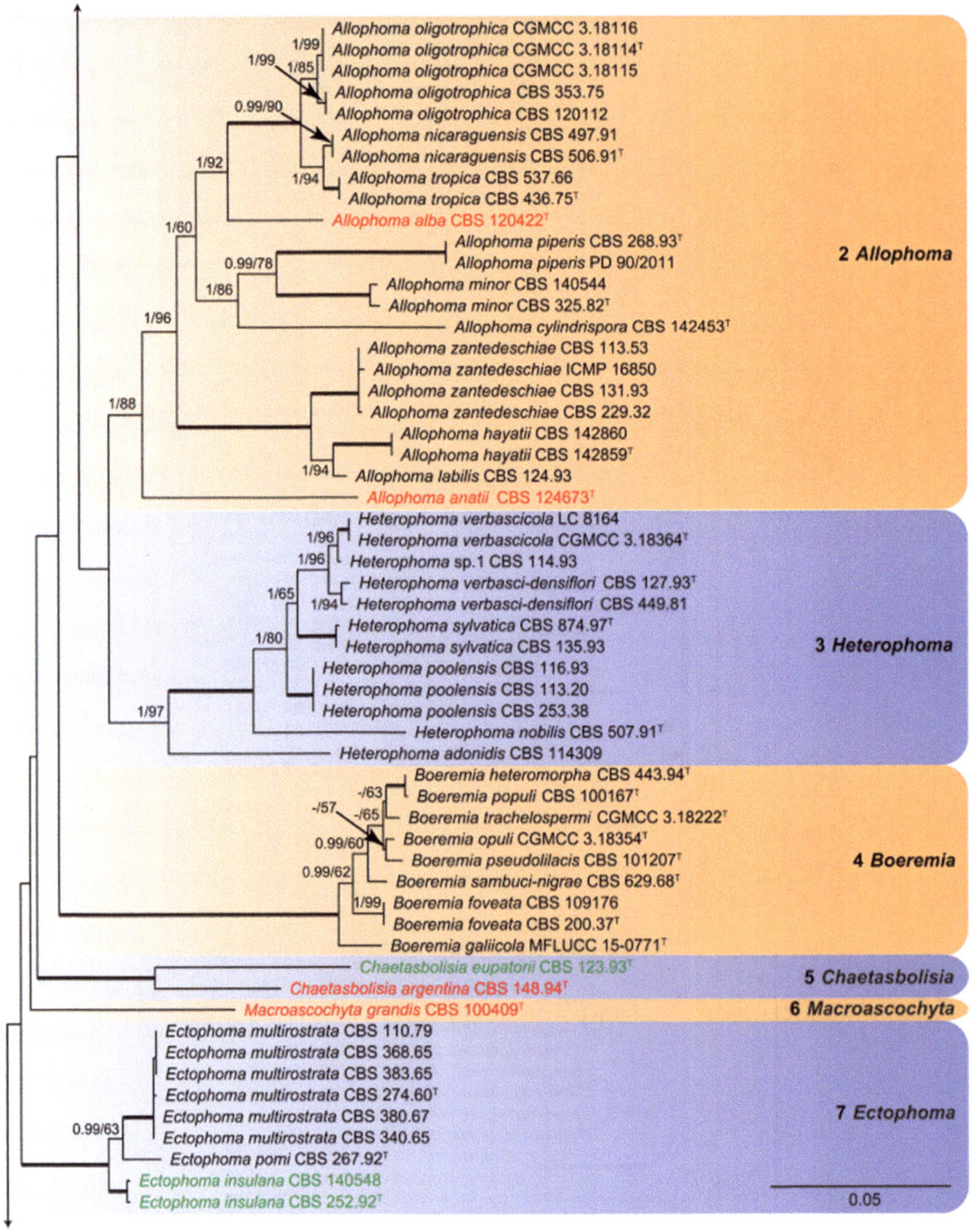

Fig. 2.5 (continued)

2.4 Use of Metabolomics for Chemotaxonomy of *Phoma* and Allied Genera

A polyphasic approach combining morphological and physiological features with DNA sequence data is now considered as the comprehensive line to classify *Phoma* and allied genera. Chemotaxonomy is one method of biological classification based on the similarity of chemical compounds such as carbohydrates, lipids, amino acids or secondary metabolites (Benedict 1970; Velázquez et al. 2006; Rai et al. 2009, 2014). Filamentous fungi, especially ascomycetes, are

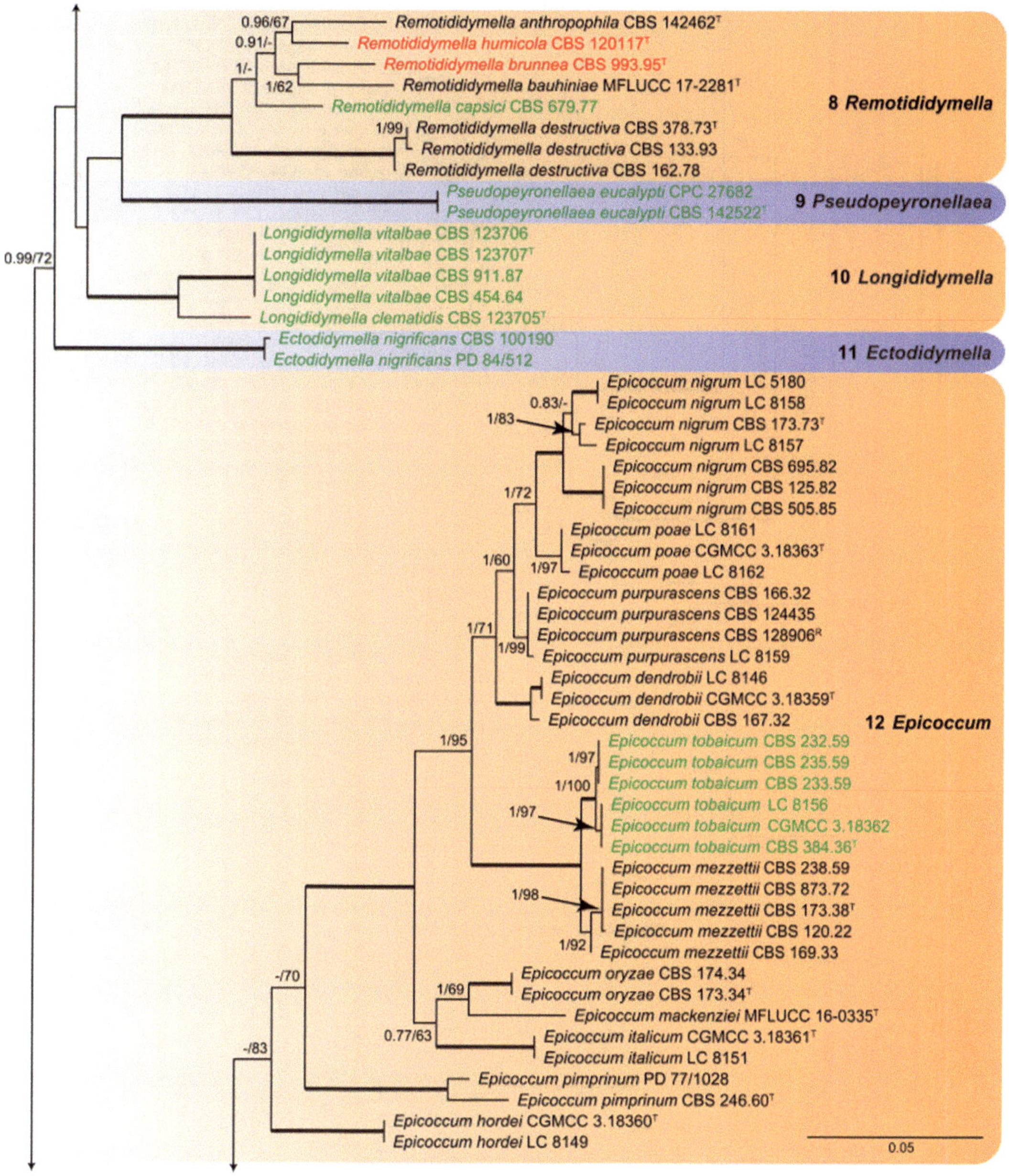

Fig. 2.5 (continued)

known to produce a vast array of secondary metabolites, such as non-ribosomal peptides, terpenes, polyketides as well as many other small organic compounds of mixed biosynthetic origin. Extensive genome sequencing efforts have revealed that genomic differences between fungal species are often related to the number and similarity of the genes involved in synthesis of secondary metabolites like polyketide synthases and non-ribosomal peptide synthetizes, which result in the production of a unique set of secondary metabolites (Smedsgaard and Nielsen 2005). Therefore, currently, it is much more common to include the use of secondary metabolite profiles as a taxonomic diagnostic tool.

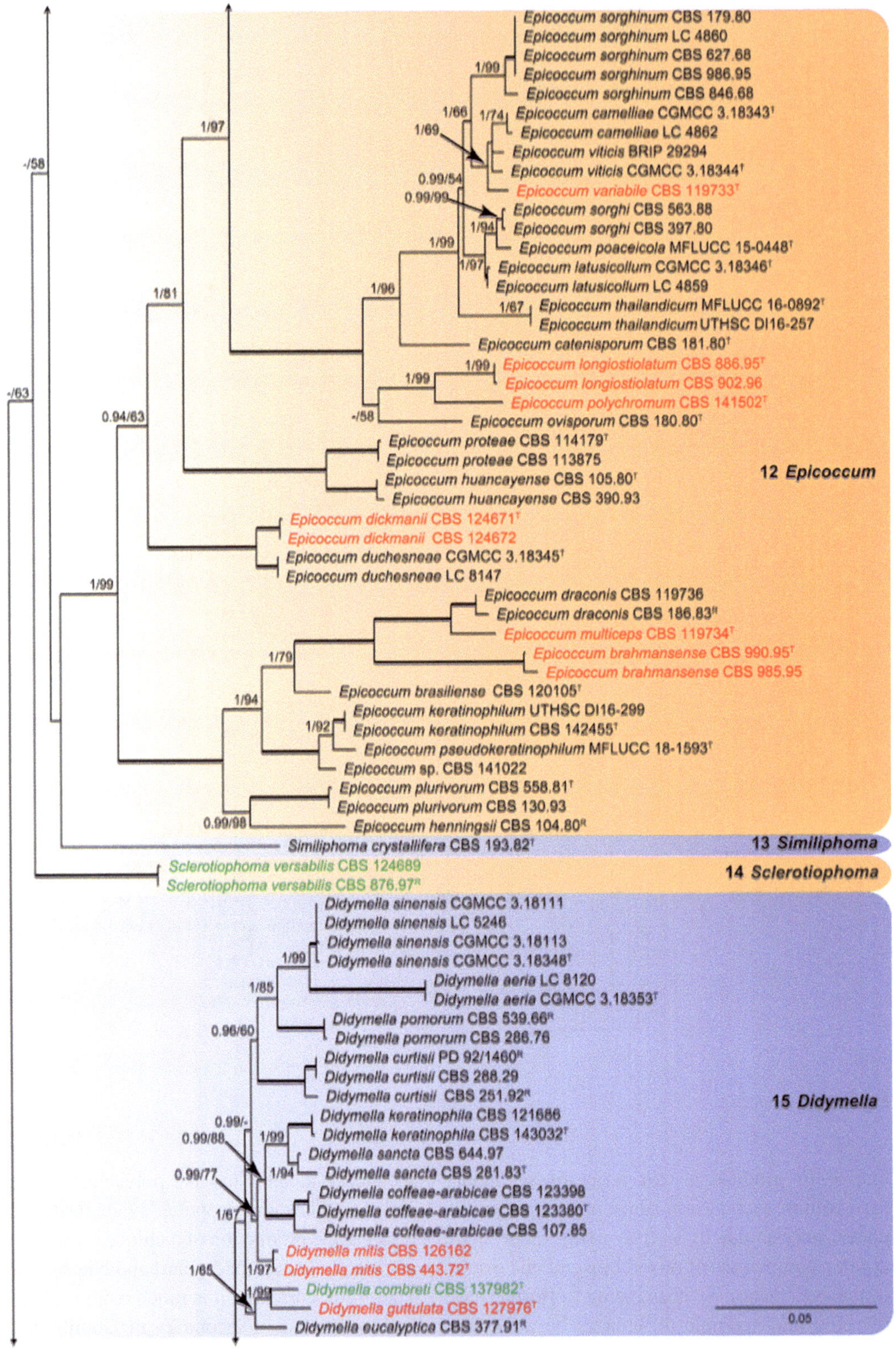

Fig. 2.5 (continued)

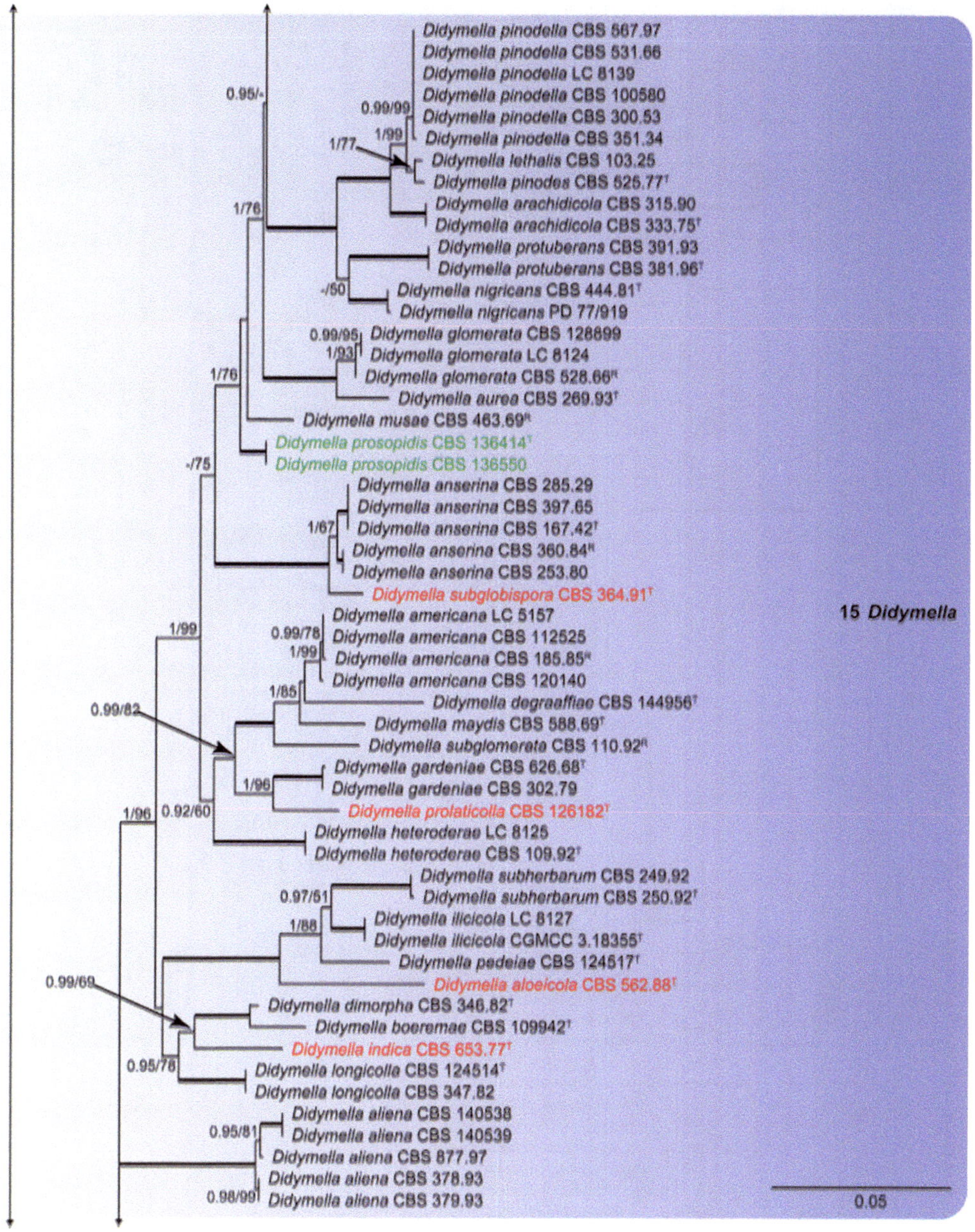

Fig. 2.5 (continued)

The most commonly used method for screening metabolomics profile was paper chromatography and later thin layer chromatography (TLC). TLC is still considered as a good method for confirming outcomes obtained by other chromatographic separation methods and is occasionally used for chemotaxonomic studies in conjunction with high-pressure liquid chromatography (HPLC). For filamentous fungi, including *Phoma* and allied genera, a very simple agar plug standardized TLC application technique was developed in 1980, allowing examination of a large number of isolates (Filtenborg et al. 1983). This standardized TLC method was later complemented with HPLC and standardized HPLC with diode array detection (DAD). Later a direct inject

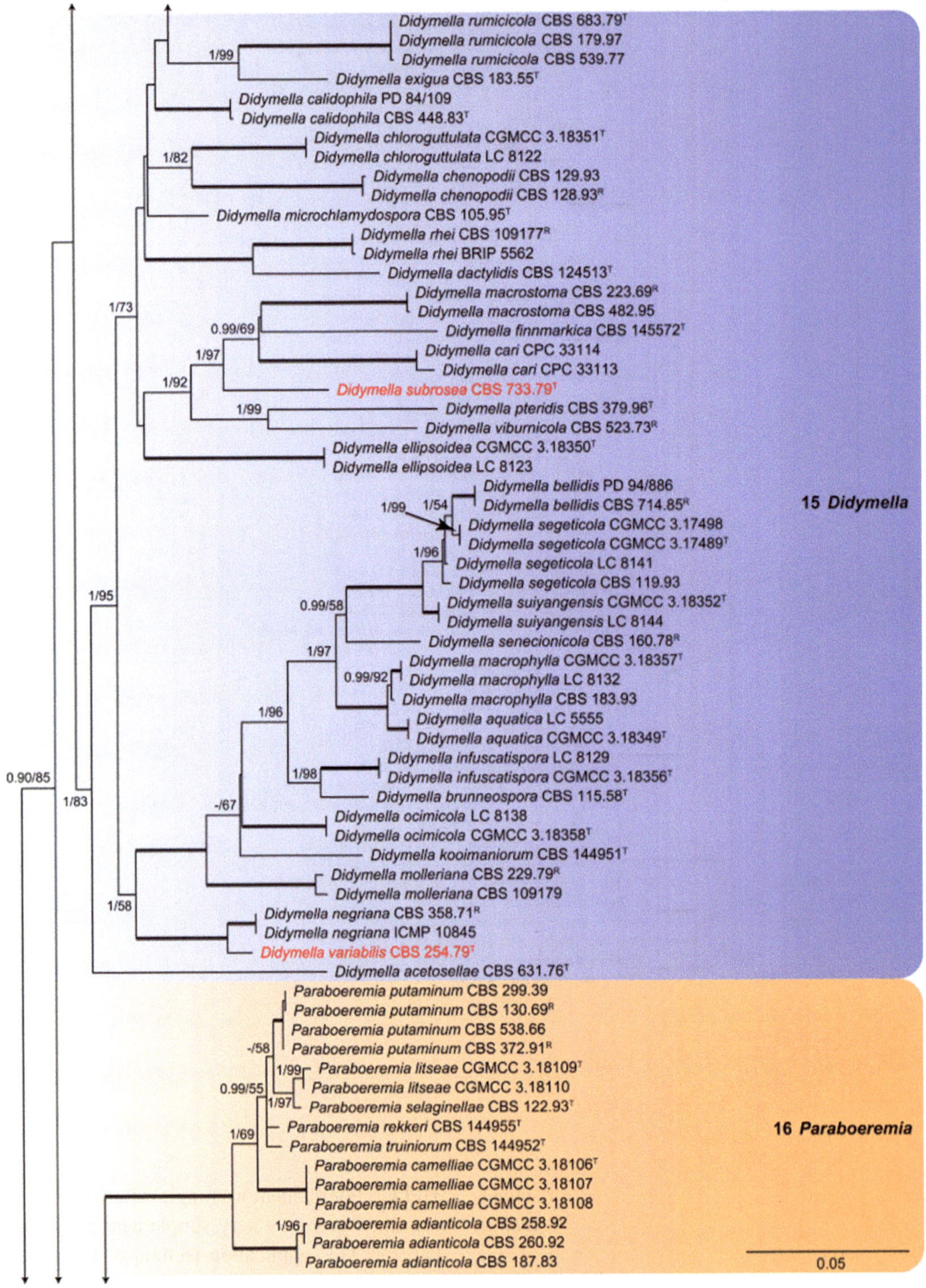

Fig. 2.5 (continued)

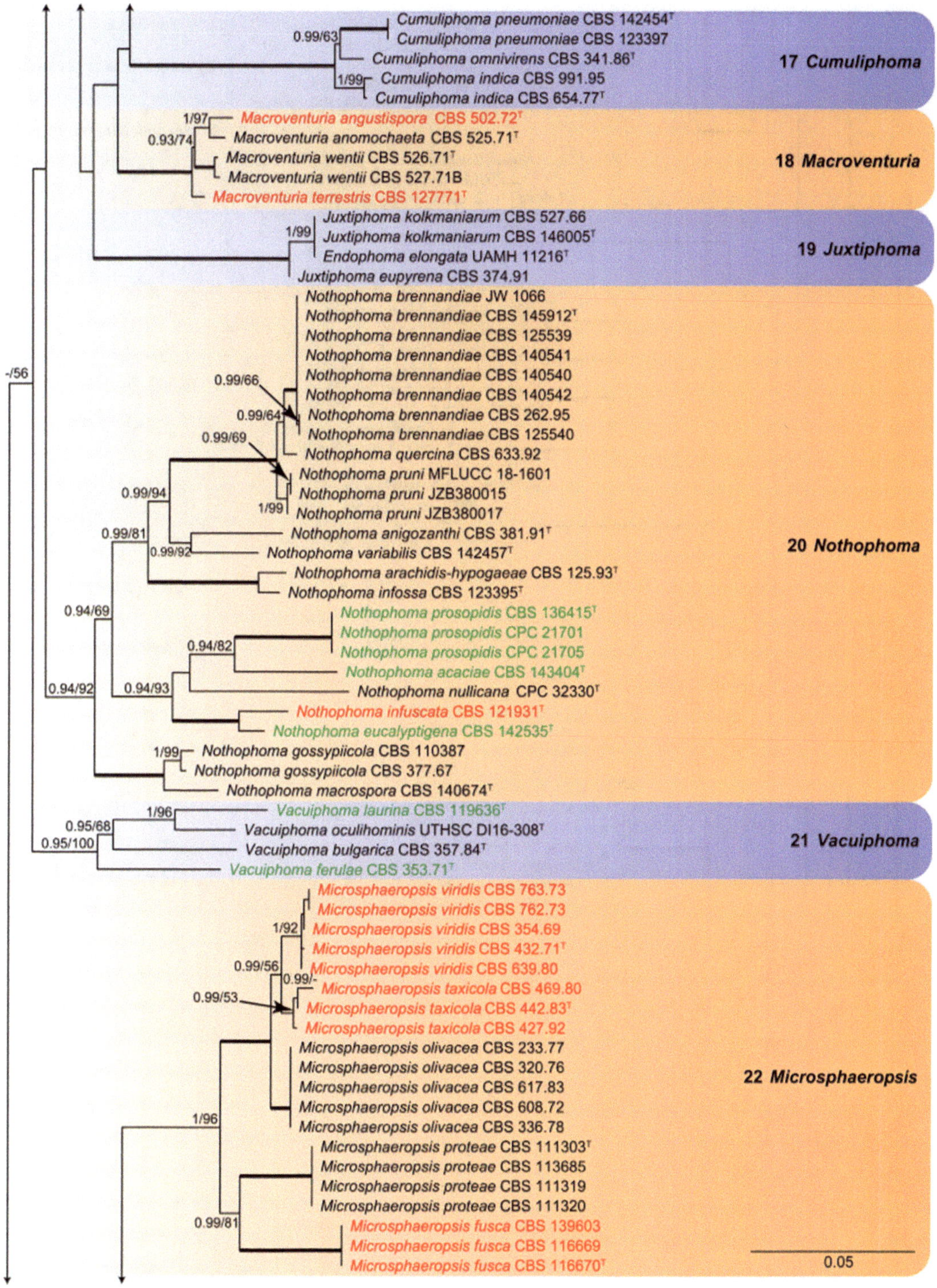

Fig. 2.5 (continued)

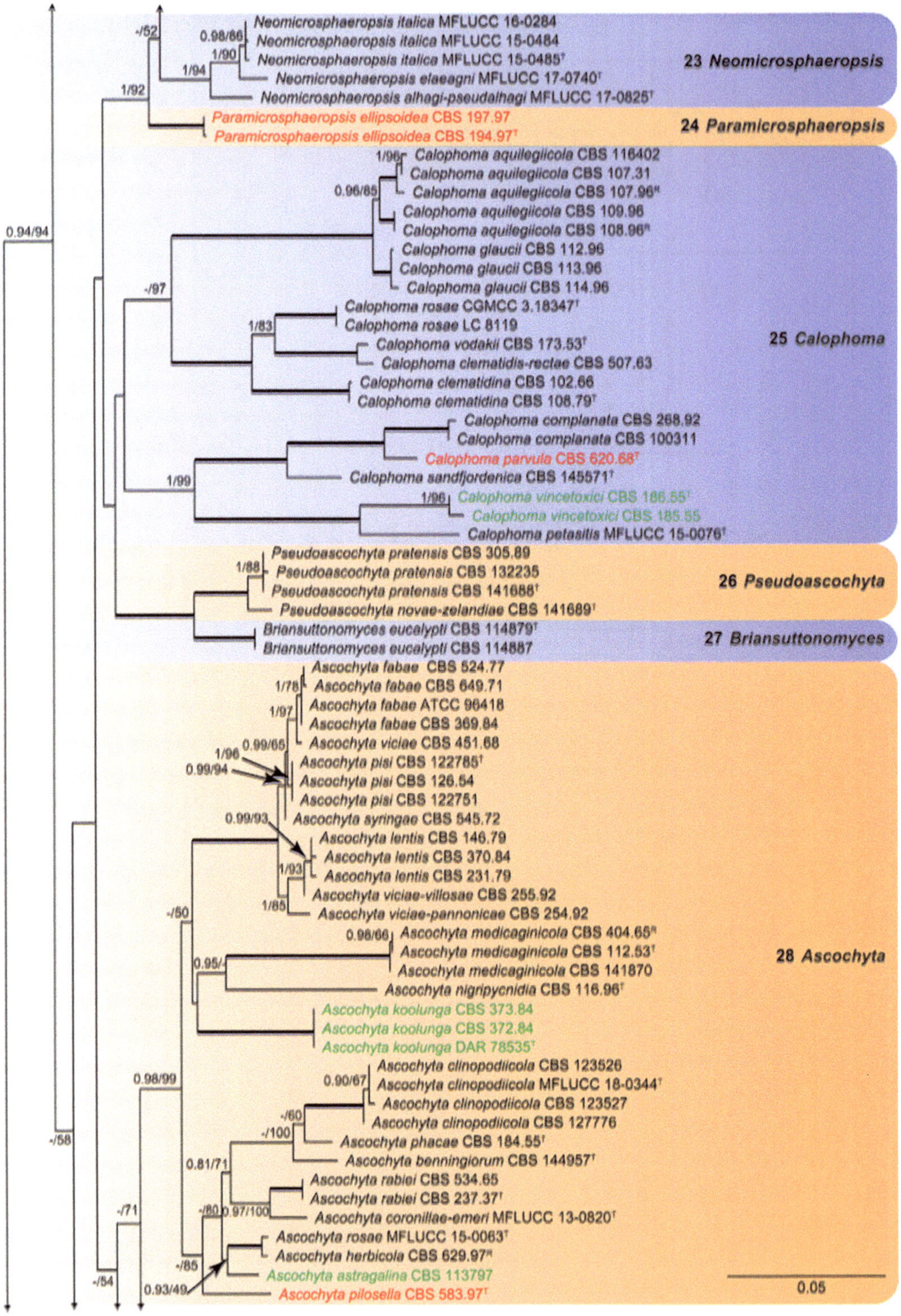

Fig. 2.5 (continued)

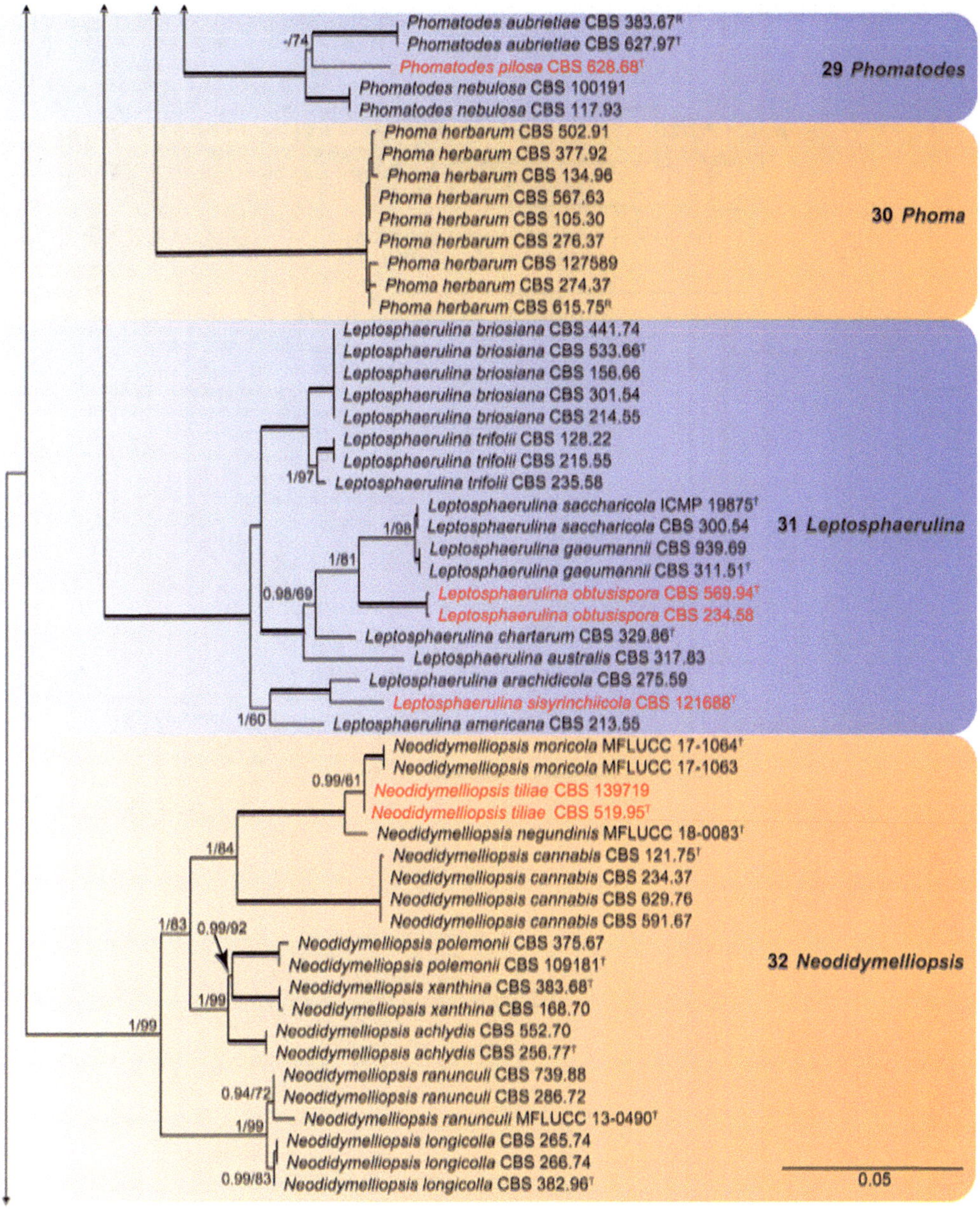

Fig. 2.5 (continued)

electrospray method for mass spectrometric analysis of fungal extracts was introduced, allowing database search as an identification tool, also using few agar plugs from fungal cultures (Smedsgaard 1997). For chemotaxonomic purposes, LC methods were later combined with both DAD and mass spectrometry (MS) detection, especially using electrospray MS. The chemotaxonomic results obtained are often treated using multivariate data analysis such as hierarchical cluster analysis, principal component analysis, correspondence analysis and fuzzy cluster analysis (Nielsen and Smedsgaard 2003).

Although the molecular techniques have greatly contributed to identification and taxonomy of *Phoma* species, the secondary metabolite profiling should also be evaluated as one of the relevant factors which could be useful to distin-

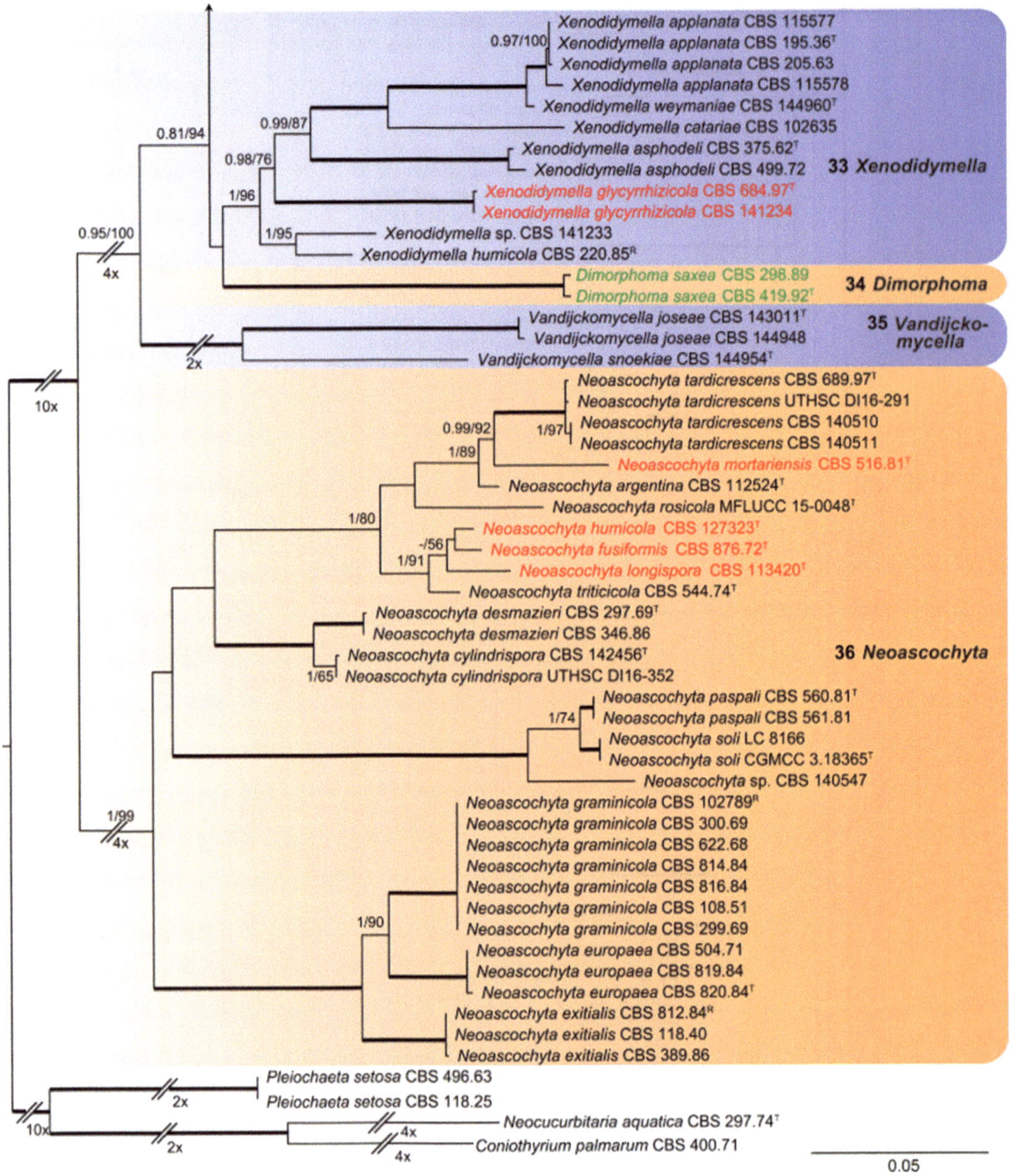

Fig. 2.5 (continued)

guish fungi within *Phoma* complex (Rai et al. 2014). A great diversity of metabolites, which includes alkaloids, peptides, steroids, terpenoids, phenols, quinones and flavonoids which can act as phytotoxins, antimicrobials and mycoherbicides from both terrestrial and marine origin *Phoma* spp., have already been described (Pedras and Biesenthal 2000; Cimmino et al. 2008; Wang et al. 2012; Rai et al. 2014, 2018). Increasing knowledge about the structure and functions of *Phoma* secondary metabolites profile is fundamental in the aim to correct identification as well as for chemotaxonomy of fungal taxa lacking information on their metabolomics content.

2.5 Future Considerations and Conclusions

Since Saccardo (1980) officially introduced the genus *Phoma*, remarkable progress has been made in taxonomy of *Phoma* and allied genera to obtain their better understanding. The extensive application of molecular tools and data resolved the phylogenetic relationships for taxa with similar morphological characteristics in *Didymellaceae* (Aveskamp et al. 2010; Chen et al. 2015; Valenzuela-Lopez et al. 2018). The currently recognized phylogenetic relationships of *Didymellaceae* were mainly based on the ex-type strains and specimens examined by Boerema and his co-workers, but still a great number of species required to be re-examined, including numerous *Phoma*-like species accepted by Boerema et al. (2004) and Aveskamp et al. (2010). However, a large number of unknown *Didymellaceae* species still exist in nature, especially from previously ignored ecosystems and diverse substrates. The majority of isolates represent 301 taxa currently incorporated in *Didymellaceae* that occurred on plants, either as pathogens, endophytes or saprobes (Hou et al. 2020). Increasing knowledge about diversity on the distribution and species from other substrates such as air, soil, water, limestone in caves and inorganic materials including asbestos, cement and paint or saline-alkali soils, deep-sea sediments and oligotrophic environments is fundamental to better understand their taxonomical status, as well as biology and environmental significance. To achieve this goal, global community-wide efforts to investigate species diversity of *Didymellaceae* are needed, by the use of comprehensive morphological, phylogenetic and chemotaxonomy tools.

References

Ariyawansa HA, Hyde KD, Jayasiri SC, et al (2015) Fungal diversity notes taxonomic and phylogenetic contributions to fungal taxa. Fungal Divers 75: 27–274.

Aveskamp MM, de Gruyter J, Crous PW (2008) Biology and recent developments in the systematics of *Phoma*, a complex genus of major quarantine significance. Fungal Divers 31: 1–18.

Aveskamp MM, Verkley GJM, de Gruyter J, Murace MA, Perelló A, Woudenberg JHC, Groenewald JZ, Crous PW (2009a) DNA phylogeny reveals polyphyly of *Phoma* section *Peyronellaea* and multiple taxonomic novelties. Mycologia 101: 363–382. doi: https://doi.org/10.3852/08-199

Aveskamp MM, Woudenberg JHC, de Gruyter J, Turco E, Groenewald JZ, Crous PW (2009b) Development of taxon-specific sequence characterized amplified region (SCAR) markers based on actin sequences and DNA amplification fingerprinting (DAF): a case study in the *Phoma exigua* species complex. Mol Plant Pathol 10: 403–414. https://doi.org/10.1111/j.1364-3703.2009.00540.x

Aveskamp MM, de Gruyter J, Woudenberg JHC, Verkley GJM, Crous PW (2010) Highlights of the *Didymellaceae*: A polyphasic approach to characterise *Phoma* and related pleosporalean genera. Stud Mycol 65: 1–60.

Balmas V, Scherm B, Ghignone S, Salem AOM, Cacciola SO, Migheli Q (2005) Characterisation of *Phoma tracheiphila* by RAPD-PCR, microsatellite-primed PCR and ITS rDNA sequencing and development of specific primers for *in planta* PCR detection. Eur J Plant Pathol 111: 235-247.

Benedict R (1970) Chemotaxonomic relationships among basidiomycetes. Adv Appl Microbiol 13: 1–23

Bennett A, Ponder MM, Garcia-Diaz J (2018) *Phoma* Infections: Classification, Potential Food Sources, and Their Clinical Impact. Microorganisms , 6: 58; doi:https://doi.org/10.3390/microorganisms6030058.

Boerema GH (1964) *Phoma herbarum* Westend., the type-species of the form-genus *Phoma* Sacc. Persoonia 3: 9-16.

Boerema GH (1970) Additional notes on *Phoma herbarum*. Persoonia 6: 15-48.

Boerema GH (1976) The species studied in culture by Dr Dennis. Trans Br Mycol Soc 67: 289-319.

Boerema GH (1993) Contributions towards a monograph of *Phoma* (Coelomycetes) – II. Section *Peyronellaea*. Persoonia 15: 197–221.

Boerema GH (1997) Contributions towards a monograph of *Phoma* (Coelomycetes) – V. Subdivision of the genus in sections. Mycotaxon 64: 321–333.

Boerema GH (2003) Contributions towards a monograph of *Phoma* (Coelomycetes) – X. Section *Pilosa* (taxa with a *Pleospora* teleomorph) and nomenclatural notes on some other taxa. Persoonia 18: 153–161.

Boerema, G.H. and Bollen, G.J. (1975). Conidiogenesis and conidial septation ads differentiating criteria between *Phoma* and *Ascochyta*. Persoonia 8: 111-144

Boerema GH, de Gruyter J (1998) Contributions towards a monograph of *Phoma* (Coelomycetes) – VII. Section *Sclerophomella*: Taxa with thick-walled pseudoparenchymatous pycnidia. Persoonia 17: 81–95.

Boerema GH, de Gruyter J (1999) Contributions towards a monograph of *Phoma* (Coelomycetes) –

III-Supplement: Additional species of section *Plenodomus*. Persoonia 17: 273–280.

Boerema GH, Dorenbosch MMJ (1973) The *Phoma* and *Ascochyta* species described by Wollenweber and Hochapfel in their study on fruit-rooting. Stud Mycol 3: 1-50.

Boerema GH, de Gruyter J, van Kesteren HA (1994) Contributions towards a monograph of *Phoma* (Coelomycetes) – III-1. Section *Plenodomus*: Taxa often with a *Leptosphaeria* teleomorph. Persoonia 15: 431–487.

Boerema GH, Loerakker WM, Hamers MEC (1996) Contributions towards a monograph of *Phoma* (Coelomycetes) – III-2. Misapplications of the type species-name and the generic synonyms of section *Plenodomus* (Excluded species). Persoonia 16: 141–190.

Boerema GH, de Gruyter J, Noordeloos ME (1997) Contributions towards a monograph of *Phoma* (Coelomycetes) – IV. Section *Heterospora*: Taxa with large sized conidial dimorphs, *in vivo* sometimes as *Stagonosporopsis* synanamorphs. Persoonia 16: 335–371.

Boerema GH, De Gruyter J, Noordeloos ME, Hamers MEC (2004) Phoma *identification manual. Differentiation of specific and infraspecific taxa in culture*. CABI Publishing, United Kingdom.

Brewer JG, Boerema GH (1965) Electron microscope observations on the development of pycnidiospores in *Phoma* and *Ascochyta* spp. Proceedings of the Koninklijke Nederlandse Akademie van Wetenschappen C. 68: 86-97.

Chen Q, Jiang JR, Zhang GZ, et al. (2015) Resolving the *Phoma* enigma. Stud Mycol 82: 137–217.

Chen Q Hou LW, Duan WJ, et al. (2017) *Didymellaceae* revisited. Stud Mycol 87: 105–159.

Chilvers MI, Rogers JD, Dugan FM, Stewart JE, Chen WD, Peever L (2009) *Didymella pisi sp. nov.*, the teleomorph of *Ascochyta pisi*. Mycological Research 113: 391–400. doi: https://doi.org/10.1016/j.mycres.2008.11.017

Cimmino A, Andolfi A, Berestetskiy A, Evidente A (2008) Production of phytotoxins by *Phoma exigua* var. *exigua*, a potential mycoherbicide against perennial thistles. J Agric Food Chem 56: 630-634.

Crous PW, Gams W (2000) *Phaeomoniella chlamydospora* gen. et comb. nov., a causal organism of Petri grapevine decline and esca. Phytopathol Mediterr 39: 112- 118.

Cullen DW, Toth IK, Boonham N, Walsh K, Barker I, Lees AK (2007) Development and validation of conventional and quantitative polymerase chain reaction assays for the detection of storage rot potato pathogens, *Phytophthora erythroseptica*, *Pythium ultimum*, *Phoma foveata*. J Phytopathol 155: 309-315.

de Gruyter J (2002). Contributions towards a monograph of *Phoma* (Coelomycetes) – IX. Section *Macrospora*. Persoonia 18: 85–102.

de Gruyter J, Boerema GH (2002) Contributions towards a monograph of *Phoma* (Coelomycetes) – VIII. Section *Paraphoma*: Taxa with setose pycnidia. Persoonia 17 ("2001"): 541–561.

de Gruyter J, Noordeloos ME (1992) Contributions towards a monograph of *Phoma* (*Coelomycetes*) – I. 1. Section *Phoma*: Taxa with very small conidia *in vitro*. Persoonia 15: 71-92.

de Gruyter J, Boerema GH, van der Aa HA (2002) Contributions towards a monograph of *Phoma* (Coelomycetes) – VI-2. Section *Phyllostictoides*: Outline of its taxa. Persoonia 18: 1–53.

De Gruyter, J., Noordeloos, M.E. and Boerema, G.H. (1993). Contributions towards a monograph of *Phoma*: (*Coelomycetes*) - I. 2. Section *Phoma*: Additional taxa with very small conidia and taxa with conidia up to 7 μm long. Persoonia 15: 369–400.

De Gruyter J, Noordeloos ME, Boerema GH (1998). Contributions towards a monograph of *Phoma* (*Coelomycetes*) – I. 3. Section *Phoma*: taxa with conidia longer than 7 μm. Persoonia 16: 471–490.

de Gruyter J, Aveskamp MM, Woudenberg JHC, Verkley GJM, Groenewald JZ, Crous PW (2009) Molecular phylogeny of *Phoma* and allied anamorph genera: towards a reclassification of the *Phoma* complex. Mycological Research 113: 508–519.

de Gruyter J, Woudenberg JHC, Aveskamp MM, Verkley GJM, Groenewald JZ, Crous PW (2010) Systematic reappraisal of species in *Phoma* section *Paraphoma*, *Pyrenochaeta* and *Pleurophoma*. Mycologia 102: 1066–1081.

de Gruyter J, Woudenberg JHC, Aveskamp MM, Verkley GJM, Groenewald JZ, Crous PW (2012) Redisposition of *Phoma*-like anamorphs in *Pleosporales*. Stud Mycol 75: 1–36.

de Gruyter J, Woudenberg JHC, Aveskamp MM, et al (2013) Redisposition of Phoma-like anamorphs in Pleosporales. Stud Mycol 75: 1–36.

Dennis RWG (1946) Notes on some British fungi ascribed to *Phoma* and related genera. T Brit Mycol Soc 29: 11-42.

EPPO Global Database (2019). https://gd.eppo.int/.

Filtenborg O, Frisvad JC, Svendsen JA (1983) Simple screening method for moulds producing intracellular mycotoxins in pure cultures. Appl Envitron Microbiol 45:581–585.

Goidànich G (1937) Studi sulla microflora fungina della pasta di legno destinata alla fabbricazione della carte. cont. Bolletino della Stazione di Patologia Vegetale di Roma 17: 405–474.

Grodona I, Monte E, Garcia-Acha I, Sutton BC (1997) *Pyrenochaeta dolichi*: an example of a confusing species. Mycological Research 101: 1404–1408.

Hou LW, Groenewald Z, Pfenning LH, Yarden O, Crous PW, Cai L (2020) The Phoma-like dilemma. Stud Mycol 96: 309-396.

Hughes SJ (1953) Conidiophores, conidia and classification. Can J Botany 31:577-599.

Irinyi L, Kövics GJ, Sándor E (2009) Taxonomical re-evaluation of *Phoma*-like soybean pathogenic fungi. Mycological Research 113: 249–260. doi: https://doi.org/10.1016/j.mycres.2008.11.003

Ito S, Iwadare S (1934) Studies on the red blotch of rice grains. Hokkaido Agricultural Experiment Station 31: 1–84.

Koch CA, Utkhede RS (2004) Development of a multiplex classical polymerase chain reaction technique for detection of *Didymella bryoniae* in infected cucumber tissues and greenhouse air samples. Can J Plant Pathol 26: 291-298.

Licciardella G, Grasso FM, Bella P, Cirvilleri G, Grimaldi V, Catara V (2006) Identification and detection of *Phoma tracheiphila*, causal agent of Citrus Mal Secco disease, by real-time polymerase chain reaction. Plant Dis 90: 1523-1530.

Mendes-Pereira E, Balesdent MH, Brun H, Rouxel T (2003) Molecular phylogeny of the *Leptosphaeria maculans* - *L. biglobosa* species complex. Mycological Research 107: 1287–1304.

Monte E, Bridge PD, Sutton BC (1991) An integrated approach to *Phoma* systematics. Mycopathologia 115: 89–103

Morgan-Jones G (1988) Studies on the genus *Phoma*. XIV. Concerning *Phoma herbarum*, the type species, a widespread saprophyte. Mycotaxon 33: 81-90.

Morgan-Jones G, White JF (1983) Studies in the genus *Phoma*. I. *Phoma americana*. Mycotaxon 16: 403-413.

Nielsen KF, Smedsgaard J (2003) Fungal metabolite screening: database of 474 mycotoxins and fungal metabolites for dereplication by standardized liquid chromatography-UV-mass spectrometry methodology. J Chromatogr A 1002:111–136.

Pedras M, Biesenthal CJ (2000) HPLC analyses of cultures of Phoma species: differentiation among groups and species through secondary metabolite profiles. Can J Microbiol 46: 685–691.

Punithalingam E (2004) Book review: Conidial fungi, *Phoma* Identification Manual: Differentiation of specific and infra-specific taxa in culture. Mycotaxon 90: 487–493.

Rai MK, Deshmukh P, Gade A, Ingle A, Kövics GJ, Irinyi L (2009) *Phoma* Saccardo: Distribution, secondary metabolite production and biotechnological applications. Crit Rev Microbiol 35: 182–196.

Rai MK, Tiwari VV Irinyi L, Kövics GJ (2014) Advances in taxonomy of genus *Phoma*: Polyphyletic nature and role of phenotypic traits and molecular systematics. Indian J Microbiol 54: 123-128.

Rai M, Gade A, Zimowska B, Ingle PA, Ingle P (2018) Marine-derived *Phoma*- the gold mine of bioactive compounds. Appl Microbiol Biot 102: 9053-9066. https://doi.org/10.1007/s00253-018-9329-2

Saccardo PA (1880) *Conspectus generum fungorum Italiae inferiorum nempe ad Sphaeropsideas, Melanconieas et Hyphomyceteas pertinentium systemate sporologico dispositorum*. Michelia 2: 1-38.

Saccardo PA (1884) *Sylloge Sphaeropsidearum et Melanconiearum omnium hucusque cognitorum*. Sylloge Fungorum 3: 1-860.

Schoch CL, Seifert KA, Huhndorf S, Robert V, Spouge JL, Levesque CA, Chen W (2012) Nuclear ribosomal internal transcribed spacer (ITS) region as a universal DNA barcode marker for Fungi. Proceedings of the National Academy of Sciences of the United States of America 9(16): 6241–6246. doi: https://doi.org/10.1073/pnas.1117018109

Smedsgaard J (1997) Micro-scale extraction procedure for standardized screening of fungal metabolite production in cultures. J Chromatogr A 760:264–270.

Smedsgaard J and Nielsen J (2005) Metabolite profiling of fungi and yeast: from phenotype to metabolome by MS and informatics. J Exp Bot 56: 273–286.

Somai BM, Keinath AP, Dean RA (2002) Development of PCR-ELISA for detection and differentiation of *Didymella bryoniae* from related *Phoma* species. Plant Dis 86: 710- 716.

Sutton BC (1964) *Phoma* and related genera. Trans Brit Mycol Soc 47: 497- 509.

Sutton BC (1977) Coelomycetes VI. Nomenclature of generic names proposed for Coelomycetes. Mycological Papers 141: 1–253.

Sutton BC (1980) *The Coelomycetes. Fungi Imperfecti with Pycnidia, Acervuli and Stromata.* 1st edn. Commonwealth Mycological Institute, United Kingdom.

Thambugala KM, Daranagama DA, Phillips AJL, et al. (2016) Microfungi on Tamarix. Fungal Divers 82: 239–306.

Torres M, White JF, Cazares G, Bergen M, Bischoff JF, Sullivan RF (2005) A new species and its phylogenetic placement in the *Didymella/Phoma* complex (*Phaeosphaeriaceae*, *Pleosporale*s). Mycotaxon 93: 297–308.

Valenzuela-Lopez N, Cano-Lira JF, Guarro J, et al. (2018) *Coelomycetous Dothideomycetes* with emphasis on the families *Cucurbitariaceae* and *Didymellaceae*. Stud Mycol 90: 1–69.

van der Aa HA, Noordeloos ME, de Gruyter J (1990) Species concepts in some larger genera of the *Coelomycetes*. Stud Mycol 32: 3-19.

van der Aa HA, Boerema GH, de Gruyter J (2000) Contributions towards a monograph of *Phoma* (Coelomycetes) – VI. 1. Section *Phyllostictoides*: characteristics and nomenclature of its type species *Phoma exigua*. Persoonia 17: 435-456.

Velázquez E, del Villar, M, Grondona I, Monte E, González-Villa T (2006) Ultrastructural and chemotaxonomic analysis of a xylanolytic strain of *Cryptococcus adeliensis* isolated from sheep droppings in Spain. Arch Microbiol 186: 195–202.

Von Szilvinyi A (1936) Archiv für Hydrobiologie Supplement 14. Tropische Binnengewasser 6a: 519.

Wang LW, Xu BG, Wang J., Su ZZ, Lin FC, Zhang CL, Kubicek CP (2012) Bioactive metabolites from *Phoma* species, an endophytic fungus from the Chinese medicinal plant *Arisaema erubescens*. Appl Microbiol Biot 93:1231-1239.

Wijayawardene NN, Hyde KD, Wanasinghe DN, et al. (2016) Taxonomy and phylogeny of dematiaceous coelomycetes. Fungal Divers 77: 1–316.

Wollenweber HW, Hochapfel H (1936) Beiträge zur Kenntnis parasitärer und saprophytotischer Pilze. I.

Phomopsis, *Dendrophoma*, *Phoma* und *Ascochyta* und ihre Beziehung zur Fruchtfäule. Z Parasitenk 8: 561-605.

Woudenberg JHC, Aveskamp MM, de Gruyter J, Spiers AG, Crous PW (2009) Multiple *Didymella* teleomorphs are linked to the *Phoma clematidina* morphotype. Persoonia 22: 56–62. doi: https://doi.org/10.3767/003158509X427808

Zimowska B (2011) Conidiogenesis of *Phoma strasseri* the fungus responsible for black stem and rhizomes rot in peppermint (*Mentha piperita*). Acta Sci. Pol. Hortorum Cultus, 10 (4): 171–178.

Zimowska B, Król ED, Furmańczyk A, Abramczyk B, Okoń S (2018) Molecular characterization of *Boeremia strasseri* the causal agent of black stems and rhizomes rot of peppermint. J Plant Pathol 100: 13–24, doi: https://doi.org/10.1007/s42161-018-0003-4

Part II

Diversity, Diseases and Management

Phoma on Medicinal and Aromatic Plants

3

Beata Zimowska

3.1 Introduction

Within *Didymellaceae* family, *Phoma* genus is the most polyphyletic one, including species worldwide recorded, with differentiated biological specialization (Chen et al. 2015). The genus consists of pathogens, opportunists, saprobic species from a wide range of substrates, endophytes, and human pathogens (Rai 1989; Aveskamp et al. 2008; de Gruyter et al. 2009). The majority of *Phoma* species are plant pathogens causing mainly leaf and stem spots on land plants. Many are recognized as important quarantine species, posing a threat to a diverse range of crops (Aveskamp et al. 2010). Terrestrial fungi belonging to *Phoma sensu lato* are well-known phytopathogens of medicinal and aromatic plants (MAPs), causing significant loses in commercial crops (Zimowska 2007a, b, 2008, 2012). They also inhabit tissue of the mentioned group of plants as saprotrophic and endophytic species (Nicoletti and Fiorentino 2015; Chen et al. 2020). *Phoma sensu lato* are considered as important yield-reducing factors, which may severely curtain biomass production. Pathogenic species could negatively affect assimilation rate due to changes in physiological parameters, as it happens when leaves, roots, or vessels are infected, but also competition may exert remarkable effect on plant metabolism when a shared use of environmental resources is of concern (Kegge and Pierik 2010).

Medicinal and aromatic plants (MAPs) are considered as the minor crops; that's why research about issue concerning plant diseases is generally lacking compared with economically strategic crops, which the bigger attention of research is addressed to. However, the increasing rate of cultivation of MAPs observed recently due to the relevance of this group of plants and rapid spread of MAPs monocultures over wider areas will result in a major concern about pathogens, including *Phoma sensu lato.* Nowadays, mainly in Europe, Asia, and America, an increasing interest on MAPs cultivation is observed resulting from the consumer demand towards the healthy diet and natural products (Rodeva et al. 2016). Cultivation of MAPs in Poland reached 20,000 ha last year, which puts this country at the second place as regards the crop area in Europe. At present, more than 70 species of medicinal and aromatic plants are cultivated in this European country. During the last two decades, in southeastern Poland different species belonging to Lamiaceae (thyme, lemon balm, sage, peppermint, oregano, motherworts) have been cultivated. The interest in the cultivation of these species is primarily determined by their content of essential oils or other secondary metabolites.

B. Zimowska (✉)
Department of Plant Protection, Institute of Plant Pathology and Mycology, University of Life Sciences in Lublin, Lublin, Poland
e-mail: beata.zimowska@up.lublin.pl

M. Rai et al. (eds.), *Phoma: Diversity, Taxonomy, Bioactivities, and Nanotechnology*,
https://doi.org/10.1007/978-3-030-81218-8_3

The concentration of the MAPs crops in this region led to an increase of pathogen occurrence, incorporating *Phoma sensu lato* fungi, which results in a decrease of the yields but also the quality of raw materials (Zimowska 2007a, b, 2008, 2015; Machowicz-Stefaniak et al. 2008).

Nowadays, although most of MAPs cultivation is still mainly concentrated in certain world regions, herbs and their derivatives are used throughout the world, also due to human migration arising from the globalization process. Therefore, the knowledge on the "new" pathogens detected within the last 5 years; the emerging ones, which have always been present but whose importance has grown in recent years; and finally the threatening ones, which have not been noted in MAPs cultivation areas so far or whose occurrence has been limited is promptly needed. This chapter is devoted to pathogenic, saprotrophic, and endophytic *Phoma sensu lato* fungi reported in recent years from MAPs species, summarizing the effort of scientific community concerning this particular topic.

3.2 Diversity of Pathogenic and Saprotrophic *Phoma sensu lato* on Medicinal and Aromatic Plants

Pathogenic species of *Phoma* infect MAPs hosts usually by wounds that are resulted by cultivation practices, interaction with other organisms, or weather conditions. In case of some species like *Boeremia strasseri*, direct infection through epidermis also may occur (Zimowska 2012). Primarily, the fungal hyphae grow intercellularly through plant tissues; after this symptomless stage, pathogen becomes necrotrophic, due to production of phytotoxins or to hypersensitive response, after which the pathogen has access to the resources of the dead plan tissue. At this stage, etiological symptoms in form of globose or flask-shaped pycnidial conidiomata are usually observed, with pale white, cream, or pinkish colored exudate of conidial matrix (Zimowska 2007a). Conidia, and in some cases also mycelial fragments, dispersed easily by water-splash, misting, or wind, constitute the secondary source of infection of new host plants. Also birds and insects may act as vectors. In absence of a suitable host, due to crop harvesting, for example, most species survive as saprobes on decaying organic material in the soil. In most cases, this material is the residue of plants that were previously infected. In this way, the hyphae may survive periods of stress, such as drought or extreme cold. The structures which are most suitable for long-term survival are the conidiomata and the uni- or multicellular chlamydospores, which could be formed by some *Phoma* species (Aveskamp et al. 2008; Zimowska 2011c).

The analysis of recent literature as well as studies carried out by the author of this chapter shows that different species of MAPs are infected or inhabited by distinct *Phoma sensu lato* species (Table 3.1). *Boeremia exigua* var. exigua (syn. *Phoma exigua* var. *exigua*) is the worldwide recorded wound and weak parasitic soil fungus, which in Eurasia has been isolated from more than 200 different host genera. The species is an opportunistic plant pathogen which may cause necrosis of leaves and stems or roots rot (Zimowska 2007b, 2008; Machowicz-Stefaniak et al. 2008). *B. exigua* var. *exigua* isolates turned out to be pathogenic for seedlings and schizocarps of lemon balm (*Melissa officinalis*) and thyme (*Thymus vulgaris*) in Poland (Machowicz-Stefaniak et al. 2008), as well as for dill (*A. graveolens*) (Fig. 3.1).

B. exigua var. *exigua* has been commonly isolated from roots and stem base of many diverse MAPs species with symptoms of necrosis and rot (Table 3.1). This species was often obtained from the inner tissues of the MAPs seeds decreasing the ability of seeds to germinate and causing the symptoms of pre-emergence and post-emergence damping off of the seedlings. It must be highlighted that sowing material of MAPs plays the significant role as the source of primary infection for the following generation of plants (Machowicz-Stefaniak and Zimowska 2000). Black stem and rhizomes rot of peppermint (*Mentha piperita*), also called phomosis of mint, is considered as the most harmful disease of this plant species and is caused by *Boeremia strasseri*

Table 3.1 Pathogenic and saprotrophic *Phoma sensu lato* reported from medicinal and aromatic plants

Phoma sp.[a]	Plant species/disease or (and) organ	Country	Reference
Boeremia exigua var. *exigua*	*Melissa officinalis*/leaf and stem	Poland	Wielgusz and Seidler-Łożykowska (2017)
Boeremia exigua var. *exigua*	*Hypericum perforatum*/leaf and stem	Poland	Wielgusz and Seidler-Łożykowska (2017)
Didymella cari	*Carum carvi*/blossom blight	Canada	Armstrong-Cho and Banniza (2020)
Didymella cari	*Coriandrum sativum*/blossom blight	Canada	Armstrong-Cho and Banniza (2020)
Phoma anethi	*Carum carvi*/damping off seedlings	Lithuania	Mačkinaitė (2012)
Phoma anethi	*A. graveolens*/stem	India	Pandotra and Sastry (1969)
Phoma betae	*Aloe vera*/leaf spot	India	Avasthi et al. (2013)
Phoma exigua var. *exigua*	*Carum carvi*/stem necrosis	Poland	Machowicz-Stefaniak and Zalewska (2008)
Phoma exigua var. *exigua*	*A. graveolens*/stem necrosis	Poland	Machowicz-Stefaniak and Zalewska (2008)
Phoma exigua var. *exigua*	*Trichosanthes cucumerina*/seed-borne	India	Akhtar et al. (2017)
Phoma exigua var. *exigua*	*Thymus vulgaris*/root and stem base rot	Poland	Machowicz-Stefaniak et al. (2002)
Phoma exigua var. *exigua*	*Hypericum perforatum*/root and stem base rot	Poland	Zimowska (2004)
Phoma exigua var. *exigua*	*Leonurus cardiaca*/root, stem base rot and leaf	Poland	Zimowska (2008)
Phoma exigua var. *exigua*	*Thymus vulgaris*/root, stem, leaf, seeds	Poland	Machowicz-Stefaniak et al. (2008)
Phoma exigua var. *exigua*	*Melissa officinalis*/root, stem, leaf and seed	Poland	Machowicz-Stefaniak et al. (2008)
Phoma exigua var. *exigua*	*Mentha piperita*/root, stem base rot and leaf	Poland	Machowicz-Stefaniak et al. (2008)
Phoma exigua var. *exigua*	*Carum carvi*/stem, leaf and seed	Poland	Machowicz-Stefaniak et al. (2008)
Phoma exigua var. *exigua*	*A. graveolens*/stem, leaf and seed	Poland	Machowicz-Stefaniak et al. (2008)
Phoma exigua var. *exigua*	*Pastinaca sativa*/root, stem and seed	Poland	Machowicz-Stefaniak et al. (2008)
Phoma exigua var. *exigua*	*Coriandrum sativum*/seed	Poland	Machowicz-Stefaniak et al. (2008)
Phoma exigua var. *exigua*	*Achillea millefolium*/seed	Poland	Machowicz-Stefaniak et al. (2008)
Phoma exigua var. *exigua*	*Satureja hortensis*/root, stem base rot and leaf	Poland	Zimowska (2010a)
Phoma cannabis comb. nov. (≡ *Depazea cannabis*)	*Cannabis sativa*/stem and leaf	USA	McPartland (1994)
Phoma capitulum	*Thymus vulgaris*/root	Poland	Machowicz-Stefaniak et al. (2002)
Phoma capitulum	*Hypericum perforatum*/root and stem	Poland	Zimowska (2004)
Phoma capitulum	*Leonurus cardiaca*/root	Poland	Zimowska (2007a)
Phoma capitulum	*Satureja hortensis*/root	Poland	Zimowska (2010a)
Phoma capitulum	*Plantago lanceolata*/root, stem and leaf	Poland	Zimowska (2013)
Phoma capitulum	*Perilla frutescens*/seed-borne	India	Akhtar et al. (2017)

(continued)

Table 3.1 (continued)

Phoma sp.[a]	Plant species/disease or (and) organ	Country	Reference
Phoma capitulum	*Tagetes patula*/seed-borne	India	Akhtar et al. (2017)
Phoma capitulum	*Tagetes cucumerina*/seed-borne	India	Akhtar et al. (2017)
Phoma capitulum	*Tagetes bracteata*/seed-borne	India	Akhtar et al. (2017)
Phoma complanata	*Archangelica officinalis*/stems, leaf, and umbels spot	Poland	Machowicz-Stefaniak et al. (2014)
Phoma dictamnicola	*Thymus vulgaris*/leaf	Poland	Machowicz-Stefaniak et al. (2002)
Phoma destructiva	*Ziziphus jujuba*/fruit shrink disease	Iran	Mirzaee (2014)
Phoma glaucii	*Thymus vulgaris*/leaf	Poland	Machowicz-Stefaniak et al. (2002)
Phoma glomerata	*Catharanthus roseus*/leaf	India	Nagaraja et al. (2010)
Phoma glomerata	*Foeniculum vulgare*/crown rot of seedlings	Italy	Lahoz et al. (2007)
Phoma glomerata	*Coriandrum sativum*/leaf spot	Bulgaria	Rodeva et al. (2013)
Phoma glomerata	*Satureja hortensis*/root and stem	Poland	Zimowska (2010a)
Phoma glomerata	*Zanthoxylum armatum*/leaf	India	Gupta et al. (2013)
Phoma glomerata	*Plantago lanceolata*/root, stem, and leaf	Poland	Zimowska (2013)
Phoma glomerata	*Thymus vulgaris*/root	Poland	Machowicz-Stefaniak et al. (2002)
Phoma glomerata	*Hypericum perforatum*/leaf	Poland	Zimowska (2004)
Phoma glomerata	*Leonurus cardiaca*/root and stem	Poland	Zimowska (2008)
Phoma glomerata	*Cannabis sativa*/stem and leaf	USA	McPartland (1994)
Phoma herbarum	*Salvia nemorosa*/leaf spot	Italy	Bertetti et al. (2018)
Phoma herbarum	*Acalypha indica*/leaf spot	India	Saranraj et al. (2016)
Phoma herbarum	*Phyllanthus amarus*/leaf spot	India	Saranraj et al. (2016)
Phoma herbarum	*Fagonia cretica*/leaf spot	Egypt	Abdel-Hafez and El Naggar (2006)
Phoma herbarum	*Peganum harmala*/leaf spot	Egypt	Abdel-Hafez and El Naggar (2006)
Phoma herbarum	*Origanum vulgare*/leaf spot	Iran	Basavand et al. (2019)
Phoma herbarum	*Scorzonera undulata*/leaf	Egypt	Abdel-Hafez and El Naggar (2006)
Phoma herbarum f. sp. *cannabis*	*Cannabis sativa*/leaf	USA	McPartland (1994)
Phoma heteroderae	*Plantago lanceolata*/root, stem, and leaf	Poland	Zimowska (2013)
Phoma labilis	*Leonurus cardiaca* L./root and stem	Poland	Zimowska (2008)
Phoma lavandulae	*Lavandula officinalis*/stem and leaf spot	Bulgaria	Radev (2020)
Phoma leonuri	*Origanum vulgare*/stem and leaf spot	Poland	Zimowska (2015)
Phoma ligulicola var. *inoxydabilis*	*Tanacetum cinerariifolium*/leaf and stem ray blight	Australia	Pethybridge Sarah et al. (2012)

Phoma sp.[a]	Plant species/disease or (and) organ	Country	Reference
Phoma medicaginis	*Medicago truncatula*/spring black stem and leaf spot	Tunisia	Djebali (2013)
Phoma medicaginis	*Medicago ciliaris*/spring black stem and leaf spot	Tunisia	Djebali (2013)
Phoma medicaginis	*Medicago sativa*/spring black stem and leaf spot	USA	Akamatsu et al. (2008)
Phoma multirostrata	*Coriandrum sativum*/stem necrosis	India	Khare et al. (2017)
Phoma multirostrata var. *macrospora*	*Origanum vulgare*/root, stem, and leaf	Poland	Zimowska (2015)
Phoma multirostrata var. *microspora*	*Thymus vulgaris*/stem	Poland	Machowicz-Stefaniak et al. (2002)
Phoma nepeticola	*Leonurus cardiaca*/stem and leaf spot		Zimowska (2008)
Phoma odoratissimi	*Polygonatum odoratum*/leaf spot	Italy	Karimi et al. (2016)
Phoma pinodella	*Trigonella foenum-graecum*/black stem and leaf spot	Australia	Bretag and Cunnington (2005)
Phoma pinodella	*Trigonella foenum-graecum*/black stem and leaf spot	India	Acharya et al. (2014)
Phoma piskorzii	*Cannabis sativa*/leaf spot	USA	McPartland (1994)
Phoma sabdariffae	*Hibiscus sabdariffa*/leaf spot	Malaysia	Ansari et al. (2013)
Phoma salviae	*Salvia officinalis*/leaf spot	Serbia	Pavlović et al. (2009)
Phoma salviae	*Salvia officinalis*/leaf and stalk spot	Serbia	Bokor et al. (2008)
Phoma septicidalis	*Leonurus cardiaca*/root and leaf	Poland	Zimowska (2007a)
Phoma sclerotioides	*Medicago sativa*/brown root rot	Canada	Wang et al. (2004)
Phoma sp.	*Carum carvi*/seed-borne	Lithuania	Mačkinaitė (2012)
Phoma sp.	*Trigonella foenum-graecum* /seed borne	Pakistan	Tariq et al. (2016)
Phoma sp.	*Pimpinella anisum*/seed-borne	Serbia	Pavlović (2015)
Phoma sp.	*Rosmarinus officinalis*/seed-borne	Brazil	Kruppa and Russomanno (2008)
Phoma sp.	*Origanum majorana*/seed-borne	Brazil	Kruppa and Russomanno (2008)
Phoma sp.	*Thymus vulgaris*/seed-borne	Brazil	Kruppa and Russomanno (2008)
Phoma sp.	*Salvia officinalis*/seed-borne	Brazil	Kruppa and Russomanno (2008)
Phoma sp.	*Leuzea carthamoides*/seed-borne	Serbia	Pavlović et al. (2012)
Phoma sp.	*Coriandrum sativum*/leaf spot	Bulgaria	Atanasova et al. (2013)
Phoma sp.	*Althea officinalis*/stem and leaf spot	Serbia	Pavlović et al. (2007)
Phoma sp.	*Gentiana lutea*/leaf spot	Serbia	Pavlović et al. (2011)
Phoma sp.	*Echinacea angustifolia*/stem	Serbia	Pavlović et al. (2009)

(continued)

Table 3.1 (continued)

Phoma sp.[a]	Plant species/disease or (and) organ	Country	Reference
Phoma sp.	*Echinacea purpurea*/stem	Serbia	Pavlović et al. (2009)
Phoma strasseri	*Mentha piperita*/rhizome and stem black rot	Poland	Zimowska (2007b)
Phoma strasseri	*Mentha piperita*/rhizome and stem black rot	Hungary	Paizs and Naggy (1975)
Phoma strasseri	*Mentha piperita*/rhizome and stem black rot	India	Kalra et al. (2005)
Phoma strasseri	*Mentha piperita*/rhizome and stem black rot	USA	Horner (1971)
Phoma subglomerata	*Leonurus cardiaca*/rot and stem	Poland	Zimowska (2008)
Phoma versabilis	*Origanum vulgare*/stem, leaf	Poland	Zimowska (2015)

[a]Species are reported according to the name used in the corresponding reference

Fig. 3.1 Necrosis of *Carum carvi* shoots caused by *Boeremia exigua* var. exigua

(syn. *Phoma strasseri*). The disease symptoms on the plants cultivated in the field are visible on the stems, first in the form of necrotic, slightly hollow spots enfolding the stem around. With time, the tissue in the place of the spots gets rotten. Such symptoms are most often formed just under the ground surface to the height up to 10 cm from the base. The secondary symptom is the reddening or (and) reduction of the leaf blades. Very young stems usually die out without any secondary symptoms of the disease. The rot proceeds very fast on the rhizomes. Young rhizomes rot away wholly, while the bark layer often comes off on older ones. The symptoms on peppermint cultivated in a glasshouse are similar to those that are observed in the field, with no secondary symptoms in the form of the reddening and reduction of the leaves, however. The yield losses as a result of plant infection can even reach 90%. In Poland the occurrence of the disease is connected with the spread of the pathogen by the infected rhizomes, which play the role of the propagation material and which constitute the most important source of primary infection for new plants on peppermint plantations. The mycelium living on the lower parts of 2-year-old plants can also be the source of infection (Zimowska 2007b). *Boeremia strasseri* makes infection through the epidermis of mint stems and rhizomes, omitting the stomata. Many of *Phoma sensu lato* pathogens penetrate the tissues of their hosts in this way. Such a model of infection is most frequently accompanied by the formation of appressorium, the adhesive structure which enables a pathogen to be attached to a plant tissue. A significant role in the initial stage of infection by *B. strasseri* is played by the mucous substance which is produced during the process of conidiogenesis at the final stage of differentiation of the conidium wall to its final structure. Thanks to it, the spore is attached to the surface of the tissue of mint stems and rhizomes, owing to which successive infection can take place (Zimowska 2011a). It is known that direct penetration of pathogenic fungi takes place as a result of cooperation of two factors, namely, the pressure of the fast-growing infection hypha and enzymes decomposing the cutin and next pectin compounds and cellulose composing the cell wall. In the case of *B. strasseri*, pectolytic enzymes produced by the pathogen take part in the process of active penetration. Among the numerous compounds produced by pathogenic species belonging to *Phoma sensu lato* group, enzymes play an essential role in the process of pathogenesis (Okafar et al. 2010). They act as catalyzers of the reaction of plant structure decomposition, thus making the processes of infection and colonization of the host tissues easier. In both processes the most important barriers to overcome are pectin compounds building the cell walls and the middle lamella (Gummadi and

Panda 2003). The enzymes produced by *B. strasseri* which are responsible for tissue maceration of mint stems and rhizomes mainly concerned β-glucosidase, the enzyme which catalyzes the reactions of cellobiose decomposition to glucose (Zimowska and Targoński 2015). Melouk and Horner (1973) pointed to the role of β-glucosidase and pectinolytic enzymes in the process of *B. strasseri* pathogenesis, and they showed that the optimal pH for the activity of β-glucosidase is within the range from 4.5 to 5.5. Zimowska and Targoński (2015) proved that pectinolytic enzymes of *B. strasseri* show the greatest activity at the temperature above 30 °C, which together with the eurythermal character of the fungus suggests that increased occurrence of black rot of stems and rhizomes should be taken into consideration during hot periods of vegetation (Fig. 3.2).

The occurrence of *Phoma sensu lato* fungi, including pathogenic as well as saprotrophic, has been noted on motherwort (*Leonurus cardiaca*). They included *Westerdykella capitulum* (syn. *Phoma capitulum*), *Phoma septicidalis*, *Allophoma labilis* (syn. *Phoma labilis*), and *Phoma nepeticola* (syn. *Ascochyta nepeticola*). *Phoma labilis* strains were obtained from the roots and the stems of motherwort, which points to the presence of this species in the cultivated environment of *Leonurus cardiaca* (Zimowska 2008). Cultures of the fungus were earlier obtained from the soil and the aboveground parts of other MAPs species in Italy, Turkey, and Israel. This species is considered as a frequently occurring necrotroph, especially in the regions of warm climate (Boerema et al. 2004). *Phoma nepeticola* was considered as particularly harmful pathogen. It caused symptoms in the form of small, regular, necrotic spots on motherwort leaves and stems, thus contributing to the worse quality of the raw material of Herba Leonuri cardiaceae (Zimowska 2008). This species is described in the literature as a commonly occurring pathogen of catmint (*Nepeta cataria*) and other species from the *Nepeta* genus (Boerema et al. 2004). A monograph on fungi of *Ascochyta* genus prepared by Mel'nik (2000) includes information that *P. nepeticola* may infect other plants from the *Labiatae* family, for example, peppermint *(Mentha piperita, Leonurus cardiaca)* (Fig. 3.3).

Calophoma complanata (syn. *Phoma complanata)* based on pathogenicity tests carried out in accordance with Koch's postulates was found to be a facultative pathogen of garden angelica (*Archangelica officinalis* Hoffm.) in Poland (Machowicz-Stefaniak et al. 2014; Zimowska et al. 2017). Favorable abiotic conditions, mainly high humidity, lead to the development of a variety of MAPs diseases caused by *Phoma sensu lato*. Periods of leaf wetness affected by the levels of relative humidity have been found to play important roles in various pathosystems at different stages of fungal growth. Measures of relative humidity can be used as a predictor to forecast diseases like spring black stem and leaf spot of

Fig. 3.2 Black rot of stems and rhizomes of *Mentha piperita* caused by *Boeremia strasseri*

Fig. 3.3 Necrotic spots on the leaf of *Leonurus cardiaca* casued by *Phoma nepeticola*

Medicago spp. caused by *P. medicaginis*. It has been found that increasing the duration of humid conditions leads to higher infections (Wang et al. 2004). *Phoma betae* is highly favored by rainfall and periods of high humidity. Pycnidia of the fungus exude abundant spores in gelatinous masses, which are readily spread by splashing rain or during sprinkler irrigation. Rainy and cloudy days before harvesting favor the inoculum buildup and spread of the pathogen to *A. vera* plants, causing infection of leaves. Rainfall period stimulates the sporulation of pycnidia and consequent spread of spores of the pathogen (Avasthi et al. 2013). As regards temperature requirements, *Phoma sensu lato* pathogens present diverse needments. *Phoma sclerotioides* favors a lower temperature range of 15–16 °C to optimally grow on alfalfa. Being inactive during hot summer months, the pathogen can grow at a temperature range of 0–27 °C. *P. sclerotioides* does not attack seedlings, and the disease invariably occurs in 4- to 6-month-old plants (Gray et al. 2004). In the case of *B. strasseri*, the growth of the pathogen colonies is possible at the temperature from 5 to 28 °C, with the optimum from 16 to 28 °C, which provided the basis to recognize *B. strasseri* as a eurythermal species. The fungus isolates formed the infection material within the temperature range from 5 to 28 °C; however, at the temperature from 24 to 28 °C, pycnidia were already formed after 4 days (Roustaee et al. 2000; Zimowska 2011b). Another factor that affects the infections of MAPs by *Phoma sensu lato* pathogens is soil, as a dynamic ecological habitat contains diverse groups of active fungal flora. Survival of soil-borne *Phoma* pathogens and their ability to cause disease in the following season depend primarily on soil characteristics such as temperature and moisture (Raaijmakers et al. 2009). Soil with high moisture content harbors significantly higher numbers of fungal propagules than soil with lower moisture-holding capacities. Moist soil conditions are necessary for favorable growth and survival, for example, opportunistic pathogen like *P. exigua* var. *exigua*. Occurrence of root rot and base stem rot of MAPs is highest in areas having highest soil moisture content (Zimowska 2007b, 2008).

The pathogenic *Phoma sensu lato* species which occur in the cultivation environment of MAPs crops can seriously affect the plant's healthiness. The quality of the raw material of MAPs is largely determined by the content of volatile oils and secondary metabolites produced in plant tissues. Pathogens infect the plant parts and disturb their metabolism through changes in physiological processes, which results in a decreased synthesis of volatile oils and modification of the composition of plants' volatile fractions (Zechini et al. 1995; Carrubba et al. 2015). The special use of raw materials obtained from this group of plants and restrictions in pesticide application point to the necessity of employing integrative methods of protection, where the priority should be the agrotechnical methods considering plant rotation, healthy propagation material, as well as early and effective discovery of first disease symptoms at plantations (WHO 2003).

3.3 Diversity of Endophytic *Phoma sensu lato* Species Associated with Medicinal and Aromatic Plants and Their Bioactivity

Fungal endophytes are the organisms which colonize the internal plant tissues without causing any apparent symptoms and provide resistance against abiotic and/or biotic stress (Larran et al. 2016). This group of microorganisms are now considered as an outstanding source of bioactive natural products, because they occupy unique biological niches as they grow in many unusual environments (Strobel and Daisy 2003; Strobel et al. 2004). They are known to produce metabolites that mimic the structure and function of host compounds (Strobel 2002), which can be provided directly to their hosts, thereby contributing to their chemical defense, or they might have transferred the corresponding genes to the host genome or vice versa. Endophytes must synthesize metabolites for the competence with co-occurring endophytes, host, and pathogens in order to colonize the host and for the nutrition (Wink 2008). They live primarily in the intercellular spaces of the hosts' tissue; acquire support, protection, and food from the nutrient-rich phloem; and, at the same time, produce one or more metabolomics that enhance protection from pathogens' infection to the hosts. Exciting possibilities for exploiting endophytic fungi for the production of a plethora of known and novel biologically active secondary metabolites provide the impetus for a number of investigations and encourage scientists all over the world to figure out how can endophytes be exploited for large-scale in vitro production of high-value biological compounds. The importance of bioactive natural compounds, which make them relevant for both agricultural and pharmaceutical applications, stimulates the scientific community to explore new environmental contexts and the associated microbial diversity. As the one of the largest frontiers in endophyte discovery, medicinal and aromatic plants (MAPs) represents a significant source of fungal endophytes, including *Phoma sensu lato* (Table 3.2), producing novel secondary metabolites with interesting bioactivities (Gunatilaka 2006; Chen et al. 2020).

The endophytic *Phoma* sp. strain producing anticancer compound has been isolated from *Cinnamomum mollissimum.* The anticancer compound was identified as 5-hydroxyramulosin, a polyketide produced via the action of pentaketide synthase. The production of the polyketide derives from the synthesis of melanin, a metabolic precursor, which increases the potential of virulence in fungal endophytes and improves its survivability under stressful condition. The result showed that the extracted 5-hydroxyramulosin demonstrated the significant inhibitory effect on the P388 murine leukemia cells (Santiago et al. 2012). *Phoma* sp. isolated from *Catharanthus roseus* demonstrated strong antioxidant potential. Total antioxidant capacity (TAC) of the extracts was estimated using DPPH in vitro method, and TAC IC50 values reached 2686.0 ± 51.7 mg/mL (Khiralla et al. 2015). The endophytic fungus *Phoma* sp. ZJWCF006 which was isolated from *Arisaema erubescens* has been found to produce different varieties of metabolite classes that showed strong or moderate antifungal and cytotoxic activities. (3S)-3,6,7-Trihydroxy-α-tetralone (1) was discovered as a new member of α-tetralone derivatives with antifungal activities. Cercosporamide demonstrated strong antifungal and antitumor activities in vitro. β-Sitosterol had antifungal activities, while trichodermin showed the broad spectrum of antifungal and antitumor activities (Wang et al. 2012). The endophytic strain of *P. herbarum* isolated from *Aegle marmelos* exhibited an impressive antibacterial activity against *Salmonella boydii* and *S. flexnii:* moreover this strain also demonstrated antifungal, antimalarial, and antioxidant activity (Kharwar et al. 2014). Caffeic acid was found in the metabolome profiles of isolate of *Paraphoma radicina*

Table 3.2 Endophytic *Phoma sensu lato* species reported from medicinal and aromatic plants

Phoma sp.[a]/strain	Plant species	Country	Reference
Didymella glomerata	*Salvia miltiorrhiza*	China	Jingfeng et al. (2013)
Didymella pedeiae	*Salvia miltiorrhiza*	China	Jingfeng et al. (2013)
Nothophoma multilocularis	*Rhazya stricta*	Saudi Arabia	Abdel-Wahab et al. (2017)
Paraphoma radicina	*Salvia abrotanoides*	Iran	Teimoori-Boghsani et al. (2020)
Phoma betae	*Costus spicatus*	India	Palanichamy et al. (2018)
Phoma betae	*Lepisanthes tetraphylla*	India	Palanichamy et al. (2018)
Phoma betae	*Kalidium foliatum*	China	Tan et al. (2018)
Phoma betae	*Artemisia nilagirica*	India	Parismita et al. (2016)
Phoma capitulum	*Justicia gendarussa*	China	Gangadevi and Muthumary (2008)
Phoma chrysanthemicola	*Phyllanthus amarus*	India	Kandasamy et al. (2015)
Phoma chrysanthemicola	*Bauhinia racemosa*	India	Kandasamy et al. (2015)
Phoma chrysanthemicola	*Artemisia nilagirica*	India	Parismita et al. (2016)
Phoma chrysanthemicola	*Calotropis procera*	India	Khan et al. (2007)
Phoma eupyrena	*Artemisia nilagirica*	India	Parismita et al. (2016)
Phoma eupyrena	*Euphorbia geniculata*	Egypt	Kamel et al. (2020)
Phoma eupyrena	*Azadirachta indica*	India	George et al. (2005)
Phoma eupyrena	*Azadirachta indica*	India	Verma et al. (2007)
Phoma epicoccinia	*Phyllanthus amarus*	India	Kandasamy et al. (2015)
Phoma exigua	*Artemisia nilagirica*	India	Parismita et al. (2016)
Phoma glomerata	*Solanum cernuum*	Brazil	Vieira et al. (2012)
Phoma glomerata	*Artemisia nilagirica*	India	Parismita et al. (2016)
Phoma glomerata D14	*Salvia miltiorrhiza*	China	Li et al. (2016)
Phoma herbarum	*Aegle marmelos*	India	Kharwar et al. (2014)
Phoma herbarum	*Curcuma longa*	India	Gupta et al. (2016)
Phoma herbarum	*Panax notoginseng*	China	Zheng et al. (2017)
Phoma herbarum	*Ruta graveolens*	Bulgaria	Rusanova et al. (2019)
Phoma herbarum D603	*Salvia miltiorrhiza*	China	Chen et al. (2020)
Phoma hedericola	*Mentha viridis*	India	Kumar et al. 2016
Phoma hedericola	*Calotropis procera*	India	Khan et al. (2007)
Phoma leveillei	*Catharanthus roseus*	India	Krishnamurthy et al. (2008)
Phoma leveillei	*Rauwolfia serpentina*	India	Krishnamurthy et al. (2008)
Phoma macrostoma	*Justicia gendarussa*	India	Palanichamy et al. (2018)
Phoma macrostoma	*Pueraria lobata*	China	Wan et al. (2014)
Phoma medicaginis	*Solanum torvum*	India	Kannan and Muthumary (2012)
Phoma medicaginis	*Solanum trilobatum*	India	Kannan and Muthumary (2012)
Phoma medicaginis	*Medicago* sp.	Italy	Weber et al. (2004)
Phoma moricola	*Solanum cernuum*	Brazil	Vieira et al. (2012)
Phoma multirostrata	*Withania somnifera*	India	Krishnamurthy et al. (2008)
Phoma multirostrata	*Catharanthus roseus*	India	Krishnamurthy et al. (2008)
Phoma multirostrata	*Cassia tora*	India	Krishnamurthy et al. (2008)
Phoma multirostrata	*Hemidesmus indicus*	India	Krishnamurthy et al. (2008)
Phoma nebulosa	*Mimusops elengi*	India	Kannan et al. (2017)
Phoma radicina	*Panax ginseng*	Korea	Park et al. (2012)
Phoma sorghina	*Tithonia diversifolia*	West Africa	Borges and Pupo (2006)
Phoma sp.	*Catharanthus roseus*	Sudan	Khiralla et al. (2015)
Phoma sp.	*Ocimum sanctum*	India	Rajagopal et al. (2010)
Phoma sp.	*Ocimum basilicum*	India	Rajagopal et al. (2010)
Phoma sp.	*Leucas aspera*	India	Rajagopal et al. (2010)
Phoma sp.	*Aconitum vilmorinianum*	China	Liu et al. (2019)

(continued)

Table 3.2 (continued)

Phoma sp.[a]/strain	Plant species	Country	Reference
Phoma sp.	*Cinnamomum mollissimum*	Malaysia	Santiago et al. (2012)
Phoma sp.	*Tylophora asthmatica*	India	Nalini et al. (2014)
Phoma sp.	*Rubia cordifolia*	India	Nalini et al. (2014)
Phoma sp.	*Viola odorata*	India	Gautam (2014)
Phoma sp.	*Cannabis sativa*	India	Gautam (2014)
Phoma sp.	*Glycyrrhiza glabra*	India	Arora et al. (2019)
Phoma sp.	*Ginkgo biloba*	Canada	Ateah (2019)
Phoma sp.	*Salvia officinalis*	Bulgaria	Rusanova et al. (2019)
Phoma sp.	*Panax ginseng*	China	Park et al. (2017)
Phoma sp.	*Mentha canadensis*	Brazil	Herrmann (2019)
Phoma sp.	*Cissus quadrangularis*	India	Tenguria and Firodiya (2015)
Phoma sp.	*Asclepias curassavica*	India	Tenguria and Firodiya (2015)
Phoma sp.	*Cassia fistula*	India	Tenguria and Firodiya (2015)
Phoma sp.	*Carissa carandas*	India	Tenguria and Firodiya (2015)
Phoma sp.	*Ocimum basilicum*	India	Tenguria and Firodiya (2015)
Phoma sp.	*Leucas aspera*	India	Tenguria and Firodiya (2015)
Phoma sp.	*Dysosma versipellis*	China	Tan et al. (2018)
Phoma sp.	*Vernonia amygdalina*	Indonesia	Praptiwi et al. (2020)
Phoma sp.	*Securinega suffruticosa*	China	Du et al. (2020)
Phoma sp.	*Acanthospermum australe*	Argentina	Giusiano et al. (2010)
Phoma sp.	*Centella asiatica*	Madagascar	Rakotoniriana (2012)
Phoma sp.	*Nothapodytes foetida*	India	Samaga and Rai (2016)
Phoma sp.	*Hypericum mysorense*	India	Samaga and Rai (2016)
Phoma sp.	*Hypericum japonicum*	India	Samaga and Rai (2016)
Phoma sp.	*Huperzia serrata*	Vietnam	Le et al. (2019)
Phoma sp.	*Tripterygium wilfordii*	China	Kumar and Hyde (2004)
Phoma sp.	*Stephania dielsiana*	China	Wu et al. (2020)
Phoma sp.	*Mimusops elengi*	India	Kannan et al. (2017)
Phoma sp.	*Salicornia europaea*	Korea	Park et al. (2016)
Phoma sp.	*Fagonia cretica*	Spain	Krohn et al. (2007)
Phoma sp.	*Saurauia scaberrinae*	Australia	Hoffman et al. (2008)
Phoma sp.	*Melastoma malabathricum*	India	Mishra et al. (2016)
Phoma sp. PF2	*Artemisia princeps*	China	Kim et al. (2019)
Phoma sp./ PG23	*Taraxacum mongolicum*	China	Zhang et al. (2013)
Phoma ZJWCF006	*Arisaema erubescens*	China	Wang et al. (2012)
Setophoma terrestris	*Panax ginseng*	Korea	Park et al. (2012)

[a]Species are reported according to the name used in the corresponding reference

endophytic in *Salvia abrotanoides*. Caffeic acid showed antibacterial, antifungal, and modulatory effects. Cryptotanshinone, another nor-abietanoid diterpenoid, which is a main bioactive compound of *S. abrotanoides* known for leishmanicidal, antiplasmodial, and cytotoxic activity (Sairafianpour et al. 2001) has been found to be produced in roots of *S. yangii* (Ślusarczyk et al. 2015). Very recently, this compound has been reported as a secondary metabolite of endophytic strains of *P. radicina*, independently of the host plant. Moreover, the effect of exogenous gibberellin (GA3) on *S. abrotanoides* and endophytic fungi was shown to have a positive effect on increasing the cryptotanshinone production in the plant as well as in endophytic fungi cultivated under axenic conditions (Teimoori-Boghsani et al. 2020). Salvianolic acid C was obtained from both mycelia and fermentation broth of *P. glomerata* D14 endophytic strain isolated from *S. miltiorrhiza*. This compound consists of two units of tashinol and one unit of caffeic acid, which exhibited firm anti-platelet aggregation effect (Chen et al. 2016), and thus it is recommended for cardiovascular and cerebrovascular diseases. It can also effectively inhibit human HMG-CoA reductase (hHMGR), a key enzyme in the cholesterol biosynthesis pathway. A high level of cholesterol can lead to atherosclerosis. So salvianolic acid C may exhibit hypolipidemic activity through the inhibition of hHMGR. In addition, it displayed significant antiproliferative effect on cancer cells, which blocks cell cycle through inhibiting tubulin polymerization and then induces apoptosis (Yang et al. 2010). In recent years, due to overcollection of *S. miltiorrhiza*, this medicinal plant is getting to exist in short supply in the natural environment; thus finding another alternative source of salvianolic acid C is urgently necessary. Such breeder can form an endophytic microbial community associated with *S. miltiorrhiza.* A new rare 14-nordrimane-type sesquiterpenoid, named phomanolide, together with five known compounds, (−)-6-methoxymellein, 7-hydroxy-3, 5-dimethyl-isochromen-1-one, norlichexanthone, 6-methylsa-licylic acid, and gentisyl alcohol, was obtained from the culture broth of an endophytic fungus *Phoma* sp. isolated from the roots of *Aconitum vilmorinianum.* Phomanolide and (−)-6-methoxymellein exhibited strong antiviral activities against influenza A virus (Liu et al. 2019). Antibacterial activity was displayed by 2-hydroxy-6-methylbenzoic acid isolated from endophytic *Phoma* sp./PG23 which was obtained from the leaves of *Taraxacum mongolicum* (Zhang et al. 2013), while antifungal activity against pathogens like *Alternaria panax*, *Botrytis cinerea*, *Cylindrocarpon destructans*, *Pythium* sp., and *Rhizoctonia solani* demonstrated endophytic *Phoma* sp. associated with *Panax ginseng* (Park et al. 2017). Phenolic compounds, as well as flavonoids exhibiting antioxidant, antibacterial, and antifungal properties, have been extracted from *Phoma* sp. endophyte inhabiting *Vernonia amygdalina* (Praptiwi et al. 2020). *P. herbarum* associated with *Curcuma longa* was found to produce 2,5-dihydroxy benzyl alcohol which is commonly known as gentisyl alcohol. The purified compound showed strong antagonistic activity against *Colletotrichum gloeosporioides* causing leaf spot of turmeric and started to be considered as a biocontrol agent for the host-specific phytopathogen (Gupta et al. 2016). Two isochromanone derivatives, (3*S*,4*S*)-3,8-dihydroxy-6-methoxy-3,4,5-trimethylisochroman-1-one and methyl (*S*)-8-hydroxy-6-methoxy-5-methyl-4a-(3-oxobutan-2-yl)benzoate, were isolated from the cultures of an endophytic fungus *Phoma* sp. PF2 obtained from *Artemisia princeps.* These bioactive secondary metabolites have been found to demonstrate antimicrobial activities (Kim et al. 2019). Endophytic fungus *P. betae* inhabiting in plant *Kalidium foliatum* was found to produce unique products spiciferone A, F, G, and H and spiciferol A demonstrating cytotoxic activity against cancer cells (Tan et al. 2018). Endophytic strain *Nothophoma multilocularis* yielded 55 active biological compounds, but 10 major products, di-n-octyl phthalate, 2-allyl-3,4-dimethoxybenzaldehyde, maltol, cetene, 1-tetradecene, E-15-heptadecenal, 2,5-cyclohexadien-1-one, 1-octadecene, diethyldithiophosphinic acid, and phenol, 2,4-di-t-butyl-6-nitrophenol, showed strong antimicrobial activity (Abdel-Wahab et al. 2017). Three known anthraquinones (1,7-dihydroxy-3-methyl-9,10-anthraquinone, 1,6-dihydroxy-3-methyl-9,10-anthraquinone, and 1-hydroxy-3-methyl-9,10-anthraquinone), one new anthraquinone (1,7-dihydroxy-3-hydroxymethyl-9,10-anthraquinone), and two new hexahydro anthraquinone derivatives, dendryols E and F, were isolated from the culture of the endophytic fungus *P. sorghina*, found in association with *Tithonia diversifolia* (Borges and Pupo 2006). The great number of biologically active compounds synthetized by endophytic *Phoma* species presents antifungal and antibacterial activities. The former aptitude depicted two compounds pyrenophorol and pyrenophorolic acid, obtained from endophytic *Phoma* strain associated with *Fagonia cretica.* However, very rarely secondary metabolites show the algaecide properties. The last metabolite regardless of antifungal activity against *Microbotryum violaceum* effectively inhabited the growth of *Chlorella* *fusca* (Krohn et al. 2007). Phomodione and cercosporamide derivatives were isolated from culture broth of a *Phoma* species, discovered as an endophyte on *Saurauia scaberrinae*. Phomodione exhibited a minimum inhibitory concentration of 1.6 lg/mL against *Staphylococcus aureus* and was found to be active against *P. ultimum*, *S. sclerotiorum*, and *R. solani* (Hoffman et al. 2008). Endophytic fungus *Phoma* species ZJWCF006 in *Arisaema erubescens* afforded the α-tetralone, (3S)-3,6,7-trihydroxy-α-tetralone, cercosporamide, β-sitosterol, and trichodermin compounds. The first one has been marked by antifungal activity against *F. oxysporum* and *R. solani*, while the second one exhibited cytotoxic aptitudes towards tumor cells (Wang et al. 2012).

It is well-known that fungal endophytes pose the source of profound compounds usually easier and more economical to produce, than from plants, considering the threat of depletion of natural resources. Therefore, if an endophytic fungus isolated from MAPs may participate in metabolic paths and boost their own natural biosynthetic activity, or may gain some genetic information to synthesize biologically active compounds closely related to those produced by the host plants, should be screened in view of assessing their biotechnological potential.

3.4 Conclusions and Future Prospects

The knowledge regarding MAPs diseases in their cultivation conditions represents a meaningful fund of experience also in the environments, in which these plants are growing. *Phoma sensu lato* constitutes the important and common group of pathogens affecting the quality and quantity of herbs, and due to the globalization of plant markets and increasing rate of MAPs production, it is possible to foresee that new, threatening pathogenic species which have not been noted in MAPs cultivation areas so far or whose occurrence was limited are going to be spread. Therefore, further investigations and active cooperation among researchers currently working on this topic worldwide should be carried out to understand various aspects of pathogenicity and mutual interaction between different members of *Phoma sensu lato* in causing a particular disease of MAPs. Besides the negative aspect, *Phoma* constitutes the huge diversity of endophytic fungi associated with medicinal and aromatic plants of outstanding importance. The available literature specifically concerning endophytic *Phoma sensu lato* highlighted that research in this field is promptly needed to avoid continuous depletion of natural resources of MAPs and achieve goals of biotechnological applications of bioactive compounds in pharmaceutical, agricultural, and other industries. A set of genetic metabolic and bioprocess engineering should be used to sustain and enhance production of high-value secondary metabolites from endophytic *Phoma sensu lato* strains, whose biosynthetic abilities can be improved through physical and chemical mutagenesis or various methods for genetic transformation. Therefore, due to an enormous potential to explore *Phoma sensu lato* endophytes from the diverse species of MAPs, a growing interest of researchers all over the world should be focused on the bioprospecting of endophytic fungal communities inhabiting this peculiar group of plants.

References

Abdel-Wahab MA, Bahkali AH, El-Gorban AM, Hodhod MS (2017) Natural products of *Nothophoma multilocularis* sp. nov. an endophyte of the medicinal plant *Rhazya stricta*. Mycosphere 8(8): 1185-1200.

Abdel-Hafez SII, El Naggar SM (2006) Morphological, reproductive and mycobiota characters of three wild medicinal plants inhabiting Western Mediterranean coastal land, Egypt. Feddes Repertorium: Zeitschrift für botanische Taxomonie und Geobotanik 117: 240-249.

Acharya K, Chakraborty N, Chatterjee S, Basu SK (2014) Fungal diseases of fenugreek. American Journal of Social Issues and Humanities ISSN 2276-6928.

Akamatsu HO, Chilvers MI, Peever TL (2008) First report of spring black stem and leaf spot of alfalfa in Washington State caused by *Phoma medicaginis*. Plant Dis 92: 833-833.

Akhtar J, Singh B, Aravindaram PK, Kumar A (2017) Status of seed-borne fungi in some indigenous medicinal and aromatic plants conserved in National Gene Bank, India. Indian Phytopathology 70: 206-215.

Ansari M, Eslaminejad T., Sarhadynejad Z, Eslaminejad T (2013) An overview of the roselle plant with particular reference to its cultivation, diseases and usages. Eur J Med Plants 135–145.

Arora P, Wani ZA, Ahmad T, Sultan P, Gupta S, Riyaz-Ul-Hassan S (2019) Community structure, spatial distribution, diversity and functional characterization of culturable endophytic fungi associated with *Glycyrrhiza glabra* L. Fungal Biol 123: 373-383.

Armstrong-Cho C, Banniza S (2020) Pathogens of caraway and coriander blossoms in western Canada. Can J Plant Pathol 42: 367-376.

Atanasova D, Maneva V, Dacheva S, Atanassov A (2013) Phytosanitary monitoring of coriander crops in south-east Bulgaria. Povrćarstvo, ukrasno, aromatično i ljekovito bilje 48.

Ateah D (2019) *A survey of fungal endophytes from healthy branches of Ginkgo biloba* (Doctoral dissertation).

Avasthi S, Gautam AK, Bhadauria R (2013) First report of *Phoma betae* on *Aloe vera* in India. Arch Phytopathol Plant Prot 46: 1508-1511.

Aveskamp MM, De Gruyter J, Crous PW (2008) Biology and recent developments in the systematics of *Phoma*, a complex genus of major quarantine significance. Fungal Divers 31: 1-18.

Aveskamp MM, de Gruyter J, Woudenberg JHC, Verkley GJM, Crous PW (2010) Highlights of the *Didymellaceae*: A polyphasic approach to characterise *Phoma* and related pleosporalean genera. Stud Mycol 65: 1–60.

Basavand E, Babaeizad V, Mirhosseini HA, Niri MD (2019) Occurrence of leaf spot disease caused by *Phoma herbarum* on oregano in Iran. J Plant Pathol 1-2.

Bertetti D, Gilardi G, Gullino ML, Garibaldi A (2018) Nuovi parassiti fungini su colture ornamentali comparsi in Italia settentrionale negli ultimi mesi. Protezione delle colture 118: 2-6 (in Italian).

Boerema GH, De Gruyter J, Noordeloos ME, Hamers MEC (2004) Phoma *identification manual. Differentiation of specific and infraspecific taxa in culture.* CABI Publishing, United Kingdom.

Bokor P, Tancik J, Habán M, Marinković BJ, Poláček M (2008) The occurrence of pests on lemon balm (*Melissa officinalis*) and garden sage (*Salvia officinalis*). Proc Nat Sci Matica Srpska Novi Sad 115: 59–64.

Borges WDS, Pupo MT (2006) Novel anthraquinone derivatives produced by *Phoma sorghina*, an endophyte found in association with the medicinal plant *Tithonia diversifolia* (Asteraceae). J Braz Chem Soc 17: 929-934.

Bretag TW, Cunnington JH (2005) First report of spring black stem and leaf spot in fenugreek (*Trigonella foenum-graecum*) caused by *Phoma pinodella* in Australia. Australasian Plant Pathology 34: 619-620.

Carrubba, A., Verde, G. L., & Salamone, A. (2015). Sustainable weed, disease and pest management in medicinal and aromatic plants. In Medicinal and Aromatic Plants of the World (pp. 205–235). Springer, Dordrecht.

Chen Q, Jiang J, Zhang G, Cai L, Crous PW (2015) Resolving the *Phoma* enigma. Stud Mycol 82: 137–217.

Chen Y, Zhang N, Ma J, Zhu Y, Wang M, Wang X, Zhang P (2016) A Platelet/CMC coupled with offline UPLC-QTOF-MS/MS for screening antiplatelet activity components from aqueous extract of Danshen. Journal of Pharmaceutical and Biomedical Analysis 117: 178-183.

Chen HM, Wu HX, He XY, Zhang HH, Miao F, Liang ZS (2020) Promoting tanshinone synthesis of *Salvia miltiorrhiza* root by a seed endophytic fungus, *Phoma herbarum* D603. China J. Chinese Mater. Medica 45: 65–71 doi:https://doi.org/10.19540/j.cnki.cjcmm.20191113.101.

Djebali N (2013) Aggressiveness and host range of *Phoma medicaginis* isolated from *Medicago* species growing in Tunisia. Phytopathol Mediterr 3-15.

Du W, Yao Li J, Sun C, Xia J, Wang B, Ren L (2020) Diversity and antimicrobial activity of endophytic fungi isolated from *Securinega suffruticosa* in the Yellow River Delta. PloS one 15 e0229589.

Gangadevi V, Muthumary J (2008) A simple and rapid method for the determination of taxol produced by fungal endophytes from medicinal plants using high performance thin layer chromatography. Chin J Chromat 26: 50-55.

Gautam AK (2014) Diversity of fungal endophytes in some medicinal plants of Himachal Pradesh, India. Archives of Phytopathology and Plant Protection 47: 537-544.

George KJ, GangaG, Varma RS, Sasikumar B, Saji KV (2005) Endophytic mycoflora of inner bark of *Azadirachta indica* A. Juss. Curr Sci 88: 218.

Giusiano G, Rodolfi M, Mangiaterra M, Piontelli E, Picco AM (2010) Endophytic fungi in medicinal plants of northeast of Argentina. I: Morphotaxonomic approach of their foliar community. Bol Micol 25: 15-27.

Gray F, Hollingsworth C, Groose R, Reedy C, Larsen R (2004) Brown root rot resistance *Phoma sclerotioides*. Available at: http://www.naaic.org/stdtests/brrootrot.pdf [Accessed 18 May 2020].

de Gruyter J, Aveskamp MM, Woudenberg JHC, Verkley GJM, Groenewald JZ, Crous PW (2009) Molecular phylogeny of *Phoma* and allied anamorph genera: towards a reclassification of the *Phoma* complex. Mycological Research 113: 508–519.

Gummadi SN, Panda T (2003) Purification and biochemical properties of microbial pectinases: a review. Process Biochem 38: 987-996.

Gunatilaka AL (2006) Natural products from plant-associated microorganisms: distribution, structural diversity, bioactivity, and implications of their occurrence. J Nat Prod 69:509–526.

Gupta D, Sharma S, Sharma YP (2013) Mycoflora and natural aflatoxin contamination in some traditional medicinal plants from Jammu, India. J Mycol Plant Pathol 43: 360.

Gupta S, Kaul S, Singh B, Vishwakarma RA, Dhar MK (2016) Production of gentisyl alcohol from *Phoma herbarum* endophytic in *Curcuma longa* L. and its antagonistic activity towards leaf spot pathogen *Colletotrichum gloeosporioides*. App Biochem Biotech 180: 1093-1109.

Herrmann LW, Poitevin CG, Schuindt LC, De Almeida AB, Pimentel IC (2019) Diversity of fungal endophytes in leaves of *Mentha piperita* and *Mentha canadensis*. Int J Botany Stud 4: 44-49.

Hoffman AM, Mayer SG, Strobel GA, Hess WM, Sovocool GW, Grange AH, Kelley-Swift EG (2008) Purification, identification and activity of phomodione, a furandione from an endophytic *Phoma* species. Phytochemistry 69: 1049-1056.

Horner CE (1971) Rhizome and stem rot of peppermint caused by *Phoma strasseri*. Plant Dis Rep 55: 814–816.

Jingfeng L, Linyun F, Ruiya L, Xiaohan W, Haiyu L, Ligang Z (2013) Endophytic fungi from medicinal herb *Salvia miltiorrhiza* Bunge and their antimicrobial activity. African J Microbiol Res 7: 5343–5349. doi: https://doi.org/10.5897/ajmr2013.6358.

Kalra A, Singh HB, Pandey R, Samad A, Patra NK, Kumar S (2005) Diseases in mint: causal organisms, distribution, and control measures. J Herbs Spices Med Plants 11: 71-91.

Kannan KP, Muthumary J (2012) Comparative analysis of endophytic mycobiota in different tissues of medicinal plants. Afr J Microbiol Res 6: 4219-4225.

Kannan KP, Abdul Basheed MI, Kannadhasan S, Pondurai S, Dhakshinamoorthy M (2017) Mycoendophytes Isolated from *Mimusops elengi*. L-A First report. Int Biol Biomed J 3: 25-29.

Kamel NM, Abdel-Motaal FF, El-Zayat SA (2020) Endophytic fungi from the medicinal herb *Euphorbia geniculata* as a potential source for bioactive metabolites. Arch Microbiol 202: 247-255.

Kandasamy P, Manogaran S, Dhakshinamoorthy M, Kannan KP (2015) Evaluation of antioxidant and antibacterial activities of endophytic fungi isolated from *Bauhinia racemosa* Lam and *Phyllanthus amarus* Schum and Thonn. J Chem Pharm Res 7: 366-379.

Karimi K, Khodaei S, Rota-Stabelli O, Arzanlou M, Pertot I (2016) Identification and characterization of two new fungal pathogens of *Polygonatum odoratum* (Angular Solomon's seal) in Italy. J Phytopathol 164: 1075-1084.

Kegge W, Pierik R (2010) Biogenic volatile organic compounds and plant competition. Trends Plant Sci 15: 126-132.

Khan R, Shahzad S, Choudhary MI, Khan SA, Ahmad A (2007) Biodiversity of the endophytic fungi isolated from *Calotropis procera* (Ait.) R. Br. Pak J Botany 39: e2239.

Kharwar RN, Upadhyay RS, Dubey NK, Raghuwanshi R (2014) Microbial diversity and biotechnology in food security. Springer India.

Khare MN, Tiwari SP, Sharma YK (2017) Disease problems in the cultivation of coriander (*Coriandrum sativum* L.) and their management leading to production of high quality pathogen free seed. International J Seed Spices 7: 1-7.

Khiralla A, Mohamed I, Thomas J, Mignard B, Spina R, Yagi S, Laurain-Mattar D (2015) A pilot study of antioxidant potential of endophytic fungi from some Sudanese medicinal plants. Asian Pac J Trop Med 8: 701-704. doi:https://doi.org/10.1016/j.apjtm.2015.07.032.

Kim JW, Choi HG, Song JH, Kang KS, Shim SH (2019) Bioactive secondary metabolites from an endophytic fungus *Phoma* sp. PF2 derived from *Artemisia princeps* Pamp. J Antibiot 72: 174-177.

Krishnamurthy YL, Naik SB, Jayaram S (2008) Fungal communities in herbaceous medicinal plants from the Malnad region, Southern India. Microbes Environ 23: 24-28.

Krohn K, Farooq U, Flörke U, Schulz B, Draeger S, Pescitelli G, Kurtán T (2007) Secondary metabolites isolated from an endophytic *Phoma* sp.–absolute configuration of tetrahydropyrenophorol Using the Solid-State TDDFT CD Methodology. Eur J Org Chem 19: 3206-3211.

Kruppa PC, Russomanno OM (2008) Ocorrência de fungos em sementes de plantas medicinais, aromáticas e condimentares da família Lamiaceae. Tropical Plant Pathology 33: 72-75.

Kumar S, Upadhyay R, Aharwal RP, Sandhu SS (2016) Antibacterial activity of some isolated endophytic fungi from *Mentha viridis*. Int J Appl Biol Pharmac Technol 7: 239-249.

Kumar DSS, Hyde KD (2004) Biodiversity and tissue-recurrence of endophytic fungi in *Tripterygium wilfordii*. Fungal Divers 17:69-90.

Lahoz E, Caiazzo R, Fanigliulo A, Comes S, Crescenzi A (2007) *Phoma glomerata* as causal agent of crown rot disease of fennel in southern Italy. Commun Agri Appl Biol Sci 72: 875-878.

Larran S, Simón MR, Moreno MV, Siurana MPS, Perelló A (2016) Endophytes from wheat as biocontrol agents against tan spot disease. Biol Control 92: 17–23.

Le TTM, Hoang ATH, Le TTB, Vo TTB, Van Quyen D, Chu HH (2019) Isolation of endophytic fungi and screening of Huperzine A–producing fungus from *Huperzia serrata* in Vietnam. Sci Rep 9: 1-13.

Li X, Zhai X, Shu Z, Dong R, Ming Q, Qin L, Zheng C (2016) *Phoma glomerata* D14: An endophytic fungus from *Salvia miltiorrhiza* that produces salvianolic acid C. Current Microbiology 73: 31-37.

Liu SS, Jiang JX, Huang R, Wang YT, Jiang BG, Zheng KX, Wu SH (2019) A new antiviral 14-nordrimane sesquiterpenoid from an endophytic fungus *Phoma* sp. Phytochem Lett 29: 75-78.

Machowicz-Stefaniak Z, Zimowska B (2000). Grzyby przenoszone przez materiał siewny roślin zielarskich [Fungi transmitted by saw material of herbs]. Acta Agrobot 2: 25-38.

Machowicz-Stefaniak Z, Zimowska B, Zalewska E (2002) Grzyby zasiedlające różne organy tymianku właściwego *Thymus vulgaris* L. uprawianego na Lubelszczyźnie [Fungi colonizing various organs of thyme *Thymus vulgaris* L. cultivated in the region of Lublin]. Acta Agrobot 55: 185-197.

Machowicz-Stefaniak Z, Zalewska E (2008) Biodiversity of fungi colonizing different part of caraway (*Carum carvi* L.). Electronic J Polish Agricult Univer 11: 21. http://www.ejpau.media.pl/volume11/issue1/art-21.html.

Machowicz-Stefaniak Z, Zimowska B, Zalewska E (2008) The occurrence and pathogenicity of *Phoma exigua* Desm. var. *exigua* for selected species of herbs. Acta Agrobot 61:157-166.

Machowicz-Stefaniak Z, Zalewska ED, Zimowska B, Król ED (2014) Pathogenicity of *Phoma complanata* (Tode) Desm. towards angelica (*Archangelica officinalis* Hoffm.). Acta Sci Pol Hortorum Cultus 13: 45-58.

Mačkinaitė R (2012) Potential pathogens of common caraway (*Carum carvi* L.) seeds and search for measures suppressing their spread. Žemdirbystė–Agriculture 99: 179-88.

McPartland JM (1994) Cannabis pathogens X: *Phoma, Ascochyta* and *Didymella* species. Mycologia 86: 870-878.

Mel'nik VA (2000) Key to the fungi of the genus *Ascochyta* Lib.(*Coelomycetes*). Parey. VA Mel'nik, U Braun, G Hagedorn (Eds.), Mitteilungen aus der Biologischen Bundesanstalt fuür Land- und Forstwirtschaft. Heft 379, Perley Buchverlag, Berlin Germany.

Melouk HA, Horner CE (1973) β-glucosidase from *Phoma strasseri* and its possible role in a disease of peppermint. Phytopathology 63: 973-975.

Mishra VK, Singh G, Passari AK, Yadav MK, Gupta VK, Singh BP (2016) Distribution and antimicrobial potential of endophytic fungi associated with ethnomedicinal plant *Melastoma malabathricum* L. J Environ Biol 37: 229.

Mirzaee MR (2014) An overview of jujube (*Ziziphus jujuba*) diseases. Arch Phytopathol Plant Protect 47: 82-89.

Nagaraja O, Mahishi P, Krishnappa M (2010) Mycoflora associated with periwinkle (*Catharanthus Roseus* L.), pathogens city and their effect on primary metabolites. J Plant Dis Sci 5: 173-178.

Nalini MS, Sunayana N, Prakash HS (2014) Endophytic fungal diversity in medicinal plants of Western Ghats, India. Int J Biodivers 1-9 doi: https://doi.org/10.1155/2014/494213.

Nicoletti R, Fiorentino A (2015) Plant bioactive metabolites and drugs produced by endophytic fungi of Spermatophyta. Agriculture 5: 918-970.

Okafar UA, Okachi VI, Chinedu SN, Ebuehi OAT, Onygeme-Okerenta BM (2010) Pectinolytic activity of wild-type filamentous fungi fermented on agrowastes. Afr J Microbiol Res 24: 2729-2734.

Paizs L, Naggy F (1975) *Phoma strasseri*: A new pathogen of mint in Hungary. Herba Hungarica 14: 65-75.

Palanichamy P, Krishnamoorthy G, Kannan S, Marudhamuthu M (2018) Bioactive potential of secondary metabolites derived from medicinal plant endophytes. Egypt J Basic Appl Sci 5: 303-312.

Park YH, Kim YC, Park SU, Lim HS, Kim JB, Cho BK, Bae H (2012) Age-dependent distribution of fungal endophytes in *Panax ginseng* roots cultivated in Korea. J Ginseng Res 36: 327-333.

Park JM, Park JH, Kim JG, You YH (2016) Securing and analysis of fungal endophytic diversity from roots of *Salicornia europaea* L. J Life Sci 50: 89-98.

Park YH, Kim Y, Mishra RC, Bae H (2017) Fungal endophytes inhabiting mountain-cultivated ginseng (*Panax ginseng* Meyer): Diversity and biocontrol activity against ginseng pathogens. Sci Rep 7: 1-10.

Pandotra VR, Sastry KSM (1969) Fungi on medicinal and aromatic plants in the North-West Himalayas-V. In Proceedings of the Indian Academy of Sciences-Section B Springer India 70:88–97.

Parismita S, Manoj K, Highland K, Ruth L, Richa R (2016) Endophytic fungi associated with the medicinally important aromatic plant *Artemisia nilagirica* (Clarke) Pamp. and antimicrobial activity of selected endophytic fungi against *Rhizoctonia solani.* Asian J Biol Life Sci 5 (2).

Pavlović SĐ, Stojšin VB, Stojanović SD (2007) Mycopopulation of marshmallow (*Althaea officinalis* L). Zbornik Matice srpske za prirodne nauke 113: 193-202.

Pavlović S, Stojšin V, Stojanović S, Starović M (2009) Mycopopulation in five important cultivated medicinal plants in Serbia. Planta Med 75: 1005-1005.

Pavlović DS, Stević RT, Pljevljakušić SD, Stojanović DS, Starović SM, Jošić LD (2012) Mycopopulation of *Leuzea carthamoides* DC. cultivated in Serbia. In Proceedings of the Seventh Conference on Medicinal and Aromatic Plants of Southeast European Countries (Proceedings of the 7th CMAPSEEC) Subotica Serbia 27-31 May, pp. 315-320). Institute for Medicinal Plant Research and Association for Medicinal and Aromatic Plants of Southeast European Countries (AMAPSEEC).

Pavlović SĐ, Stojanović SD, Starovic MS, Jošić DL, Menković NR (2011) Parasitic mycobiota of yellow gentian (*Gentiana lutea* L.). Zbornik Matice srpske za prirodne nauke 120: 177-182.

Pavlović DS, Stević RT, Ristić D, Starović M, Aleksić G, Kuzmanović S, Stojanović S (2015) The first report of *Bipolaris/Drechlera sorociniana* (Sacc in sorok.) on anise seeds in Serbia. In Sixth International Scientific Agricultural Symposium "Agrosym 2015", Jahorina, Bosnia and Herzegovina, October 15-18, 2015. Book of Proceedings (pp. 1101-1105). University of East Sarajevo.

Pethybridge Sarah J, Jason B. Scott, Frank S. Hay (2012) Lack of evidence for recombination or spatial structure in *Phoma ligulicola* var. *inoxydabilis* populations from Australian pyrethrum fields. Plant Dis 96: 746-751.

Praptiwi P, Fathoni A, Ilyas M (2020) Diversity of endophytic fungi from *Vernonia amygdalina*, their phenolic and flavonoid contents and bioactivities. Biodiversitas 21: 436-441.

Radev Z (2020) Study on the use of honey bees (*Apis mellifera* L.) for biological control against pests in lavender (*Lavandula officinalis* L). New knowledge Journal of science, 9: 165-168.

Rai MK (1989) *Phoma sorghina* infection in human being. Mycopathologia 105:167-170.

Rajagopal K, Kalavathy S, Kokila S, Karthikeyan S, Kathiravan G, Prasad R, Balasubraminan P (2010) Diversity of fungal endophytes in few medicinal herbs of South India. Asian J Exp Biol Sci 1: 415-418.

Raaijmakers JM, Paulitz TC, Steinberg C, Alabouvette C, Moënne-Loccoz Y (2009) The rhizosphere: a playground and battlefield for soilborne pathogens and beneficial microorganisms. Plant Soil 321: 341-361.

Rakotoniriana EF (2012) Biodiversity and antifungal properties of endophytes from the madagascar medicinal plants *Centella asiatica* (L.) Urb. and *Catharanthus roseus* (L.) G. Don. Disertasi. Universite Catholique de Louvain. Louvain.

Rodeva R, Carrieri R, Stoyanova Z, Dacheva S, Lahoz E, Fanigliulo A, Crescenzi A (2013) New report of *Phoma glomerata* on *Coriandrum sativum* L. Commun Agricult App Biol Sci 78: 617-620.

Rodeva R, Gabler J, Machowicz-Stefaniak Z, Kačergius A, Zimowska B, Zalewska E, Stoyanova Z (2016) New, emerging and re-emerging fungal diseases on medicinal and aromatic plants in European domain. Proc. 6th International Sympozium Breeding Research on Medicinal and Aromatic Plants, BREEDMAP 6, Quedlinburg, Germany: 33-39.

Roustaee A, Dechamp-Guillaume G, Gelie B, Savy C, Dargent R Barrault G (2000) Ultrastructural studies of the mode of penetration by *Phoma macdonaldii* in sunflower seedlings. Phytopathology 90: 915-920.

Rusanova M, Rusanov K, Butterweck V, Atanassov I (2019) Exploring the capacity of endophytic fungi isolated from medicinal plants for fermentation and phenolics biotransformation of rose oil distillation wastewater. Biotechnology & Biotechnological Equipment 33: 651-663.

Sairafianpour M, Christensen J, Stærk D, Budnik BA, Kharazmi A, Bagherzadeh K, Jaroszewski JW (2001) Leishmanicidal, antiplasmodial, and cytotoxic activity of novel diterpenoid 1, 2-quinones from *Perovskia abrotanoides*: new source of tanshinones. J Nat Prod 64: 1398-1403.

Samaga PV, Rai VR (2016) Diversity and bioactive potential of endophytic fungi from *Nothapodytes foetida*, *Hypericum mysorense* and *Hypericum japonicum* collected from Western Ghats of India. Ann Microbiol 66: 229-244.

Santiago C, Fitchett C, Munro MH, Jalil J, Santhanam J (2012) Cytotoxic and antifungal activities of 5-hydroxyramulosin, a compound produced by an endophytic fungus isolated from *Cinnamomum mollisimum*. Evid Based Complement Alternat Med. doi:https://doi.org/10.1155/2012/689310.

Saranraj P, Sivasakthivelan P, Sivasakthi S (2016) Prevalence of fungal diseases in medicinal plants of *Vellore district* of Tamil Nadu in India. Int J Adv Multidiscip Res 3: 49-63.

Strobel GA (2002) Microbial gifts from the rainforest. Can J Phytopathol 24: 14-20.

Strobel G, Daisy B (2003) Bioprospecting for microbial endophytes and their natural products. Microbiol Mol Biol Rev 67:491–502.

Strobel G, Daisy B, Castillo U, Harper J (2004) Natural products from endophytic microorganisms. J Nat Prod 67:257–68.

Ślusarczyk S. Topolski J, Domaradzki K, Adams M, Hamburger M., Matkowski A (2015) Isolation and fast selective determination of nor-abietanoid diterpenoids from *Perovskia atriplicifolia* roots using LC-ESI-MS/MS with multiple reaction monitoring. Nat Prod Commun 10 (7). 1934578X1501000703.

Tan XM, Zhou YQ, Zhou XL, Xia XH, Wei Y, He LL, Yu LY (2018) Diversity and bioactive potential of culturable fungal endophytes of *Dysosma versipellis*; a rare medicinal plant endemic to China. Sci Rep 8: 1-9.

Tariq M, Dawar S, Tufail S (2016). Identification and isolation of fungi from *Trigonella foenum-graecum* L. Int. J Biol Res 4: 47-50.

Teimoori-Boghsani Y, Ganjeali A, Cernava T, Müller H, Asili J, Berg G (2020) Endophytic fungi of native *Salvia abrotanoides* plants reveal high taxonomic diversity and unique profiles of secondary metabolites. Front Microbiol 10: 3013.

Tenguria RK, Firodiya A (2015) Occurrence of endophytic fungi in leaves of medicinal plants from central region of Madhya Pradesh, India. World J Pharmacy Pharma Sci 4: 1921-1934.

Wan SN, Ma, YP, Wang JW, Wei XY (2014) Identification and antimicrobial activity of endophytic fungi from *Pueraria lobata*. Nat Prod Res Develop 11: 34.

Wang HSF, Hwang KF, Chang BD, Gossen GD, Turnbull RJ, Howard (2004) Assessing resistance to spring black stem and leaf spot of alfalfa caused by *Phoma* spp. Can J Plant Sci 84: 311-317.

Wang LW, Xu BG, Wang JY, Su ZZ, Lin FC, Zhang CL, Kubicek CP (2012) Bioactive metabolites from *Phoma* species, an endophytic fungus from the Chinese medicinal plant *Arisaema erubescens*. App Microbiol Biotechnol 93: 1231-1239.

Weber RW, Stenger E, Meffert A, Matthias HAHN (2004) Brefeldin A production by *Phoma medicaginis* in dead pre-colonized plant tissue: a strategy for habitat conquest? Mycol Res 108: 662-671.

WHO (2003) WHO guidelines on good agricultural and collection practices (GACP) for medicinal plant. Geneva, Switzerland: World Health Organization.

Wielgusz K, Seidler-Łożykowska K (2017) Fungi colonizing and damaging different parts of some medicinal plants. Herba Pol 63: 18-26.

Wink M (2008) Plant secondary metabolism: diversity, function and its evolution. Nat Prod Commun 3:1205–1216.

Wu H, Yan Z, Deng Y, Wu Z, Xu X, Li X, Luo H (2020) Endophytic fungi from the root tubers of medicinal plant *Stephania dielsiana* and their antimicrobial activity. Acta Ecologica Sinica 5: 383-387.

Verma VC, Gond SK, Kumar A, Kharwar RN, Strobel G (2007) The endophytic mycoflora of bark, leaf, and stem tissues of *Azadirachta indica* A. Juss (Neem) from Varanasi (India). Microb Ecol 54: 119-125.

Vieira ML, Hughes AF, Gil VB, Vaz AB, Alves TM, Zani CL, Rosa LH (2012) Diversity and antimicrobial activities of the fungal endophyte community associated with the traditional Brazilian medicinal plant *Solanum cernuum* Vell. (Solanaceae). Can J Microbiol 58: 54-66.

Yang H, Cheng JJ, Liang G (2010) Induction of mitotic arrest and apoptosis by salvianolic acid C in human hepatoma HepG2 cells. Chin Pharmacol Bull 26: 1208-1212.

Zechini A, D'auerio, Zambonelli A, Bianch A, Alabasini A (1995) Micro morphological and chemical investigation into the effects of fungal diseases on *Melissa officinalis* L.. *Mentha x piperita* L., *Salvia officinalis L.* Phytopathology 143: 179-183.

Zhang H, Xiong Y, Zhao H, Yi Y, Zhang C, Yu C, Xu C (2013) An antimicrobial compound from the endophytic fungus *Phoma* sp. isolated from the medicinal plant *Taraxacum mongolicum*. J Taiwan Inst Chem Eng 44: 177–181. doi: https://doi.org/10.1016/j.jtice.2012.11.013.

Zheng YK, Miao CP, Chen HH, Huang FF, Xia YM, Chen YW, Zhao LX (2017) Endophytic fungi harbored in *Panax notoginseng*: diversity and potential as biological control agents against host plant pathogens of root-rot disease. J Ginseng Res 41: 353-360.

Zimowska B (2004) Fungi threatening with cultivation of St. John's wort (*Hypericum perforatum* L.) in particular with regard to *Seimatosporium hypericinum* (Ces.) Sutton (*Melanconiales, Deuteromycotina*). Phytopathol Pol 31: 79-80.

Zimowska B (2007a) New *Phoma* species on *Leonurus cardiaca*. Acta Mycol 42: 119-123.

Zimowska B (2007b) Fungi colonizing and damaging different parts of peppermint (*Mentha piperita* L.) cultivated in south-eastern Poland. Herba Pol 53: 97-105.

Zimowska B (2008) Biodiversity of fungi colonizing and damaging selected parts of motherwort (*Leonurus cardiaca* L.). Herba Pol 54: 30-40.

Zimowska B (2010a) Diversity of fungi occurring on savory (*Satureja hortensis* L.). Herba Pol 56: 29-37.

Zimowska B (2011a) Conidiogenesis of *Phoma strasseri* the fungus responsible for black stem and rhizomes rot in peppermint (*Mentha piperita*). Acta Sci Pol Hortorum Cultus 10: 171-178.

Zimowska B (2011b) Biotic activity of *Phoma strasseri* and the effect of thermal conditions on the growth and formation of the pathogen's infectious material. Acta Sci Pol Hortorum Cultus 10: 213-224.

Zimowska B (2011c) Biotic activity of *Phoma strasseri* and the effect of thermal conditions on the growth and formation of the pathogen's infectious material. Acta Sci Pol Hortorum Cultus 10: 259-271.

Zimowska B (2012) Pathogenicity and ultrastructural studies of the mode of penetration by *Phoma strasseri* in peppermint stems and rhizomes. Pol J Microbiol 61: 273-279.

Zimowska B (2013) Diversity of fungi colonizing and damaging selected parts of ribwort (*Plantago lanceolata* L.). Acta Sci Pol Hortorum Cultus 12: 91-103.

Zimowska B (2015) Fungi threatening the cultivation of oregano (*Origanum vulgare* L.) in south-eastern Poland. Acta Sci Pol Hortorum Cultus 14: 65-78.

Zimowska B, Targoński Z (2015) Pectinolytic activity of *Boeremia strasseri* the causal agent of black stem and rhizomes rot of peppermint. Acta Sci Pol Hortorum Cultus 14: 19-28.

Zimowska B, Zalewska ED, Krol ED, Furmańczyk A (2017) Morphological and molecular characterization of *Phoma complanata*, a new causal agent of *Archangelica officinalis* Hoffm. in Poland. Pol J Microbiol 66: 281-285.

Diseases of Fruits, Tubers, and Seeds Caused by *Phoma* sensu lato Species Complex

4

Graciela Dolores Avila-Quezada and Mahendra Rai

4.1 Introduction

Fruits are a rich source of vitamins and minerals for human nutrition, and this is the reason why the Food and Agriculture Organization (FAO) declared 2021 as the international year of fruits and vegetables. In addition, the production of fruits contributes strongly to the economy of nations. Since the world population has been increasing, food production must increase in the same proportion. The quality of fruits and nuts and consumer preferences are determined by the size, shape, color, flavor, aroma, texture, and health benefits (Bernardez et al. 2004; Gardea et al. 2017; Yahia et al. 2019). These are requirements to ensure profitable income for producers. Fruits, due to their sugar content, are very attractive to insects and pathogens.

The pathogens induce diseases in fruits reducing their quality, causing them to be sold in local markets or not to be marketable, which generates economic losses (Prusky et al. 2009). Fruit pathogens can attack fruits from pre-harvest, and then, with the primary inoculum on them, they can infect fruits postharvest (Pétriacq et al. 2018). The well-known famines are examples of how devastating a plant pathogen could be (Avila-Quezada et al. 2018). The losses of agricultural products related to phytopathogens are attributed to the lack of plant material free of pathogens, monoculture, the lack of phytosanitary measures, or emerging diseases (Maresi et al. 2013; García-González et al. 2018). To avoid devastating losses in fruit and vegetable production, the World Trade Organization (WTO) has imposed conditions for the export and import of fresh fruits and vegetables. These actions have minimized, to some extent, the dispersal of quarantined pathogens. Among fungal diseases, fruit rot is very destructive, causing considerable losses from the field to storage and marketing.

The genus *Phoma* has been classified as the phylum *Ascomycota*, class *Dothideomycetes*, order *Pleosporales*, and family *Didymellaceae* (Bennett et al. 2018). The genus is characterized by the formation of flask-shaped pycnidia with a clear ostiole (Fig. 4.1). These pycnidia may be simple flask-shaped, setose to pilose or pycnosclerotia-like (Aveskamp et al. 2008). Inside pycnidia, conidia are produced, which are hyaline, ellipsoidal with or without guttules. Sometimes, different kinds of chlamydospores are formed such as in *Phoma sorghina*, *P. glomerata*, and *Phoma arachidicola* (Rai and Rajak 1993).

The impact of *Phoma* sensu lato species complex on agriculture products is significant; they can cause yield loss of up to 25% (Fitt et al.

G. D. Avila-Quezada (✉)
Facultad de Ciencias Agrotecnológicas, Universidad Autónoma de Chihuahua, Chihuahua, Mexico

M. Rai
Department of Biotechnology, SGB Amravati University, Amravati, Maharashtra, India

M. Rai et al. (eds.), *Phoma: Diversity, Taxonomy, Bioactivities, and Nanotechnology*,
https://doi.org/10.1007/978-3-030-81218-8_4

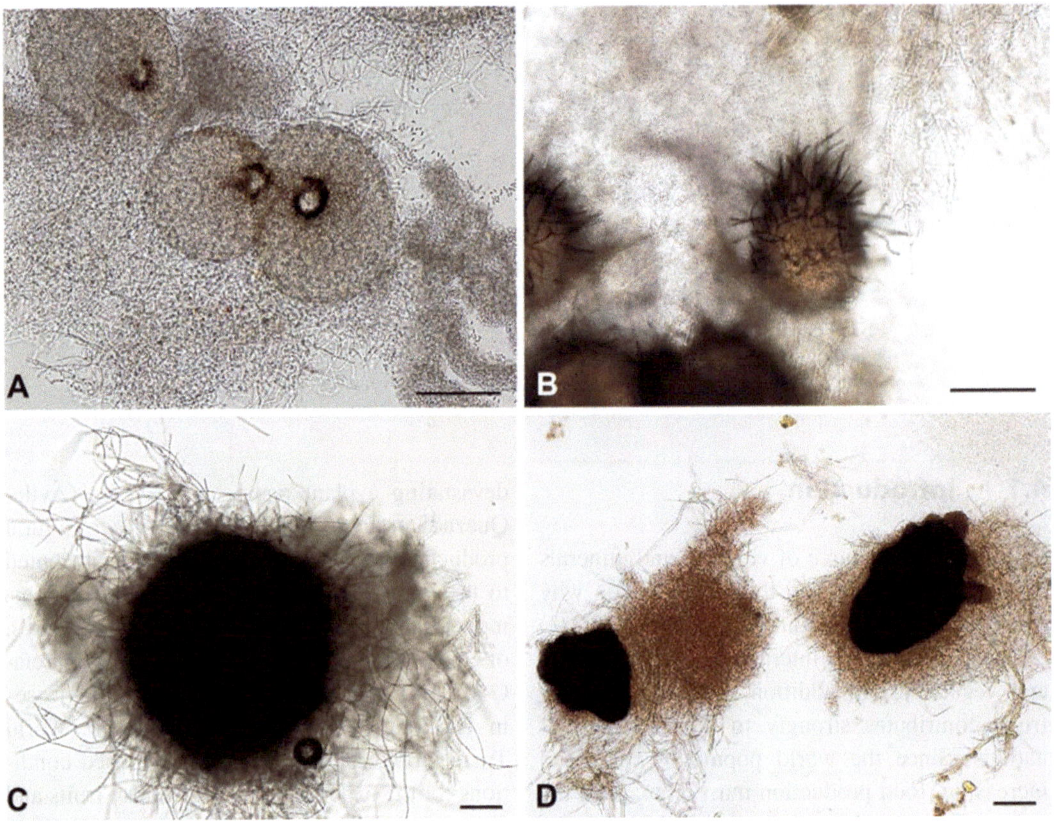

Fig. 4.1 (**a–d**) Different types of pycnidia of *Phoma*. (**a**) Regular glabrous as in *P. herbarum* with clear ostioles. (**b**) Setose as in *P. carteri*. (**c**) Pilose as in *P. betae*. (**d**) Pycnosclerotia-like pycnidia as in *P. incompta*. Scale bars = 100μm. Reproduced from Aveskamp et al. 2008 under Creative Commons Attribution 4.0 International license

2006). The disease incidence by *Phoma exigua* causing fruit rot of brinjal (eggplant) was ranging from 30% to 65% in different fields (Gogoi and Senapoty 2018). Moreover, some species of *Phoma* produce mycotoxins. For instance, *P. sorghina* produces tenuazonic acid (Steyn and Rabie 1976). This mycotoxin is a potent inhibitor of protein biosynthesis in animal tests (Smith et al. 1968). In humans, tenuazonic acid can cause bleeding disorder when *Phoma*-infected grains are ingested (Oliveira et al. 2017). Therefore, food safety protocols must be carried out.

The chapter is focused on different diseases caused by *Phoma* sensu lato species complex on fruits, seeds, and tubers. Moreover, it discusses general disease management strategies.

4.2 Diseases on Fruit, Tubers, and Seeds by *Phoma* spp.

Different species of the genus *Phoma* cause fruit rot or spots on fruit and seeds. For example, the symptoms on fruits of pepper and tomato are large spreading spots forming concentric, brown round lesions with black fruiting bodies (Jackson 2017; Kubota et al. 2000). *Phoma* sensu lato species complex is a pathogen of many crops that can affect the whole plant. The pycnidia and pseudothecia that occur in the plant are the source of inoculum for the same or the crop next cycle and affect the fruits in postharvest (Fig. 4.2).

Phoma infects seedlings causing symptoms such as leaf spots and persists through winter as infections in plant tissue. Spots appear on the

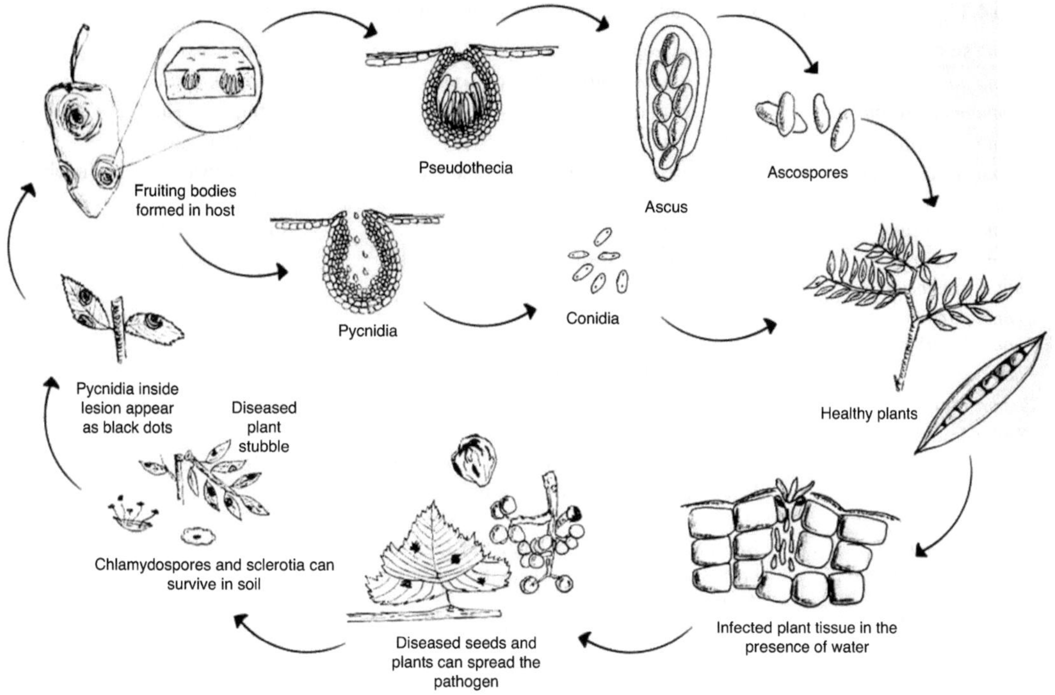

Fig. 4.2 Schematic representation of the life cycle of *Phoma* spp. in fruits. Conidia are formed by phialidic ontogeny, in basipetal succession of the phialides with prominent collarettes around the conidiogenetic locus of the conidiophore (Handmade drawing by Carolina Alvarado)

spring shoots; later, lesions appear on the stems and seeds. During the time of rainfall or high relative humidity, the pycnidia of *Phoma* exude conidia that may be in mucilaginous masses. These conidia are easily spread by splashes of rain or by the wind. When the conidia come into contact with developing flowers, the pathogen infects the seeds (Leach and Macdonald 1976).

The main source of infection by *Phoma* spp. is the stubble of diseased plants where chlamydospores, dormant mycelia, sclerotia, or pseudosclerotia can be found. The pycnidia which are also hibernating structures, under favorable conditions, expel the conidia. When the conidium reaches the host, it forms germ tubes; these break the host cuticle by direct penetration or by appressoria (Deb et al. 2020).

Stomata, being natural openings, facilitate the entry of fungi. *P. exigua* var. *foveata* (=*Boeremia foveata*) enters through stomata or lenticels of young potato tubers (Umaerus et al. 1989). The pathogen produces phytotoxins that emphasize symptoms. Some of the most devastating *Phoma* species of fruits and seeds includes *P. destructiva*, *P. glomerata*, and *P. exigua*. *P. destructiva* causes fruit spot in bell pepper (*Capsicum annuum*), whilst *P. exigua* causes the potato gangrene (Boerema 1967; Didvania et al. 2012). *Phoma* sensu lato includes several important plant pathogens affecting fruits, tubers, and seeds (Table 4.1).

4.3 Incidence and Losses

The seeds have been considered an important method for the spread of the disease; although in some occasions the percentage of diseased seed is low, the disease can be spread by means of diseased seedlings. In the study of Lloyd (1959), less than 0.1% of brassica seeds were infected by *Phoma lingam*. Maširević et al. (2014) reported that sunflower hybrids had seed infection from 1 to 3.5% by *Phoma macdonaldi*. However, there is a significant loss due to infection by *Phoma* spp. For instance,

Table 4.1 *Phoma* sensu lato species complex, causing disease on fruit, tubers, and seeds

Phoma species	Crop/symptoms and diseases	References
Phoma pomorum	*Physalis pubescens*/postharvest fruit rot	Liang et al. (2020)
Phoma multirostrata	Mango/mature fruit rot	Fatima and Khot (2017)
Phoma tracheiphila	Citrus fruit/[b]	Ezra et al. (2007)
Phoma caricae-papayae	Papaya = pawpaw/rot on fruits and seeds	Echerenwa and Umechuruba (2004)
Phoma herbarum	Maize/infected maize kernels[b]	Pitt et al. (1993)
Phoma herbarum *Phoma macrostoma*	Pecans/infected freshly harvested pecans[b]	Huang and Hanlin (1975)
Phoma glomerata and *Phoma hibernica*	Groundnut kernels/[b]	Joffe (1969)
Phoma sp.	Cashew nut/black spots	Wonni et al. (2017)
Phoma endogena	Nuts of *Castanea sativa*/brown rot	Maresi et al. (2013)
Phoma destructiva	Pepper/fruit spot	Didvania et al. (2012)
Phoma valerianellae	Lamb's lettuce (*Valerianella olitoria* L. Poll. syn. *V. locusta* L. Betcke)/Blackening and necrosis of cotyledons	Pellegrino et al. (2010)
Phoma lingam	*Brassica* (swede, turnip, chou moellier, rape, and kale) seeds/infection starts in the hypocotyl. Abundant pycnidia on seeds	Lloyd (1959)
Phoma sp.	Aonla (*Phyllanthus emblica*)/fruit rot	Jamaluddin et al. (1975)
Phoma sojicola comb. nov. (syn. *Ascochyta sojicola)*	Soybean seeds/infected seeds are brownish, shrunken, or they dont develop	Kővics et al. (1999)
Phoma sorghina	Sorghum grains/infected sorghum grains[a]	Oliveira et al. (2017)
Phoma lycopersici	Tomato *Lycopersicon esculentum* mill/rot, brown and round lesions on fruit	Boerema (1973), Morgan-Jones and Burch (1988), Kubota et al. (2000)
Phoma exigua	Tomato/fruit rot	Laundon (1971), Boerema (1973)
Phoma exigua	Brinjal (eggplant) (*Solanum melongena* L/fruit rot	Gogoi and Senapoty (2018)
Phoma exigua	Capsicum/large spots on fruit	Jackson (2017)
Phoma destructiva	Tomato/fruit rot	Laundon (1971), Boerema (1973)
Phoma aliena	Pomegranate/fruit rots	Palavouzis et al. (2015)
Phoma foveata, *Phoma exigua* var. *exigua*, *Phoma eupyrena*	Potato/gangrene or dry rot in storage potato	EPPO (1997), Yang et al. (2012), A'Hara (2015)
Phoma cucurbitacearum	*Momordica charantia* L./black rot on fruit	Choi et al. (2015)
Phoma macdonaldii Boerema, teleomorph *Leptosphaeria lindquistii* Frezzi	Sunflower seeds/pycnidia and pycniospores on the surface of seeds[b]	Maširević et al. (2014)
Phoma koolunga	Pea seed/small necrotic lesions on seeds	Khani (2015)
Phoma betae	Beet seeds/[b]	Leach and Macdonald (1976)

[a]The name of *Phoma* species are mentioned as given in their respective references
[b]Does not specify the symptom

Ascochyta blight or pea black spot (*Pisum sativum*) caused by *Phoma koolunga*, which has a worldwide distribution, causes grain production losses of 15% per year in Australia, since infected seeds transmit the fungus to seedlings (Khani 2015).

There are several reports of *Phoma* causing seed infection, for instance, pathogen of citrus *Phoma tracheiphila* (Ezra et al. 2007), *Phoma sorghina* transmitted by Brazilian sorghum grains (Oliveira et al. 2017), and *Phoma valerianellae* in lamb's lettuce transmitted by seeds (Pellegrino et al. 2010). The reappearance of this pathogen on lamb's lettuce (*Valerianella olitoria* L.) in many farms of Lombardy, Italy, caused losses of 15–35% (Pellegrino et al. 2010).

4.4 Quarantined *Phoma* Species and Diagnosis

The genus includes several plant pathogens of quarantine importance that pose serious problems in plant health quarantine regulation. Quarantined species, due to their condition, cause restrictions in international trade or can cause denial or destruction of lots suspected of contamination with the quarantined pathogen. Many species of *Phoma* are of quarantine importance because they are pathogens of potential economic importance for an agricultural area. Species of quarantine importance are *Phoma exigua* and *P. foveata* (*P. exigua* var. *foveata*) in South America (Mendes et al. 2006).

The global market and commercial exchange led to the mobilization of fruits and can disperse pathogens (Avila-Quezada et al. 2018). In this context, the fast identification of *Phoma* spp. is very important and requires speed and accuracy for global markets. The morphological identification of a pathogen such as *Phoma* to distinguish species is difficult since it requires highly qualified technicians. Moreover, several *Phoma* spp. which are similar morphologically are different on the basis of molecular markers. Therefore, as per current trends, molecular markers are used for the identification of the species.

The recent molecular advance in our knowledge to detect and diagnose pathogens in products, even at a very low level, made it quite mandatory to produce exportable products free of pathogens. Molecular methods are reliable for the rapid detection of *Phoma* spp. To identify *Phoma* species, transcribed internal spacer regions 1 and 2 and 5.8S nrDNA (ITS), large partial subunit 28S nrDNA region (LSU), β-tubulin gene (TUB) (Woudenberg et al. 2009; Aveskamp et al. 2010; Rai et al. 2014), the RNA polymerase II gene (RPB2) (Chilvers et al. 2009), actin (ACT) (Aveskamp et al. 2009; de Gruyter 2012), chitin synthase 1 (CHS-1) (de Gruyter 2012), the translation elongation factor (TEF1) gene (Irinyi et al. 2009), and LEPB/LEPF (Chong et al. 2015) have been used.

In addition, multiple locus sequences from ITS, LSU, TUB, and RPB2 have been used to determine their phylogenetic relationships (Chen et al. 2015). This, in combination with morphological characteristics, has helped to identify *Phoma* spp.

4.5 Management Strategies

Since there is weak, opportunistic, or wound pathogen such as *Phoma exigua* that does not cause major disease (Jackson 2017), there are also quarantined *Phoma* spp. in some countries. In trees such as chestnut (*Castanea sativa*), *Phoma endogena* and *Phomopsis endogena* are endophytic in bark tissues (Maresi et al. 2013); therefore, their control becomes more difficult.

The pathogen *Phoma* sensu lato is commonly controlled by fungicides such as benzimidazoles. Thiophanate methyl controls *P. exigua* in the field (Schmitz et al. 2006). Khani et al. (2016) mentioned that the fungicides thiabendazole and thiram are effective against *P. koolunga* and *P. exigua* var. *foveata* in artificially inoculated seedlings. Also, the fungicide prochloraz has been effective against *P. exigua* var. *foveata* (Maughan et al. 1991). The disadvantages of the continued use of benzimidazoles are that they can generate resistant strains. In another example, flax foot rot caused by *P. exigua* var. *linicola*, controlled by thiabendazole and carbendazim, produced resistant strains (Deb et al. 2020).

Chlorothalonil and mancozeb inhibit the germination of conidia of *P. complanata*. The former is a fungicide that alters the cellular metabolism of fungi by deactivating glutathione. Being electrophilic, it inhibits thiol enzymes and sulfhydryl groups (Krieger 2001). Moreover, mancozeb inhibits conidium germination (Shin et al. 2017). Mancozeb, thiram, and captafol provided effective control of *P. destructiva*, a chili pepper pathogen that is transmitted by seeds (Dhyani et al. 1990). Benomyl and propiconazole also inhibited the growth of *P. medicaginis* and *P. sclerotioides* in vitro and in the field (Deb et al. 2020). Copper treatments are effective against *Phoma* spp. as reported by Belisario (1996) who found this treatment effective to control *Phoma juglandina* in *Juglans regia* and *J. nigra*.

Among the cultural measures, the following are some of the main measures applied for opportunistic *Phoma* spp. and quarantined *Phoma* spp. In the field, it is suggested the use of certified seeds (Okello et al. 2017), since infected seeds can be a source of initial inoculum for seedlings. Also avoid damaging the fruits, since *Phoma* spp. enter through the wounds (Aveskamp et al. 2008), as well as the elimination of weeds to avoid being an inoculum source. Besides, soil solarization can be applied for *Phoma* spp. (Mihajlović et al. 2017) when the crop and conditions allow it. In addition, it is possible that pathogens develop heat tolerance due to repeated solarizations (Garibaldi and Gullino 1991).

According to Mihajlović et al. (2017), compost provides effective control of soil pathogens such as *Phoma* spp. Davidson et al. (2012) recommend rotation with non-host crops for more than 4 years to prevent the survival of soil pathogens. For the infection of *Phoma valerianellae* on lamb's lettuce seeds, besides chemical treatment, aerated steam, hot water, treatment with electrons, and thyme oil have been effective (Nega et al. 2003; Schmitt et al. 2009).

4.6 Conclusions

Phoma spp. affect fruits, seeds, and tubers causing losses up to 25%, reducing quality, and having the risk of spreading the pathogens to the seedlings. Besides, some species produce mycotoxins, so food safety protocols must be followed. The determination of identity is very important to deal with *Phoma* spp. The current approach of identification is based on molecular tools which have made it possible to separate species that cause similar symptoms in fruits. In this context, more research is needed to design control strategies that minimize contamination and reduce production costs, particularly for more aggressive species of *Phoma* which should be regulated or quarantined.

Acknowledgments We are indebted to Carolina Alvarado Gonzalez for the artwork of Fig. 4.2.

References

A'Hara D (2015) Detection and identification of *Phoma* pathogens of potato. In: Plant Pathology. Humana Press, New York, NY. pp. 17-27.

Aveskamp MM, De Gruyter J, Crous, PW (2008) Biology and recent developments in the systematics of *Phoma*, a complex genus of major quarantine significance. Fungal Divers 31:1-18.

Aveskamp MM, Woudenberg JHC, de Gruyter J, Turco E, Groenewald JZ, Crous PW (2009) Development of taxon-specific sequence characterized amplified region (SCAR) markers based on actin sequences and DNA amplification fingerprinting (DAF): a case study in the *Phoma exigua* species complex. Mol Plant Pathol 10:403-414.

Aveskamp MM, de Gruyter J, Woudenberg JHC, Verkley GJM, Crous PW (2010) Highlights of the Didymellaceae: A polyphasic approach to characterise *Phoma* and related pleosporalean genera. Stud Mycol 65:1-60.

Avila-Quezada GD, Esquivel JF, Silva-Rojas HV, Leyva-Mir SG, Garcia-Avila C, Noriega-Orozco L, Rivas-Valencia P, Ojeda-Barrios D, Melgoza-Castillo A (2018) Emerging plant diseases under a changing climate scenario: Threats to our global food supply. Emirates J Food Agriculture 30:1-10.

Belisario A (1996) The principal diseases of walnut in Italy. Infor Fitopatol 46: 20-25.

Bennett A, Ponder MM, Garcia-Diaz J (2018) *Phoma* Infections: Classification, potential food sources, and its clinical impact. Microorganisms 6:58.

Bernardez MM, De la Montana MJ, Garcia QJ (2004) HPLC determination of sugars in varieties of chestnut fruits from Galicia (Spain). J Food Compos Anal 17:63-67.

Boerema GH (1967) The *Phoma* organisms causing gangrene of potatoes. Neth J Plant Pathol 73:190-192.

Boerema GH (1973) The *Phoma* and *Ascochyta* species described by Wollenweber and Hochapfel in their study on fruit-rotting. Stud Mycol 3:1-50.

Chen Q, Zhang KE, Zhang G, Cai L (2015) A polyphasic approach to characterise two novel species of *Phoma* (Didymellaceae) from China. Phytotaxa 197:267-281.

Chilvers MI, Rogers JD, Dugan FM, Stewart JE, Chen WD, Peever L (2009) *Didymella pisi* sp. nov., the teleomorph of *Ascochyta pisi*. Mycol Res 113:391-400.

Choi IY, Kim JH, Lee WH, Park JH, Shin HD (2015) First report of black rot caused by *Phoma cucurbitacearum* on *Momordica charantia* in Korea. Plant Dis 99:727.

Chong W, Xiaoju Z, Xianglin Z, Wei Z, Yike Q, Yawei L, Yu Z, Yanfei S, Entry Y (2015) Quarantine and identification of *Phoma helianthi* Taberosi in sunflower imported from Kazakhstan. Plant Quarantine 02.

Davidson JA (2012) Epidemiology and management of ascochyta blight of field pea (*Pisum sativum*) in South Australia. PhD thesis. In: https://digital.library.adelaide.edu.au/dspace/bitstream/2440/77182/8/02whole.pdf

de Gruyter J (2012) Revised taxonomy of *Phoma* and allied genera. PhD thesis. Wageningen University, Wageningen, the Netherlands, 180 pp

Deb D, Khan A, Dey N (2020) *Phoma* diseases: Epidemiology and control. Plant Pathol 69:1203-1217.

Dhyani A, Sati M, Khulbe R (1990) Seed-borne fungi of chilli with special reference to pathogenicity and control of *Phoma destructiva*. Indian Phytopathol 43:110-113.

Didvania S, Shah R, Jadon KS (2012) A new disease of bell pepper (*Capsicum annuum* var. *grossum*) caused by *Drechslera bicolor*, its pathophysiology, efficacy of fungicides and botanicals. Plant Pathol J 11:68-72.

Echerenwa MC, Umechuruba CI (2004) Post-harvest fungal diseases of pawpaw (*Carica papaya* L.) fruits and seeds in Nigeria. Glob J Pure Appl Sci 10:69-73.

EPPO (1997) *Phoma exigua* var. *foveata* data sheets on quarantine pests. Prepared by CABI and EPPO for the European Union. Quarantine pests for Europe, University Press, Cambridge. pp. 865-871.

Ezra D, Kroitor T, Sadowsky A (2007) Molecular characterization of *Phoma tracheiphila*, causal agent of Mal secco disease of citrus, in Israel. Eur J Plant Pathol 118:183-191.

Fatima S, Khot YC (2017) Isolation of post harvest fungi from mango (*Mangifera indica*) fruits. Epitome J 3. http://epitomejournals.com/VolumeArticles/FullTextPDF/253_Research_Paper.pdf

Fitt BD, Brun H, Barbetti M, Rimmer S (2006) World-wide importance of phoma stem canker (*Leptosphaeria maculans* and *L. biglobosa*) on oilseed rape (*Brassica napus*). In: Sustainable strategies for managing *Brassica napus* (oilseed rape) resistance to *Leptosphaeria maculans* (phoma stem canker). Springer. pp 3-15.

García-González T, Sáenz-Hidalgo HK, Silva-Rojas HV, Morales-Nieto C, Vancheva T, Koebnik R, Ávila-Quezada GD (2018) *Enterobacter cloacae*, an emerging plant-pathogenic bacterium affecting chili pepper seedlings. Plant Pathol J 34:1-10.

Gardea AA, García-Bañuelos ML, Orozco-Avitia JA, Sánchez-Chávez E, Sastré-Flores B, Ávila-Quezada G (2017) Cacao (*Theobroma cacao* L.). In: Fruit and Vegetable Phytochemicals: Chemistry and Human Health. 2nd Edition Wiley Blackwell. pp. 921-940.

Garibaldi A, Gullino ML (1991) Soil solarization in southern European countries, with emphasis on soilborne disease control of protected crops. Soil solarization. CRC Press, Boca Raton, FL, USA. pp. 227-235.

Gogoi S, Senapoty D (2018) First report of *Phoma exigua* causing fruit rot of brinjal in northeast India (Assam) with a new pathogenicity test method. Int J Curr Microbiol Appl Sci 7:2593-2596.

Huang LH, Hanlin RT (1975) Fungi occurring in freshly harvested and in-market pecans. Mycologia 67:689-700.

Irinyi L, Kövics GJ, Sándor E (2009) Taxonomical re-evaluation of *Phoma*-like soybean pathogenic fungi. Mycol Res 113:249-260.

Jackson G (2017) Pacific Pests and Pathogens – Fact Sheets. Capsicum Phoma blight (239) https://www.pestnet.org/fact_sheets/capsicum_phoma_blight_239.htm

Jamaluddin M, Tandon P, Tandon RN (1975) A fruit rot of aonla (*Phyllanthus emblica* L.) caused by *Phoma* sp. Proc Nat Acad Sci India 45:75-76.

Joffe AZ 1969 The mycoflora of fresh and stored groundnut kernels in Israel. Mycopathol Mycol Appl 39:255-264.

Khani M (2015) Aspects of epidemiology of *Phoma koolunga* (ascochyta blight of field pea) PhD thesis. The University of Adelaide. Australia.

Khani M, Davidson J, Sosnowski M, Scott E (2016) Survival, transmission and control of *Phoma koolunga* in field pea seed and reaction of field pea genotypes to the pathogen. Australas Plant Pathol 45:91-102.

Kövics GJ, de Gruyter J, van der Aa HA (1999) *Phoma sojicola* comb. nov. and other hyaline-spored coelomycetes pathogenic on soybean. Mycol Res 103:1065-1070.

Krieger R (2001) Handbook of Pesticide Toxicology: Principles and Agents. 2nd edition. Academic Press.

Kubota M, Kishi K, Abiko K (2000) *Phoma* leaf spot, stem and fruit rot of tomato caused by *Phoma lycopersici* Cooke in Japan. Jpn J Phytopathol 66:12-17.

Laundon GF (1971) Records of fungal plant disease in New Zealand. N Z J Bot 9:610-624.

Leach LD, Macdonald JD (1976) Seed-borne phoma betae as influenced by area of sugarbeet production, seed processing and fungicidal seed treatments. J Amer Soc Sugarbeet Technologists 19: 4-15.

Liang Y, Zhang R, Liu H, Hou Y (2020) First report of postharvest fruit rot of *Physalis pubescens* in China caused by *Phoma pomorum*. J Plant Pathol 102:979-979.

Lloyd AB (1959) The transmission of *Phoma lingam* (Tode) Desm. in the seeds of swede, turnip, chou moellier, rape, and kale. N Z J Agric Res 2:649-658.

Maresi G, Longa O, Turchetti T (2013) Brown rot on nuts of *Castanea sativa* Mill: an emerging disease and its causal agent. iForest 6:294

Maširević SN, Medić-Pap SS, Terzić AN, Dedić BP, Balalić ID (2014) *Phoma Macdonaldi* on seed and its importance in etiology of Phoma black stem in sunflower. Zb Matitse srp prir nauke 126:57-65.

Maughan J, Shanmuganathan N, Hepworth G (1991) Fungicide treatments for the control of storage rots of seed potatoes. Australas Plant Pathol 20:142-145.

Mendes MAS, Urben AF, Oliviera AS, Marhinho VLA (2006) Interceptação de *Phoma exigua* var. *foveata*, praga exótica e quarentenária para o Brasil, em germoplasma de batata procedente da França. Fitopatol Bras 31:601-603.

Mihajlović M, Rekanović E, Hrustić J, Grahovac M, Tanović B (2017) Methods for management of soil-borne plant pathogens. Pestic fitomed 32:9-24.

Morgan-Jones G, Burch KB (1988) Studies in the genus *Phoma*. XI. Concerning *Phoma lycopersici*, the anamorph of *Didymella lycopersici*, causal organism of stem canker and fruit rot of tomato. Mycotaxon 32:133-142.

Nega E, Ulrich R, Werner S, Jahn M (2003) Hot water treatment of vegetable seed: an alternative seed treatment method to control seed-borne pathogens in organic farming. Z Pflanzenk Pflanzen 110:220-234.

Okello JJ, Zhou Y, Kwikiriza N, Ogutu S, Barker I, Schulte-Geldermann E, Atieno E, Ahmed JT (2017) Productivity and food security effects of using of certified seed potato: the case of Kenya's potato farmers. Agric Food Secur 6:1-9.

Oliveira RC, Goncalves SS, Oliveira MS, Dilkin P, Mallmann CA, Freitas RS, Bianchi P, Correa B (2017) Natural occurrence of tenuazonic acid and *Phoma sorghina* in Brazilian sorghum grains at different maturity stages. Food Chem 230:491-496.

Palavouzis S, Tzamos S, Paplomatas E, Thomidis T (2015) First report of *Phoma aliena* causing fruit rots of pomegranates in northern Greece. J Plant Pathol 97:215.

Pellegrino C, Gilardi G, Gullino ML, Garibaldi A (2010) Detection of *Phoma valerianellae* in lamb's lettuce seeds. Phytoparasitica 38:159-165.

Pétriacq P, López A, Luna E (2018) Fruit decay to diseases: can induced resistance and priming help?. Plants 7:77.

Pitt JI, Hocking AD, Bhudhasamai K, Miscamble BF, Wheeler KA, Tanboon-Ek P (1993) The normal mycoflora of commodities from Thailand. 1. Nuts and oilseeds. Int J Food Microbiol 20:211-226.

Prusky D, Kobiler I, Miyara I, Alkan N (2009) Fruit diseases. In: The mango, botany, production and uses. CAB International, 2nd edition. pp. 210-231.

Rai M, Rajak RC (1993) Distinguishing characteristics of some *Phoma* species. Mycotaxon 48:389-414.

Rai M, Tiwari VV, Irinyi L, Kövics GJ (2014) Advances in taxonomy of genus *Phoma*: Polyphyletic nature and role of phenotypic traits and molecular systematics. Indian J Microbiol 54:123-128.

Schmitt A, Koch E, Stephan D, Kromphardt C, Jahn M, Krauthausen HJ, Forsberg G, Werner S, Amein T, Wright SAI, Tinivella F, van der Wolf J, Groot SPC (2009) Evaluation of non-chemical seed treatment methods for the control of *Phoma valerianellae* on lamb's lettuce seeds. J Plant Dis Protec 116:200-207.

Schmitz S, Zini J, Etienne M, Moreau J-M, Chandelier A, Cavelier M, 2006. Effectiveness of thiophanate-methyl, trifloxystrobin and vinclozolin on canker caused by *Phoma exigua* Desm. on ash tree seedlings. BASE. Liege Université.

Shin JH, Fu T, Park KH, Kim KS (2017) The effect of fungicides on mycelial growth and conidial germination of the ginseng root rot fungus, *Cylindrocarpon destructans*. Mycobiology 45:220-225.

Smith ER, Fredrickson TN, Hadidian Z (1968) Toxic effects of the sodium and the N, N'-dibenzylethylenediamine salts of tenuazonic acid (NSC-525816 and NSC-82260). Cancer Chemother Rep 52:579-585.

Steyn PS, Rabie CJ (1976) Characterization of magnesium and calcium tenuazonate from *Phoma sorghina*. Phytochemistry 15:1977-1979.

Umaerus V, Scholte K, Turkensteen L (1989) Crop rotation and the occurrence of fungal diseases in potatoes. In: Effects of Crop Rotation on Potato Production in the Temperate Zones. Springer. pp. 171-189.

Wonni I, Sereme D, Ouedraogo I, Kassankagno A, Dao I, Ouedraogo L, Nacro, S (2017) Diseases of cashew nut plants (*Anacardium Occidentale* L.) in Burkina Faso. Adv Plants Agric Res 6:00216.

Woudenberg JHC., Aveskamp MM, de Gruyter J, Spiers AG, Crous PW (2009) Multiple *Didymella* teleomorphs are linked to the *Phoma clematidina* morphotype. Persoonia 22:56-62.

Yahia EM, Gardea-Béjar A, Ornelas-Paz J, Maya-Meraz IO, Rodríguez-Roque MJ, Rios-Velasco C, Salas-Marina MA (2019) Preharvest factors affecting postharvest quality. In: Postharvest Technology of Perishable Horticultural Commodities. Woodhead Publishing. pp. 99-128.

Yang C-D, Chen X-R, Jiang H-X, Pu C.-J (2012) First report of potato gangrene caused by *Phoma foveata* in China. Plant Dis 96:1698.

Overview of *Phoma*-Like Fungi on Important Legumes (Papilionaceous Plants)

5

György János Kövics, László Irinyi, and Mahendra Rai

5.1 Introduction

Phoma-like fungi (phomoid fungi) are a large, controversial group of anamorphic ascomycetes that inhabit a wide range of ecosystems (Aveskamp et al. 2008), from soil to water and from plants to animals (Aveskamp et al. 2010a, b). The biology and taxonomic boundaries among *Phoma*-like genera and species, as well as the ecological aspects of these fungi, remain confused, troublesome and controversial more than 130 years after the first *Phoma* descriptions (Aveskamp et al. 2009). The conventional system of species identification based on morphological features and host plant associations is not sufficient and has serious limitations due to the similarity of morphological and cultural features and the wide host range of many species (Aveskamp et al. 2010a, b; Rai et al. 2014; Chen et al. 2015a). It is critical to note that the identification of these *Phoma*-like fungi was carried out mainly by morphological features or only 'based on the host'. However, such identification of the pathogen could be incorrect, since it is possible to encounter a similar but widely specialized saprotrophic or weakly pathogenic species (Gomzhina et al. 2020). The role of host plant associations and morphological or cultural features on different substrates has not have specified importance any more in the identification of *Phoma*-like fungi. A recently published review (Deb et al. 2020) has summarized the epidemiology and control of *Phoma* diseases. Both *Ascochyta* and *Didymella* and the polyphyletic genus *Phoma* of the family *Didymellaceae* cause devastating diseases on vegetables, oilseed crops, forage crops, horticultural plants, including papilionaceous plants, and many more. Distributed across many continents, *Phoma* attacks several economically important plants causing serious yield losses (Chen et al. 2015b, 2017). The phomoid fungi on legume hosts are based on phylogenetic analyses of DNA sequences from the ribosomal internal transcribed spacer regions (ITS), as well as portions of three protein-coding genes: *glyceraldehyde-3-phosphate-dehydrogenase* (*G3PD*), *translation elongation factor 1-α* (*TEF1α*) and *chitin syn-*

G. J. Kövics (✉)
Plant Protection Institute, University of Debrecen, Debrecen, Hungary

L. Irinyi
Molecular Mycology Research Laboratory, Centre for Infectious Diseases and Microbiology, Faculty of Medicine and Health, Sydney Medical School, Westmead Clinical School, The University of Sydney, Sydney, NSW, Australia

Westmead Institute for Medical Research, Westmead, NSW, Australia

M. Rai
Department of Microbiology, Nicolaus Copernicus University, Torun, Poland
e-mail: mahendra.rai@v.umk.pl

M. Rai et al. (eds.), *Phoma: Diversity, Taxonomy, Bioactivities, and Nanotechnology*,
https://doi.org/10.1007/978-3-030-81218-8_5

thase 1 (*CHS1*). All legume-associated *Ascochyta* species had nearly identical ITS sequences and are clustered with other *Ascochyta*, *Phoma* and *Didymella* species from legume and non-legume hosts (Peever et al. 2007).

Despite above contradictions, nevertheless, it seems reasonable to overview briefly the history of these pathogenic fungi on papilionaceous (Fabaceae or Leguminosae) plants. Their occurrence are more or less regular, causing serious losses as harmful necrotrophic or semibiotrophic pathogens, which sometimes appear weak or opportunistic parasites if environmental factors are unfavourable to the host development. The taxonomical positions of these species have been changed during the last 30 years applying molecular tools. The taxonomical revisions based on molecular results have been attached to most taxa, including *Phoma*-like fungi.

Phoma and its relatives are the most common pleosporalean asexual morphs and are characterized by the presence of pycnidia (globose to pyriform conidiomata from which the conidia arise throughout an apical opening). *Pleosporales* are mainly saprobic on plant debris; epiphytic, endophytic, or parasitic on living plants, fungi, and insects; or mycobionts in lichens (Aveskamp et al. 2008, 2010a, b; De Gruyter et al. 2009). These fungi can also infect humans (Punithalingam 1979; Rai 1989; Ahmed et al. 2014; Valenzuela-Lopez et al. 2016, 2018).

Until 2013, polymorphic higher fungi (*Dikarya*) were allowed to carry multiple names to describe their teleomorphic and anamorphic stages. These forms occur independently, and their relationship was not always established. The wide use of molecular methods has facilitated the establishment of relationships between these various stages in the life cycles of fungi and has led to the abolition of the dual nomenclature system of fungi and the adoption of the One Fungus – One Name system, as provided in the new 'International Code of Nomenclature for Algae, Fungi, and Plants' (Melbourne Code) (Hawksworth et al. 2011; McNeill et al. 2012).

According to the most recent taxonomical rules established in Shenzhen Code (Turland et al. 2018) and its addition (May et al. 2019), moreover, some practical interpretations summarized in a recent proposal that *all scientific names and all taxonomic ranks, including the fungal families, should be written in italics* in the scientific papers (Thines et al. 2020) and should complement the available fungus data with DNA barcoding (Lücking et al. 2020). To promote the better temporary orientation of the generations which are not professionally skilled in fungal taxonomy, we indicated the morphs after the binominal fungus name in brackets: (a) means anamorphic name(s), and (t) means teleomorphic name(s) in the former Saccardo's systems.

5.2 Alfalfa (*Medicago* L.)

5.2.1 *Ascochyta medicaginicola*

Qian Chen & L. Cai (a) in Chen, Jiang, Zhang, Cai & Crous *Stud. Mycol.* **82**: 187 (2015), Fam. *Didymellaceae* Gruyter, Aveskamp & Verkley (2009).

Syn. *Phoma medicaginis* (a) (1886), *Ascochyta imperfecta* (a) (1912), *Phoma medicaginis* var. *medicaginis* (a) (1965), *Phoma medicaginis* var. *macrospora* (a) (1993) and some more.

Spring black stem and leaf spot disease of alfalfa and clovers or 'Phoma'/'Ascochyta' (black) stem and leaf spot disease (Kövics et al. 2007; Deb et al. 2020) is a cool-season foliar disease. Symptoms include small, black-to-dark brown spots on lower leaves, petioles and stems. The lesions are irregularly to triangularly shaped. As they increase in size, lesions coalesce and become light brown. Affected leaves turn yellow and often wither before falling. Lesions on stems and petioles enlarge, causing large areas near the base of the plant to turn black. Young shoots are often girdled and killed. Most damage occurs before the first cutting.

Probably a misinterpretation of exact identification (Deb et al. 2020) that causative agent of 'Ascochyta blight of field pea' should be '*Phoma medicaginis* (syn. *Ascochyta medicaginicola*)' without any indication of '*varietas*', neither *Phoma medicaginis* var. *medicaginis* (1965), autonym (= *Ascochyta medicaginicola*), nor *P. medicaginis* var.

macrospora (1993) (= *Ascochyta medicaginicola*) both pathogenic to alfalfa and listed among the synonyms above. However, *Phoma medicaginis* var. *pinodella* (1965) is the former, synonymous name of the well-known pea pathogen *Didymella pinodella* (2015) and differs in numerous features (Boerema et al. 2004) from the mentioned vars.

5.2.2 *Leptosphaeria pratensis*

Sacc. & Briard (t) *Revue mycol., Toulouse* **7**(27): 209 (1885), Fam. *Leptosphaeriaceae* M.E. Barr (1987).

Syn. *Septoria medicaginis* (a) (1847), *Stagonospora trifolii* (a) (1891), *Phoma meliloti* (a) (1892), *Stagonosporopsis trifolii* (a) (1932) and some more.

[Not *Stagonospora meliloti* (Lasch.) Petr. (= *Longiseptatispora meliloti* (Lasch ex Rabenh.) L.W. Hou & Crous et al. (2020); not *Leptosphaeria weimeri* Shoemaker, C.E. Babc. & J.A.G. Irwin (1991)].

'Stagonospora' leaf spot and root rot of alfalfa: The causative agents of disease are specified as *Leptosphaeria weimeri* Shoemaker, C. E. Babc. & J. A. G. Irwin (syn. *Leptosphaeria pratensis* Sacc. & Briard) and *Stagonospora meliloti* (Lasch) Petr. *sensu* Samac et al. (2014).

Formerly, and even later also, the *L. pratensis* confused with *L. weimeri* and *L. viridella* (Boerema et al. 2004).

Kövics et al. (2007) referred the disease as alfalfa (*Medicago sativa*) leaf spot and root rot and, moreover, disease on red clover (*Trifolium pratense*), yellow sweet clover (*Melilotus officinalis*) and vetch species (*Vicia* spp.).

5.2.3 *Leptosphaeria weimeri*

Shoemaker, C.E. Babc. & J.A.G. Irwin (t) *Can. J. Bot.* **69**(3): 572 (1991), Fam. *Leptosphaeriaceae* M.E. Barr (1987).

Syn. Not known.

The Index Fungorum considered this species as a legitim, independent species without any synonyms and/or anamorphs. In contrast with Schoemaker et al. (2011), who described a new species *Massarina walkeri*, the teleomorph of *Acrocalymma medicaginis* (= *Acrocalymma walkeri*) from *Medicago sativa* contrasted with *Leptosphaeria pratensis*, the newly established *L. weimeri* and, moreover, *L. viridella* (= *Phaeosphaeria viridella*, 1984).

The above statement also conflicts to Boerema et al. (2004) who synonymized *L. weimeri* with *Phoma meliloti* Allesch (1892) and *Stagonospora meliloti* (Lash) Petr. (1920) synanamorphs. They adopted Jones and Weimer (1938. p. 807) and, moreover, Lucas and Webster (1967) statements who both confirmed the anamorph-teleomorph connection between *Leptosphaeria weimeri* and *Phoma meliloti* (= *Longiseptatispora meliloti*, 2020).

5.2.4 *Leptosphaeria sclerotioides*

(Preuss ex Sacc.) Gruyter, Aveskamp & Verkley (t) (2012) in Gruyter, Woudenberg, Aveskamp, Verkley, Groenewald & Crous, *Stud. Mycol.* **75**: 19, Fam. *Leptosphaeriaceae* M.E. Barr (1987).

Nom. inval. *Plenodomus sclerotioides* Preuss (a) (1849) without description (Boerema et al. 2004).

Nom. inval. *Plenodomus meliloti* Dearn. & G.B. Sanford (a) (1930) Art. 53.1, *sensu* Mycobank and Index Fungorum.

Syn. *Phoma sclerotioides* (a) (1892).

The frequently used name is *Phoma sclerotioides*, the causal agent of brown root rot of alfalfa, which causes severe root and crown lesions on alfalfa and other perennial forage legumes in regions with harsh winters, called even winterkill of alfalfa (Deb et al. 2020). Isolates of *Phoma sclerotioides* (= *Leptosphaeria sclerotioides*, 2012) exhibit diverse cultural morphologies on potato dextrose agar (PDA), suggesting that they may exhibit a high degree of genetic diversity (Wunsch and Bergstrom 2011).

The pathogen occurs in the northern parts of Eurasia and North America (Hollingsworth and Gray 1999), common on the roots and occasional on lower stem parts of various herbaceous plants, particularly destructive on sweet clover (*Melilotus alba*) and alfalfa (*Medicago sativa*) following winter dormancy. This fungus may also be pathogenic on Poaceae species, exposed to low temperature (Boerema et al. 2004).

5.2.5 *Acrocalymma walkeri*

(Shoemaker, C.E. Babc. & J.A.G. Irwin) Crous & Trakun. (a) in Trakunyingcharoen, Lombard, Groenewald, Cheewangkoon, Toanun, Alfenas & Crous, *IMA Fungus* **5**(2): 407 (2014), Fam. *Acrocalymmaceae* (*sensu* Hawksworth and David 1989)

Syn. *Acrocalymma medicaginis* (a) (1987), *Massarina walkeri* (t) (2011).

Acrocalymma walkeri is a plant pathogen that causes root and crown rot in alfalfa, this pathogen being referred in Australia. The teleomorph of *Acrocalymma medicaginis* (= *A. walkeri*) was formed in pure culture. It was described as a new species as *Massarina walkeri* Shoemaker, C.E. Babc. & J.A.G. Irwin (t) (Shoemaker et al. 2011). This finding was contrasted with three somewhat similar species that parasitize legumes: *Leptosphaeria pratensis* (t) (1875), *Leptosphaeria weimeri* (t) (1991) (anamorph *Stagonospora meliloti* (a) (1920) (= *Longiseptatispora meliloti*, 2020) (see more below) and *L. viridella* (Peck) Saccardo (Shoemaker et al. 2011). However there is no recent data about its occurrence except the legitimate description (= *Phaeosphaeria viridella* / Peck/ Leuchtm., *Sydowia* **37**: 108) in 1984, accepted by Index Fungorum.

5.2.6 *Longiseptatispora meliloti*

(Lasch ex Rabenh.) L.W. Hou & Crous (a) in Crous, Wingfield, Schumacher, Akulov, Bulgakov, Carnegie, Jurjević, Decock, Denman, Lombard, Lawrence, Stack, Gordon, Bostock, Burgess, Summerell, Taylor, Edwards, Hou, Cai, Rossman, Wöhner, Allen, Castlebury, Visagie & Groenewald, *Fungal Systematics and Evolution* **6**: 199 (2020), Fam. *Phaeosphaeriaceae* M.E. Barr (1979).

Syn. *Depazea meliloti* (a) (1844), *Septoria meliloti* (a) (1890), *Stagonospora meliloti* (a) (1920).

[Not *Leptosphaeria pratensis* (t) L.W. Hou & Crous et al. 2020; not *Leptosphaeria weimeri* Shoemaker, C.E. Babc. & J.A.G. Irwin (1991)].

The most frequently used name of alfalfa leaf spot and root rot disease is *Stagonospora meliloti* (= *Longiseptatispora medicaginis*), the causal agent of a leaf spot, crown and root rot of alfalfa, in North America, Australia and New Zealand. It has also been recorded on a range of forage legumes in Europe and South America. Symptoms are produced on alfalfa roots and leaves with the most characteristic on crown and root infection being the presence of a diffuse pink to red rot (blotching) in the internal, woody tissue, with reddish flecking at the extremity of the rotted (blotched) area (Irwin et al. 2004). The recent change in the taxonomical position was made by Crous et al. (2020).

5.3 Beans (*Phaseolus* L.)

'Ascochyta' blight symptoms on common bean can be caused by three fungus species with similar symptoms, viz. *Ascochyta boltshauseri*, *Boeremia exigua* and *Boeremia diversispora*.

5.3.1 *Boeremia exigua*

(Desm.) Aveskamp, Gruyter & Verkley (a) in Aveskamp, Gruyter, Woudenberg, Verkley & Crous, *Stud. Mycol.* **65**: 37 (2010), Fam. *Didymellaceae* Gruyter, Aveskamp & Verkley (2009).

Syn. This fungus has numerous, about 150, synonyms, invalid later homonyms and *nomina nuda* (for details see in Boerema et al. 2004).

Phoma exigua var. *exigua* (basionym) (1849), *Ascochyta phaseolorum* (1878).

It is a common fungus found in soils, causing a minor leaf spot disease on bean leaves and pods. Usually, it is either a wound pathogen following damage by insects and weather events or a secondary invader on weakened or diseased plants. It may be pathogenic on beans causing 'Ascochyta' leaf spot, on tobacco 'Phoma' blight, on soybean 'Phoma' or 'Ascochyta' leaf and pot spots and many different hosts mainly as a weak parasite. The pathogen causes typical blight symptoms on beans and was observed occasionally on dark-red kidney, brown kidney, and black bean plants. It rarely was destructive unless the crop had been weakened by some unfavourable environmental conditions. Infected leaves had brown-to-black lesions that developed concentric zones 10–30 mm in diameter and also contained small, black pycnidia. Concentric dark grey-black lesions also appeared on branches, stems, nodes and pods. Infected seeds turned brown to black. Plants sometimes showed defoliation and pod drop (Bardas et al. 2008). Spread of the fungus occurs in wet weather which helps the discharge of spores. As the conidia ooze out of the pycnidia through ostiole at the top, they are splashed in the rain (or irrigation water). There is a possibility that the spores of the fungus can be carried on seed. The fungus survives on dead plant material in the soil.

See more synonyms and data at soybean diseases (Sect. 5.11.2).

5.3.2 *Ascochyta boltshauseri*

Sacc. apud Boltshauser (a) *Z. Pfl.Krankh.* **1**: 136 (1891), *sensu* Index of Fungi, Fam. *Didymellaceae* Gruyter, Aves-kamp & Verk-ley (2009).

Syn. *Hendersonia hortensis* (a) (1882), *Stagonospora hortensis* (a) (1884), *Stagonosporopsis hortensis* (a) (1921), *Ascochyta hortensis* (a) (1945), *Stagonosporopsis boltshauseri* (a) (1975), *Phoma boltshauseri* (1993), *Phoma subboltshauseri* (a) (1997) and some more. Remark: *Phoma subboltshauseri* is a synanamorph of *Stagonosporopsis hortensis sensu* Farr and Rossman (2017).

The Index Fungorum disagrees with Mycobank relating to legitim name of fungus; the latest proposes *Stagonosporopsis hortensis* (Sacc. & Malbr.) Petr., Annales Mycologici 19 (1-2): 21 (1921).

The symptoms have been associated with 'Ascochyta' blight described on *Ascochyta boltshauseri, Phoma exigua* var. *exigua* (= *Boeremia exigua*), and *P. exigua* var. *diversispora* (= *Boeremia diversispora*), respectively (Boerema et al. 1981; Schwartz 2005; Ferreira et al. 2016). The typical on leaves are well-delimited black lesions that may develop in concentric zones. Severe attack on stems or branches can lead to premature defoliation and in extreme cases death of plants, particularly if the attack is on the first node of the main stem. The pathogens can also cause lesions on pods and damage seeds. Infected seeds can have a range of discolouration, resulting in reduced seed quality and decreased yield (Ferreira et al. 2016). See more at faba bean diseases (Sect. 5.6.2).

5.3.3 *Boeremia diversispora*

(Bubák) Aveskamp, Gruyter & Verkley (a) in Aveskamp, Gruyter, Woudenberg, Verkley & Crous, *Stud. Mycol.* **65**: 37 (2010), Fam. *Didymellaceae* Gruyter, Aveskamp & Verkley (2009).

Syn. *Phoma diversispora* (a) (1905), *Phoma exigua* var. *diversispora* (a) (1980).

'Ascochyta' leaf blight of common bean appears the most destructive enemy of beans in Andean highlands of Colombia and Peru. Infection of the fungus is being favoured by the high relative humidity and the cold to moderate temperatures of the regions. On *Phaseolus vulgaris,* early attack of *Phoma exigua* var. *diversispora (= B. diversispora)* can destroy leaves and pods completely and cause 74% yield losses in Andean highlands (Baudoin et al. 1997).

Recently in Spain it was examined the causative agent of 'Ascochyta' blight of bean, based on amplification for the SCAR markers actin and H-CCT, specific for species *Boeremia exigua* and *Boeremia diversispora*, respectively, and revealed a unique fragment to marker actin. In consequence, all evidence suggested that both isolates were close to *Boeremia diversispora* (Ferreira et al. 2016).

5.3.4 *Setophoma terrestris*

(H. N. Hansen) Gruyter, Aveskamp & Verkley (a) in de Gruyter, Woudenberg, Aveskamp, Verkley, Groenewald & Crous, *Mycologia* **102**(5): 1077 (2010), Fam. *Phaeosphaeriaceae* M.E. Barr (1979).

Syn. *Phoma terrestris* (a) (1929), *Pyrenochaeta terrestris* (a) (1948).

Setophoma terrestris is present worldwide but is most common in subtropical and tropical regions. Optimum temperatures for growth of the pathogen and disease development are 24–28 °C. Fungus is able to attack the roots of vegetable crops besides Amaryllidaceae (onions), Amaranthaceae (spinach), Apiaceae (carrot), Brassicaceae (cauliflower), Cucurbitaceae (cantaloupe, cucumber, pumpkin, squash and muskmelon), Fabaceae (bean, cowpea, lima bean and soybean), Solanaceae (eggplant, pepper, potato and tomato) and even Poaceae (sweet corn) (Kreutzer 1941; Punithalingam and Holiday 1973; Hoinacki 2004). However, the most references relating to pink root disease of onions were reported *Setophoma terrestris as causative agent.*

5.3.5 *Mycosphaerella phaseolorum*

Siemaszko (t) Mater. Mikol. Fitopat. Ross. 1(3): 25 (1915), Fam. *Mycosphaerellaceae* Lindau (1897) ***Syn.*** *Sphaerella phaseolorum (Siemaszko) Trotter (t) (1928).*

Since dry beans are produced in South Africa areas and *Mycosphaerella phaseolorum* is common on this crop, it was believed that the pathogen might have moved to soybean. In cross inoculations, both soybean and dry bean isolates were pathogenic to both hosts. However, *M. phaseolorum* isolates were more aggressive to both hosts than the *Ascochyta* sp. (van Wyk et al. 2001).

5.4 Chickpea (*Cicer arietinum* L.)

5.4.1 *Mycosphaerella rabiei*

Kovatsch. ex Gruyter (t) in Boerema et al. *Phoma* identification manual: *Differentiation of specific and infra-specific taxa in culture.* 398 (2004), Fam. *Mycosphaerellaceae* Lindau (1897).

Nom. inval. *Didymella rabiei* Kovatsch. ex Arx (t) in Müller & von Arx *Beitr. Kryptfl. Schweiz* **11**(2): 364 (1962) Art. 39.1 (Melbourne Code).

Nom inval. *Phoma rabiei* (Pass.) Khune (1979) (a).

Nom. inval. *Phoma rabiei* (Pass.) Khune & J.N. Kapoor (1980) (a).

Syn. *Ascochyta rabiei* (a) (1931), *Mycosphaerella rabiei* (t) (2002).

This is a noxious fungus on chickpea (*Cicer arietinum,* Fabaceae). The names of diseases are anthracnose, chickpea blight, 'Ascochyta' blight, chickpea 'Ascochyta' blight fungus which causative agent is the most important pathogen on chickpea-growing areas. In phytopathological literature the disease is commonly called just as '*Ascochyta rabiei*' and 'the Ascochyta pathogen of chickpea'. Hungary is not among the most significant chickpea-growing countries, however it caused a serious disease in the mid-1980s (Kövics et al. 1986). Being seed-borne, the mycelium can be present and contaminate seeds both outside and inside. A range of transmission frequencies up to 90% for *A. rabiei* (= *Mycosphaerella rabiei*) on chickpea (Kaiser et al. 1973) have been reported. It seems to be heterothallic because compatible mating types are required for ascospore development (Boerema et al. 2004) . Detailed information about the lifecycle of pathogen is available in publication of Kaiser (1997).

5.4.2 *Ectophoma multirostrata*

(P.N. Mathur, S.K. Menon & Thirum) Valenz.-Lopez, Cano, Crous, Guarro & Stchigel (a) in Valenzuela-Lopez, Cano-Lira, Guarro, Sutton, Wiederhold, Crous & Stchigel, *Stud. Mycol.* **90**: 34 (2017), Fam. *Pleosporales* Luttr. ex M.E. Barr (1987).

Syn. *Phoma decorticans* var. *microspora* (a) (1898), *Phoma lucknowensis* (a) (1968), *Phoma multirostrata* (1973), *Phoma multirostrata* var. *macrospora* (1986), *Phoma multirostrata* var. *microspora* (1986).

This fungus is a halophilic, soil-borne saprophyte or opportunistic pathogen, which may cause dry and brittle root rot of chickpea (Chobe et al. 2020; Deb et al. 2020) and, moreover, blight and dieback symptoms in subtropical regions recorded in south-west Asia (India) in the soil of mangrove vegetation, northern Africa (Egypt) (Boerema et al. 2004) and Australia (Golzar et al. 2015). Gilardi et al. (2010) reported a leaf spot disease on *Fuchsia* × *hybrida* which was observed for the first time in Northern Italy and worldwide. The fungus was identified on ITS-based molecular method as *Phoma multirostrata* (= *Ectophoma multirostrata*).

The fungus is rather variable both in size of pycnidia and even in wide range of conidial dimensions which led to the differentiation of the former *Phoma multirostrata* vars. *macrospora* and *microspora,* respectively (Boerema et al. 2004); however, intermediate variants also commonly occur based on the size of conidia, depending on pH and media (Rai 1998).

In recent disease notes (Chobe et al. 2020), the occurrence of the pathogen was described in chickpea-growing soils in central India. Besides the most frequent *Macrophomina phaseolina*, a few samples (10 to 15%) recurrently yielded another fungus which was identified morphologically, and the ITS region was also sequenced, and a BLASTn search revealed 100% sequence similarity (524 bp nucleotide similarity) to *Ectophoma multirostrata* (Valenzuela-Lopez et al. 2018) isolate (accession no. KM659039). The sequence was submitted to the GenBank database maintained by the National Center for Biotechnology Information (accession no. MG897497).

5.4.3 *Ascochyta medicaginicola*

Qian Chen & L. Cai (a) in Chen, Jiang, Zhang, Cai & Crous *Stud. Mycol.* **82**: 187 (2015), Fam. *Didymellaceae* Gruyter, Aveskamp & Verkley (2009).

Syn. *Phoma medicaginis* (a) (1886), *Ascochyta imperfecta* (a) (1912), *Phoma medicaginis* var. *medicaginis* (a) (autonym) (1965), *Phoma medicaginis* var. *macrospora* (a) (1993) and some more.

The main host of the pathogen is alfalfa, causing spring black stem and leaf spot disease; however, it can occur on chickpea, causing 'Phoma' blight, or spot disease (as *Phoma medicaginis*) and lentil. More details are accessible in the Compendium of Chickpea and Lentil Diseases and Pests (Chen et al. 2011), and see alfalfa diseases (Sect. 5.2.1).

5.5 Clovers (*Trifolium* L.)

5.5.1 *Ascochyta caulicola*

Laubert (a) *Art. on Biol. Abtheil am Gesundheit*: 443 (1903), Fam. *Didymellaceae* Gruyter, Aveskamp & Verkley (2009).

Syn. *Ascochyta caulicola* var. *lupini* (a) (1935).

Ascochyta stem and leaf spot, stem canker or gooseneck disease of *Melilotus* species, including the yellow sweet clover, *M. officinalis*, and white clover (*Trifolium repens*). The silvery-white lesions of variable size with a brown halo, located on the stem, petioles and leaf, spots enlarge, increasing blights sometimes coalescence. Numerous black pycnidia give the lesions a stippled appearance. The affected leaves turn yellow, and the heavily infected stems are swollen, retarded in their development, so plants remain in backward (Gorz 1955; Kövics 1983).

There are no recent reports about the pathogen.

5.5.2 *Ascochyta trifolii*

Siemaszko (a) *Acta Soc. Scient. Warszav.* 7(3): 8 (extr.) (1914), Fam. *Didymellaceae* Gruyter, Aveskamp & Verkley (2009).

Inc. sed. *Ascochyta trifolii* Bondartsev & Trusova (a) *Zh. Bolêz. Rast.* **7**: 215 (1913).

Nom. illeg. *Ascochyta trifolii-montani* Bond.-Mont. (a) *Gribnye bolezni klevera i lucerny:* 118 (1954) According to Art. 39.1 (Melbourne Code).

In a survey about the distribution and prevalence of pathogens in a naturally infected field of white clover (*Trifolium repens*) grown for seed in Poland, white clover was infected by a number of pathogens (Nadolnik 1981). The most prevalent attacking leaves were *Peronospora trifoliorum*, *Pseudopeziza trifolii* and *Ascochyta trifolii* (*cit.* Clements 1996). Besides the species description of the fungus on *Trifolium* sp. in Poland, the fungus occurs (mostly) in Estonia; however, there are no recent references, except some Species Catalogue items (Roskov et al. 2013; PlutoF 2017). Although Gafforov (2017) refers in his list about the presence of '*A. trifolii* Bondartsev & Trusova, *Zh. Bolêz. Rast.* 7: 215 (1913)' in Uzbekistan, cited in a PhD thesis from 1989 (in Russia), but this data might be questionable on both validity of the scientific name and the form of the used author names. However, the CABI Datasheet (No 7345, 2019) suggests the scientific name as '*Ascochyta trifolii* Bondartsev & Trus'. The Mycobank database manages both "*A. trifolii*" names considered valid, but after the latest descriptors, it uses a question mark (?).

5.5.3 *Ascochyta trifolii-alpestris*

Dominik (a) *Acta Societatis Botanicorum Poloniae* **11**: 242 (1934), Fam. *Didymellaceae* Gruyter, Aveskamp & Verkley (2009).

It occurs on *Trifolium alpestris* described in Poland. Pycnidia are epiphyllous, brown, globose or almost globose-depressed, 180 × 120 μm. Conidia are fusiform, both ends acute, straight, slightly constricted, 9–15 × 2–3 μm. Occurs on living leaves of *Trifolium alpestris*. Distribution: Europe (Poland) (Melnik 2000). There is no recent data about the occurrence of the pathogen.

5.5.4 *Leptosphaeria pratensis*

Sacc. & Briard (t) *Revue mycol., Toulouse* **7**(27): 209 (1885) *sensu* Index Fungorum, Fam. *Leptosphaeriaceae* M.E. Barr (1987).

Syn. *Phoma meliloti* (a) (1892).

The pathogen occurs in Australia, Europe and North America on various papilionaceous plants. *Leptosphaeria pratensis* formerly was confused with *L. weimeri* Schoemaker, C.E. Babc. & J.A.G. Irwin (t) (1991) and *L. viridella* (Peck) Sacc. (1883) (= *Phaeosphaeria viridella* /Peck/ Leuchtm. (t) (1984) *sensu* Boerema et al. (2004).

Leptosphaeria pratensis occurs on different Fabaceae, e.g. *Medicago, Melilotus, Trifolium* and *Ulex* spp. It causes leaf spot, stem canker, and root rot of forage legumes. Low temperature (about 8°C) induces to develop the *Phoma* anamorph (Boerema et al. 2004).

5.5.5 *Stagonospora recedens*

(C. Massal.) F.R. Jones & Weimer (a) *J. Agric. Res., Washington* **57**: 807 (1938), Fam. *Didymellaceae* Gruyter, Aveskamp & Verkley (2009).

Syn. *Phloeospora trifolii* var. *recedens* (a) (1889), *Stagonospora recedens* f. *trifolii-alexandrini* (1953).

Stagonospora recedens is a fungal plant pathogen infecting clovers, including red (*Trifolium repens*) and Alexandrian clovers (*Trifolium alexandrinum*). There is no recent reference about this pathogen.

5.6 Faba Bean, Broad Bean and Horse Bean (*Vicia faba* L.)

The causative agents of the 'Ascochyta' leaf and pod blight of horse bean include three different *Phoma*-like fungi, viz. *Didymella fabae*, *Didymella lethalis* (= *Peyronellaea lethalis*), *Phoma subboltshauseri* (= *Didymella subboltshauseri*) and *Ascochyta viciae* (= *Septoria viciae*). Host plant of *D. fabae* is only the broad bean (*Vicia faba*), *Ascochyta viciae* (= *Septoria viciae*) pathogenic on sweet clover (*Melilotus*) and vetch (*Vicia*) species including *V. faba and Phoma subboltshauseri* (= *Ascochyta subboltshauseri*) pathogenic on beans (*Phaseolus* spp.), faba bean (*Vicia faba*) and cowpea (*Vigna unguic-*

ulata). The *Ascochyta pisi* (= *Didymella pisi*) is a multi-host pathogen, including grass pea (*Lathyrus*), lupines (*Lupinus*), sweet clovers (*Melilotus*), clovers (*Trifolium*), pea (*Pisum*), faba bean (*Vicia faba*), vetches (*Vicia* spp.) and cowpea (*Vigna*) and, moreover, some other papilionaceous plant species (Lenti et al. 2008).

5.6.1 *Didymella fabae*

G.J. Jellis & Punith. (t) *Pl. Path.* **40**(1): 151 (1991), Fam. *Didymellaceae* Gruyter, Aveskamp & Verkley (2009).

Syn. *Ascochyta fabae* (a) (1899).

The pathogen occurs worldwide (Punthalingam 1993). Lesions can be formed on leaves, stems and pods of infected faba bean. On both the top and bottom sides of leaves, small, circular, dark-brown spots first appear which enlarge and turn light to dark grey as the disease develops. They become irregular and often zonate and may coalesce to cover most of the leaf surface. Leaf tissue next to the lesions may become black and necrotic. Within the lesions, numerous black fruiting bodies (pycnidia) of the fungus develop (Simay 1988; Lenti et al. 2008). The transmission frequency of *Ascochyta fabae* (= *Didymella fabae*) on faba bean (Hewett 1973) was less than 10%. The teleomorph *D. fabae*, produced over winter on infected bean straw in stubble fields, is now considered to be a major source of infection through air-borne dispersal of the sexual spores (Jellis and Punithalingam 1991; Rubiales and Trapero-Casas 2002).

In a recent host pathogenicity test also, it is confirmed that faba bean isolates of *D. fabae* were specific to *Vicia faba* (Sillero and Rubiales 2014).

5.6.2 *Ascochyta boltshauseri*

Sacc. apud Boltshauser (a) *Z. Pfl.Krankh.* **1**: 136 (1891), *sensu* Index Fungorum, Fam. *Didymellaceae* Gruyter, Aveskamp & Verkley (2009).

Syn. *Hendersonia hortensis* (a) (1882), *Stagonospora hortensis* (a) (1884), *Stagonosporopsis hortensis* (a) (1921), *Ascochyta hortensis* (a) (1945), *Stagonosporopsis boltshauseri* (a) (1975), *Phoma boltshauseri* (1993), *Phoma subboltshauseri* (a) (1997) and some more. Remark: *Phoma subboltshauseri* is a synanamorph of *Stagonosporopsis hortensis sensu* Farr and Rossman (2017).

The *Ascochyta boltshauseri* can be pathogenic on beans, cowpea and broad bean and was described first in Hungary in the middle of 1960. Since the pathogen is seed-borne and survive on plant residues, it can spread easily (Békési 1965). In a keynote lecture at the international workshop on faba bean, W. Link from Germany (2006) referred that *Ascochyta* resistance was developed in faba bean which was confirmed by molecular markers. The line 29H, among others, was repeatedly used to improve *Ascochyta* resistance; however, *Phoma* (= *Ascochyta boltshauseri*?) and mildew cause further, less well-studied foliar diseases in *Vicia faba* (Link 2006).

See more about this pathogen in bean diseases (Sect. 5.3.2).

5.7 Lentil (*Lens culinaris* Medik)

5.7.1 *Didymella lentis*

W.J. Kaiser, B.C. Wang & J.D. Rogers (t) *Plant Dis.* **81**(7): 815 (1997), Fam. *Didymellaceae* Gruyter, Aveskamp & Verkley (2009).

Syn. *Ascochyta lentis* (a) (1938).

Disease name(s): 'Ascochyta' blight, 'Ascochyta' leaf blight; ascochytosis; lentil blight; lentil stain. The global cool-season grain legume production is largely represented by chickpea (*Cicer arietinum* L.), pea (*Pisum sativum* L.) and cultivated lentil (*Lens culinaris* Medik. ssp. *culinaris*) (Khazaei et al. 2016). Environmental conditions during germination influence the frequency of transmission of the pathogen from seed (Ahmed and Beniwal 1998) to the aerial portions of seedlings. *Ascochyta fabae* (= *Didymella fabae*) (Yu 1947), *A rabiei* (= *Mycosphaerella rabiei*) (Kaiser 1973) and *A. len-*

tis (= *D. lentis*) Bondartzeva-Monteverde and Vassiljevski (1940) survive at least one winter season at the soil surface.

Ascospores of *Didymella lentis* were trapped from lentil stubble that was weathered under natural conditions in the field in Western Australia and Victoria; naturally occurring pseudothecia were present on the stubble. This was the first report of the teleomorph of *Ascochyta lentis* on lentil stubble from the field in Australia. The presence of the teleomorph has implications in the long-distance dispersal of *D. lentis* and on resistance breeding programs (Galloway et al. 2004).

5.7.2 *Ascochyta medicaginicola*

Qian Chen & L. Cai (a) in Chen, Jiang, Zhang, Cai & Crous *Stud. Mycol.* **82**: 187 (2015), Fam. *Didymellaceae* Gruyter, Aveskamp & Verkley (2009).

Syn. *Phoma medicaginis* (a) (1886), *Ascochyta imperfecta* (a) (1912), *Phoma medicaginis* var. *medicaginis* (a) (autonym) (1965), *Phoma medicaginis* var. *macrospora* (a) (1993) and some more.

The main host of the pathogen is alfalfa, causing spring black stem and leaf spot disease, but it also occurs on lentil and chickpea causing 'Phoma' leaf spot diseases, under frequently used '*Phoma medicaginis*' name. More details are available in the Compendium of Chickpea and Lentil Diseases and Pests (Chen et al. 2011), and alfalfa diseases (Sect. 5.2.1).

5.8 Lupines (*Lupinus* L.)

'Ascochyta stem canker' causes infected lupine seedlings to become girdled by a dark brown to black canker. Affected plants of any age can die from this fungus, and plants entering maturity may wilt and die suddenly. The fungus that causes ascochyta stem canker is the same fungus that causes blight on cotton (Putnam et al. 1989). Presumably this is a false statement, because the cotton ascochytosis is caused by *Ascochyta gossypii* (= *Nothophoma gossypiicola* /Gruyter/ Qian Chen & L. Cai (a) in Chen, Jiang, Zhang, Cai & Crous, *Stud. Mycol.* 82: 213, 2015). The only methods of control for lupine ascochyta are planting resistant varieties, when available, and planting lupine in areas with good drainage to avoid soggy soil. Also, avoid planting lupine in soil with a high pH level (Putnam et al. 1989).

However, the exact causative agent of described lupine 'stem canker' symptoms and control measurements have remained unclear. Kövics (1995a, b) listed *Ascochyta pisi* (= *Didymella pisi*), *Phoma lupini* (= *Plenodomus lupini*), *Phoma lupinicola* and *Phoma pedicularis* (= *Leptosphaeria pedicularis*) among potential pathogens on lupines in Hungary.

5.8.1 *Didymella pisi*

Chilvers, J.D. Rogers & Peever (t) *Mycol. Res.* **113**(3): 396 (2009), Fam. *Didymellaceae* Gruyter, Aveskamp &Verkley (2009).

Nom. rej. *Sphaeria pinodes* Berk. & A. Bloxam (1861).

Syn. *Ascochyta pisi* (a) (1830), *Ascochyta pisi* var. *lupini* (a) (1920), *Ascochyta pisi* var. *medicaginis* (a) (1920), *Ascochyta pisi* var. *onobrychidis* (a) (1920) *Peyronellaea pinodes* (a) (2010).

Occurrence of the *Didymella pisi* on *Lupinus angustifolius* and *L. arboreus* are reported from New Zealand (Pennycook 1989), South Africa (Crous et al. 2000), moreover on *L. luteus* and *L. perennis* appeared in some lists of US pathogens (cit. Farr and Rossman 2017).

For more details of pathogen, see pea diseases (Sect. 5.9.1).

5.8.2 *Plenodomus lupini*

(Ellis & Everh) Gruyter, Aveskamp & Verkley (a) in Gruyter, Woudenberg, Aveskamp, Verkley, Groenewald & Crous, *Stud. Mycol.* **75**: 22 (2012) [2013], Fam. *Pleosporales* Luttr. ex M. E. Barr (1987).

Syn. *Phoma lupini* (a) (1884), *Sphaeropsis lupini* (1898), *Stictochorella lupini* (1924), *Asteromella lupini* (1955).

This pathogen was isolated from various *Lupinus* spp., indigenous to North and South America. The fungus can infect all aboveground parts: on leaves, stems and pods causing spots. In the Andean highlands, above 3800 m, a common disease on *Lupinus mutabilis*, a protein source of the Indians (Boerema et al. 2004).

5.8.3 *Phoma schneiderae*

Boerema, Gruyter & P. Graa (a) *Persoonia* **17**(2): 282 (1999).

Remarks: Nom. inval., Art. 40.3 (Melbourne Code) *sensu* Index Fungorum in contrast to Mycobank, which acknowledges this as a valid species. *Stagonosporopsis lupini* is a synanamorph of *Phoma schneiderae* (Farr and Rossman 2017).

Syn. *Ascochyta lupini* (1984), *Stagonosporopsis lupini* (*stagonosporoid* synanamorph) (1999).

The pycnidia of pathogen are dimorph: with large *stagonosporoid* conidia, described as *Stagonosporopsis lupini* (Boerema & R. Schneid.) Boerema et al., and one-celled *phomoid* conidia of *Phoma schneiderae* (Boerema et al. 1999).

Earlier, based on one (*phomoid*) form of conidia was described with diagnostic data (Latin description, conidial variability in vivo and in vitro, drawings from R. Schneider (Boerema 1984), disease symptoms, reproduced from Frey and Yabar (1983); differentiation of the *phomoid* phenotype as *Ph. schneiderae* (Boerema et al. 2004).

The pathogen is responsible for leaf spot and blight of lupine diseases. The fungus is a specific, seed-borne pathogen of lupines (*Lupinus* spp.), causing spots on leaves, stems and pods. In South America it causes serious problem on the cultivated *Lupinus mutabilis*; however, in North America it has been found on the wild *Lupinus perennis*. In Europe, the pathogen is known with occasional findings on *L. arborescens* in England and was recently isolated from cultivated *Lupinus albus* in Poland and the UK (Boerema et al. 2004). More recently the occurrence was reported from Australia (Bretag et al. 2006). The pathogen seems dangerous, because its hosts can be *Lupinus albus* (white lupine), *Lupinus mutabilis* (sweet lupine), *Lupinus angustifolius* (blue lupine, narrow leaf lupine or narrow-leaved lupine) and *Lupinus luteus* (yellow lupine). Western Australia cultivates over 800,000 hectares of *Lupinus angustifolius* with a market value of approximately AU$200 million. There are also large areas of lupine crops in New South Wales, Victoria and South Australia (over 100,000 hectares in total). This explains that the pathogen is on the members of Australian Plant Pest Alert List (CCEPP 2007).

5.8.4 Rare or Dubious *Phoma*-Like Fungi Described on Lupines

Ascochyta lupinicola

Petr. (a) *Annls mycol.* **19**(5-6): 281 (1921), Fam. *Didymellaceae* Gruyter, Aveskamp & Verkley (2009).

Syn. Not known.

The pathogen was described exactly 100 years ago, according to the valid rules of description; however, there was no additional data about its occurrence.

Phoma lupinicola

Earle (a) *Greene, Plant. Bak.* **2**(1): 27 (1901), Fam. *Didymellaceae* Gruyter, Aveskamp & Verkley (2009).

Syn. Not known.

Besides the species description and an item on list of potential lupine *Phoma*-like diseases (Kövics 1995a, b), there is no reference about the occurrence of the pathogen.

Leptosphaeria pedicularis

(Fuckel) Gruyter, Aveskamp & Verkley (t) in Gruyter, Woudenberg, Aveskamp, Verkley, Groenewald & Crous, *Stud. Mycol.* **75**: 19 (2012) [2013], Fam. *Leptosphaeriaceae* M.E. Barr (1987).

Syn. *Phoma pedicularis* [as '*pedicularidis*'] (a) (1874), *Diplodina pedicularis* (a) (1924), *Ascochyta pedicularis* (a) (1963) and some more. [Not *Phoma pedicularis* (a) (1946), which equals with '*Phoma herbicola*' (Boerema et al. 2004) (= *Ascochyta herbicola* (Wehm.) Qian Chen & L. Cai (a) (2015)].

This fungus occurs on dead stems and leaves of various herbaceous plants, but is not pathogenic on lupines (*Lupinus* spp.) (Boerema et al. 2004).

5.9 Pea (*Pisum* L.)

There are three fungi which contribute to the 'pea "Ascochyta" blight complex'. As the fungi are seed-borne, nowadays distributed worldwide in arable soils. Because *Didymella pisi*, *Didymella pinodes* and *Didymella pinodella* are similar in host range and disease symptoms, including seedling blights, foot rots, stem, pod and leaf blights, so frequently treated as 'pea "Ascochyta" blight diseases' which occur throughout the world and can be of significant economic importance.

5.9.1 *Didymella pisi*

Chilvers, J.D. Rogers & Peever (t) *Mycol. Res.* **113**(3): 396 (2009), Fam. *Didymellaceae* Gruyter, Aveskamp & Verkley (2009).

Syn. *Ascochyta pisi* (a) (1830), *Sphaeria pisicola* (1859).

Didymella pisi, the legitim (teleomorph) name of the former anamorphic *Ascochyta pisi* pathogen, is a well-known fungus of papilionaceous plants, especially on *Pisum sativum*, causing black stem, foot rot and leaf spot. Lesions start as small dark flecks on the lower parts of the plant and enlarge under humid conditions, often creating a target effect with alternating rings of grey and brown. The pathogens overwinter on pea stubble and residue, the primary source of inoculum, and can survive on stubble or in the soil for many years. Airborne ascospores form perithecia and conidia from pycnidia are released and spread by rain splash to plants nearby, or by wind to plants up to several kilometres away. This creates a disease risk even in fields where no field peas have been grown previously.

The cultural characteristics of *Didymella pisi* are different from the other relative species on pea host. *Ascochyta pisi* Lib., the type species of *Ascochyta*, and its conidia always septate (ascochytoid) both in vivo and in vitro, described as 'wall-thickening septation' (Boerema 1984).

5.9.2 *Didymella pinodes*

(Berk. & A. Bloxam) Petr. (t) *Annls mycol.* **22**(1/2): 16 (1924), Fam. *Didymellaceae* Gruyter, Aveskamp & Verkley (2009).

Syn. *Sphaeria pinodes* (a) (basionym) (1861), *Mycosphaerella pinodes* (t) (1912), *Ascochyta pinodes* (a) (1927), *Peyronellaea pinodes* (a) (2010) and some others.

A phylogenetic comparison has proved that *Ascochyta pinodes* (= *Didymella pinodes*) clustered with *Didymella* species and not with the well-characterized *Mycosphaerella* species from other hosts, and it was proposed that the name *Didymella pinodes* (Berk. & A. Bloxam) Petrak be used to describe this fungus. Analysis of G3PD revealed two major clades among legume-associated *Ascochyta* fungi with members of both clades infecting pea ('Ascochyta complex') (Peever et al. 2007).

5.9.3 *Didymella pinodella*

(L.K. Jones) Qian Chen & L. Cai (t) in Chen, Jiang, Zhang, Cai & Crous, *Stud. Mycol.* **82**: 178 (2015), Fam. *Didymellaceae* Gruyter, Aveskamp & Verkley (2009).

Syn. *Ascochyta pinodella* (a) (1927), *Phoma medicaginis* var. *pinodella* (a) (1965), *Phoma pinodella* (a) (1987), *Peyronellaea pinodella* (a) (2010) and some more.

All these three *Didymella* species, mentioned till 2009, were the causative agents of the pea 'Ascochyta' blight diseases, respectively. An additional *Phoma* fungus was reported from

Australia, viz. *Phoma koolunga* (Davidson et al. 2009). Among the well-known three pathogenic fungi of pea, *Didymella pinodes* is the most important (Ali et al. 1978). *Didymella pinodella* is one of the causative agents of soybean *Phoma*-like fungi as well. More synonyms of *Didymella pinodella*, detailed cultural, morphological and pathological features, and molecular data can be found in soybean diseases (Sect. 5.11.1).

5.9.4 *Boeremia exigua*

(Desm.) Aveskamp, Gruyter & Verkley (a) in Aveskamp, Gruyter, Woudenberg, Verkley & Crous, *Stud. Mycol.* **65**: 37 (2010) Fam. *Didymellaceae* Gruyter, Aveskamp & Verkley (2009).

Syn. *Phoma exigua* (basionym) (1849), *Phoma exigua* var. *exigua* (autonym) (1849), *Ascochyta phaseolorum* (a) (1878), *Boeremia exigua* var. *heteromorpha* (a) (2010). This fungus has numerous, about 150, synonyms, invalid later homonyms and *nomina nuda* (for details see in Boerema et al. 2004).

This ubiquitous fungus is discussed at soybean diseases (Sect. 5.11.2).

5.9.5 *Phoma koolunga*

J.A. Davidson, D. Hartley, Priest, S. Kaczm., Herdina, A. McKay & E.S. Scott *Mycologia* **101**(1): 124 (2009), Fam. *Didymellaceae* Gruyter, Aveskamp & Verkley (2009).

This fungus was described in Australia as causative agent of 'Ascochyta blight' on pea. 'Ascochyta blight' (synonym: blackspot) was first recorded in Australian field pea crops in the 1960s (Ali and Dennis 1992). In Western Australia, isolates of *P. koolunga* are more virulent and cause significant disease symptoms compared to *Didymella pinodes* and *Phoma pinodella* (= *Didymella pinodella*) (Tran et al. 2015). It was recently described in host range experiments which were conducted in a growth room using 41 legumes comprising Australian native, naturalised, crop and pasture legumes and wild-type *Pisum*, *Lathyrus* and *Vicia* spp. All legumes tested, except *Cicer arietinum* (chickpea), developed leaf lesions and some also had stem lesions. Disease incidence and severity differed significantly among species but not consistently between the *P. koolunga* isolates (Keirnan et al. 2020).

Additionally, *Phoma herbarum* Westend., 1852 (Li et al. 2011) and *Phoma glomerata* (Tran et al. 2014) have also been shown to be associated with the 'Ascochyta blight complex' on dry pea in Australia.

5.9.6 *Setophoma terrestris*

(H.N. Hansen) Gruyter, Aveskamp & Verkley (a) in de Gruyter, Woudenberg, Aveskamp, Verkley, Groenewald & Crous, *Mycologia* **102**(5): 1077 (2010), Fam. *Phaeosphaeriaceae.*

Syn. *Phoma terrestris* (a) (1929), *Pyrenochaeta terrestris* (a) (1948).

This fungus is widely distributed in soil, and it is the causal organism of onion pink root, referred from North and South America, and common inhabitant in Europe, Africa and Australia associated with the cultivation of *Allium* species (garlic, leek, shallot and chive) (Boerema et al. 2004) and, moreover, *Carthamus*, *Solanum* and *Spinacia* spp. (Kövics 2000). The fungus is frequently isolated from roots of different Poaceae plants, without any disease symptoms. The fungus produces red pigments in the mycelium, and this easily distinguishes from morphologically similar *Phoma*-like fungi (Boerema et al. 2004). See additional details in soybean diseases (Sect. 5.11.3).

5.9.7 *Didymella glomerata*

(Corda) Qian Chen & L. Cai (t) in Chen, Jiang, Zhang, Cai & Crous, *Stud. Mycol.* **82**: 176 (2015), Fam. *Didymellaceae* Gruyter, Aveskamp & Verkley (2009).

Syn. *Coniothyrium glomeratum* (a) (basionym) (1840), *Peyronellaea glomerata* (a) (1946), *Phoma glomerata* (a) (1936) and many more synonyms; see Boerema et al. (2004).

The former name of the fungus, *Phoma glomerata*, is still widely used. *Didymella glomerata* is a worldwide distributed soil fungus, which has been isolated from various plants (belonging to more than 100 host plant genera), as well as association with animal and human mycoses. Generally it is considered as secondary invader or opportunistic pathogen (Boerema et al. 2004). It was also reported as a mycoparasite of powdery mildew (Sullivan and White Jr. 2000). *Phoma glomerata* (= *Didymella glomerata*) and *P. calorpreferens* (= *Peyronellaea calorpreferens*) were recorded as endophyte fungi from Korean pine (*Pinus koraiensis*) leaves (Deng et al. 2011). It was also associated with "Pea 'Ascochyta blight complex" in Australia (Tran et al. 2014).

5.10 Vetch Species (*Vicia* L. spp.)

Common vetch (*Vicia sativa*) is one of the most important species cultivated in the genus *Vicia*. There were five *Ascochyta* species described in family *Didymellaceae* and one *Septoria* in family *Mycosphaerellaceae* which does not belong to phomoid fungus group; however, its basionym was described as *Ascochyta viciae* Lib. (= *Septoria viciae*).

5.10.1 *Septoria viciae*

(Lib.) Westend. (a) *Herb. crypt. Belg.*: no. **1151** (1857), Fam. *Mycosphaerellaceae* Lindau (1897).

Syn. *Ascochyta viciae* (a) (1837), *Phyllosticta viciae* (a) (1871).

Although the morphology of *Septoria*, especially the shape of conidia, is significantly distinct from *Phoma*-like species, from the middle of the twentieth century, it was mentioned only in different checklists as 'leaf spot disease' on *Vicia sativa* (Weiss 1946; Anonymous 1960; Shaw 1973; Kövics 1995a, b). The teleomorphic connection with *Mycosphaerella lethalis* has been referred by Melnik (2000). However, this fungus is synonymous with *Peyronellaea lethalis sensu* Index Fungorum.

5.10.2 *Ascochyta punctata*

Naumov (a) *Bull. Angew. Bot. St. Petersburg* **6**(3): 204 (1913), Fam. *Didymellaceae* Gruyter, Aveskamp & Verkley (2009).

Described on *Vicia sativa*. There are no recent data about this pathogen.

5.10.3 *Ascochyta viciae-villosae*

Ondřej (a) *Biológia, Bratislava* **23**: 815 (1968), Fam. *Didymellaceae* Gruyter, Aveskamp & Verkley (2009).

It is isolated from *Vicia villosa*. Watson (1971) listed this species among fungi from former Czechoslovakia. Although there is no recent record about its occurrence, an isolate (CCM 65) has been preserved in the Czech Collection of Microorganisms at Masaryk University.

5.10.4 *Ascochyta viciae-lathyroidis*

Syd. & P. Syd. (a) *Hedwigia* **39** (Beibl.): (3) (1900), Fam. *Didymellaceae* Gruyter, Aveskamp & Verkley (2009).

Although this species was listed in the book of foreign fungus diseases in the USA (Watson 1971) from Germany, it was neither included in Index Fungorum nor Mycobank databases because presumably of its invalidity.

5.10.5 *Ascochyta viciae-pannonicae*

Ondřej (a) *Biológia, Bratislava, séria A – Botanica* **25**(10): 685 (1970), Fam. *Didymellaceae* Gruyter, Aveskamp & Verkley (2009).

It was described on *Vicia pannosa*. There is no recent reference about its occurrence on fields. Although one isolate was deposited to CBS-KNAW collection, backbone molecular data of *Ascochyta viciae-pannonicae* strain CBS 254.92 was compared to similar species shown in *Ascochyta* clade (No. 28) by Hou et al. (2020).

5.10.6 *Ascochyta viciae-pisiformis*

Bubák (a) *Annls mycol.* **2**(5): 397 (1904), Fam. *Didymellaceae* Gruyter, Aveskamp & Verkley (2009).

Only limited research has been published about foliar diseases of *Vicia* species, excluding *Vicia faba*, and there is insufficient information on their potential as a green bridge for dissemination of faba bean diseases (Sillero and Rubiales 2014).

5.10.7 *Peyronellaea lethalis*

(Ellis & Barthol.) Aveskamp, Gruyter & Verkley (a) in Aveskamp, Gruyter, Woudenberg, Verkley & Crous, *Stud. Mycol.* **65**: 32 (2010), Fam. *Didymellaceae* Gruyter, Aveskamp & Verkley (2009).

Syn. *Ascochyta lethalis* (a) 1903, *Mycosphaerella lethalis* (t) (1912), *Didymellina lethalis* (t) (1918), *Sphaerella lethalis* (t) (1928), *Didymella lethalis* (t) (1984).

At least 18 fungal leaf spots of *Vicia* species have been reported, including those caused by *Ascochyta fabae* Speg., *A. pinodes* Jones (= *Didymella pinodes*), *A. pisi* Lib. (= *Didymella pisi*), *A. viciae* Lib. (= *Septoria viciae*) and an unidentified species of *Ascochyta*. These fungi are similar in morphology and have overlapping host ranges. *A. viciae* (= *Septoria viciae*) is a questionable species, as has been assigned synonym with *A. lethalis* (= *Peyronellaea lethalis*) and *A pisi* (= *Didymella pisi*) as well (Leth et al. 1994).

Pey. lethalis fungus (as *Ascochyta lethalis*) was listed among sweet clover (*Melilotus officinalis*) diseases, being presumed as the pathogen of sweet clover 'black stem disease' (Honey et al. 1945). Recently, the *Pey. lethalis* was described as a pathogen of pea 'Ascochyta' blight complex in Serbia (Petrović et al. 2017).

5.11 Soybean (*Glycine max* (L.) Merr)

Different species of hyaline-spore pycnidial fungi have been associated with leaf and pod spot diseases of soybean (*Glycine max*(L.) Merr.) from different parts of the world. These diseases might periodically occur, causing yield losses under favourable environmental conditions. The causative agents of these diseases with rather similar symptoms of all *Phoma*-like species, are geographically widespread and are found in numerous ecological niches. These fungi are soil-borne group inhabiting organic debris and water, as well as species that parasitize other fungi, lichens, insects and vertebrates. Specimens have been isolated mainly from soil and from a wide range of plant hosts where they reside as primary pathogens, opportunists, saprobes or endophytes (Aveskamp et al. 2008; Kövics et al. 2014). Recently, in a diversity study of endophytic fungi of soybean leaves and roots, 229 items were isolated and identified. *Phoma*-like fungi had been isolated from soybean leaves only, which represented *Phoma glomerata* (2.5%) (= *Didymella glomerata*) and *Phoma pomorum* (0.7%) (= *Didymella pomorum*) and in a very high rate of *Phoma* sp. (29.4%) without precisely identification (Fernandes et al. 2015). Besides the pathogenic, weak or opportunistic parasite, saprophyte and endophyte roles of fungi, it would be reasonable considering more accurately their complexity in a niche of growing plants.

Molecular datasets have provided a better outlook towards the phylogenetic and evolutionary trends of phomoid fungi (Aveskamp et al. 2008, 2010a, b; Gomzhina et al. 2020).

The family *Didymellaceae* was recently revised based on morphological and phylogenetic analyses of ex-type strains subjected to DNA sequencing of partial gene data of the *LSU, ITS, RPB2* and *TUB2* loci. A robust, multi-locus reference phylogenetic tree of *Didymellaceae* was generated. In addition, *RPB2* was found to be the most effective locus for the identification of Fam. *Didymellaceae* at species level and is proposed as a secondary DNA marker for the family Hou et al. (2020). In a recent re-evaluation of *Phoma*-like fungi of soybean was summarized and clarified of relevant species, including their historical, pathological, morphological and taxonomical aspects, and confirmed by molecular analyses (Kövics et al. 2014).

5.11.1 *Didymella pinodella*

(L.K. Jones) Qian Chen & L. Cai (t) in Chen, Jiang, Zhang, Cai & Crous, *Stud. Mycol.* **82**: 178 (2015), Fam. *Didymellaceae* Gruyter, Aveskamp & Verkley (2009).

Nom. inval. *Ascochyta sojina* Bedlan (a) *J. Kulturpfl.* 66(9): 320 (2014).

Syn. *Ascochyta pinodella* (a) (1927), *Ascochyta sojicola* (as '*sojaecola*') Abramov (a) (Abramov 1931), *Phoma medicaginis* var. *pinodella* (a) (1965), *Phoma pinodella* (a) (1987), *Phoma sojicola* (1999), *Peyronellaea pinodella* (a) (2010).

The fungus '*Ascochyta sojicola*' (as '*sojaecola*') was the first description of a soybean pathogen (probably as weak parasite) in Far Eastern region in 1931 by Abramov. Later in Europe, Böning (1939) referred a soybean pathogen as '*Phyllosticta sojaecola*' (= *Boeremia exigua*), and then Frandsen (1953) in Germany and Hungary (Tóth and Kövics 1978) referred '*Ascochyta sojaecola*' with similar, indistinguishable symptoms. Based on detailed cultural, morphological and pathological tests made by Bondartzeva-Monteverde and Vassiljevskiy (1940), *Ascochyta pisi* (= *Didymella pisi*) could not be shown to be pathogenic to soybean in contrast with *A. pinodella* (= *Didymella pinodella*), and they sharply differentiated *Ascochyta phaseolorum* (= *Boeremia exigua*) and *A. sojaecola* (now synonymous with *Didymella pinodella*) as distinct species and the former to possess higher pathogenicity on soybean. The pathogenicity of *Phoma pinodella*, *A. sojicola* (= *Didymella pinodella*) and *P. exigua* var. *exigua* (= *Boeremia exigua*) isolates was also confirmed on the basis of a wide range of pathogenicity tests made on soybean, and *Phoma sojicola* was described as a new species (Kövics et al. 1999).

Later, in a comprehensive phylogenetic analysis, sequences from *Phoma pinodella* (= *Didymella pinodella*), *P. exigua* var. *exigua* (= *Boeremia exigua*) and *P. glomerata* (= *Didymella glomerata*) isolated from soybean and other hosts as well as non-soybean pathogenic *Phoma* and *Ascochyta* species were also used. Two protein-encoding genes, tef1α (Fig. 5.1) and β-tubulin (Fig. 5.2), together with the complete ITS sequences (Fig. 5.3) yielded consensus results in species delimitations and confirmed the previous pathogenicity and morphology data. However, *Phoma sojicola* isolates always formed one clade with all the examined *P. pinodella* isolates, so they had to synonymize with *P. pinodella* (= *Didymella pinodella*) (Irinyi et al. 2009). This establishment was also confirmed by Aveskamp et al. (2010a, b).

The *Ascochyta sojina* was collected on soybean (*Glycine max* /L./ Merr.) plants and described as a 'new species' in Upper Austria which differed in some morphological characteristics from other '*Ascochyta* species' (Bedlan 2014). According to the valid taxonomic rules, the fungus does not appropriate for validity. The author did not study morphological, cultural or pathological characteristics, and it has not been placed in any public collections either. Despite these missings, the Index Fungorum considers it as a synonym (?) of *Didymella pinodella*.

5.11.2 *Boeremia exigua*

(Desm.) Aveskamp, Gruyter & Verkley (a) in Aveskamp, Gruyter, Woudenberg, Verkley & Crous, *Stud. Mycol.* **65**: 37 (2010), Fam. *Didymellaceae* Gruyter, Aveskamp & Verkley (2009).

Nom. illeg. *Phyllosticta glycines* (as '*glycineum*') Tehon & E.Y. Daniels in Mycologia 19: 117-118. 1927. According to Art. 53.1.

Syn. This fungus has numerous, about 150, synonyms, invalid later homonyms and *nomina nuda* (for details see in Boerema et al. 2004).

Phoma exigua var. *exigua* (basionym) (1849), *Ascochyta phaseolorum* (1878), *Phyllosticta sojicola* (as '*sojaecola*') Massal. in Atti del Reale Istituto Veneto di Scienze, Lettere ed. Arti 59: 688. (1900). The latest name was suggested to synonymize in the Index Fungorum based on the publication of Kövics et al. (2014).

The *Boeremia exigua* fungus is a cosmopolitan, ubiquitous species on diseased and dead plant materials (Morgan-Jones and Burch 1988). The recently established genus

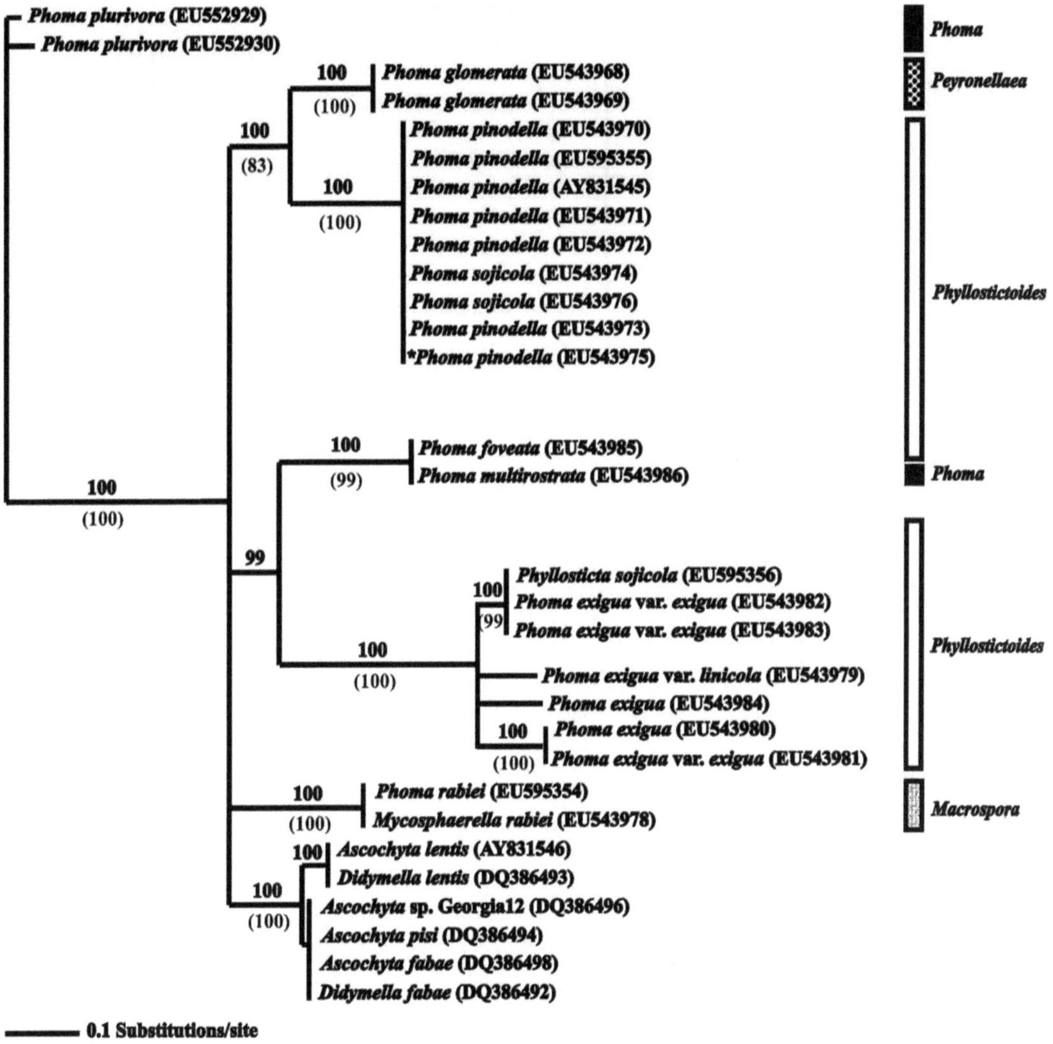

Fig. 5.1 Phylogenetic relationships of *Phoma*-like strains inferred by the Bayesian analysis of *tef1* sequences. Numbers above lines indicate Bayesian posterior probability values. Numbers in brackets represent the bootstrap values from 1000 bootstrap samples in Parsimony analysis. *Phoma pinodella* cluster (= *Didymella pinodella*) includes '*Phoma sojicola*' (EU543974 & 976), and *Phoma exigua* cluster (= *Boeremia exigua*) includes '*Phyllosticta sojicola*' (EU595357). **Phoma pinodella* (EU543975) was formerly misidentified as '*Phoma exigua*' (Irinyi et al. 2009)

(Aveskamp et al. 2010a, b) was named after Gerhard H. Boerema (1945–2019) who made great contributions to our understanding of the taxonomy of phomoid (*Phoma*-like) fungi. Within the *Phoma*-like *Boeremia exigua* species, at least nine varieties of subtaxa (varietas, var.) were admitted (Boerema et al. 2004) and confirmed by molecular methods (Abeln et al. 2002). In addition to type species, *B. exigua* (Aveskamp et al. 2010a, b), two novel varieties were recently described, namely, *B. exigua* var. *gilvescens* and *B. exigua* var. *pseudolilacis*; however, recently both of them were raised to species rank as *B. gilvescens* and *B. pseudolilacis*, respectively (Jayawardena et al. 2019).

This seed-borne fungus appeared to be the most common *Phoma*-like species involved with

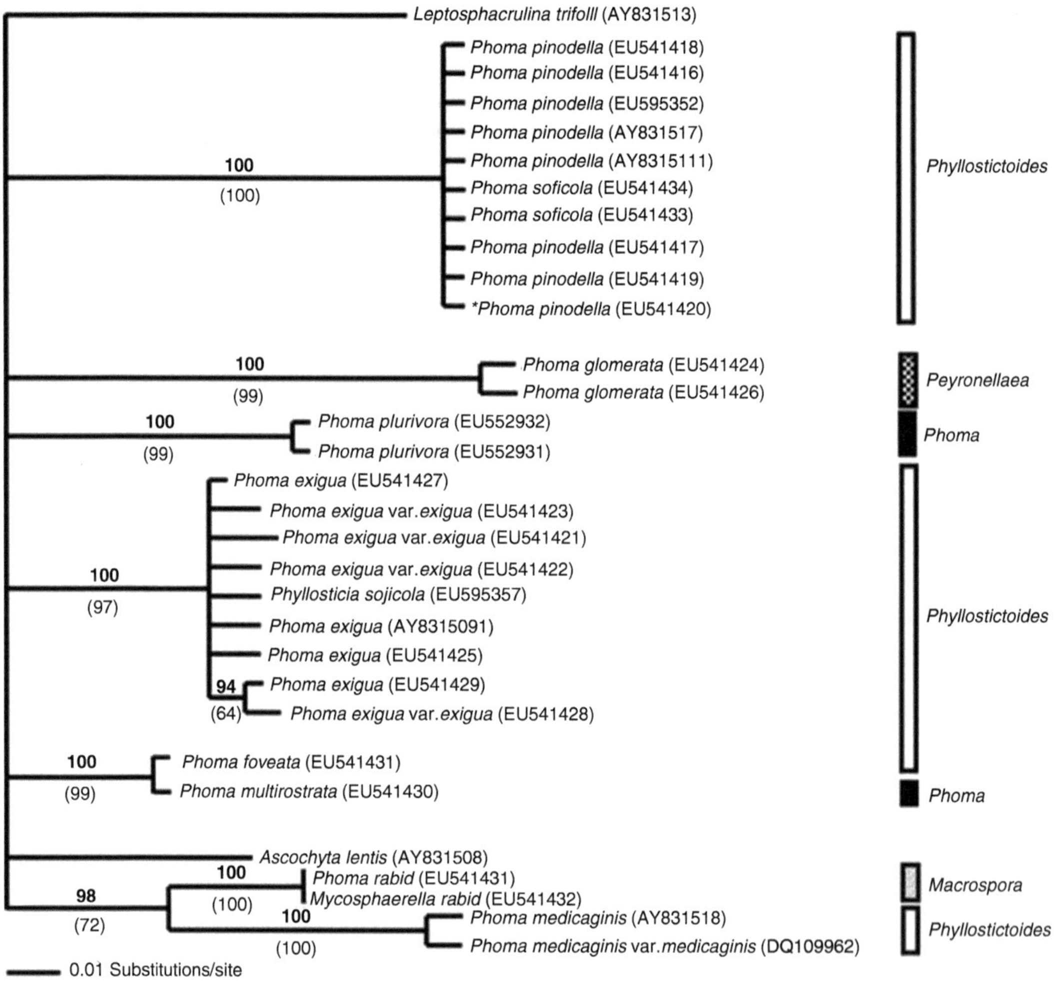

Fig. 5.2 Phylogenetic relationships of *Phoma*-like strains inferred by the Bayesian analysis of *β-tubulin* sequences. Numbers above lines indicate Bayesian posterior probability values. Numbers in brackets represent the bootstrap values from 1000 bootstrap samples in Parsimony analysis. *Phoma pinodella* cluster (= Didymella pinodella) includes 'Phoma sojicola' (EU541433 & 434), and Phoma exigua cluster (= Boeremia exigua) includes 'Phyllosticta sojicola' (EU595357). **Phoma pinodella* (EU541420) was formerly misidentified as '*Phoma exigua*' (Irinyi et al. 2009)

numerous papilionaceous plant pathogens, including soybean in Eurasia.

Phyllosticta sojicola (as '*sojaecola*') isolate (CBS 310.39) is collected and deposited by Böning (1939). Examined type material of *Phyllosticta sojicola*, on leaves of *Glycine max* (Massalongo 1900), however, seems to represent the plurivorous *P. exigua* var. *exigua*, so it was proposed to synonymize with this species (= *Boeremia exigua*) (Kövics et al. 1999). Later it was complemented by molecular features of Böning's isolate and confirmed that *Phyllosticta sojicola* was synonymous with *Ph. exigua* var. *exigua* (= *Boeremia exigua*) (Irinyi et al. 2009).

According to a recent viewpoint (Hou et al. 2020), the proper delimitation of *Boeremia* species should apply a combined phylogeny rather than that from any single locus. The proposed five loci are *actin*, *β-tubulin*, *calmodulin*, *translation elongation factor 1-α*, and the ITS region is recommended to identify species in *Boeremia.*

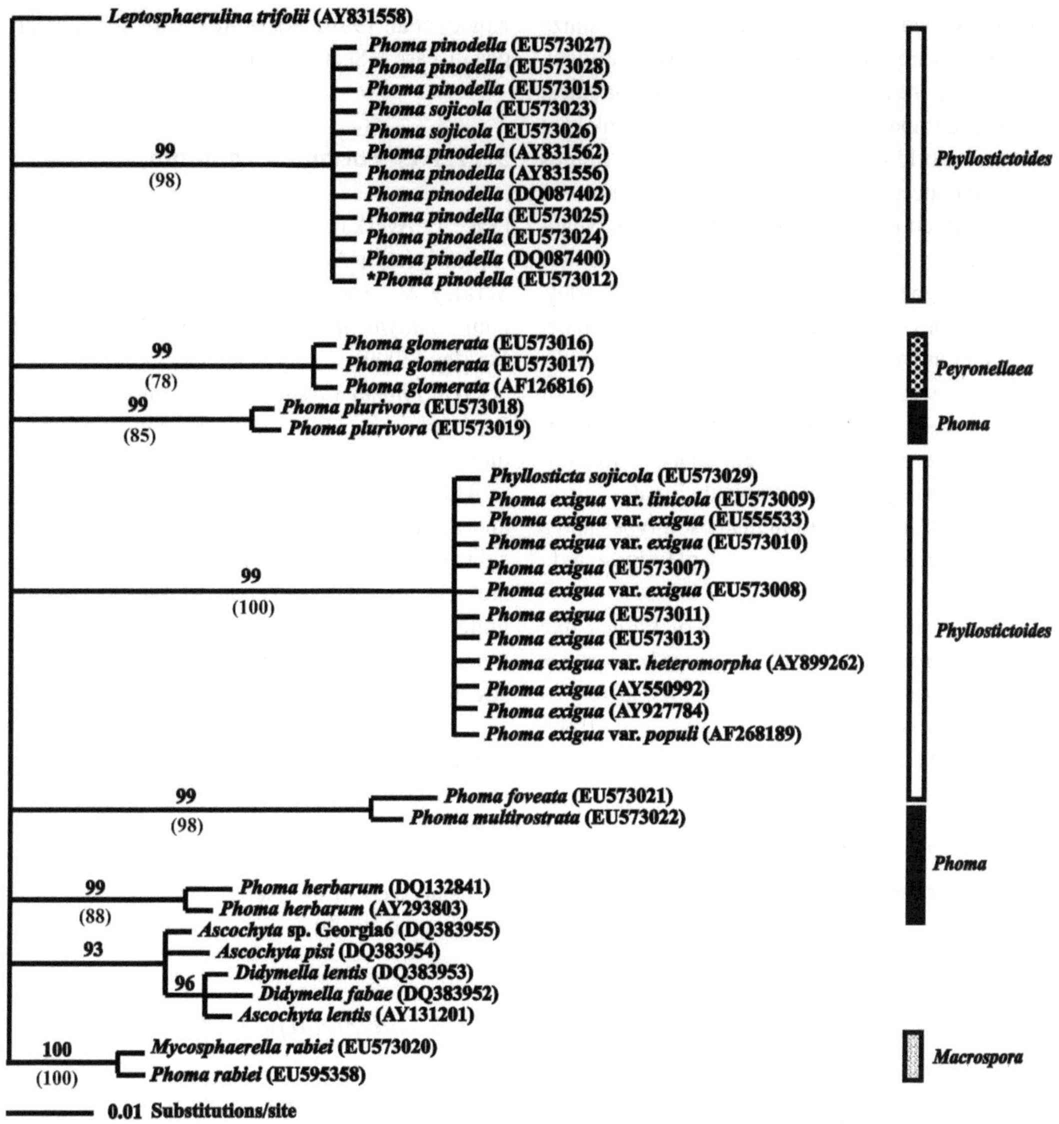

Fig. 5.3 Phylogenetic relationships of *Phoma*-like strains inferred by the Bayesian analysis of ITS sequences. Numbers above lines indicate Bayesian posterior probability values. Numbers in brackets represent the bootstrap values from 1000 bootstrap samples in Parsimony analysis. *Phoma pinodella* cluster (= *Didymella pinodella*) includes '*Phoma sojicola*' (EU573023 & 026), and *Phoma exigua* cluster (= *Boeremia exigua*) includes '*Phyllosticta sojicola*' (EU573029). **Phoma pinodella* (EU573012) was formerly misidentified as '*Phoma exigua*' (Irinyi et al. 2009)

5.11.3 *Setophoma terrestris*

(H.N. Hansen) Gruyter, Aveskamp & Verkley (a) in de Gruyter, Woudenberg, Aveskamp, Verkley, Groenewald & Crous, *Mycologia* **102**(5): 1077 (2010), Fam. *Phaeosphaeriaceae* M.E. Barr (1979).

Syn. *Phoma terrestris* (a) (1929), *Pyrenochaeta terrestris* (a) (1948).

Setophoma terrestris is reported as a widespread saprophyte and a weak parasite of many hosts. The most devastating losses from this pathogen in many cases are caused from a disease complex with other species. Pink root of onion (*Allium cepa*) and red root rot of maize (*Zea mays*) are caused by the soil-borne pathogen *Setophoma terrestris*. Historically the greatest economic impact has been shown with pink root of onion worldwide. *Setophoma terrestris* fungus has been successfully isolated from hosts of 45 genera which include many of the cereals, vegetables and grasses. In 1941 it was reported to survive on soybean, pea, cane, millet, oats, barley, wheat, corn, squash, cucumber, cantaloupe, muskmelon, tomato, pepper, eggplant, cauliflower, carrot, spinach and onion with reports of high disease to no disease damage (Kreutzer 1941). The pathogen is well adapted to subtemperate, temperate and tropical climates due to its ability to survive well in many soil types and a wide range of temperature and pH (Frye n.d., internet).

Differential in vitro characters of *Phoma*-like species were compared to the recorded species from *Glycine max*, including *Setophoma terrestris* (as '*Phoma terrestris*') (Kövics et al. 1999; Boerema et al. 2004). See more additional details at pea diseases (Sect. 5.9.6).

5.11.4 *Ectophoma multirostrata*

(P.N. Mathur, S.K. Menon & Thirum) Valenz.-Lopez, Cano, Crous, Guarro & Stchigel (a) in Valenzuela-Lopez, Cano-Lira, Guarro, Sutton, Wiederhold, Crous & Stchigel, *Stud. Mycol.* **90**: 34 (2017), Fam. *Pleosporales* Luttr. ex M.E. Barr (1987).

Syn. *Phoma decorticans* var. *microspora* (a) (1898), *Phoma lucknowensis* (a) (1968), *Phoma multirostrata* (1973), *Phoma multirostrata* var. *macrospora* (1986), *Phoma multirostrata* var. *microspora* (1986).

Ectophoma multirostrata is a soil-borne fungus, which was isolated on necrotic leaves and stems of soybean. In vitro the fungus can be differentiated from *Phoma sojicola* (= *Didymella pinodella)* by the absence of septate conidia and the relatively fast growth of the colonies (Dorenbosch and Boerema 1973; Boerema 1986; Kövics et al. 1999). For more details see at chickpea diseases (Sect. 5.4.2).

5.11.5 *Epicoccum sorghinum*

(Sacc.) Aveskamp, Gruyter & Verkley [as '*sorghi*'] (a), in Aveskamp, Gruyter, Woudenberg, Verkley & Crous, *Stud. Mycol.* **65**: 36 (2010), Fam. *Didymellaceae* Gruyter, Aveskamp & Verkley (2009).

Syn. *Phoma sorghina* (synanamorph) (1973), *Leptosphaeria spegazzinii* (t) (1899) and many more.

This fungus occurs worldwide as soil-borne and seed-borne, particularly in the tropics and subtropics; however, it also appears in temperate regions, especially on glasshouse plants, under continental climate, too. The fungus is associated with Poaceae (e.g. sorghum, rice, sugarcane) causing spots on leaves, glumes and seeds and, moreover, dying off. However, the fungus is considered as a weak parasite and secondary invader of weakened plants (Boerema et al. 2004).

The *Epicoccum sorghinum* was isolated from soybean seed from Nepal (IMI 333584) (cit. Kövics et al. 1999).

5.11.6 *Didymella pomorum*

(Thüm.) Qian Chen & L. Cai, (t) in Chen, Jiang, Zhang, Cai & Crous, *Stud. Mycol.* **82**: 179 (2015), Fam. *Didymellaceae* Gruyter, Aveskamp & Verkley (2009).

Syn. *Phoma prunicola* (1936), *Phoma pomorum* (1879) var. *pomorum* (a), *Phoma pomorum* var. *calorpreferens* (1993) (a), *Peyronellaea pomorum* (a) (2010).

A worldwide occurring soil- and seed-borne opportunistic parasite is formerly commonly treated under *Phoma prunicola* synonym name. The fungus refers to its occurrence in association with leaf spots on rosaceous plants, like apple, pear and *Prunus* species and, moreover, on strawberry plants as causative agent of 'black root rot' complex.

The examined isolates from soybean, which were identified and deposited in ATCC 44494 (= CBS 568.97) as *Phyllosticta sojicola* (as '*sojaecola*') (Walters and Martin 1981), produced multicellular chlamydospores, and corresponds with *P. pomorum* var. *calorpreferens (= Didymella pomorum*, 2015). As '*Phyllosticta sojicola* (ATCC 44494; = CBS 568.97) isolates' (Walters and Martin 1981) probably were misidentified, Kövics et al. (1999) proposed to synonymize these isolates with *Phoma pomorum* (= *Didymella pomorum*).

5.12 Conclusions

In this chapter, the *Phoma*-like fungi that occur on papilionaceous agricultural crops were overviewed based on the current 'One fungus – One name' concept. Because these species formerly were treated on host plant base, a lot of misidentification and synonymous names were used both in teleomorph and anamorph stages. In this compilation, we collected as many phomoid fungi as possible of important agricultural papilionaceous plants. These fungi can cause well-known serious diseases; however, others occur sometimes as weak parasites, depending on special unfavourable circumstances. This review may come in handy to find the collection of potential pathogens, which appear regularly, occasionally, rarely or particularly in agricultural practice. According to the valid taxonomic rules, all the fungal species were ranked in the proper ascomycetous order, ranking the only valid name and their most synonyms and moreover indicating the formerly applied morphic stages, viz. anamorph (a) or teleomorph (t). The circle of treated plants includes ten host plant groups from alfalfa to vetches.

Acknowledgement MR is thankful to the Polish National Agency for Academic Exchange (NAWA) for financial support (Project No. PPN/ULM/2019/1/00117/A/DRAFT/00001) to visit the Department of Microbiology, Nicolaus Copernicus University, Toruń, Poland.

References

Abeln, E. C. A., Stax, A. M., de Gruyter, J. & van der Aa, H. A., (2002). Genetic differentiation of *Phoma exigua* varieties by means of AFLP fingerprints. *Mycological Research, 106*: 419-427.

Abramov, I. N. (1931). Gribnyje boleznyi sojevich bobov na Dalnem Vostoke. Boleznyi i vrediteli sojevich bobov na Dalnem Vostoke. Vladivostok. 84 pp. Sojevaja ascochyta - *Ascochyta sojaecola* sp. nov. (68) 62-70.

Ahmed, S., & Beniwal, S. P. S. (1998). Survival of *Ascochyta fabae* f. sp. *lentis* on infected lentil seeds. *LENS Newsletter, 25*, 71-74.

Ahmed, S. A., van de Sande, W. W., Stevens, D. A., et al. (2014). Revision of agents of black-grain eumycetoma in the order *Pleosporales*. *Persoonia, 33*, 141-154.

Ali, S.M. & Dennis, J. (1992). Host range and physiologic specialisation of *Macrophomina phaseolina* isolated from field peas in South Australia. *Aust. J. Exp. Agric., 32*, 1121-1125.

Ali, M., Nitschke, L., Krause, M. & Cameron, B. (1978). Selection of Pea Lines for Resistance to Pathotypes of *Ascochyta pinodes, A. pisi* and *Phoma medicaginis* var. *pinodella*, *Australian Journal of Agricultural Research, 29*(4), 841-849. https://doi.org/10.1071/ar9780841

Anonymous (1960). Index of Plant Diseases in the United States. *U.S.D.A. Agric. Handb., 165*, 1-531. (94)

Aveskamp, M. M., Gruyter, J. de & Crous, P. W. (2008). Biology and recent developments in the systematics of *Phoma*, a complex genus of major quarantine significance. *Fungal Diversity, 31*, 1-18.

Aveskamp, M. M., Verkley, G. J. M., de Gruyter, J, Murace, M. A., Perelló, A., Woudenberg, J. H. C., Groenewald, J. Z., & Crous, P. W. (2009). DNA phylogeny reveals polyphyly of *Phoma* section *Peyronellaea* and multiple taxonomic novelties. *Mycologia, 101*, 363-382.

Aveskamp, M. M., de Gruyter, J., Woudenberg, J. H. C. et al. (2010a). Highlights of the *Didymellaceae*: a polyphasic approach to characterise *Phoma* and related pleosporalean genera. *Studies in Mycology, 65*, 1-60.

Aveskamp, M. M., de Gruyter, J., Woudenberg, J. H. C., Verkley, G. J. M., Crous, P. W. (2010b). Highlights of the *Didymellaceae*: A polyphasic approach to characterise *Phoma* and related pleosporalean genera. *Studies in Mycology, 65*, 1-60.

Bardas, G. A., Tziros, G. T., & Tzavella-Klonar, K. (2008). First Report of Ascochyta Leaf Spot Caused by *Phoma exigua* var. *exigua* on Common Bean in Greece. Disease Notes. *Plant Disease 92*(4), https://doi.org/10.1094/PDIS-92-4-0653C

Baudoin, J., Camarena, F. & Lobo, M. (1997). Improving *Phaseolus* genotypes for multiple cropping systems. *Euphytica, 96*, 115-123.

Bedlan, G. (2014). *Ascochyta sojina sp. nov.*, a new pathogen on *Glycine max* (L.) Merr. *Journal für Kulturpflanzen* (formerly: Nachrichtenblatt des

Deutschen Pflanzenschutzdienstes, seit 2010 vereint mit Pflanzenbauwissenschaften), *66*(9), 319-321.

Békési, P. (1965). Occurrence of *Ascochyta boltshauseri* Sacc. in Hungary on *Vicia faba*. *Növényvédelem (Crop Protection, Budapest), 1*(3), 28-36. (in Hungarian)

Boerema, G. H. (1984). Mycologisch-taxonomisch onderzoek. Echte *Ascochyta*-soorten (met hyaline conodiën) die pathogeen zijn voor Papilionaceae. *Verslagen en Mededelingen Plantenziektenkundige Dienst Wageningen* 162 (Jaarb. 1983), 23-31.

Boerema, G. H. (1986). Een subtropische bodemschimmel die zich ook thuis voelt in onze kassen. *Phoma multirostrata* var. *macrospora* Boerema *var. nov. Verslagen en Mededelingen Plantenziektenkundige Dienst Wageningen* 164 (Jaarb. 1985), 28-32.

Boerema, G. H., Crüger, G., Gerlagh, M. & Nirenberg, H. (1981). *Phoma exigua* var. *diversispora* and related fungi on *Phaseolus beans*. *Journal Plant Disease and Protection, 88*(10), 597-607.

Boerema, G. H., de Gruyter, J., & van de Graaf, P. (1999). Contribution towards a monograph of *Phoma* (Coelomycetes) – IV. Supplement: An addition to section Heterospora: *Phoma schneiderae spec. nov.*, synanamorph *Stagonosporopsis lupini* (Boerema and E. Schneider) *comb. nov. Persoonia, 17*(2), 281-285.

Boerema, G. H., de Gruyter, J., Noordeloos, M. E. & Hamers, M. E. C. (2004). *Phoma* Identification Manual Differentiation of Species and Infra-Specific Taxa in Culture. CABI Publishing, CAB International, Wallingford, Oxfordshire, UK.

Bondartzeva-Monteverde, V. N. & Vassiljevskiy, N. I. (1940). K biologii i morfologii nekotorikh vidov *Ascochyta* na bobovikh. *Acta Instituti Botanici Acad. Sci. URSS. Ser II, (Fasc 4) 1938*, Trudy Bot. Inst. Akad. Nauk SSSR. Sporoviye rasteniya, Moskva-Leningrad, 345-376.

Böning, K. (1939). *Phyllosticta* - Fleckenkrankheit der Sojabohne. *Praktische Blätter für Pflanzenbau und Pflanzenschutz, 16*, 168-172.

Bretag, T. W., Cunnington, J. H. & de Gruyter, J (2006). First record of the lupin pathogen *Phoma schneiderae* in Australia. *Australasian Plant Disease Notes, 1*, 5–6 (47083)

CABI Datasheet (2019): *Ascochyta trifolii*. No. 7345. https://www.cabi.org/isc/datasheet/7345

CCEPP (2007). Plant Pest Alert – LOOK OUT FOR *Phoma schneiderae*, a disease of lupins. The Consultative Committee on Emergency of Australia. https://www.agriculture.gov.au/plant/health/committees/ccepp

Chen, W., Sharma, H. C., Muehlbauer, F. J. (2011). Compendium of Chickpea and Lentil Diseases and Pests. The American Phytopathological Society. 165 pp.

Chen, Q., Jiang, J. R., Zhang, G. Z. Cai, L. & Crous, P. W. (2015a): Resolving the *Phoma* enigma. *Studies in Mycology, 82*, 137-217.

Chen, Q., Zhang, K., Zhang, G. Z. & Cai, L. (2015b). A polyphasic approach to characterise two novel species of *Phoma* (*Didymellaceae*) from China. *Phytotaxa, 197*(4), 267-281. https://doi.org/10.11646/phytotaxa.197.4.4

Chen, Q., Hou, L., Duan, W., Crous, P. W. & Cai, L. (2017). *Didymellaceae* revisited. *Studies in Mycology, 87*, 105-159.

Chobe, D. R., Tarafdar, A., Sharath Chandran, U. S., Sharma, S. & Sharma, M. (2020). First report of *Ectophoma multirostrata* causing root rot in chickpea. *Plant Disease, 104*(6), 1866.

Clements, R. O. (1996). Pest and disease damage to white clover (*Trifolium repens*) in Europe. A compilation. In: Interregional Cooperative Research and Development Network on Pastures and Fodder Crops (FAO/CIHEAM) (ed. John Frame). *REU TECHNICAL SERIES*, 42. http://www.fao.org/3/v9968e/v9968e03.htm

Crous, P. W., Phillips, A. J. L., & Baxter, A. P. (2000). Phytopathogenic fungi from South Africa. *University of Stellenbosch, Department of Plant Pathology Press*, 358 pp.

Crous, P. W., Wingfield, M. J., Schumacher, R. K., Akulov, A., Bulgakov, T. S., Carnegie, A. J., Jurjević, Ž., Decock, C., Denman, S., Lombard, L., Lawrence, D. P., Stack, A. J., Gordon, T. R., Bostock, R. M., Burgess, T., Summerell, B. A., Taylor, P. W. J., Edwards, J., Hou, L. W., Cai, L., Rossman, A. Y., Wöhner, T., Allen, W. C., Castlebury, L. A., Visagie, C. M. & Groenewald, J. Z. (2020). New and Interesting Fungi. 3. *FUSE 6*, 157-231.

Davidson, J. A., Hartley, D., Priest, M., Herdina, M. K., McKay, A. & Scott, E. S. (2009). A new species of *Phoma* causes ascochyta blight symptoms on field peas (*Pisum sativum*) in South Australia. *Mycologia, 101*(1) 120-128. doi: https://doi.org/10.3852/07-199.

De Gruyter J., Aveskamp, M. M., Woudenberg, J. H. C. et al. (2009). Molecular phylogeny of *Phoma* and allied anamorph genera: towards a reclassification of the *Phoma* complex. *Mycological Research, 113*, 508-519.

Deb, D., Khan, A., & Dey, N. (2020). *Phoma* diseases: Epidemiology and control. Plant Pathology, 69, 1203-1217. https://doi.org/10.1111/ppa.13221

Deng, J. X., Paul, N. C., Li, M. J., Seo, E. Y., Sung, G. H. & Yu, S. H. (2011). Molecular characterization and morphology of two endophytic *Peyronellaea* species from *Pinus koraiensis* in Korea. *Mycobiology, 39*(4), 266-271. https://doi.org/10.5941/MYCO.2011.39.4.266

Dorenbosch, M. M. J. & Boerema, G. H. (1973). About *Phoma liliana* Chandra & Tandon II. *Mycopathologia & Mycologica Applicata, 50*, 255-256.

Farr, D. F. & Rossman, A. Y. (2017). Fungal Databases, U.S. National Fungus Collections, *ARS, USDA*. https://nt.ars-grin.gov/fungaldatabases/

Fernandes, E. G., Pereira, O. L., Silva, C. C., Bento, C. B. P. & Queiroz, M. V. (2015). Diversity of endophytic fungi in *Glycine max*. *Microbiological Research, 181*, 84-92.

Ferreira, J. J., Campa, A. & Pérez-Vega, E. (2016). Variation in the response to ascochyta blight in common bean germplasm. *Eur. J. Plant Pathol., 146*, 977-985. https://doi.org/10.1007/s10658-016-0974-y

Frandsen, N. O. (1953). *Ascochyta sojaecola* auf Sojabohne in Deutschland. *Phytopath. Z., 20* (4), 375-382.

Frey, F. & Yabar, E. (1983). Enfermedades y y plagas de lupinos en el Peru. *ScgrReihe dt. Ges. Tech. Zusammenarb. (GTZ) Eschborn*. 142.

Frye, J. (n.d.) *Phoma terrestris* E. M. Hans. (Syn. *Pyrenochaeta terrestris*) (= *Setophoma terrestris*) https://projects.ncsu.edu/cals/course/pp728/Phoma/Phoma_terrestris.html

Gafforov, Y. S. (2017). A preliminary checklist of ascomycetous microfungi from Southern Uzbekistan. *Mycosphere, 8*(4), 660-696. https://doi.org/10.5943/mycosphere/8/4/12

Galloway, J., Macleod, W. J. & Lindbeck, K. D. (2004). Formation of *Didymella lentis*, the teleomorph of *Ascochyta lentis*, on lentil stubble in the field in Victoria and Western Australia. *Australasian Plant Pathology, 33*(3), 449-450. https://doi.org/10.1071/AP04033

Gilardi, G., Luongo, I., Gullino, M. L., & Garibaldi, A. (2010). Attacks of *Phoma multirostrata* on *Fuchsia × hybrida* in Northern Italy. *Protezione delle Colture, 2010*(4), 44-45.

Golzar, H., Lanoiselet, V., Wang, C. et al. (2015). First report of *Phoma multirostrata* in Australia. *Australasian Plant Dis. Notes, 10*, 8. https://doi.org/10.1007/s13314-015-0161-6

Gomzhina, M., Gasich, E., Khlopunova, L. & Gannibal, P. (2020). New species and new findings of *Phoma*-like fungi (*Didymellaceae*) associated with some Asteraceae in Russia. *Nova Hedwigia, 111*(1-2), 131-149.

Gorz, H. J. (1955). Inheritance of reaction to *Ascochyta caulicola* in sweetclover (*Melilotus alba*). *Agronomy Journal, 47*(8): 379-383. https://doi.org/10.2134/agronj1955.00021962004700080009x

Hawksworth, D. L. & David, J. C. (1989). Index of Fungi Supplement - Family Names. 75 pp.

Hawksworth, D. L., Crous, P. W., Redhead, S. A., Reynolds, D. R., Samson, R. A., Seifert, K. A. et al. (2011). The Amsterdam declaration on fungal nomenclature. *IMA Fungus, 2*(1), 105-112.

Hewett, P. D. (1973). The field behaviour of seed-borne *Ascochyta fabae* and disease control in field beans. *Ann. Appl. Biol., 74*, 287-295.

Hoinacki, B. (2004). Root rot of sweet corn in Western Oregon. *Extension Catalog EM 8859*, 1-4. Oregon State University, Corvallis.

Hollingsworth, C. R., & Gray, F. A. (1999). First report of brown root rot on alfalfa caused by *Phoma sclerotioides* in the Continental United States. *Plant Diseases, 83*(11), 1071. https://doi.org/10.1094/PDIS.1999.83.11.1071A

Honey, E. E., Dickson, J. G., Jones, F. R., & Allison, J. L. (1945). Diseases and injuries of forage legumes and grasses in Wisconsin. *Plant Dis. Reptr. 29*(22) (603) 600-606.

Hou, L. W., Groenewald, J. Z., Pfenning, L. H., Yarden, O., Crous, P. W. & Cai, L. (2020). The *Phoma*-like dilemma. *Studies in Mycology, 21*(96), 309-396. https://doi.org/10.1016/j.simyco.2020.05.001

Irinyi, L., Kövics, G.J. & Sándor, E. (2009). Taxonomical re-evaluation of *Phoma*-like soybean pathogenic fungi. *Mycological Research, 113*(2), 249-260. https://doi.org/10.1016/j.mycres.2008.11.003

Irwin, J. A. G., Mackie, J. M., Marney, T. S. et al. (2004). Incidence of *Stagonaspora meliloti* and *Acracalymma medicaginis* in lucerne crowns and roots in eastern Australia, their comparative aggressiveness to lucerne and inheritance of reaction to *S. meliloti* in lucerne. *Australasian Plant Pathology, 33*, 61-67. https://doi.org/10.1071/AP03083

Jayawardena, R. S., Hyde, K. D., Jeewon, R., Ghobad-Nejhad, M., Wanasinghe, D. N., Liu, N. G., Phillips, A. J. L., Oliveira-Filho, J. R. C., da Silva, G. A., Gibertoni, T. B., Abeywikrama, P., Carris, L. M., Chethana, K. W. T., Dissanayake, A. J., Hongsanan, S., Jayasiri, S. C., McTaggart, A. R., Perera, R. H., Phutthacharoen, K., Savchenko, K. G. , Shivas, R. G., Thongklang, N., Dong, W., Wei, D. P., Wijayawardena, N. N. & Kang, J. C. (2019). One stop shop II: taxonomic update with molecular phylogeny for important phytopathogenic genera: 26-50. *Fungal Divers., 94*, 41-129.

Jellis, G. J. & Punithalingam, E. (1991). Discovery of *Didymella fabae sp. nov.*, the teleomorph of *Ascochyta fabae*, on faba bean straw. *Plant Pathology, 40*(1), 150-157.

Jones, F. R. & Weimer, J. L. (1938). Stagonospora leaf spot and root rot of forage legumes. *J. Agric. Res., Washington, 57*, 791-812.

Kaiser, W. J. (1973). Factors affecting growth, sporulation, pathogenicity, and survival of *Ascochyta rabiei*. *Mycologia, 65*, 444-457.

Kaiser, W. J. (1997). Inter- and intranational spread of *Ascochyta* pathogens of chickpea, faba bean, and lentil. *Canadian Journal of Plant Pathology, 19*, 215-224.

Kaiser, W. J., Okhovat, M. & Mossahebi, G. H. (1973). Effect of seed-treatment fungicides on control of *Ascochyta rabiei* in chickpea seed infected with the pathogen. *Plant Dis. Reptr., 57*, 742-746.

Keirnan, E. C., Davidson, J. A., Correll, R. L. et al. (2020). Host range investigation of *Phoma koolunga*, a causal agent of ascochyta blight of field pea. *Australasian Plant Pathol., 49*, 707-719. https://doi.org/10.1007/s13313-020-00747-z

Khazaei, H., Caron, C. T., Fedoruk, M., Diapari, M., Vandenberg, A., Coyne, C. J. et al. (2016). Genetic diversity of cultivated lentil (*Lens culinaris* Medik.) and its relation to the world's agro-ecological zones. *Front. Plant Sci., 6*, 1093. https://doi.org/10.3389/fpls.2016.01093

Kövics, G. (1983). Diseases and control of papilionaceous forage plants. In: Cultivation of papilionaceous forage plants. (ed. L. Bányai). Mezőgazdasági Kiadó (Agricultural Publishing, Budapest). 196-211. (in Hungarian)

Kövics, G. (1995a). Remarks on taxonomic positions of *Ascochyta – Phoma – Phyllosticta* plant parasitic fungi. *Növényvédelem (Crop Protection, Budapest), 31*(7), 307-315. (In Hungarian)

Kövics, G. (1995b). Remarks on the taxonomical questions of phytopathogenic *Ascochyta, Phoma,* and *Phyllosticta* species. A review. *Növényvédelem (Crop Protection, Budapest), 31*(7), 307-315. (In Hungarian)

Kövics, G. (2000). Collection of scientific names of plant pathogenic fungi. Mezőgazda Publishers, Budapest. 255 pp. ISBN 963 9239 81 X (in Hungarian)

Kövics, G., Holly, L., & Simay, E. I. (1986). An ascochytosis of the chickpea (*Cicer arietinum* L.) caused by *Didymella rabiei* (Kov.) v. Arx; Imperfect *Ascochyta rabiei* (Pass.) Lab. in Hungary. *Acta Phytopathologica et Entomologica Acad.Sci.Hung., 21*(1–2), 147-150.

Kövics, G. J., Gruyter, J. de & Aa, H. A. van der (1999). *Phoma sojicola comb. nov.* and other hyaline-spored coelomycetes pathogenic on soybean. *Mycological Research, 103*(8), 1065-1070. https://doi.org/10.1017/S0953756298

Kövics, G., Bozsik, A., Dávid, I., Szarukán, I., Radócz, L., Karaffa, E., Irinyi, L., Szarvas, P., & Tarcali, G. (2007). Protection of alfalfa Part I. Alfalfa diseases and animal pests on roots and leaves. *Növényvédelem (Crop Protection, Budapest), 43*(4), 119-137. (In Hungarian)

Kövics, G. J., Sándor, E., Rai, M. & Irinyi, L. (2014). *Phoma*-like fungi on soybeans. *Critical Reviews in Microbiology, 40*(1), 49-62. https://doi.org/10.3109/1040841X.2012.755948

Kreutzer, W. A. (1941). Host-parasite relationships in pink root of *Allium cepa* L. II. The action of *Phoma terrestris* on *Allium cepa* and other hosts. *Phytopathology, 31,* 907-915.

Lenti, I., Borbély, F., Kövics G. J., Bozsik, A. & Dávid, I. (2008). Crop protection of faba bean (*Vicia faba* L.). *Növényvédelem (Crop Protection, Budapest), 44*(8), 403-421. (in Hungarian)

Leth, K. T., DeGregorio, R. E. & Ashley, R. A. (1994). Foliar blight of bigflower vetch caused by *Ascochyta fabae* f. sp. *viciae*. *Plant Disease, 78,* 637-639.

Li, Y. P., You, M. P., Khan, T. N., Finnegan, P. M. & Barbetti, M. J. (2011). First report of *Phoma herbarum* on field pea (*Pisum sativum*) in Australia. *Plant Disease, 95,* 1590.

Link, W. (2006). Methods and objectives in faba bean breeding. *International Workshop on Faba Bean Breeding and Agronomy*. pp. 1-8. October 25-27 2006, Córdoba, Spain. https://plants.usda.gov/plantguide/pdf/pg_vifa.pdf

Lucas, M. T. & Webster, J. (1967). Conidial states of British species of *Leptosphaeria*. *Transactions of the British Mycological Society, 50*(1), 85-121. https://doi.org/10.1016/S0007-1536(67)80067-6.

Lücking, R., Aime, M. C., Robbertse, B. et al. (2020). Unambiguous identification of fungi: where do we stand and how accurate and precise is fungal DNA barcoding? *IMA Fungus, 11,* 14. https://doi.org/10.1186/s43008-020-00033-z

Massalongo, C. (1900). De nonnullis speciebus novis micromycetum agri veronensis. *Atti del Reale Istituto Veneto di Scienze, Lettere ed Arti, 59,* 688.

May, T. W., Redhead, S. A., Bensch, K. et al. (2019). Chapter F of the *International Code of Nomenclature for algae, fungi, and plants* as approved by the 11th International Mycological Congress, San Juan, Puerto Rico, July 2018. *IMA Fungus, 10,* 21. https://doi.org/10.1186/s43008-019-0019-1

McNeill, J., Barrie, F. R., Buck, W. R., Demoulin, V., Greuter, W., Hawksworth, D. L. et al. (2012). *International Code of Nomenclature for algae, fungi, and plants* (Melbourne Code). In: Regnum Vegetabile, vol. 154. Konigstein: Koeltz Sci. Books.

Melnik, V. A. (2000). *Ascochyta trifolii-alpestris* & *A. viciae* No.185 & 186 pp. 91-92. In: Key to the genus *Ascochyta* Lib. (Coelomycetes). (Ed. & transl. Vadim A. Melnik, Uwe Braun & Gregor Hagedorn). *Mitteilungen aus der Biologischen Bundesanstalt für Land- und Forstwirtschaft Berlin-Dahlem, Heft 379.* ISBN 3-8263-3355-1

Morgan-Jones, G. & Burch, K. B. (1988). Studies on the genus *Phoma* X. Concerning *Phoma eupyrena*, a ubiquitous, soilborne species. *Mycotaxon, 31,* 427-434.

Nadolnik, M. (1981). Wystepowanie i szkodliwosc chorob na plantacjack nasiennych koniczyny bialej (*Trifolium repens* L.). *Zeszyty Naukowe Akademii Rolniczej IM. H. Kollatja W Krakowie, 167,* 271-284.

Peever, T., Barve, M., Stone, L. & Kaiser, W. (2007). Evolutionary relationships among *Ascochyta* species infecting wild and cultivated hosts in the Legume Tribes *Cicereae* and *Vicieae*. *Mycologia, 99*(1), 59-77. http://www.jstor.org/stable/20444807

Pennycook, S. R. (1989). Plant diseases recorded in New Zealand. 3 Vol. *Pl. Dis. Div., D.S.I.R.,* Auckland.

Petrović, K., Riccioni, L., Valente, M. T., Đorđević, V., Živanov, D., Jevtić, R. & Malidža, G. (2017). First report of *Peyronellaea lethalis* associated with Ascochyta blight complex of field pea in Serbia. Disease Notes. *Plant Disease, 101*(8), https://doi.org/10.1094/PDIS-11-16-1668-PDN

PlutoF (2017). PlutoF platform reference-based occurrences. Version 1.11. Occurrence Dataset https://doi.org/10.15468/e15jve accessed via GBIF.org on 2018-04-21

Punithalingam, E. (1979). *Sphaeropsidales* in culture from humans. *Nova Hedwigia, 31,* 119-158.

Punithalingam, E. & Holiday, P. (1973). *Pyrenochaeta terrestris. CMI descriptions of pathogenic fungi and bacteria no. 397.*

Punthalingam, E. (1993). *IMI Descriptions of Fungi and Bacteria* No. *117* Sheet 1164. https://doi.org/10.1079/DFB/20056401164

Putnam, D. H., Oplinger, E. S., Hardman L. L. & Doll, J. D. (1989). Lupine. In: Alternative Field Crops Manual. *Series of Univ. Wisconsin and Univ. Minnesota Extension Services 1989–1992.* https://hort.purdue.edu/newcrop/afcm/index.html

Rai, M. (1989). *Phoma sorghina* infection in human being. *Mycopathologia, 105*(3), 167-170. https://doi.org/10.1007/BF00437250. PMID: 2761610

Rai, M. (1998). The genus *Phoma*. (Identity and Taxonomy). International Book Distributors, Dehra Dun, India.

Rai, M., Tiwari, V. V., Irinyi, L. & Kövics, G. J. (2014). Advances in taxonomy of genus *Phoma*: Polyphyletic nature and role of phenotypic traits and molecular systematics, *Indian J. Microbiol., 54*(2), 123-128. https://doi.org/10.1007/s12088-013-0442-8

Roskov, Y., Kunze, T., Paglinawan, L., Orrell, T., Nicolson, D., Culham, A., Bailly, N., Kirk, P., Bourgoin, T., Baillargeon, G., Hernandez, F. & De Wever, A. (red) (2013). "Species 2000 & ITIS Catalogue of Life: 2013 Annual Checklist". *Species 2000.* Reading, UK.

Rubiales, D. & Trapero-Casas, A. (2002). Occurrence of *Didymella fabae*, the teleomorph of *Ascochyta fabae*, on faba bean straw in Spain. *Journal of Phytopathology, 150*(3), 146-148.

Samac, D. A., Rhodes, L. H. & Lamp, W. O. (2014). The diseases of alfalfa (*Medicago sativa* L.). https://www.apsnet.org/edcenter/resources/commonnames/Pages/Alfalfa.aspx

Schwartz, H. F. (2005). Ascochyta leaf spot. pp. 22-28. In: Compendium of bean diseases (H.F Schwartz, J. R. Steadman, R. Hall & R. L. Forster (Eds.) Minnesota, USA, The American Phytopathology Society.

Shaw, C. G. (1973). Host fungus index for the Pacific Northwest - I. Hosts. *Washington State Univ. Agric. Exp. Sta. Bull., 765*(44), 1-121

Shoemaker, R. A., Babcock, C. E. & Irwin, J. A. G. (2011). *Massarina walkeri n. sp.*, the teleomorph of *Acrocalymma medicaginis* from *Medicago sativa* contrasted with *Leptosphaeria pratensis*, *L. weimeri n. sp.*, and *L. viridella*. *Canadian Journal of Botany, 69*(3), 569-573. https://doi.org/10.1139/b91-077

Sillero, J. C. & Rubiales, D. (2014). Response of *Vicia* species to *Ascochyta fabae* and *Uromyces viciae-fabae*. *Czech J. Genet. Plant Breed., 50*, 109-115.

Simay, E. I. (1988). Occurrence of *Ascochyta fabae* Speg. in Hungary. *Acta Phytopathologica et Entomologica Hungarica, 23*, 119-122.

Sullivan, R. F. & White, J. F. Jr. (2000): *Phoma glomerata* as a mycoparasite of powdery mildew. *Appl. Environ. Microbiol., 66*, 425-427.

Thines, M., Aoki, T., Crous, P. W. et al. (2020). Setting scientific names at all taxonomic ranks in italics facilitates their quick recognition in scientific papers. *IMA Fungus 11*, 25. https://doi.org/10.1186/s43008-020-00048-6

Tóth, O. & Kövics, G. (1978). The occurrence of *Ascochyta sojaecola* Abramov soybean pathogen in Hungary. *Növényvédelem (Crop Protection, Budapest), 24*(7), 299-304.

Tran, H. S., You, M. P., Lanoiselet, V., Khan, T. N. & Barbetti, M. J. (2014). First report of *Phoma glomerata* associated with the Ascochyta blight complex on field pea (*Pisum sativum*) in Australia. *Plant Disease, 98*, 427.

Tran, H. S., You, M. P., Khan, T. N. & Barbetti, M. J. (2015). Relative host resistance to black spot disease in field pea (*Pisum sativum*) is determined by individual pathogens. *Plant Disease, 99*, 580-587.

Turland, N. J., Wiersema, J. H., Barrie, F. R., Greuter, W., Hawksworth, D. L., Herendeen, P. S., Knapp, S., Kusber, W.-H., Li, D.-Z., Marhold, K., May, T. W., McNeill, J., Monro, A. M., Prado, J., Price, M. J. & Smith, G. F. (eds.) (2018). *International Code of Nomenclature for algae, fungi, and plants (Shenzhen Code) adopted by the Nineteenth International Botanical Congress Shenzhen, China, July 2017.* Regnum Vegetabile 159. Glashütten: Koeltz Botanical Books. https://doi.org/10.12705/Code.2018

Valenzuela-Lopez, N., Sutton, D.A., Cano-Lira, J.F. et al. (2016). Coelomycetous fungi in the clinical setting: morphological convergence and cryptic diversity. *Journal of Clinical Microbiology, 55*, 552-567.

Valenzuela-Lopez, N., Cano-Lira, J. F., Guarro, J., Sutton, D. A., Wiederhold, N., Crous, P. W. & Stchigel, A. M. (2018). Coelomycetous *Dothideomycetes* with emphasis on the families *Cucurbitariaceae* and *Didymellaceae*. *Studies in Mycology, 90*, 1-69. https://doi.org/10.1016/j.simyco.2017.11.00

van Wyk, P.S., Herselman, L. & Linde, E. J. van der (2001). First report of *Ascochyta* spp. on soybean in South Africa. *Plant Disease, 85*(3), 334. https://doi.org/10.1094/PDIS.2001.85.3.334B

Walters, H. J. & Martin, K. F. (1981). *Phyllosticta sojaecola* on pods of soybeans in Arkansas. *Plant Disease, 65*, 161-162.

Watson, A. J. (1971). Foreign bacterial and fungus diseases of food, forage, and fiber crops: an annotated list. *U.S. Agricultural Research Service, 97*, 1-111.

Weiss, F. (1946). Check list revision. *Plant Disease Reptr., 30*(7), 218-225.

Wunsch, M.J. & Bergstrom, G. C. (2011). Genetic and morphological evidence that *Phoma sclerotioides*, causal agent of brown root rot of alfalfa, is composed of a species complex. *Phytopathology, 101*(5), 594-610. https://doi.org/10.1094/PHYTO-04-10-0107.

Yu, T. F. (1947). Ascochyta blight and leaf and pod spot of broad bean in China. *Phytopathology, 37*, 207-214.

Diseases of Vegetables Caused by *Phoma* spp.

6

Farjana Sultana and Md. Motaher Hossain

Abbreviations

AP	Anilino-pyrimidines
DMI	Demethylation inhibitors
MBC	Methyl benzimidazole carbamates
MBI-D	Melanin biosynthesis inhibitors-dehydratase
PP	Phenylpyrroles
QoI	Quinone outside inhibitors
SDHI	Succinate-dehydrogenase inhibitors

6.1 Introduction

Vegetables embrace a wide range of horticultural plants from various plant families and represent one of the three major food groups (grains, fruits, and vegetables) of plant origin. Roots, leaves, stems, flowers, fruits, and/or seeds of these plants are consumed as cooked or raw. The nutritional and medicinal value of vegetables makes them absolute necessities in regular human diets. Vegetables are important sources of carbohydrates, proteins, fibers, vitamins, minerals, and non-nutritive phytochemicals such as phenolic compounds, flavonoids, and bioactive peptides. The well-known vegetable phytochemicals include polyphenols, carotenoids, organo-sulfur, and seleno-compounds. These nutritive and non-nutritive phytochemicals have synergistic health benefits. Multiple studies show that the lycopene in tomato and watermelon fruit reduces cancer risk (Tang et al. 2009; Butt et al. 2013). Beta-carotene from carrots and squash, lutein from broccoli and peas, and glucosinolate from crucifers lower blood pressure (Huang et al. 2018). Anthocyanins in beets, red cabbage, and kidney beans improve the circulatory system, and anti-oxidant lutein in spinach leaves benefits vision (Kramer 2004; Feng et al. 2019). Vitamin folate in legumes and spinach reduces depression (Gougeon et al. 2016), while allicin in onion and garlic reduces heart disease and cancer risk (Bayan et al. 2014). Thus, increasing vegetable consumption appears to be a pathway for a change towards healthier and more sustainable diets.

Globally, vegetable production has grown intensively, especially on a per capita basis, which has increased 60% over the last 20 years. This trend is particularly strong in developing countries. This phenomenal growth is seen almost in all major vegetable crops, including potato, cucurbits, tomato, eggplant, onion, pep-

F. Sultana
College of Agricultural Sciences, International University of Business Agriculture and Technology, Dhaka, Bangladesh

Md. M. Hossain (✉)
Department of Plant Pathology, Bangabandhu Sheikh Mujibur Rahman Agricultural University, Gazipur, Bangladesh
e-mail: hossainmm@bsmrau.edu.bd

M. Rai et al. (eds.), *Phoma: Diversity, Taxonomy, Bioactivities, and Nanotechnology*,
https://doi.org/10.1007/978-3-030-81218-8_6

per, spinach, lettuce, carrot, pea, and bean. Compared to vegetable production growth, over the same 30-year period, the human population has increased approximately 1.29 times, from 5.90 to 7.64 billion people. With this pace of growth, the global population is projected to reach around 9.7 billion by 2050. Although economic growth will help increase vegetable production in the future, particularly in lower-income countries, this alone will not ensure an adequate supply of vegetables. In 2050, the number of people living in countries with an insufficient supply of vegetables will increase by 1.5 billion (Mason-D'Croz et al. 2019). For ensuring adequate vegetable supply for this large population group, systematic strategies must be pursued to exterminate the production constraints. One such strategy is to counteract agents and factors that incur crop losses in the field. Efforts need to focus on improving current mitigation technologies and practices for preventing such losses.

Diseases are serious concerns for vegetable production worldwide. The perishable nature of the vegetables makes them more vulnerable to serious diseases. Diseases are primarily caused by living organisms such as fungi, bacteria, nematodes, viruses, mollicutes, and other pathogens. Plant diseases caused by fungal pathogens have detrimental effects on vegetable production, and human nutrition, besides being causes for great famines and the resultant untold human suffering (Agrios 2005). *Phoma* is among the important fungal genera that have displayed significant hostilities against vegetable crops and other economically important plants (Derevnina et al. 2016). Various species of *Phoma* may decline the leaves, roots, stems, flowers, or fruits and upset the normal productivity of the host. Some may take a heavy toll on plants and have become a limiting factor in vegetable cultivation in some regions. Genus *Phoma* is also called *Phoma sensu lato* because it involves a group of about ten different genera and some new genera (De Gruyter 2012). *Phoma sensu lato* is a highly polyphyletic genus with indistinct species boundaries (Aveskamp et al. 2008). In this context, many *Phoma* spp. are causal agents of fatal vegetable diseases and distributed across many continents. They often infect multiple plant parts of vegetable hosts, causing serious yield losses in the greenhouse and field.

This chapter has reviewed the basic information on the significance, distribution, host range, causal agent, symptomatology, and epidemiology of different vegetable diseases caused by members of *Phoma* spp. Also, we have discussed the cultural, biological, and chemical measures for controlling the devastation caused by *Phoma* spp. It is expected that this chapter will serve as a valuable document illustrating the *Phoma* etiology of vegetables.

6.2 *Phoma*: The Largest Genus of Agricultural Significance

The first *Phoma* spp. were described in 1821 (Sutton 1980), and the generic name was first introduced in 1880 by Saccardo (1880). The concept was later revised in 1975 by Boerema and Bollen (1975). In later years, Boerema and associates worked on the reclassification and other aspects of *Phoma* and *Phoma*-like fungi (reviewed in Rai et al. 2009). Ever since *Phoma* Sacc. emend. Boerema & G.J. Bollen (*Pleosporales*) has been considered a filamentous fungal genus capable of forming pycnidium and belonging to *Coelomycetes*. *Phoma* is a ubiquitously spread taxon and is always considered one of the largest fungal genera (Aveskamp et al. 2008). Initially, it included over 3000 species (Monte et al. 1991). Such many species described within *Phoma sensu lato* were related to nomenclature, based on the host plant characteristics and the marginalization of micro-morphological properties (Aveskamp et al. 2008). Dutch mycologists studied the relative morphology of different species of *Phoma* on various culture media in constant conditions since 1992, which led to their division into nine sections (Boerema 1997). Because of 40 years of taxonomic research based on the morphological and cultural characteristics of *Phoma* species, the number of species was curtailed to 223 (Boerema et al. 2004a, b). Nowadays, the genus *Phoma sensu lato* that has been parted into clades rose to the rank of genera, including species with similar relationships and placed within *Didymellaceae* (Aveskamp et al. 2010; Chen et al. 2015). Despite these amendments, the taxonomy of these genera remains confused. This is largely shown by the high number of synonyms in this complex (Aveskamp et al. 2008).

Being geographically prevalent in occurrence, fungi belonging to *Phoma sensu lato* embrace the largest number of species that are saprobic to parasitic (Entwistle 1990; Koike et al. 2006; Hossain et al. 2008; Sultana et al. 2008; Sultana et al. 2009; Hossain et al. 2017; Hossain and Sultana 2020). Approximately 110 *Phoma* species are pathogenic to plants, some of which are economically important plant pathogens and even are pathogens of quarantine importance. Pathogenic members of *Phoma sensu lato* are common inhabitants in soil and plant materials and can affect both belowground and aboveground parts of plants. *Phoma sensu lato* causes an infection that results in symptoms ranging from leaf blight to root rot, even fruit rot, and wilting of the plant. Large yield losses of the infested crop are often recorded in the field under favorable environmental conditions (Fitt et al. 2006). Substantial postharvest crop losses may occur during storage and in the supply chain. *Phoma* pathogens produce a range of phytotoxins that affect the transpiration rate, photosynthetic efficiency, and actin cytoskeleton-based functions and cause electrolyte leakage from affected plants (Cimmino et al. 2008; Möbius and Hertweck 2009; Deb et al. 2020). Minimizing the damage from these fungal pathogens is important to support grower profitability, the viability of the food and feed industries, and the integrity of the trade.

6.3 Diseases of Vegetables Caused by *Phoma* spp.

Vegetable growers face substantial losses in quality and quantity due to damage by various diseases. The polyphyletic genus *Phoma* of the family *Didymellaceae* constitutes a major problem in the production of vegetable crops. Different *Phoma sensu lato* species are associated with the various diseases of vegetables of the family Alliaceae, Apiaceae, Cucurbitaceae, Brassicaceae, Amaranthaceae, Fabaceae, Asteraceae, and Solanaceae in the field and the storage (Table 6.1). However, the relative importance of these diseases varies from crop to crop. Many of these dangerous diseases are seed-borne, while there are diseases in which soil-borne aspects of the pathogen play an important role in the disease epidemiology. Host plants show a range of symptoms that include leaf blight to root rot, even fruit rot, and wilting of the plant. Controlling the diseases caused by members of *Phoma sensu lato* is challenging. Various cultural, biological, and chemical techniques are applied to manage these diseases.

6.3.1 *Phoma* Disease of Alliaceae (Onion Family) Vegetables

Alliums were formerly classified with the Liliaceae (lilies and relatives). Now they have been assigned to their own family, the Alliaceae. The plants in this family are usually used for seasoning and cooking. The most familiar allium crops include onion (*A. cepa*) and garlic (*A. sativum*). Allium plants are monocots and may have originated in Asia. Several fungal diseases affect alliums. A *Phoma* disease, pink rot, is a significant disease of Alliaceae and other hosts.

Pink Root

Introduction, Significance, Occurrence, and Host Range

The pink root is a severe disease in onion, reducing both the bulb yield and quality. It also infects garlic but seldom causes significant crop loss. The pathogen affects many other vegetable crops, including eggplant, pepper, tomato, legumes, spinach, carrot, crucifers, and cucurbits. Even grain crops, such as maize, millet, and sorghum, are susceptible (Table 6.2). The disease was first detected in Texas (Tabenhaus 1917). Since then, the disease has been found in most onion-growing regions worldwide (Table 6.2). Although the disease is more prevalent in tropical and subtropical

Table 6.1 Important diseases of vegetables caused by *Phoma* spp.

Name of the family	Name of the vegetable	Name of the disease	Name of the pathogen	References
Alliaceae	Onion	Pink root	*Phoma terrestris*	Nischwitz and Dhiman (2012)
Apiaceae	Celery	Phoma crown and root rot	*Phoma apiicola*	Dixon 1981
	Parsnip	Phoma canker	*Calophoma complanata* (Syn. *Phoma complanata*)	Cerkauskas (1984)
Amaranthaceae	Beet	Blackleg	*Phoma betae*	Koenick et al. (2019)
Brassicaceae	Brassicas	Blackleg/Phoma leaf spot/stem canker	*Plenodomus lingam* (Syn. *Phoma lingam*)	Gabrielson (1983)
Cucurbitaceae	Cucurbits	Gummy stem blight, black rot	*Phoma cucurbitacearum*	Keinath (2013)
Fabaceae	Peas	Root and foot rot complex	*Peyronellaea pinodella* (Syn. *Phoma pinodella*)	Bretag and Ramsey (2001)
	Peas	Foot rot complex		Koike et al. (2007)
	Peas	Ascochyta blights		Bretag et al. (2006)
	Peas	Ascochyta blights	*Ascochyta koolunga* (Syn. *Phoma koolunga*)	Davidson et al. (2009)
Asteraceae	Lettuce	Phoma basal rot	*Boeremia exigua* (Syn. *Phoma exigua*)	Koike et al. (2006)
Solanaceae	Tomato	Phoma blight/Phoma rot of tomato	*Remotididymella destructiva* (Syn. *Phoma destructiva*)	Jones et al. (2014)
	Potato	Gangrene	*Boeremia foveata* (Syn. *Phoma exigua* var. *foveata*) *Boeremia exigua* var. *exigua* (Syn. *Phoma exigua* var. *exigua*)	Giebel and DopieraLa (2004)

regions, the pathogen is well adapted to a wide range of climates due to its ability to survive in many soil types at different temperatures and pH ranges.

Causal Agent

The pink root disease is caused by the fungus *Phoma terrestris* (synonym *Pyrenochaeta terrestris*) (Entwistle 1990). *P. terrestris* is a soil-borne fungus and a common saprophyte (Table 6.2). It can also be present in roots or dry outer scales of the onion bulbs. However, it does not attack the basal plate or stem of the bulb. Hence, root injury is considered the utmost concern to onion growers as it may decrease the onion yield. *P. terrestris* is a weak parasite of many hosts. It has been isolated from plants of 45 genera, including other *Allium* spp., broadleaf weeds, grasses, grains, and vegetables. *P. terrestris* produces globose, dark-brown to black pycnidia on infected roots or fleshy tissue. Setae grow on these pycnidia, especially around the ostiole, and measure 8–120μm long. Conidia are ovoid, single-celled, hyaline, and biguttulate and measure 3.7–5.8 x 1.8–2.4μm. The fungus produces chlamydospores.

Symptoms and Diagnostic Features

Infected roots are pink in color; as the disease develops, the pink coloration progresses to become more intense, and later, it turns dark or red-purple (Fig. 6.1a–c). Roots develop water-soaked rots and dry out. Such roots become brittle and papery in texture. Eventually, the roots die. New young roots also are infected as they emerge (Fig. 6.1c). In the above-ground, severely infected plants appear stunted and small. Nutrient deficiencies and stress symptoms are seen in the

Table 6.2 Host range, geographic distribution, mode of spread, and nature of infection of economically important *Phoma* pathogens in vegetable crops

Name of the pathogen	Host range	Geographic distribution	Mode of spread	Nature of infection
Phoma terrestris	Onion, garlic, eggplant, pepper, tomato, legumes, spinach, carrot, crucifers, and cucurbits. Even grain crops, such as maize, millet, and sorghum	Worldwide, prevalent in tropical and subtropical regions	Infected crop residues and soil	Field
Phoma apiicola	Celery, celeriac, fennel, carrot, caraway, parsley, and parsnip	Europe and North America	Seed-borne and infested debris	Field and storage
Calophoma complanata (Syn. *Phoma complanata)*	Parsnips and cow parsnip	Canada and UK	Seed-borne and infested debris	Field and storage
Phoma betae	Table beet, sugar beet, spinach, Swiss chard, and lambs' quarters	Most beet-producing areas especially in Europe, North America, and Africa	Seed-borne, infested debris, and airborne ascospores	Seedbed, field, and storage
Plenodomus lingam (Syn. *Phoma lingam*)	Cabbage, broccoli, Brussels sprouts, cauliflower, rutabaga, turnip, forage, and canola	Worldwide, but Europe, North America, and Australia suffers the most	Seed-borne, infected crop debris, and airborne ascospores	Field
Phoma cucurbitacearum	All commonly cultivated cucurbits including squash, pumpkin, cucumber, etc.	Subtropical, tropical, and temperate areas of the world	Seed-borne and infected crop debris	Field and storage
Peyronellaea pinodella (Syn. *Phoma pinodella*)	Legumes including peas and beans	Worldwide, including all major bean- and pea-growing areas	Seed-borne and chlamydospores in soil	Field
Ascochyta koolunga (Syn. *Phoma koolunga*)	Peas	South Australia	Infected crop debris and seed	Field
Boeremia exigua (Syn. Phoma exigua)	Lettuce	USA and UK	Soil	Field
Remotididymella destructiva (Syn. *Phoma destructiva*)	Tomato, bell pepper, and chili	Asia (India, Malaysia), Europe, and North and South America (United States and Brazil)	Seed and plant debris	Seedbed and field
Boeremia foveata (Syn. *Phoma exigua* var. *foveata*) *Boeremia exigua* var. *exigua* (Syn. *Phoma exigua* var. *exigua*)	Potato	Cold or temperate climates such as in Northern Europe	Soil	Storage

foliage. Leaf tips show dieback symptoms (Fig. 6.1d). Severely affected seedlings may die. The external bulb scales can be infected and may develop pink blemishes or water-soaked rots within 1–3 weeks of planting.

Disease Cycle

The pathogen survives in infected crop residues or soil as chlamydospores for only a few years. The fungus can stay in the soil up to 45 cm deep. Invasion of the root by the pathogen occurs just

Fig. 6.1 Onion plants infected with pink root. (**a**, **b**, and **c**) Diseased onion roots infected with pink root. (**d**) Onion plants infected with pink root showing foliar symptoms. (Image courtesy of Lindsey J. du Toit and Carrie Wohleb of Washington State University)

behind the root tip through direct penetration and then spreads upward into the main root body. Symptoms express 7–21 days after infection (Babadoost 1990). The pink root becomes severe, especially during seasons of warm soil temperatures, as the fungus is most active at soil temperatures of 24–28 °C. If soil temperatures are below 16 °C, it may cause minor infection. The disease severity is favored in fields with heavy and poorly drained soils. The pathogen can spread to unaffected fields by irrigation water.

Control

Rotation of onions with non-host crops is often suggested to reduce soil-borne inoculum (Entwistle 1990). A standard crop rotation scheme requires at least 3 years between susceptible crops. However, the extensive host range of *P. terrestris* may make this control approach unfeasible. Initial inoculum in the soil can be reduced via soil fumigation or solarization. Fumigation with dazomet (Porter et al. 1989) and metam-sodium (sodium methyldithiocarbamate) (Hartz et al. 1989) increased yields and decreased incidence of the pink root. Soil fumigation with methyl bromide +chloropicrin (67% + 33%) or chloropicrin effectively reduced the fungal population and increased marketable bulb yield (Table 6.3). However, the use of methyl bromide is not recommended due to the Montreal Protocol on ozone-depleting substances (Koike et al. 2006). Soil solarization alone or with fumigation increased onion yields in some regions (Hartz et al. 1989, Porter et al. 1989). Soil fumigation is very expensive and can only be justified in severely infested fields (Lee et al. 2007). Use of

Table 6.3 Examples of chemicals, their FARC code, mode of application, and target site used in the control of *Phoma* diseases

Pathogen.	Group name	Chemical group	Active ingredient	FARC code	Mode of application	Target site and code	References
Phoma terrestris	Heteroaromatics	Thiadiazinane	Dazomet	14	Soil fumigation	F3 Cell peroxidation	Porter et al. (1989)
	Dithiocarbamates and relatives	Dithiocarbamates and relatives	Metam-sodium	M03	Soil fumigation	Multi-site contact activity	Hartz et al. (1989)
Phoma apiicola	Dithiocarbonates and relatives	Dithiocarbonates and relatives	Thiram	M03	Seed soaking or foliar spray	Multi-site contact activity	Koike et al. (2007)
			Zineb	M03	Foliar spray	Multi-site contact activity	Koike et al. (2007)
Calophoma complanata (Syn. *Phoma complanata*)	Chloronitriles (phthalonitriles)	Chloronitriles (phthalonitriles)	Chlorothalonil	M05	Foliar spray	Multi-site contact activity	Cerkauskas and McGarvey (1987)
	Dithiocarbonates and relatives	Dithiocarbonates and relatives	Mancozeb	M03	Foliar spray	Multi-site contact activity	Cerkauskas and McGarvey (1987)
Phoma betae	Organomercurials	Organomercurials	Ethyl mercury phosphate		Seed soaking	–	Gates (1959)
	Dithiocarbonates and relatives	Dithiocarbonates and relatives	Thiram	M03	Seed soaking	Multi-site contact activity	Maude and Bambridge (1985)
	DMI-fungicides (demethylation inhibitors)	Imidazoles	Prochloraz	3	Seed soaking and foliar spray	G1 C14-demethylase in sterol biosynthesis (erg11/cyp51)	Piszczek (1997)
	DMI-fungicides (demethylation inhibitors)	Imidazoles	Spartak 45 EC	3	Foliar spray	G1 C14-demethylase in sterol biosynthesis (*erg11/cyp51*)	Piszczek (1997)
Plenodomus lingam (Syn. *Phoma lingam*)	MBC (methyl benzimidazole carbamates)	Benzimidazole	Thiabendazole	1	Seed treatment and foliar spray	B1 ß-tubulin assembly in mitosis	Ash (2000); Maude et al. (1984)
	MBI-D (melanin biosynthesis inhibitors – dehydratase)	Carboxamide	Diclocymet	16.2	Foliar spray	I2 Dehydratase in melanin biosynthesis	Ash (2000)

(continued)

Table 6.3 (continued)

Pathogen.	Group name	Chemical group	Active ingredient	FARC code	Mode of application	Target site and code	References
	Dicarboximides	Dicarboximides	Iprodione	2	Seed treatment and foliar spray	E3 MAP/histidine-kinase in osmotic signal transduction (*os-1, Daf1*)	Ash (2000); Maude et al. (1984)
	Amines ("morpholines")	Morpholines	Fenpropimorph	5	Seed treatment	G2 Δ14-reductase and Δ8→Δ^{7-} isomerase in sterol biosynthesis (*erg24, erg2*)	Ash (2000); Maude et al. (1984)
Phoma cucurbitacearum	Dithiocarbonates and relatives	Dithiocarbamates and relatives	Mancozeb	M3	Foliar spray	Multi-site contact activity	Mathews et al. (2018)
	Chloronitriles (phthalonitriles)	Chloronitriles (phthalonitriles)	Chlorothalonil	M5	Foliar spray	Multi-site contact activity	Freeman et al. (2017)
	DMI-fungicides (demethylation inhibitors)	Triazoles	Tebuconazole	3	Foliar spray	G1 C14-demethylase in sterol biosynthesis (*erg11/ cyp51*)	Freeman et al. (2017)
	SDHI (succinate-dehydrogenase inhibitors)	Pyridine-carboxamides	Boscalid	7	Foliar spray	C2 Complex II: succinate-dehydrogenase	Freeman et al. (2017)
	QoI-fungicides (quinone outside inhibitors)	Methoxy-acrylates	Azoxystrobin	11	Foliar spray	C3 Complex III: cytochrome bc1 (ubiquinol oxidase) at Qo site (*cyt b* gene)	Freeman et al. (2017)
Peyronellaea pinodella (Syn. *Phoma pinodella*)	Phthalimides	Phthalimides	Captan	M04	Seed treatment	Multi-site contact activity	Maude (1966)
	Dithiocarbonates and relatives	Dithiocarbonates and relatives	Thiram	M03	Seed treatment	Multi-site contact activity	Maude (1966)
	MBC (methyl benzimidazole carbamates)	Benzimidazoles	Benomyl	1	Seed treatment	B1 ß-tubulin assembly in mitosis	Maude et al. (1986)
			Thiabendazole	–	Seed treatment		Maude et al. (1986)
Boeremia exigua (Syn. *Phoma exigua*)	Heteroaromatics	Thiadiazinane.	Dazomet	14	Soil Fumigation	F3 Cell peroxidation	Koike et al. (2006)

	QoI-fungicides (quinone outside inhibitors)	Methoxy-acrylates	Azoxystrobin	11	Foliar spray	C3 Complex III: cytochrome bc1 (ubiquinol oxidase) at Qo site (*cyt b* gene)	Koike et al. (2006)
	SDHI (succinate-dehydrogenase inhibitors)	Pyridine-carboxamides	Boscalid	7	Foliar spray	C2 Complex II: succinate-dehydrogenase	Koike et al. (2006)
	PP-fungicides (phenylpyrroles)	Phenylpyrroles	Fludioxonil	12	Foliar spray	E2 MAP/histidine-kinase in osmotic signal transduction (*os-2*, *HOG1*)	Koike et al. (2006)
	AP fungicides (anilino-pyrimidines)	Anilino-pyrimidines	Cyprodinil	9	Foliar spray	D1 Methionine biosynthesis (proposed) (*cgs* gene)	Koike et al. (2006)
	DMI-fungicides (demethylation inhibitors)	Imidazoles	Prochloraz	3	Foliar spray	G1 C14-demethylase in sterol biosynthesis (*erg11/cyp51*)	O'Neill et al. (2000)
	QoI fungicides (quinone outside inhibitors)	Methoxy-carbamates	Pyraclostrobin	11	Foliar spray	C3 Complex III: cytochrome bc1 (ubiquinol oxidase) at Qo site (*cyt b* gene)	O'Neill et al. (2000)
	QoI fungicides (quinone outside inhibitors)	Methoxy-acrylates	Azoxystrobin	11	Foliar spray	C3 Complex III: cytochrome bc1 (ubiquinol oxidase) at Qo site (*cyt b* gene)	O'Neill et al. (2000)
	SDHI (succinate-dehydrogenase inhibitors)	Pyridine-carboxamides	Boscalid	7	Foliar spray	C2 Complex II: succinate-dehydrogenase	O'Neill et al. (2000)
Remotididymella destructiva (Syn. *Phoma destructiva*)	QoI-fungicides (quinone outside inhibitors)	Methoxy-acrylates	Azoxystrobin	11	Foliar spray	C3 Complex III: cytochrome bc1(ubiquinol oxidase) at Qo site (*cyt b* gene)	Shankar et al. (2014)

(continued)

Table 6.3 (continued)

Pathogen.	Group name	Chemical group	Active ingredient	FARC code	Mode of application	Target site and code	References
	Chloronitriles (phthalonitriles)	Chloronitriles (phthalonitriles)	Chlorothalonil	M05	Foliar spray	Multi-site contact activity	Shankar et al. (2014)
	QoI-fungicides (quinone outside inhibitors)	Methoxy-carbamates	Pyraclostrobin	11	Foliar spray	C3 Complex III: cytochrome bc1(ubiquinol oxidase) at Qo site (*cyt b* gene)	Shankar et al. (2014)
	DMI-fungicides (demethylation inhibitors)	Triazoles	Tebuconazole	3	Foliar spray	G1 C14-demethylase in sterol biosynthesis (*erg11/cyp51*)	Shankar et al. (2014)
Boeremia foveata (Syn. *Phoma exigua* var. *foveata*) *Boeremia exigua* var. *exigua* (Syn. *Phoma exigua* var. *exigua*)	DMI-fungicides (demethylation inhibitors) (SBI: Class I)	Imidazoles	Imazalil	3	Postharvest tuber treatment	G1 C14-demethylase in sterol biosynthesis (*erg11/cyp51*)	Carnegie et al. (1998)
	MBC-fungicides (methyl benzimidazole carbamates)	Benzimidazoles	Thiabendazole	1	Postharvest tuber treatment	B1 ß-tubulin assembly in mitosis	Carnegie et al. (1998)
	PP-fungicides (phenylpyrroles	Phenylpyrrole	Fenpiclonil	12	Postharvest tuber treatment	E2 MAP/histidine-kinase in osmotic signal transduction (*os-2, HOG1*)	Carnegie et al. (1998)

resistant cultivars, if such are available, is an effective method to control pink root. Planting onion during seasons of cooler soil temperatures may help reduce disease severity. Some evidence shows that reducing the interval between irrigation applications can decrease the impact of pink root on yield. The use of fungicide-treated seed or soil fungicide applications is not effective in controlling pink roots.

6.3.2 Diseases of Apiaceae Vegetables

Apiaceae (celery, carrot, or parsley family), formerly known as the Umbelliferae, is a family of mostly aromatic plants characterized by the hollow stem and distinctive umbrella-shaped flowering structures. They are an important group of vegetables having many uses. Carrot (*Daucus carota*) and parsnip (*Pastinaca sativa*) are used for their edible roots. Celery (*A. graveolens var. dulce*) is grown for thick edible petioles. The foliage of parsley (*Petroselinum crispum*) and cilantro (also known as coriander or Chinese parsley) (*Coriandrum sativum*) is used for culinary purposes. The seeds of coriander and cumin (*Cuminum cyminum*) are also widely used as specified. Some of these plants also have medicinal properties. These plants are infected by many important fungal diseases, among which, *Phoma* crown and root rot and canker are of increasing concern to growers.

Phoma Crown and Root Rot

Introduction, Significance, Occurrence, and Host Range

Phoma crown and root rot is a minor disease, primarily in celery, celeriac, and fennel. Carrot, caraway, parsley, and parsnip are also reported to be susceptible to the disease but can be attacked by another species *Calophoma complanata* (Syn. *P. complanata*) (Table 6.1). *Phoma* crown and root rot causes crop losses both in the field and in storage (Table 6.2). Although the disease causes rotting in the storage, infections are usually initiated in the field. The disease primarily occurs in Europe and North America (Table 6.2). In Europe, the disease is recorded in several host plants belonging to Apiaceae. In Great Britain and North America, it is recorded only on celery. A severe *Phoma* crown and root rot of celery causing considerable crop losses in Auckland, New Zealand, was reported in 1965 (Dingley 1970). However, infection in New Zealand has so far been confined to celery in this single outbreak.

Causal Agent

Phoma crown and root rot is caused by the fungus *Phoma apiicola*. Morphologically, the fungus is similar to *Phoma exigua* Desm (Now *Boeremia exigua*), but it grows more slowly and produces pycnidia more freely in culture. The mycelium of the fungus is initially hyaline but becomes dark with age. The very old mycelium may form torulose. Pycnidia are embedded in host tissue, brown-black, usually formed singly, occasionally in clusters, then superficially appearing as if multilocular globose, and vary in size, measuring up to 250μm in diameter. Conidia are hyaline and almost oblong or rod-shaped that measure 3–14 × 1–2μm. The temperature range of fungal growth is −5–30 °C, with an optimum of 16–18 °C. *P. apiicola* can be seed-borne and carried in debris in the soil.

Symptoms and Diagnostic Features

Light brown lesions develop on the roots and crown of celery. The lesions later turn black and spread to the petiole bases, resulting in yellowing and death of the outer leaves. Death of the epidermal tissue may lead to cracking of the crown. Early root infection may lead to stunted growth and the possible death of the plants. On celeriac, a dry brown rot develops on the roots. In moist conditions, numerous black dot-like pycnidia are seen in the rotted tissues.

Disease Cycle

The pathogen is seed-borne and also persists in infested debris for several years (Table 6.2). The disease can be initiated from infected seed on transplants before being field planted. Conidia are disseminated by rain splash or overhead sprinkler irrigation and infect plants directly or through wounds (Koike et al. 2007). The optimum temperature for mycelial growth, pycnidial formation, and spore germination is 16–18 °C. However, wet conditions may favor disease development at much lower temperatures.

Control

Crops should be rotated for at least 2 years away from Apiaceae crops. Since the pathogen is seed-borne, the seed that does not have significant pathogen levels should be used. Seed treatment can be done either using hot water treatment (50 °C for 25 min) or using a thiram soak (24 h at 30 °C in a 0.2% aqueous thiram suspension). Fungicide sprays such as thiram or zineb may give some control (Table 6.3).

Phoma Canker

Introduction, Significance, Occurrence, and Host Range

This disease was first reported on parsnip (*Pastinaca saliva*) in Ontario, Canada, in 1984 (Cerkauskas 1984) and caused a sharp drop in annual parsnip production. The disease causes significant losses of harvest in the field due to the collapse of aboveground portions of affected plants and the decay of diseased roots in storage (Table 6.2). The pathogen is specific to parsnip and does not affect other vegetables. The cow parsnip (*Heracleum lanatum*) is only mildly susceptible to the pathogen. In the UK, *Phoma* canker has now become the most important disease parsnip (Koike et al. 2007).

Causal Agent

Phoma canker is caused by *Phoma complanata* (Tode: Fr.) Desmaz. However, a new generic name has been introduced for this fungus, and the current name of *P. complanata* is suggested to be *Calophoma complanata* (Chen et al. 2015). The fungus produces dense aerial mycelium and forms white to light gray or olivaceous gray colonies with even or regular margins. Pycnidia are spherical with a hard, dark brown to black outer wall and lack setae. Pycnidia in the host tissues are unilocular with one ostiole. The diameter ranges from 165 to 373μm with an average of 250μm. Pycnidia in culture media have thick walls consisting of about six layers of pseudoparenchymatous and are 35–49μm in diameter. Conidia are hyaline, guttulate, and variable in shape, such as ellipsoid, cylindrical, fusiform, or globose. Dimensions of aseptate conidia are 5–9 × 2–3.5μm (mean 7.4 × 2.4μm), though large (27 × 8μm). One-septate hyaline conidia with dimensions of 27.2 (22–34) by 8.1 (6–10) μm occur in older cultures or older cankers.

Symptoms and Diagnostic Features

Leaf spot, blight, and cankers on petioles and roots are the key symptoms. When present, leaf spots are small (1 mm or less in diameter), circular, and dark tan or brown and often have yellow-green halos. As the disease progresses, leaf spots may coalesce, leading to a general yellowing, necrosis, and blighting of foliage. Pycnidia develop in the center of leaf lesions about 2 weeks after infection. Light brown elliptical lesions may develop on petioles. About 2 weeks after infection, lesions become dark brown to black and coalesce, leading to canker formation. Leaves twist and bend over at the canker site. Large buff or dark brown to black canker forms on parsnip crowns and the sides of taproots. These cankers enlarge and result in cracking on the upper portion of the root. Black pycnidia appear on the surfaces of the decayed tissues or remain embedded in these tissues. Under moist conditions, pycnidia ooze a white or creamy spore exudate. Roots with *Phoma* canker are reported to have a sweet, distinct cinnamon-like odor, resulting in reduced commercial value.

Disease Cycle

The fungus survives for at least 5 months in infected parsnip residues, regardless of the depth of incorporation. The fungus can be seed-borne and may affect seed germination and seedling emergence (Table 6.2). The disease often occurs from infected plants. Moderate temperatures and rain or high humidity favor disease development. Secondary spread of the disease in the field occurs through wind-driven rain, heavy dew formation, and insects. Rain or heavy dew washes down large numbers of spores from infected foliage to the crown, where root infection may develop. Cankers in cold storage arise from field infections which may be small and visible only after two cuts. In the UK, root canker occurs, while foliar symptoms are absent because the soil-borne inoculum of the fungus is likely to be the cause of infection.

Control

The removal of diseased crop residues from the field and a minimum 1-year rotation with other crops can reduce the fungus levels in affected fields. Precautions should be taken to avoid direct skin contact with diseased plant tissues as such plants contain high concentrations of photocarcinogenic furocoumarins. The seed that does not have a significant level of the pathogen should be used. Good control of *Phoma* canker can be achieved by selecting and planting cultivars resistant or tolerant to canker. Application of fungicides such as chlorothalonil and mancozeb at the rate of 1.4 kg a.i./ha, respectively, significantly control foliar symptoms (Cerkauskas and McGarvey 1987) (Table 6.3); however, chemicals are not effective at preventing disease caused by soil-borne inoculum.

6.3.3 Phoma Disease of Amaranthaceae Vegetables

Amaranthaceae includes those vegetables, which are primarily grown as leafy vegetables and as root commodities. Economically important vegetables of this family include various beets (*Beta vulgaris* including table beet and chard) and spinach (*Spinacia oleracea*). They belong to the former Chenopodiaceae (goosefoot family) family, which is now a subfamily of Amaranthaceae. Blackleg caused by *Phoma betae* is one of the important fungal diseases of beet, particularly from a quarantine point of view as they are highly seed-transmitted.

Blackleg/Phoma Leaf Spot

Introduction, Significance, Occurrence, and Host Range

Blackleg is one of the common diseases of sugar beet and table beet (Table 6.1). The disease affects the plant at all stages of growth. The disease is widespread in seed crops and may cause severe losses in seedbeds. Seedling infection can result in reduced stands in the field. Foliage infection can cause rejection in fresh market sales. Root rot can reduce the yield and quality of root. It is also parasitic on spinach and has been isolated from some other plants such as Swiss chard, spinach, and lamb's quarters. The disease is found in most beet-producing areas, especially in Europe, North America, and Africa (Table 6.2).

Causal Agent

Blackleg is caused by *Phoma betae* (Oud.) Frank. (Teleomorph – *Pleospora bjoerlingii*). The teleomorph of the fungus has been revised as *Pleospora betae* (Chen et al. 2015). The pathogen is generally confronted in its asexual stage. *P. betae* produces grayish-black aerial colonies with black globose pycnidia (220–235μm) in concentric rings. The diseased tissues also often contain small black pycnidia. Conidia are hyaline, unicellular conidia that measure 4.1–7.0 × 2.0–4.3μm (Harveson 2006). The sexual form *Pleospora betae* produces black pseudothecia (200–500μm in diameter) on rotted roots at the end of the season and on crop residues that persist over the winter. Pseudothecia produce brown ascospores that measure 18–25 × 7–10μm.

Symptoms and Diagnostic Features

In seedbed, pre-emergence damping-off is seen, where seed and germinated seedlings die before growing aboveground. Young plants that emerge can develop black lesions on the lower part of the stem in contact with soil. These lower stem symptoms are the reason for the common name of blackleg. In root and seed crops, leaf infection is mostly confined to mature and older leaves, occurring as dark brown, circular to oval spots, 10–20 mm in diameter with dark concentric rings near the perimeter. Minute black raised spherical dots (pycnidia) are seen throughout the spots in concentric rings. In the presence of moisture, the pycnidia exude a cream-colored gelatinous mass of spores. In seed crops, dark streaks and lesions with gray centers develop on the stems. When the infection is severe, the stem breaks over at the crown. Pedicels on the flowering stems may also develop lesions. The infected roots appear near the crown, initially as soft and water-soaked lesions, which turn dark and become deep, sunken black lesions containing gray-white mycelium and pycnidia. Severely infected roots become spongy and are full of cavities. The pathogen often causes root rot during the storage of table beet, contributing to the death of the growing point and development of secondary roots.

Disease Cycle

The infested seed is the primary source of infection (long-distance dissemination) (Table 6.2). *Phoma betae* is introduced into the field when the infected seed is sown. Pre-emergence damping-off occurs if infested seeds are planted into cool, wet soils (4 to 12 °C). After emergence, disease development on foliage is favored by temperatures of 16–20 °C and high humidity periods (Leach 1986). The pathogen can be spread within a field (local dissemination) through water-splashed conidia, surviving in the soil on infected plant debris and alternative hosts. At the end of the season, the pathogen forms the pseudothecia and overwinters on crop residues. The wind-blown ascospores produced from pseudothecia spread the pathogen to farther distances between fields and cause leaf spots in newly planted root crops and stem infections in seed crops. At harvest, if beet foliage is closely clipped to the root or if roots are injured, root rots can develop during storage.

Control

The use of pathogen-free seeds is desirable. Seed treatments are standard practice for reducing seed-borne infection and improving germination. Seed soaking before storage in thiram or ethyl mercury phosphate significantly improves germination and reduces mean infection by *P. betae* (Maude and Bambridge 1985). Soaking sugar beet seed for 20 min in ten times of its weight of water containing 40 ppm ethyl mercury phosphate gives better seedling emergence than surface treatments with Panogen (methyl mercury dicyandiamide), thiram, Granosan M (ethyl mercury p-toluene sulphonanilide), captan, Dow 9B (zinc trichlorophenate), or Cerenox (acylphenylhydrazone). Seeds infected with *P. betae* can be disinfected by hot water treatment (Agarwal et al. 2006) (Table 6.3). Fungicide sprays are not commonly used on crops grown for root production but can be useful for seed crops. Foliar spray with Sportak 45 EC at a dose of 2 l/ha significantly decreases *P. betae* infection on treated plants compared with the infection level of control samples (Piszczek 1997) (Table 6.3). Implementing cultural practices such as practicing crop rotation, disposing of crop residues after harvest, and maintaining adequate fertilizers levels may result in less disease incidence.

6.3.4 *Phoma* Disease of Brassicaceae (Mustard Family) Vegetables

Plants of this family are among the most extensively grown vegetables worldwide. The family consists of more than 300 genera. Brassicaceae plants are also known as crucifers since their flowers have four petals that extend opposite each other and form a cruciform shape. The major Brassicaceae vegetable includes kale and collards (*Brassica oleracea* var. *acephala*), broccoli (*B. oleracea* var. *italica*), cauliflower (*B. oleracea* var. *botrytis*), cabbage (*B. oleracea* var. *capitata*), Brussels sprouts (*B. oleracea* var. *gemmifera*), kohlrabi (*B. oleracea* var. *gongylodes*), turnip (*B. rapa*), rutabaga or swede (*B. napus* var. *napobrassica*), Chinese cabbage (*B. pekinensis*), various leafy mustards (*B. juncea*, *B. campestris*, *B. alba*, *B. napus*, and *B. nigra*), and radish (*Raphanus sativus*). Disease caused by the fungus *Phoma lingam* is a major concern in Brassicaceae vegetables.

Blackleg/Phoma Leaf Spot/Stem Canker

Introduction, Significance, Occurrence, and Host Range

Blackleg is one of the most economically important diseases of *Brassica* crops. The disease infects numerous crucifer vegetables such as cabbage, broccoli, Brussels sprouts, cauliflower, rutabaga, turnip, and forage and arable crucifers, particularly oilseed rape including canola (Table 6.2). Although the disease has long been present worldwide, Australia suffers the most significant crop losses from this problem (Table 6.2). Yield losses of up to 90% can be caused by the disease (reviewed in Rouxel and Balesdent 2005). However, the disease has less potential to devastate the crops in China since the pathogen is the less aggressive B-type. Although blackleg is the most widely used name for this disease, *Phoma* leaf spot and stem canker are also used.

Causal Agent

Blackleg is a fungal disease caused by *Phoma lingam* Tode ex Fr. (Teleomorph *Leptosphaeria maculans* (Desm.) Ces. & de Not.). The generic classification of *Phoma lingam* was investigated by several investigators (Reddy et al. 1998; Chen et al. 2015), who suggested *Plenodomus* Preuss is the current genus for *Phoma lingam.* A brief incubation of infected plant tissues forms white mycelial growth around the edge of lesions or cankers, typically used to diagnose blackleg disease. *Plenodomus lingam* (Syn. *Phoma lingam*) produces two types of pycnidia in diseased tissues. Type I (sclerotioid form) pycnidia lack regular shape with narrow ostioles and walls composed of several layers of thick-walled sclerenchymatous cells. Type II are globose, dark brown to black ostiolate pycnidia and composed of a wall with a thick-walled outer layer. Pycnidia exude masses of conidia in deep pink ooze in wet weather, which is an important diagnostic key to distinguish this species from other *Phoma* species. Conidia are small (3–5 × 1–2μm), unicellular, hyaline, cylindrical, and straight, although a few can be curved. The sexual form *Leptosphaeria maculans* produces pseudothecia on infected woody residues of crucifer hosts. Pseudothecia are black globose that measure 300–350μm in diameter. The ascospores are large, cylindrical to ellipsoidal, and yellow-brown, measuring 35–70 x 5–8μm. An additional complication is that this pathogen is now considered a species complex (Voigt et al. 2005). The pathogen has two virulence groups (A and B). The A-group makes sirodesmin toxins (Tox+), and the B-group does not have sirodesmin toxins (Tox0). The highly virulent A-group or Tox + group causes the damaging blackleg, and the "weakly virulent" B-group or Tox0 group causes the less damaging blackleg (Williams and Fitt 1999; Rouxel et al. 2004). The A-types are currently named *L. maculans* and the B-type as *L. biglobosa.* Both *Leptosphaeria* species have the same anamorphic stage (*Plenodomus lingam*). Although both are often found together in the tissues of an individual infected plan, the symptoms and severity of disease they cause are often very different (West et al. 2001). *L. maculans* causes the most severe blackleg symptoms.

Symptoms and Diagnostic Features

Symptoms can develop on infected leaves, stems, petioles, or pods. Leaf symptoms appear as gray-green to ash-gray or almost white irregular spots. Spots tend to range in size from 0.5 to 3.0 cm in diameter. Many black, tiny dot-like pycnidia are scattered throughout the spots. These symptoms are considered to be caused by A-type isolates. Symptoms caused by the B-type usually have darker and smaller spots (1–3 mm) containing only a few pycnidia, which are difficult to see without a dissecting microscope.

The fungus from leaf spots spreads via leaf veins to the petiole and ultimately the main stem and develops stem canker. Cankers are also caused directly by inoculum from hypocotyl or stem, usually at the soil level. Cankers may enlarge and become sunken, elongated, and brown to black. Basal stem lesions appear to be the most damaging. Stem lesions near the soil line may extend below the soil surface, causing black rot in the lower stem. Since the woody tissues of the basal stem become black, the disease is called "blackleg." Severe infection may cause stunting, wilting, and general poor growth of the plant. Black pycnidia in the form of minute dot-like structures are scattered across the blackleg lesions. In seed crops, as plants mature, they may lodge over because of poor root anchorage. Symptoms on seed pods may lead to infection of the developing seed.

Disease Cycle

The fungus is seed-borne and acts as the source of infection in the clean field (Table 6.2). Seeds carry the fungus either by spores on the surface or by a dormant mycelium inside the seed coat. A low level of seed infection (<0.1%) may even start a significant disease outbreak. The pathogen can also remain viable in diseased crop debris as mycelium, pycnidium, or pseudothecia over a longer period. In some areas, the airborne ascospores from pseudothecia are more important than seed-borne inoculum for initiating the first cycle of disease. Upon maturity, the pseudothecia produce ascospores in large amounts, which are released after rainfall and can be dispersed by the wind for more than a kilometer. They likely land on the cotyledons or basal rosette leaves and germinate in the presence of free water. The germinating ascospores infect the cotyledons and/or leaves by penetrating tissues via stomata or wounds. The fungus becomes necrotrophic, causing leaf spots where asexual multiplication takes place. Single-cell conidia are produced in pycnidia and are dispersed at short distances by rain splash. Since conidia are not airborne to any significant extent, they are of minor importance as a primary infection source. Conidia cause only the secondary spread of the pathogen to nearby leaves or plants. The fungal growth follows the leaf spot phase of the disease in intercellular spaces in the direction of the crown at the base of the stem and the upper root. The fungus requires a minimum of 4 hours of leaf wetness for infection, but maximum infection requires up to 48 hours of leaf wetness. Optimum temperatures for ascospore infection are 15–20 °C. Symptoms appear in 3 to 4 days at 20 °C or 14 days at 5 °C.

Control

It is necessary to use seeds that do not have significant levels of the pathogen. Seed treatments with thiabendazole, iprodione, and fenpropimorph can eliminate seed transmission of the pathogen (Maude et al. 1984) (Table 6.3). Infected seeds can be treated with hot water by immersing the seeds tied up in a muslin bag in the water at a temperature of 50 °C for 25 minutes. A sample of the seeds, which has been subjected to the hot water treatment, should always be tested for germination before the bulk is sown. The destruction of the crop debris is useful in reducing a large proportion of the initial inoculum. Practicing a 4-year rotation in seedbeds and fields is recommended. Fungicide application is needed to protect susceptible cultivars. Foliar applications using benzimidazole, carboxamide, and dicarboxamide fungicides have been proved successful for controlling the disease (Ash 2000) (Table 6.3).

6.3.5 Phoma Disease of Cucurbitaceae Vegetables

The Cucurbitaceae (gourd family) encompasses a diverse group of vegetables grown primarily for their edible fruits and sometimes for immature leaves, stems, and blossoms. Important cultivated cucurbits include cucumber (*Cucumis sativus*); muskmelon (*Cucumis melo*); cantaloupe (*C. melo* var. *cantalupensis*); watermelon (*Citrullus lanatus*); summer and winter squashes, gourds, or courgette (*Cucurbita pepo*); vegetable marrow (*C. pepo* var. *medullosa*); pumpkin (*Cucurbita maxima*); Chinese winter melon (*Benincasa hispida*); bitter gourd (*Momordica cochinchinensis*); etc. Many of these cucurbits are originated from tropical regions in Africa, the Americas, and Asia. Cultivation of cucurbit species is severely affected by several pests and pathogens, including the fungal pathogen *Phoma* spp.

Gummy Stem Blight and Black Rot

Introduction, Significance, Occurrence, and Host Range

Gummy stem blight is one of the destructive diseases of cucurbits. It can occur on all commonly cultivated cucurbits, although the infection severity differs among species (Keinath 2013) (Table 6.2). The disease can affect both field- and greenhouse-grown crops. Black rot is the fruit rot phase of the gummy stem blight disease, reducing both pre- and postharvest yields. Significant damage occurs by the disease in the USA and subtropical and tropical regions (Table 6.2). The pathogen is also common in temperate areas, particularly on winter squash and pumpkin and greenhouse-grown cucumber.

Causal Agent

Gummy blight is caused by *Phoma cucurbitacearum* (Fr.:Fr.) Sacc. (Syn. *Ascochyta cucumis* Fautrey & Roum.). *Didymella bryoniae* (Auersw.) Rehm (Syn. *Mycosphaerella melonis* (Pass.) Chiu & J. C. Walker) is the fungus teleomorph. Aveskamp et al. (2010) described the new species *Stagonosporopsis cucurbitacearum* as a synonym of *D. bryoniae* infecting cucurbits. The anamorph *P. cucurbitacearum* is mostly isolated from diseased tissues (Somai et al. 2002). The fungus produces dark pycnidia, which are globose, ostiolate, and 125–185μm in diameter. The conidia are hyaline, cylindrical, non- or monoseptate, and 6–13μm long. The teleomorph *Didymella bryoniae* produces dark pseudothecia, which are 125–213μm in diameter. Ascospores are hyaline, oval, and mono-septate and measure 14–18 x 4–7μm. There is a constriction at the septa, and the upper cell is wider than the lower cell.

Symptoms and Diagnostic Features

The disease infects cucurbits at any growth stage, from seedling to mature vine with fruit. Disease symptoms appear on all parts of the plant, including leaves, petioles, vines, stems, tendrils, pedicels, flowers, peduncles, fruit, and seed, except for the roots (Keinath 2011). Leaf symptoms are noticed first and most often used to diagnose the disease. More than half of the spots initiate at the leaf margins or extend to the margins. Early symptoms consist of round or triangular water-soaked spots on cotyledons and true leaves. These spots enlarge rapidly and coalesce, resulting in the blighting of leaves. Constricting lesions on the petiole or the midrib of a leaf may cause the entire leaf to collapse. Small black fruiting bodies (pycnidia) of the fungus are often visible on the infected leaves. In melon, leaf lesions appear as olive brown discoloration between veins or circular tan to olive-brown spots (Fig. 6.2a).

The lesion may also develop on the stem that enlarges into patches and girdle the main stem (Fig. 6.1b). As the disease progresses, these lesions turn into cankers (Keinath et al. 2010). Cankers are light brown to beige to off-white, noticeably lighter in color than foliar leaf spots. Cracking is often visible on the stem, accompanied by a red to amber gummy ooze (Paret et al. 2018). Stem infections also contain numerous pycnidia as black raised specks (Fig. 6.1c). However, leaf and stem lesions occasionally contain the pseudothecia of the pathogen.

On fruit, initially, symptoms appear as water-soaked, dark green circular spots. The lesions subsequently expand, merge, and become brownish-black in color; hence the disease is also known as black rot. Black rot on watermelon appears as a large, expanding rot that starts at the fruit blossom end. The infected tissue contains blackish pycnidia from which masses of conidia are released under humid conditions.

Disease Cycle

The fungus survives in the diseased vines and on crop residue. It may also be seed-borne (Table 6.2). Dispersal of ascospores or conidia occurs by rains and winds from inoculum sources onto new crops. Free moisture is very important for disease development. A minimum of 1-hour leaf wetness is required for infection to take place. Continuous leaf wetness is also necessary for lesion expansion, while the optimum infection occurs at temperatures in the 20–24 °C range. The fungus penetrates leaves directly through the cuticle or intercellular spaces around the bases of trichomes and stems through wounds or by expanding leaf lesions. Wounds or flower scars aid in the penetration of the pathogen into the fruits. Insect feeding injury and powdery mildew infection predispose plants to gummy stem blight infection. Fruit rots within 2–3 days after infection.

Control

The use of disease-free seeds is essential. A 2- to 3-year rotation with non-cucurbit crops is a standard practice to reduce gummy stem blight incidence (Mathews et al. 2018). Destruction of infected residue also contributes to reducing disease potential in subsequent crops. In the green-

Fig. 6.2 Watermelon transplants infected with gummy stem blight. (**a**) Infected leaf of watermelon with gummy blight symptoms. (**b**) The infected stem of watermelon with gummy blight symptoms. (**c**) Black dot-like pycnidia on the infected stem of watermelon. (Image courtesy of Thomas S Isakeit, Texas A&M University)

house, ventilation and sprinkler irrigation should be managed properly to lessen free moisture on leaf surfaces. Proper care must be taken to avoid rind injuries to cucurbit fruits, as *P. cucurbitacearum* enters through wounds. Regular application of protectant fungicides is useful, especially to control the infection through the wound. Chlorothalonil and mancozeb are effective contact fungicides that are suggested to apply as preventive measures. A wide range of other active gradients, such as tebuconazole, boscalid, azoxystrobin, and fluazinam, is also available for use on cucurbits to control gummy stem blight (Freeman et al. 2017) (Table 6.3). There are reports of the detection of benzimidazole-resistant isolates of the pathogen in greenhouse cucumber culture in Greece and the Netherlands (Malathrakis and Vakalounakis 1983) and cucurbit fields in the eastern USA (Keinath et al. 1995). For avoiding the development of fungicide resistance, it is necessary to alternate various fungicides during the spray program.

6.3.6 Phoma Disease of Fabaceae Vegetables

Fabaceae (pea family) is the second-largest flowering plant family with worldwide distribution. It represents more than 16,000 species in 750 genera. The member plants are commonly known as legumes or pulses. Legumes are second next to cereals in terms of their economic and nutritional importance in the human diet. Several Fabaceae are important vegetable plants, especially beans and peas. The beans (*Phaseolus* spp.) and peas (*Pisum* spp.) are primarily grown for their edible pods and seeds. They can suffer from several *Phoma* diseases that can affect the roots, leaves, stems, pods, or seeds.

Root and Foot Rot Complex

Introduction, Significance, Occurrence, and Host Range

Foot and root rot is extremely a common disease in most crops. In legumes, the foot rot complex is also one of the most important disease problems (Table 6.1). The disease is prevalent in all major bean- and pea-growing areas (Table 6.2).

Causal Agents

The most frequent pathogens include *Fusarium solani* f.sp. *pisi*, *Fusarium oxysporum*, *Aphanomyces euteiches*, *Peyronellaea pinodella* (basionym *Phoma pinodella*), *Mycosphaerella pinodes*, *Thielaviopsis basicola*, *Rhizoctonia solani*, and *Pythium* species. This extensive array of pathogens may lead to complex interactions.

Each pathogen has its definite biology and shows physiological specialization at species or even cultivar level. The pathogen *P. pinodella* was previously named *Ascochyta pinodella*. *Phoma pinodella* is morphologically similar to *P. medicaginis* that it was once regarded as a variety of this species (*P. medicaginis* var. *pinodella*). The variety was later elevated to species rank after careful observation, but the varietal name is still common (Fatehi et al. 2003; Taylor and Ford 2007). Aveskamp et al. (2010) included *P. pinodella* in the re-instated genus *Peyronellaea*. The fungus was found predominantly on the roots of field beans and peas (Oyarzun et al. 1993).

Symptoms and Diagnostic Features

The pathogens may rot seeds and cause damping-off of seedlings. Dark brown sunken lesions may develop on plant hypocotyls and taproots below the soil surface. As the lesions continue to develop, the stem base and roots may show dry necrosis, often accompanied by collapsed root tissues. The infection may extend from the taproots to the lateral roots and up to the base of stems at the cotyledons level, causing a reduction in shoot and root development. Affected plants may be stunted and unproductive and in some cases can wither and die.

Disease Cycle

Pathogens reside in the soil in various survival structures, spores, or mycelium (Table 6.2). Germinating bean seeds and roots release nutrients and exudates that stimulate pathogen activity and germinate chlamydospores, oospores, and spores. The disease can occur at a wide range of temperatures. Disease development is enhanced in excessively wet and poorly draining soils. Additionally, compacted soils, acid soils, low soil fertility, and frequent plantings of legumes may increase the problem.

Control

A minimum of 4-year crop rotation with non-host crops is essential to reduce soil inoculum. Bean cultivation should be avoided in fields having histories of severe foot rot. Soil drainage should be improved, and soil compaction problems need to be corrected. Irrigation is necessary, and excess water should not be applied to the fields. Seed treatment with fungicides is useful in reducing damping-off and improving plant establishment; however, they have little effect on foot rot.

Black Spot/Foot Rot Complex of Peas

Introduction, Significance, Occurrence, and Host Ranges

This is a complex disease, affecting pea and causing both root and foliar symptoms (Table 6.1). Leaf and pod spot is more commonly known as "black spot disease." Foot rot is a common problem in most pea crops, while vining pea is usually more susceptible to foot rot than other types of peas (Table 6.2). Foot rot is a more severe problem than leaf and pod disease. The disease can be responsible for a general depression in yield and lower monetary return from the crop. Yield losses caused by this disease have been estimated to be 10–60% each year (Bretag et al. 2006). Yield losses can be high in varieties grown for dry pea products. The disease is distributed worldwide and found in almost all major pea-growing areas (Bretag et al. 2006), while the severe occurrence has been known in temperate areas (Table 6.2).

Causal Agents

The disease is caused by *Phoma medicaginis* Malbr. & Roum. var. *pinodella* (Jones) Boerema (synonyms *Ascochyta pinodella* L. K. Jones, *Phoma pinodella* (L. K. Jones) Morgan-Jones & K. B. Burch). Several other pathogens may express either similar leaf and pod spot or foot rot symptoms on the pea. However, only one pathogen *Phoma medicaginis* var. *pinodella* can cause both foliar symptoms and potentially the more destructive foot rot symptoms that may include lesions at the stem and hypocotyl base. Greater damage usually occurs to the roots compared to less damage to leaves, stems, and pods.

Symptoms and Diagnostic Features

Seed death and poor plant emergence due to seed-borne infection reduce pea stands. Seedlings that emerge from infected seeds become infected, beginning at the point of seed attachment and extending up the stem and down the taproot. Infected plants grow poorly and become stunted that results in fields having patches of uneven plants. Severe infection can kill or stunt young pea plants. As plants progress to maturity, infection leads to senescence of all lower leaves and gives a blackening appearance at the stem base and roots. Severe foot rot infection may cause death or stunt of young plants or lodging of a mature crop.

Leaf infection usually appears as many small irregularly shaped flecks that are purple or dark brown in color. In the presence of moist conditions, these lesions on leaves enlarge and turn brown to black spots that are generally round to oval in shape with distinct margins and a zonate appearance (concentric rings of alternating shades of brown). Affected leaves may be killed and dried out but remain attached to the plant. Lesions on the stem are elongate and purple in color, usually occurring at the point of attachment of an infected leaf. The lesions progressively extend and coalesce to completely girdle stems giving the base of the stem a blackening appearance. Flower infection is seen as small, pinpoint lesions on the flower parts causing the flower or immature pod to drop or limiting seed production in the infected pod. The infected seed may show shrinkage and dark-brown discoloration.

Disease Cycle

P. pinodella can survive on seed, pea debris, and soil for long periods (Table 6.2). Plants produced from infected seed may develop foot rot lesions, and a secondary spread can occur from infected seedlings to adjacent healthy plants under favorable conditions for infection. However, seed infection is usually less important than infected pea debris with *P. pinodella*. *P. pinodella* can survive as mycelium on infected pea debris or as sclerotia and chlamydospores in the soil. When peas are grown in fields, many plants are likely to be attacked by the pathogen and develop severe root rot. The disease can also occur when pycnidiospores produced on old pea trash are carried by rain splash into nearby plants. After primary infection, symptoms appear in 2–4 days. When sufficient moisture is available, pycnidia develop in new lesions and produce abundant pycnidiospores. Pycnidiospores are released during wet weather and rain-splashed onto the lower leaves of adjacent plants, resulting in a secondary infection

Control

The disease can be avoided by planting infection-free seeds. Production of seed in low rainfall areas reduces the risk of seed infection by the fungus. The seed-borne inoculum level can be reduced significantly without damaging the seed by immersing it in hot water for a short period. Where fungal infection is confirmed in the seed coat, seed treatment with captan and thiram is very effective (Table 6.3). These fungicides have limited penetration of seed as well as activity against internal fungal infection. Another problem with captan is that it is harmful to rhizobia and affects the nodulation of grain legumes. The benzimidazole group fungicides, especially benomyl and thiabendazole, effectively eradicate the seed-borne infection and are now used widely in most pea-growing regions (Maude et al. 1986). Practical disease control can also be achieved by crop rotation and the burial and destruction of infected pea debris. A 3–6-year crop rotation with non-legumes effectively minimizes losses from root rot (Davidson and Ramsey 2000; Bretag and Ramsey 2001).

Ascochyta Blights

Introduction, Significance, Occurrence, and Host Range

Ascochyta blights are of major production constraints and consistently affecting grain legume production worldwide. The diseases attack cool season food and feed legumes (pea, lentil, chickpea, cowpea, common bean, and faba bean) in all countries where they are cultivated, resulting in complete crop loss in severe cases (Table 6.2). In recent years, ascochyta blight incidence has significantly been increased on bean crops in northern Spain due to conditions favorable to these fungi: moderate temperature, high humidity, and crop management, including unusual crop rotation (Ferreira et al. 2016).

Causal Agent

Three diseases of peas, each caused by different *Ascochyta* fungi, were discovered in the late 1920s. These fungi, commonly referred to as the *Ascochyta* complex, often occur together and show similar morphological characteristics. The pathogen names comprising the original Ascochyta complex have later been changed. The ascochyta blight of pea (*Pisum sativum*) is caused by *Ascochyta pisi*, *Ascochyta pinodes* (teleomorph: *Mycosphaerella pinodes*), and *Phoma medicaginis* var. *pinodella*, formerly known as *Ascochyta pinodella*. The fungus *A. pisi* causes leaf and pod spot, while *A. pinodes* and *P. medicaginis* var. *pinodella* cause foot rot and identical symptoms on leaves, stems, pods, and seeds. These diseases occur together and are difficult to differentiate. *Mycosphaerella pinodes* produces dark brown, spherical perithecia on senescent tissues that measure 90–180μm in diameter and contain bitunicate asci. Ascospores are two-celled and 12–18 x 4–8μm in size. The asexual stage *A. pinodes* produces spherical pycnidia with hyaline, ellipsoid, two-celled conidia measuring 8–18 x 3–5μm. Conidia are exuded out of pycnidia in yellowish (buff-colored) masses. *P. medicaginis* var. *pinodella* also produces spherical pycnidia that exude hyaline conidia. The conidia of *P. medicaginis* var. *pinodella* are usually single-celled and slightly smaller than those of *M. pinodes*. In recent years, *Phoma koolunga* Davidson et al. sp. nov. has been recognized as an important element of the blight complex on field pea (*Pisum sativum* L.) in South Australia (Davidson et al. 2009). Based on ITS and *rpb2* sequences, a new combination was also proposed for this species as *Ascochyta koolunga* (Hou et al. 2020).

Symptoms and Diagnostic Features

Foliar symptoms caused by *M. pinodes* are difficult to differentiate from those caused by *P. medicaginis* var. *pinodella*. They cause lesions on leaves, stems, and pods. Lesions initially appear as small, pinpoint, irregularly shaped flecks that are dark brown in color. Lesions on leaves and pods enlarge, generally are round to oval in shape, and contain concentric rings of alternating shades of brown. Lesions on stems may become purple in color and can enlarge and girdle the stem. If conditions favor disease development, the lesions can coalesce and cause significant blighting of pea foliage. However, the leaf and pod spot pathogen *A. pisi* causes tan to light brown spots, with darker brown to red-brown borders.

Mycosphaerella pinodes and *P. medicaginis* var. *pinodella* can cause seedling damping-off and lesions on hypocotyls and roots if the seed were infested. Therefore, the planting of infested seed can result in poor stands. However, the foot rot pathogen *P. medicaginis* var. *pinodella* is known to cause much more severe symptoms on belowground parts than aboveground parts.

Disease Cycle

Seed infection is widespread and is the key source of primary inoculum (Bretag et al. 2006) (Table 6.2). Ascospores produced by *M. pinodes* are an additional source of primary inoculum that can be dispersed long distances by the air current. Chlamydospores of *M. pinodes* and *P. medicaginis* var. *pinodella* can survive in the soil for many years and serve as primary inoculum as well. Their conidia produced inside pycnidia are dispersed by rain splash within the canopy and are an important secondary inoculum source. The optimum temperature for disease development is about 20 °C, but infection can occur in temperatures ranging from 5 to 30 °C. Prolonged leaf wetness periods (minimum of 2 hours) are very crucial for spore germination and infection.

Control

Since the ascochyta blight pathogens can persist in the soil, pea should not be planted in fields with a known history of ascochyta blights. All three pathogens are seed transmitted. The seed that does not have significant levels of the pathogen should be used. Appropriate seed treatments can also contribute to the management of seedborne inoculum. Plants used as seed crops should be grown in dry areas to enhance the production of clean seeds. Foliar sprays with broad-spectrum organic fungicides, e.g., mancozeb (Warkentin et al. 2000), thiram, and chlorothalonil, are used successfully to control ascochyta blight (Table 6.3). Fungicides with systemic activity, particularly those in the benzimidazole group, e.g., benomyl, carbendazim, and thiabendazole, are also used to control ascochyta blight (Table 6.3). Spraying tebuconazole, boscalid, and iprodione, at the rate of 125 g/ha, carbendazim at the rate of 900 g/ha, and fludioxonil at the rate of 40 g/ha in a water volume of 700 L/ha are found effective against ascochyta blight (Liu et al. 2016). These fungicides have preventative and eradicative properties against the ascochyta blight fungi, coupled with good residual activity.

6.3.7 *Phoma* Disease of Asteraceae Vegetables

Asteraceae (aster family) is one of the largest angiosperm families. The family represents more than 23,000 herb, shrub, and tree species in 1620 genera distributed worldwide. Lettuce (*Lactuca sativa*) is a very popular fresh salad vegetable in the Asteraceae family and is consumed worldwide. Several important diseases attack lettuce. Disease caused by a *Phoma* pathogen is among the principal ones.

Phoma Basal Rot of Lettuce

Introduction, Significance, Occurrence, and Host Range

Phoma basal was first reported on greenhouse lettuce in the UK in 1990 (Table 6.1). The disease caused significant damage to romaine lettuce in California from 2000 to 2004 (Koike et al. 2006) (Table 6.2). Most lettuce types show some susceptibility to the disease, while romaine cultivars remain the most sensitive to damage.

Causal Agent

Phoma basal rot is caused by a soil-borne species *Phoma exigua*, the current synonym of what is now known as *Boeremia exigua* (Aveskamp et al. 2010). The fungus produces dark, ostiolate pycnidia on leaf spot but seldom in the crown rot phase. Conidia are hyaline, mostly single-celled, straight or slightly curved, ellipsoid or cylindrical conidia. The size of the conidia measures 5–10 x 2–4μm.

Symptoms and Diagnostic Features

Early *Phoma* basal rot symptoms on romaine lettuce start with yellowing and wilting of lower leaves (Koike et al. 2006). Infected plants may exhibit an uneven appearance because one side of the foliage normally grows while the diseased side appears stunted. As plants mature, the entire plant becomes stunted and eventually wilts and collapses. Examination of affected plants reveals distinct dark brown to black sunken cavities on the crown and upper taproot. The cavities extend deep into the crowns and roots, resulting in the weakening of the plant. Such plants may break at a ground level easily. The black cavities are characteristically firm and do not show fungal growth, sclerotia, or spores. In greenhouse-grown lettuce in Europe, leaves show additional symptoms that consist of circular, dark gray to black spots expanding up to 3 cm in diameter. The inner leaf spot tissue dries out and falls out, giving a shot-hole appearance. Black dot-like pycnidia can be observed in leaf spot tissues. These leaf symptoms have not been detected in California.

Disease Cycle

The pathogen can persist in soil. Laboratory tests indicate that subspecies of *Boeremia exigua* exist, some of which are weak and go to plant stems. The epidemiology of this disease has not yet been documented. For greenhouse-grown lettuce in the UK, the disease is more common during the winter months.

Control

Planting of susceptible romaine cultivars should be avoided in fields having histories of this problem. After a severe outbreak, growers may need to treat the soil with steam or dazomet (Basamid) (Koike et al. 2006). Application of protectant fungicides after thinning the crop may reduce *Phoma* basal rot problems. Fungicide should be applied before plants become too large, and direct applications to the young plant base are required. Greenhouse growers in the UK manage *Phoma* basal rot by avoiding prolonged periods of leaf wetness in the greenhouse, improving soil drainage, and applying sprays of prochloraz (Scotts Octave), pyraclostrobin + boscalid (Signum), or azoxystrobin (Amistar) (O'Neill et al. 2000). In the USA, azoxystrobin, boscalid, and cyprodinil + fludioxonil fungicides effectively control the disease in the field (Koike et al. 2006) (Table 6.3).

6.3.8 Phoma Diseases of Solanaceae Vegetables

Solanaceae (nightshade family) takes a very significant position among cultivated species. Many widely consumed and high-value vegetables such as potato, pepper, tomato, chili, and eggplants are members of Solanaceae. These vegetables play a crucial role in human nutrition and the economy of the growing countries. Diseases have a serious and even destructive impact on these vegetables. The production of some of these crop plants can be limited in the field and storage by *Phoma* spp.

Phoma Blight/Phoma Rot

Introduction, Significance, Occurrence, and Host Range

Phoma blight has been an important tomato disease, although the sporadic problem is reported in some countries such as the USA and India (Jones et al. 2014). In 2013, tomato plants cultivated in the Cameron Highlands, Pahang State, Malaysia, had extensive fruit, stem, and leaf spots (Rashid et al. 2016). Phoma was recently recorded, causing severe blight symptoms on greenhouse-grown tomatoes in Brazil (Colmán et al. 2018) (Table 6.2). In addition to Phoma blight in tomato, the pathogen also causes fruit rot and small black spots on the leaves and fruits of bell pepper (*Capsicum annuum*) and sunken leaf spot of chili.

Causal Agent

Phoma destructiva Plowr. (Syn. *Diplodina destructiva* [Plowr.] Petr.), a coelomycete fungus, is responsible for leaf spots, but mostly fruit rot (Phoma rot). The fungus was first described by Plowright (1881), infecting fruits of *Lycopersicon esculentum* in the UK. Later, the fungus was detected in similar hosts in other countries of Europe and in North and South America (De Gruyter et al. 2002; Boerema et al. 2004a, b; Colmán et al. 2018). The fungus produces olivaceous black, globose to irregular, ostiolate pycnidia of 60–150 × 70–180μm. Conidia are hyaline, aseptate, subglobose to ellipsoidal, 4.5–8 × 2.5–3μm, mainly aseptate but occasionally larger 1-septate. Chlamydospores are absent. Based on differing conidial characteristics, De Gruyter et al. (2002) differentiated *P. destructiva* into two varieties, *destructiva* and *diversispora.* However, Valenzuela-Lopez et al. (2018) did not accept these varieties, as both varieties were found morphologically and phylogenetically very similar. Recently, *P. destructiva* has been revisited as *Remotididymella destructiva* (Hou et al. 2020).

Symptoms and Diagnostic Features

Initially, on fruits, lesions of limited size, slightly sunken, appear at the sites of peduncular scars, growth cracks, mechanical injuries, or insect feeding wounds. They gradually enlarge and become distinctive dark brown to the black sunken lesion with clearly defined perimeters. The lesions become zonate, where brown to black globular structures, pycnidia, can be observed. The fruits show internal rot that may extend to the seeds. The rot eventually covers a large proportion of ripe or unripe fruits.

Although the fungus is primarily known for causing post-harvest disease of tomato fruits, it also affects stems and leaves in the field. On leaves, the lesions are initially small (1–2 mm), irregular, and dark-colored on both the upper and lower leaf surfaces. As the lesions enlarge, they become necrotic, round to irregular (1 to 2 cm diameter) in shape, and slightly sunken. The tissue around the lesions may become yellow or necrotic. Concentric patterns are also visible and tiny brownish globular structures – the fungal fructifications (pycnidia). In severely infected plants, lesions may coalesce, resulting in typical blight symptoms and premature defoliation. Lesions on stems are long and dark with visible "zonate" (an appearance of concentric rings) patterns. Stem lesions may completely girdle the stems resulting in death. Numerous brown to black fruiting bodies, pycnidia, are scattered and embedded in the damaged tissues.

Disease Cycle

The pathogen is a seed-borne fungus that can affect the germination and final crop stand in the field (Table 6.2). It reduces tomato germination by 58% (Mehrotra and Agarwal 2003) and transmits from infected seeds to germinating seedlings and/or from seedlings to growing plants. Seed transmission of *Phoma* is particularly important in paprika and tomato, where infected seeds and seedlings are considered to be involved in the long-distance pathogen dissemination (Jones et al. 2014). The fungus also survives in the diseased plant debris in the soil. The fungus is a wound pathogen and infects through the wounds, peduncular scar, and growth cracks. The mycelium of *P. destructiva* colonizes the fruits radially and produces pycnidia that are randomly distributed and embedded in the tissues. Conidia are released from pycnidia in wet periods, during or after rainfall or sprinkler irrigation, during splashing, or during cultural operations and cause secondary infections. Infection can also begin on older leaves and progress from the bottom-up infecting leaves. The disease is favored by wet weather and mild temperature.

Control

Seeds that have significant levels of infection should be avoided. Burial or burning of diseased crop debris may reduce inoculum in the soil. The insect should be controlled, and care should be taken during intercultural operations to avoid wounding plants. Cultural practices such as proper spacing and irrigation are essential to reduce humidity and leaf wetness. Tolerant tomato cultivars, if available, are recommended in areas where this disease is common. When conditions are highly favorable for disease development (temperatures of 29 °C to 32 °C and relative humidity of above 55%), fungicides may help reduce the disease severity. Applying sprays of Azoxystrobin + chlorothalonil (0.5 ml/liter +2 g/L), chlorothalonil (2 g/L), copper hydroxide (2 g/L), pyraclostrobin (1 g/L), and tebuconazole (1 ml/liter) is effective in controlling the disease (Shankar et al. 2014) (Table 6.3).

Gangrene

Introduction, Significance, Occurrence, and Host Range

Gangrene is an important dry rot of potato tubers in storage. The disease has a substantial impact on cold or temperate climates, such as in Northern Europe (Table 6.2). The use of machinery for harvesting, handling, and grading of potatoes may increase the incidence of the disease. Inadequate storage conditions may also cause increased incidence and severity of the disease. For seed potato production, this fungal group may be a significant cause of quality degradation.

Causal Agent

Gangrene is mainly caused by the fungus *Phoma exigua* Desm. var. *foveata* (Foister) Boerema or by the species *P. exigua* Desm. var. *exigua* (Syn. *Phoma solanicola* Prill. & Delacr.), the current preferred name of what are *Boeremia foveata* and *Boeremia exigua* var. *exigua*, respectively (Aveskamp et al. 2010). Among the two, *Boeremia foveata* (Syn. *P. exigua* var. *foveata*) is the most aggressive type and poses a great risk to the potato industry (De Gruyter 2012).

Symptoms and Diagnostic Features

Both pathogens incite disease on stems and tubers of the potato. However, attacks on growing crops often go undetected. No distinct symptoms are noticed, except black dot-like pycnidia may appear at the stem base at the end of the growing season or even later, after haulm destruction. The first symptoms are observed on tubers as small dark, depressed areas frequently found around the wounds, the eyes, the lenticels, or the heel end. These grow to resemble thumb mark spots. When the tuber is cut across the lesion, the infected flesh is seen as very dark brown with a sharp border against the healthy tissues. White or yellow mycelial growth can be seen in the inner tissues of the cavities. Clusters of small black pycnidia may develop in the lining of the cavities.

Disease Cycle

Gangrene is a soil-borne disease (Table 6.2). The pathogens survive in the soil as saprophytes. There have been no reports that *Boeremia exigua* varieties have been isolated from the air. Usually, *Phoma* spp. form only a small proportion of the total air-flora fungi. Infected soil adhering to the tuber acts as a propagator of the fungus. Storage areas and equipment can also be sources of contamination. Gangrene is encouraged by damage at or after harvest and by storing, handling, and moving tubers in cold conditions. Infection is initiated by *P. exigua* thriving in the wounds. Inoculum infects the tuber surface through wounds made during harvest or subsequent handling of tubers. Low temperatures, immediately after harvesting or packing, slow the healing of tuber wounds and encourage the development of *Phoma* spp. Moderate storage temperatures favor tuber rot. The pathogen can sometimes move internally into growing stems from infected seed tubers.

Control

Good farm hygiene needs to be maintained. Tolerant cultivars, if available, should be chosen.

Harvesting is desired within 3–4 weeks of haulm destruction when tuber skin is sufficiently mature. Care should be taken to minimize tuber wounds at harvest and during subsequent handling operations. Careful inspection is needed to eliminate infected tubers before storage. Adequate time and conditions (temperature and ventilation) should be allowed to favor the healing of tuber surfaces and wounds caused by handling. Appropriate long-term storage conditions should be maintained. In seed production only, chemical fungicides can be applied after harvesting on clean soil-free tubers. Fumigation of potatoes with 2-aminobutane controls gangrene. Use of 2-aminobutane for dipping or spraying of potato tubers at the rate of 2 liters t^{-1} effectively controls gangrene. The application of benzimidazole (MBC) fungicides, notably thiabendazole and/or the sterol demethylation inhibitor (DMI) imazalil, to tubers going into the store is shown to be effective in reducing gangrene. However, preplanting seed treatment does not consistently reduce disease in progeny tubers (Carnegie et al. 1998) (Table 6.3). Carnegie et al. (1998) demonstrated that fungicides including imazalil and the phenylpyrrole fenpiclonil effectively reduced gangrene when applied to tubers in store and that a mixture of thiabendazole and imazalil was more effective than imazalil alone (Table 6.3).

6.4 Conclusion

Vegetables are among the most important food crops. Vegetable production worldwide is constrained by diseases. Diseases cause great loss and hamper the efforts made to increase the yield of vegetables. Disease menace takes many different forms, depending on the vegetable crops and the pathogens. *Phoma*, a dreadful fungal genus, may affect vegetables at all stages of their development and processing, right from the plants in the nursery to the harvest, transit, or storage. Most of the *Phoma* species have a wide host range and geographic occurrence. Because the fungus survives for long periods in the absence of a host plant, minimizing the damages caused by *Phoma* diseases is challenging. The chemical-based strategies have been so far dominating for managing *Phoma* diseases in vegetables, but it has caused a serious imbalance in THE agro-ecosystem. Some problems like the emergence of fungicide-resistant *Phoma* isolates are being reported. A shift towards non-chemical strategies is likely to correct the imbalance.

Acknowledgments The authors express their deepest appreciation to Prof. Dr. Lindsey J. du Toit and Dr. Carrie Wohleb of Washington State University for sharing the images of the pink root disease of onion. We are also highly indebted to Prof. Dr. Thomas S. Isakeit of Texas A & M University for providing the images of gummy stem blight disease of watermelon transplant.

References

Agarwal PC, Singh B, Dev U, Rani I, Chand D, Khetarpal RK (2006) Seed-borne fungi detected in sugar beet seeds imported into India during last three decades. Online. Plant Health Progress doi:https://doi.org/10.1094/PHP-2006-1211-01-RS.

Agrios GN (2005) Plant Pathology. Fifth Edition. Academic Press

Ash G (2000) Blackleg of oilseed rape. 2000. The Plant Health Instructor. DOI: https://doi.org/10.1094/PHI-I-2000-1111-01

Aveskamp MM, de Gruyter J, Crous PW (2008) Biology and recent developments in the systematics of *Phoma*, a complex genus of major quarantine significance. Fungal Diversity 31:1-18.

Aveskamp MM, de Gruyter J, Woudenberg JHC, Verkley GJM, Crous PW (2010) Highlights of the Didymellaceae: A polyphasic approach to characterise *Phoma* and related pleosporalean genera. Stud Mycol 65:1-60.

Babadoost M (1990) Pink root disease. University of Illinois Extension RPD No. 932

Bayan L, Koulivand PH, Gorji A (2014) Garlic: a review of potential therapeutic effects. Avicenna J Phytomed 4:1–14.

Boerema GH (1997) Contributions towards a monograph of *Phoma* (Coelomycetes)-V subdivision of the genus in sections. Mycotaxon 64:321–333

Boerema GH, Bollen GJ (1975) Conidiogenesis and conidial septation as differentiating criteria between *Phoma* and *Ascochyta*. Persoonia 8(2): 111–444.

Boerema GH, De Gruyter J, Noordeloos ME, Hamers MEC (2004a) Phoma identification manual. Differentiation of specific and infraspecific taxa in culture. CABI Publishing, United Kingdom.

Boerema GH, de Gruyter J, Noordeloos ME, Hamers MEC (2004b) The methods used for differentiation and identification. In: Boerema GH, de Gruyter J, Noordeloos ME, Hamers MEC (Eds.), *Phoma* identification manual. Differentiation of Specific and infra-specific taxa in culture (pp. 14–18). Wallingford: CABI Publishing.

Bretag TW, Ramsey MD (2001) Foliar diseases caused by fungi: *Ascochyta* spp. In: Kraft JM, Pfleger FL (Eds.), Compendium of Pea Diseases and Pests, APS Press: The American Phytopathological Society, St. Paul, MN, pp. 24-28

Bretag TW, Keane PJ, Price TV (2006) The epidemiology and control of ascochyta blight in field peas: a review. Aust J Agr Res 57:883–902. doi: https://doi.org/10.1071/AR05222

Butt MS, Naz A, Sultan MT, Qayyum MMN (2013) Anti-oncogenic perspectives of spices/herbs: a comprehensive review. EXCLI J 12:1043–1065.

Carnegie SF, Cameron AM, Lindsay DA, Sharp E, Nevison IM (1998) The effect of treating seed potatoes with benzimidazole, imidazole and phenylpyrrole fungicides on control of rot and blemish diseases. Annals of Applied Biology 133: 343– 63.

Cerkauskas RF (1984) Parsnip petiole canker caused by *Phoma complanata*. Phytopathology 74: 829-830. (Abstr.)

Cerkauskas RF, McGarvey BD (1987) Chemical control of phoma canker of parsnip. Can J Plant Pathol 9: 274. (Abstr.)

Chen Q, Zhang K, Zhang GZ, Cai L (2015) A polyphasic approach to characterize two novel species of *Phoma* (Didymellaceae) from China. Phytotaxa 197:267–281

Cimmino A, Andolfi A, Berestetskiy A, Evidente A (2008) Production of phytotoxins by *Phoma exigua* var. *exigua*, a potential mycoherbicide against perennial thistles. Journal of Agricultural and Food Chemistry 56:6304-9.

Colmán AA, Alves JL, da Silva M, Barreto RW (2018) *Phoma destructiva* causing blight of tomato plants: a new fungal threat for tomato plantations in Brazil? Trop Plant Pathol 43:257-262

Davidson JA, Ramsey MD (2000) Pea yield decline syndrome in South Australia: the role of diseases and the impact of agronomic practices. Australian Journal of Agricultural Research 51:347– 54.

Davidson JA, Hartley D, Priest MK, Krysinska-Kaczmarek M, McKay A., Scott E.S. (2009). A new species of *Phoma* causes ascochyta blight symptoms on field peas (*Pisum sativum*) in South Australia. Mycologia 101:120-128

De Gruyter, J. (2012) Revised taxonomy of *Phoma* and allied genera. (PhD Thesis) Wageningen University, Wageningen, the Netherlands, 180 pp.

De Gruyter J, Boerema GH, van der Aa HA (2002) Contributions towards a monograph of *Phoma* (Coelomycetes)-VI-2. Section Phyllostictoides: outline of its taxa. Persoonia 18:1–53

Deb D, Khan A, Dey N (2020) Phoma diseases: Epidemiology and control. Plant Pathol 69: 1203-1217. https://doi.org/10.1111/ppa.13221

Derevnina L, Dagdas YF, De la Concepcion JC, Bialas A, Kellner R, Petre B, et al. (2016). Nine things to know about elicitins. New Phytol 212:888–895. doi: https://doi.org/10.1111/nph.14137

Dingley JM (1970) Records of fungi parasitic on plants in New Zealand 1966–68. New Zealand Journal of Agricultural Research 13:2:325-337

Dixon G.R. (1981) Pathogens of Umbelliferous Crops. In: Vegetable Crop Diseases. Palgrave Macmillan, London. https://doi.org/10.1007/978-1-349-03704-9_8

Entwistle AR (1990) Root diseases. In: Rabinowitch HD, Brewster JL (eds.). Onion and allied crops. Vol II. CRC Press, Boca Raton, FL. p. 103-154

Fatehi J, Bridge PD, E. Punithalingam (2003) Molecular relatedness within the "*Ascochyta pinodes*-complex". Mycopathologia 156: 317–327.

Feng L, Nie K, Jiang H, Fan W (2019) Effects of lutein supplementation in age-related macular degeneration. PLoS ONE 14(12): e0227048

Ferreira JJ, Campa A, Perez-Vega E (2016) Variation in the response to ascochyta blight in common bean germplasm. Eur J Plant Pathol 146:977–985

Fitt BDL, Brun H, Barbetti MJ, Rimmer SR (2006) Worldwide importance of phoma stem canker (*Leptosphaeria maculans* and *L. biglobosa*) on oilseed rape (*Brassica napus*). European Journal of Plant Pathology 114:3–15

Freeman JH, EJ McAvoy, Boyd NS, Ozores-Hampton M, Paret M, Wang Q, Miller CF, Noling JW, Martini X (2017) Cucurbit Production. In: Vallad GE, Smith HA, Dittmar PJ, Freeman JH (eds) Vegetable Production Handbook for Florida. IFAS Extension. University of Florida. http://edis.ifas.ufl.edu/pdf-files/cv/cv29200.pdf.

Gabrielson RL (1983) Blackleg disease of crucifers caused by *Leptosphaeria maculans* (*Phoma lingam*) and its control. Seed Science and Technology 11: 749–780.

Gates LF (1959) Further experiments on black-leg disease of sugar-beet seedlings. Annals of Applied Biology - AAB. 47 (3), pp. 502–510. https://doi.org/10.1111/j.1744-7348.1959.tb07282.x.

Giebel J, DopieraLa U (2004) Pathogenesis of potato gangrene caused by *Phoma exigua* var. *foveata*: II. Activities of some hydrolases and dehydrogenases. J Phytopathol 152:399–403

Gougeon L, Payette H, Morais J et al. (2016) Intakes of folate, vitamin B6 and B12 and risk of depression in community-dwelling older adults: the Quebec Longitudinal Study on Nutrition and Aging. Eur J Clin Nutr 70:380–385.

Hartz TK, Bogle CR, Bender DA, Avila FA (1989) Control of pink root disease in onion using solarization and fumigation. J Am Soc Hortic Sci 114:587-59.

Harveson RM (2006) Identifying and distinguishing seedling and root rot diseases of sugar beets. Online. Plant Health Progress doi:https://doi.org/10.1094/PHP-2006-0915-01-DG.

Hossain MM, Sultana F (2020) Application and mechanisms of plant growth promoting fungi (PGPF) for phytostimulation. In: Das SK (Ed) Organic Agriculture, ISBN 978-1-78984-669-0. IntechOpen, London, UK, https://doi.org/10.5772/intechopen.92338

Hossain MM, Sultana F, Kubota M, Koyama H, Hyakumachi M (2008) Systemic resistance to bacterial leaf speck pathogen in *Arabidopsis thaliana* induced by the culture filtrate of a plant growth-promoting fungus (PGPF) *Phoma* sp. GS8-1. Journal of General Plant Pathology 74 (3):213-221.

Hossain MM, Sultana F, Islam S (2017) Plant growth-promoting fungi (PGPF): Phytostimulation and induced systemic resistance. In: Singh D, Singh H, Prabha R (eds) Plant-Microbe Interactions in Agro-Ecological Perspectives, Volume 2: Microbial Interactions and Agro-Ecological Impacts. Singapore: Springer. pp. 135-191. DOI: https://doi.org/10.1007/978-981-10-6593-4

Hou LW, Groenewald JZ, Pfenning LH, Yarden O, Crous PW, Cai L (2020) The phoma-like dilemma. Studies in Mycology 96:309-396.

Huang J, Weinstein SJ, Yu K, Mannisto S, Albanes D (2018) Serum Beta Carotene and Overall and Cause-Specific Mortality. Circ Res 123:1339–1349.doi: https://doi.org/10.1161/CIRCRESAHA.118.313409

Jones JB, Zitter TA, Momol TM, Miller SA (2014) Compendium of tomato diseases and pests, 2nd edn. APS Press, St. Paul

Keinath AP (2011) From native plants in Central Europe to cultivated crops worldwide: The emergence of *Didymella bryoniae* as a cucurbit pathogen. Hort Science 46:532-535.

Keinath AP (2013) Diagnostic guide for gummy stem blight and black rot on cucurbits. Online. Plant Health Progress doi: https://doi.org/10.1094/PHP-2013-1024-01-DG.

Keinath AP, Farnham MW, Zitter TA (1995) Morphological, pathological and genetic differentiation of *Didymella bryoniae* and *Phoma* spp. isolated from cucurbits. Phytopathology 85:364-9.

Keinath AP, Fillippeli EL, Baccari GV, DuBose VB (2010) Susceptibility of 15 cultivars of watermelon, melon, squash, pumpkin, and bottle gourd to gummy stem blight, 2009. Plant Dis. Managem. Rep. 4: V120. doi:https://doi.org/10.1094/PDMR04.

Koenick LB, Vaghefi N, Knight NL, Du Toit LJ, Pethybridge SJ (2019) Genetic diversity and differentiation in *Phoma betae* populations on table beet in New York and Washington States. Plant Disease 103(10): 9–18.

Koike ST, Subbarao KV, Verkley GJM, Fogle D, O'Neill TM (2006) Phoma basal rot of romaine lettuce in California caused by *Phoma exigua*: Occurrence, characterization, and control. Plant Dis 90:1268-1275.

Koike ST, Gladders P, Paulus AO (2007) Vegetable Diseases. A Color Handbook. San Diego, CA, USA: Academic Press.

Kramer JH (2004) Anthocyanosides of *Vaccinium myrtillus* (bilberry) for night vision-a systematic review of placebo-controlled trials. Surv Ophthalmol 49:618.

Leach LD (1986) Seedling diseases. In: Whitney ED, Duffus JE (eds) Compendium of Beet Diseases and Insects. The American Phytopathological Society, St. Paul, MN. Pages 4-8

Lee CJ, Lee JT, Moon JS, Ha LI, Kim HD, Kim W, Cheon MG (2007) Effects of solar heating for control of pink root and other soil-borne diseases of onions. Plant Pathol 23:295–299.

Liu N, Xu S, Yao X, Zhang G, Mao W, Hu Q, Feng Z, Gong Y (2016) Studies on the control of Ascochyta blight in field peas (*Pisum sativum* L.) caused by *Ascochyta pinodes* in Zhejiang Province, China. Front. Microbiol 7:481.

Malathrakis NE, Vakalounakis DJ (1983) Resistance to benzimidazole fungicides in the gummy stem blight pathogen *Didymella bryoniae* on cucurbits. Plant Pathol. 32:395–399.

Mason-D'Croz D, Bogard J R, Sulser T B, Cenacchi N, Dunston S, Herrero M et al. (2019) Gaps between fruit and vegetable production, demand, and recommended consumption at global and national levels: an integrated modelling study. Lancet Planet Health 3 e318–29

Mathews L, Paret NSD, Freeman JH, Olson SM (2018) Integrated Management of White Mold on Vegetables in Florida. Gainesville, FL: IFAS Extension, University of Florida.

Maude RB (1966) Pea seed infection by *Mycosphaerella pinodes* and *Ascochyta pisi* and its control by seed soaks in thiram and captan suspensions. Annals of Applied Biology 57:193-200.

Maude RB, Bambridge JM (1985) Effects of seed treatment and storage on the incidence of *Phoma betae* and the viability of infected red beet seeds. Plant Pathology 34:435-437.

Maude RB, Humpherson-Jones FM, Shuring CG (1984) Treatments to control *Phoma* and *Alternaria* infections of brassica seeds. Plant Pathology 33: 525-535. doi:https://doi.org/10.1111/j.1365-3059.1984.tb02877.x

Maude RB, Bambridge JM, Spencer A (1986) Tests of fungicide seed treatments to eliminate seed-borne *Ascochyta pisi* (leaf and pod spot of peas). Annals of Applied Biology 108 (supplement Tests of Agrochemicals and Cultivars No. 7):70–71.

Mehrotra RS, Agarwal A (2003) Plant pathology, 2nd edn. Tata McGraw-Hill Publishing Company, New Delhi, India

Möbius N, Hertweck C (2009) Fungal phytotoxins as mediators of virulence. Current Opinion in Plant Biology 12:390-8.

Monte E, Bridge PD, Sutton BC (1991) An integrated approach to *Phoma systematics*. Mycopathologia 115:89–103.

Nischwitz C, Dhiman C (2012) UTAH Pest Fact Sheet. PLP017. Utah State University Extension and Utah Plant Pest Diagnostic Laboratory.

O'Neill TM, Bennison J, Gaze RH (2000) Pests and diseases of protected vegetables and mushrooms. In: DV Alford (ed) Pest and Disease Management Handbook. Blackwell Science, Oxford. pp. 317-373.

Oyarzun P, Gerlagh M, Hoogland AE (1993) Pathogenic fungi involved in root rot of peas in the Netherlands and their physiological specialization. Netherlands Journal of Plant Pathology 99:23–33.

Paret ML, Dufault NS, Newark M, Freeman JH (2018) Management of Gummy Stem Blight (Black Rot) on Cucurbits in Florida. UF/IFAS Extension. https://edis.ifas.ufl.edu/pp280

Piszczek J (1997) Protection of sugar beet seed plants against *Phoma betae* (Frank). Part I. Effect on seed health and quality. Plant Breeding and Seed Science 41:61-73

Plowright CB (1881) On the fungoid diseases of the tomato Gardeners' Chronicle, 16, pp. 620-622

Porter IJ, Merriman PR, Keane PJ (1989) Integrated control of pink root (*Pyrenochaeta terrestris*) of onions by dazomet and soil solarization. Aust. J. Agric. Res. 40:861-869

Rai M, Deshmukh P, Gade A, Ingle A, Kövics GJ, Irinyi L (2009) *Phoma saccardo*: Distribution, secondary metabolite production and biotechnological applications. Critical Reviews in Microbiology 35:182-196.

Rashid TS, Sijam K, Nasehi A, Kadir J (2016) Occurrence of Phoma blight caused by *Phoma destructiva* on tomato (*Solanum lycopersicum*) in Malaysia. Plant Disease 100:1241–1242.

Reddy PV, Patel R, White JF Jr (1998) Phylogenetic and developmental evidence supporting reclassification of cruciferous pathogens *Phoma lingam* and *Phoma wasabiae* in Plenodomus. Can J Bot 76:1916–1922

Rouxel T, Balesdent MH (2005) The stem canker (blackleg) fungus, *Leptosphaeria maculans*, enters the genomic era. Mol Plant Pathol 6:225–241.

Rouxel T, Mendes-Pereira E, Brun H, Balesdent MH (2004) Species complex of fungal phytopathogens: the *Leptosphaeria maculans*–L. biglobosa case study. In: Sharma AK, Sharma A (eds) Plant Genome: Biodiversity and Evolution, Vol. 2: Lower Groups. Enfield: Science Publishers Inc., pp. 33–75.

Saccardo PA (1880) Fungi Gallici ser. II. Michelia 2: 39–135.

Shankar R, Harsha S, Bhandary R (2014) A practical guide to identification and control of tomato diseases. P 24 Tropica Seeds Pvt Ltd|No 54, South End Road, 1st Floor, Nama Aurore Building, Basavangudi, Bangalore 560004, India

Somai BM, Dean RA, Farnham MW, Zitter TA, Keinath AP (2002) Internal transcribed spacer regions 1 and 2 and random amplified polymorphic DNA analysis of *Didymella bryoniae* and related *Phoma* species isolated from cucurbits. Phytopathology 92:997-1004.

Sultana F, Hossain MM, Kubota M, Hyakumachi M (2008) Elicitation of systemic resistance against the bacterial speck pathogen in *Arabidopsis thaliana* by culture filtrates of plant growth-promoting fungi. Canadian Journal of Plant Pathology 30 (2):196-205.

Sultana F, Hossain MM, Kubota M, Hyakumachi M (2009) Induction of systemic resistance in *Arabidopsis thaliana* in response to culture filtrate from a plant growth-promoting fungus, Phoma sp. GS8-3. Plant Biology 11 (1):97-104.

Sutton BC (1980) The Coelomycetes. Fungi imperfecti with pycnidia, acervuli and stromata. 1st edn. Commonwealth Mycological Institute, Kew, UK.

Tabenhaus JJ (1917) Pink root, a new disease of onions in Texas. Phytopathology 7:217 (abstr.).

Tang FY, Cho HJ, Pai MH, Chen YH (2009) Concomitant supplementation of lycopene and eicosapentaenoic acid inhibits the proliferation of human colon cancer cells. J Nutr Biochem. 20:426–434.

Taylor PWJ, Ford R (2007) Diagnostics, genetic diversity and pathogenic variation of ascochyta blight of cool season food and feed legumes. European Journal of Plant Pathology 119: 127–133.

Valenzuela-Lopez N, Cano-Lira JF, Guarro J, Sutton DA, Wiederhold N, Crous PW, Stchigel AM (2018)

Coelomycetous Dothideomycetes with emphasis on the families Cucurbitariaceae and Didymellaceae. Studies in Mycology 90:1-69.

Voigt K, Cozijnsen AJ, Kroymann J, Pöggeler S, Howlett BJ (2005) Phylogenetic relationships between members of the crucifer pathogenic *Leptosphaeria maculans* species complex as shown by their mating type (MAT1-2), actin, and β-tubulin sequences. Molecular Phylogenetics and Evolution 37: 541-557.

Warkentin TD, Xue AG, McAndrew DW (2000) Effect of mancozeb on the control of Mycosphaerella blight of field pea. Can. J. Plant Sci. 80:403–406.

West JS, Kharbanda PD, Barbetti MJ, Fitt BDL (2001) Epidemiology and management of *Leptosphaeria maculans* (phoma stem canker) on oilseed rape in Australia, Canada, and Europe. Plant Pathology 50:10-27.

Williams RH, Fitt BDL (1999) Differentiating A and B groups of *Leptosphaeria maculans* causal agent of stem canker (blackleg) of oilseed rape. Plant Pathol 48:161–175.

Phoma Diseases: Identification, Epidemiology, and Strategies for Management

7

Omar Atik

7.1 Introduction

Phoma is a varied genus of coelomycetous fungi. Almost 2000 *Phoma* species have been recorded (Boerema 2004). *Phoma* belongs to the order *Pleosporales* and is the most widespread and largest genus of this order, with around 3000 defined taxa. Of those described, almost 110 *phoma* species are pathogenic and present in a range of ecological places. Earlier, the imperfect *Phoma* genus was placed in the *Sphaeropsidaceae* family (order *Sphaeropsidales*). The family is not monophyletic and is described by dark-pigmented pycnidia that generally holds an ostiole. The genus is not monophyletic and is basically definited by hyaline conidia that are shaped by conidiogenous cells in separate, dark, and thin-walled pycnidia, not grouped on a common stroma (Wostemeyer 2002).

Phoma is polyphyletic genus with controversial taxonomy (Chen et al. 2015; Deb et al. 2020). *Phoma* genus contains many saprobic and parasitic species, and it includes many pathogens which affect the crop health (Hutchison et al. 1994; Sullivan and White 2000). *Phoma* species are abundant and are common inhabitants of soil (Aveskamp et al. 2008) and occasionally may affect the plants by causing root infections and diverse spot diseases. Additionally, *Phoma'* species have been reported as opportunistic aggressive pathogens for humans (Rai et al. 2014).

The taxonomy of the *Phoma* group is difficult where the subgenus contains more than 3000 diverse fungi (Monte et al. 1991), about half of them being related phytopathogens (Aveskamp et al. 2008). Actually, proof of identity of the species is not continuously forthright because of absence of enough and adequately perfect morphological features and a high degree of intraspecific colony difference among isolates in culture, even within species (Wostemeyer 2002). The common species *P. medicaginis* (Broscious and Kirby 1988) and causal agent of the disease "Black Leg" in Brassicaceae (*P. lingam* and to a lesser extent the unnamed *Phoma* anamorph of *Leptosphaeria biglobosa*) are among the most important pathogens (Gugel and Petrie 1992; Fitt et al. 2006). *Phoma lingam* was observed as the most significant pathogen of oilseed rape (Fitt et al. 2006).

Species of *Phoma* genus have been recorded as pathogens from plants (Irinyi et al. 2006), as soil-borne diseases (Wostemeyer 2002), and commonly as saprophytic species (Boerema 2004; Irinyi et al. 2006). Many *Phoma* species cause diseases to many crops, such as potatoes, beets, legumes, and citrus. The related species *P. lingam*, *P. siliquastrum*, *P. oleracea*, *P. brassicae*, *P. napobrassicae*, *P. wasabiae*, *P. nigrificans*, and *P. eupyrena* affect the crucifers and

O. Atik (✉)
Department of Plant Protection, Aleppo Center, General Commission for Scientific Agricultural Research, Aleppo, Syria

M. Rai et al. (eds.), *Phoma: Diversity, Taxonomy, Bioactivities, and Nanotechnology*,
https://doi.org/10.1007/978-3-030-81218-8_7

regularly seem in mixed infections (Wostemeyer 2002). The pathogens of this genus cause significant yield losses up to 25% (Fitt et al. 2006). *Phoma* attack numerous important plants causing serious yield losses diseases on vegetables, oilseed plants, forage crops and trees (Chen et al. 2015). In addition to the direct costs because of yield loss, moreover indirect losses are caused through preventing any introductions of probable pathogenic or quarantine relevant *Phoma* species, through application of extreme importation and exportation restrictions (Smith et al. 1992; Miric et al. 1999).

It is important to minimize the damage caused by *Phoma* spp. in order to keep profitability of farmers, the sustainability of the food and food processing, and the honesty of exports (Wostemeyer 2002). The integrated management of stem canker disease (caused by *Phoma*) includes cultivating of resistant varieties, applying proper crops rotation, debris removal, and usage of chemical control via chemical fungicides. Furthermore, Yan et al. (2020) reported that planting resistant cultivars and burying the infected debris could be the best method to manage the black stem disease of sunflower, while Lo et al. (2002) recorded that using healthy seedlings and inhibiting secondary infection by covering with polyethylene sheets might be active in the management of the black rot diseases of wasabi in the field. Seed-borne *P. ligulicola* disease can be controlled through regular use of foliar fungicides to reduce the intensity of the pathogen before harvesting of seed fields. Moreover, using fludioxonil and thiabendazole to treat the seeds also significantly minimized the severity of *P. ligulicola* disease and improved the germination ratio and survival of the seedlings (Pethybridge et al. 2005).

This chapter of the book aims to explain and discuss the importance of *Phoma* spp. diseases and to understand better about identification, epidemiology, and the best integrated management of the disease.

7.2 Identification of *Phoma* Diseases

Phoma genus is highly polyphyletic with uncertain species limits. *Phoma* is considered to be a filamentous fungal genus accomplished of developing pycnidia. *Phoma* spp. usually have high level of genetic variation although reproducing asexually (Khon 1995; Talhinhas et al. 2002). The classification methodology is functional, but it has its limitations (Rai et al. 2014). Many of *Phoma* spp. and other taxa have been identified according to the morphological, biological, and phylogenetic recognition (Harrington and Rizzo 1999; Irinyi et al. 2006).

Identification of *Phoma* is mostly dependent on growth characteristics in the laboratories, where several studies have anticipated the growth features in lab, simple physiology, and morphological and biochemical trials for fast identification of *Phoma* spp. pathogens (Aveskamp et al. 2008; Chen et al. 2015). Identification of *Phoma* strains is not easy and needs a lot of experience, and mistakes in morphological taxonomy of *Phoma* species might occur especially when isolates may not express all features of species (Bridge et al. 2003). Depending on the shape and size of pycnidia and pycnidiospores is enough to identify the *Phoma* strains up to species level. There are shared characters among *Phoma* sections which make the morphological identification sometimes not accurate (Aveskamp et al. 2009); thus, molecular DNA-based markers have been broadly used to define many *Phoma* species. However, the aforementioned methodology helps taxonomists to understand more about *Phoma* taxonomy which still require more focus from the pathologists' side (Rai et al. 2014).

The advanced biochemical and molecular studies are considered the most helpful tools for accurate identification of the fungal taxa (Harrington and Rizzo 1999; Taylor et al. 2000). Molecular markers were used in the 1990s for the identification of *Phoma*, and some isozyme analyses were used to separate morphologically undistinguishable *Phoma* species (Kövics and de Gruyter 1995; Irinyi et al. 2006). DNA polymorphisms that are based on variations in DNA sequences have three significant advantages over protein polymorphisms. Both morphological features and molecular tools are essential in the identification of *Phoma* species (Chen et al. 2000, 2002; Weekers et al. 2001; Oliverio et al. 2002; Irinyi et al. 2006).

Phoma genus usually faces taxonomy difficulties because of lack of morphological standards and undefined phylogenetic attractions. The genus is now related to three diverse teleomorph genera which exist in *Didymella*, in *Leptosphaeria*, or *infrequently* in *Pleospora* (Mostert et al. 2006). This difficulty in taxonomy makes differentiation of members of the genus *Phoma* so complicated (Yan et al. 2020). *Phoma* with its uncertain species limitations still remains a provocative genus, and it can support the fact that a multiple approach, including molecular tools, would help to understand the continuous and reliable identification of *Phoma* species (Rai et al. 2014). For example, the first reported infection of *Phoma macdonaldii* Boerema that caused the disease of the sunflower black stem was in China in 2008. Both morphological and molecular identification designated that the isolated pathogens were reported as *P. macdonaldii*. The phylogenetic examination that used ITS1, 5.8 s RNA, and ITS2 regions ensured validation of the identification (Yan et al. 2020). DNA-based methods can significantly support the classification of the fungal species (Shoemaker and Brun 2001; Bridge et al. 2003, 2004; Torres et al. 2005).

Molecular technique helped in differentiating two species of *P. lingam* (teleomorph: *L. maculans*), which could be separated through the morphological characterization approach; moreover, the weakly aggressive variant (the teleomorph) was defined as *L. biglobosa*, while the anamorph is unnamed in the *Phoma* genus (Shoemaker and Brun 2001; Voigt et al. 2005).

Mycosphaerella, *Didymella*, *Pleospora*, and *Leptosphaeria* are reported as the teleomorphs of *Phoma* genus (Boerema 1997). *Phoma* genus has close similarity with *Ascochyta* genus, and the related species of both genera were linked to the teleomorphs of *Didymella* which is considered as the main concern about taxonomy, and only simple differences in the incidence and origin of conidia such as septate conidia (Aveskamp et al. 2008) , conidiogenesis, and the conidial septation can contribute to discriminate between both of them. In *Ascochyta*, the conidia septation is an important stage of the holoblastic conidial progress, while it is a secondary process in *Phoma* (Aveskamp et al. 2010). Moreover, the pycnidia in *Phoma* may contain pseudo-Ascochyta type of septate conidia, while the true *Ascochyta* species practically continuously produces two or more celled conidia in the laboratory; this simply will help in classifying the species of these fungi (Boerema 1973; Aveskamp et al. 2010).

7.3 Epidemiology of *Phoma* Diseases

7.3.1 *Phoma* spp. Life Cycle

The infection by *Phoma* will happen after entering through the epidermis. Conidia and mycelium can spread by splash of water, wind, and misting to cause infection in the new plants; furthermore, both insects and birds can transfer this infection to healthy hosts (Rodriguez and Leath 1992; West et al. 2001; Van de Graaf et al. 2002; Castell-Miller et al. 2007). Zeilinger et al. (2015) reported that stomata are the common path of fungal entrance, where entering through the stomata has been detected for many pathogens like *P. macdonaldii*, *P. complanata*, *P. exigua* var. *foveata*, *P. exigua* var. *linicola*, *P. koolunga*, and *P. lingam* (Castell-Miller et al. 2007; Machowicz-Stefaniak et al. 2014).

There are several factors that increase the spread of disease caused by *Phoma* such as, ascospores that are produced on the infected stubble and root debris considered as the initial inoculum for the pathogen continuation. Moreover, diseased seeds are playing role as source of inoculum that should cause systemic infection for the merged seedlings (Jacobsen 2006; Deb et al. 2020). Diseased plant stubble is considered as the key source of infection by *P. koolunga*, *P. medicaginis* var. *pinodella*, and *P. macdonaldii*. In *P.*

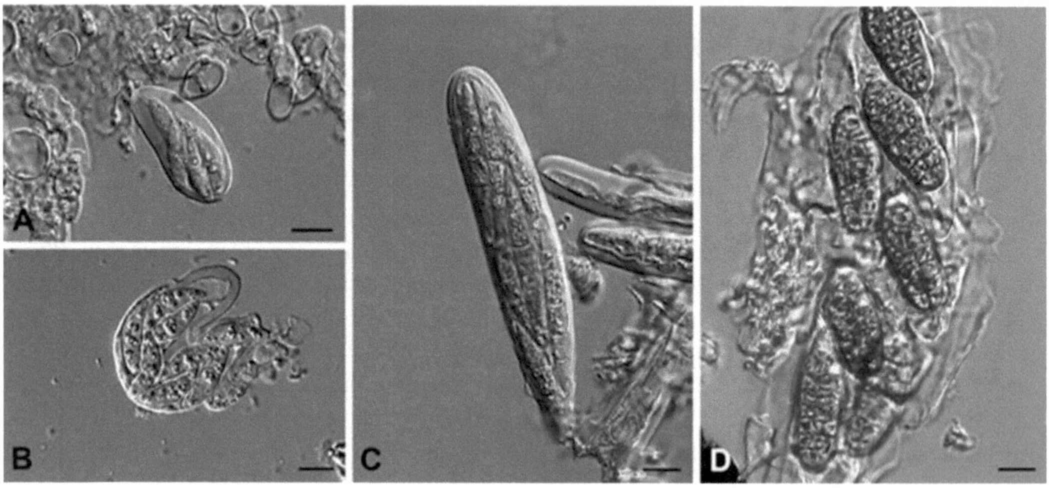

Fig. 7.1 (**a–d**) Teleomorphs of *Phoma*. Asci and ascospores of (**a**, **b**). *Didymella zeae-maydis* (anam. *P. zeae-maydis*). (**c**) *Leptosphaeria maculans* (anam. *P. lingam*). (**d**) *Pleospora herbarum*, lectotype of the genus *Pleospora*. Scale bars = 10 μm. Reproduced from Aveskamp et al. (2008) under Creative Commons Attribution 4.0 International licence

medicaginis var. *pinodella*, the chlamydospores, mycelia, and sclerotia that continue in soil for 1 year help the pathogen to survive. Similarly, pseudosclerotia help *P. ligulicola* and *P. koolunga* to survive in the soil for long time (Deb et al. 2020). The second source of infection is the pycnidia which enable *P. lycopersici* and *P. macdonaldii* to overwinter, while *P. lingam* and *P. macdonaldii* sustain as pseudothecia and mycelium created on diseased debris (Maric et al. 1981; Roustaee et al. 2000; Schwanck et al. 2016).

After penetration of the plant, *Phoma* produces mycelia, pycnidia, and pseudothecia. Ascospores (teleomorph of *Phoma*) and conidia (Fig. 7.1) may be dispersed by wind or rain splash. Kaczmarek and Jedryczka (2011) showed that the conidia that are produced by infected plant will be the secondary inoculum when these crops will be planted again in the same area and season.

Most of species remain as saprobes on decomposing organic material in the soil in case of no appropriate host plant. Few *Phoma* species produce conidiomata and chlamydospores which are the most suitable structures for long-standing (Fig. 7.2) (Liu et al. 2019); these material might cause the infection in recently planted crops, and new conidia can appear to enable forward dispersal (Aveskamp et al. 2008; Deb et al. 2020). It is reported that the alternate host plants give chances to *P. glomerata* and *P. medicaginis* var. *pinodella* to survive when the host crops are unavailable (Khani 2014). Pseudosclerotia of *P. koolunga* continue to be infective for 14 months, and new infection may happen in small planted crops in the next season (Khani et al. 2016).

7.4 *Phoma* spp. Growth and Favorable Conditions

Cool weather, high humidity, and rainy seasons are the suitable environmental conditions for developing the symptoms of *Phoma* diseases (Deb et al. 2020) and spread of the pathogen to the other parts (Leach and MacDonald 1976). As an example, rainfall and high humidity are the favored conditions for sugar beet root rot that is caused by *P. betae* (Leach and MacDonald 1976). Similarly, rain significantly affects inoculum dissemination of *P. tracheiphila*, and then temperature will help for disease development (Nigro et al. 2011; Descorps et al. 2012). Therefore, forecast for the ascospores that are released

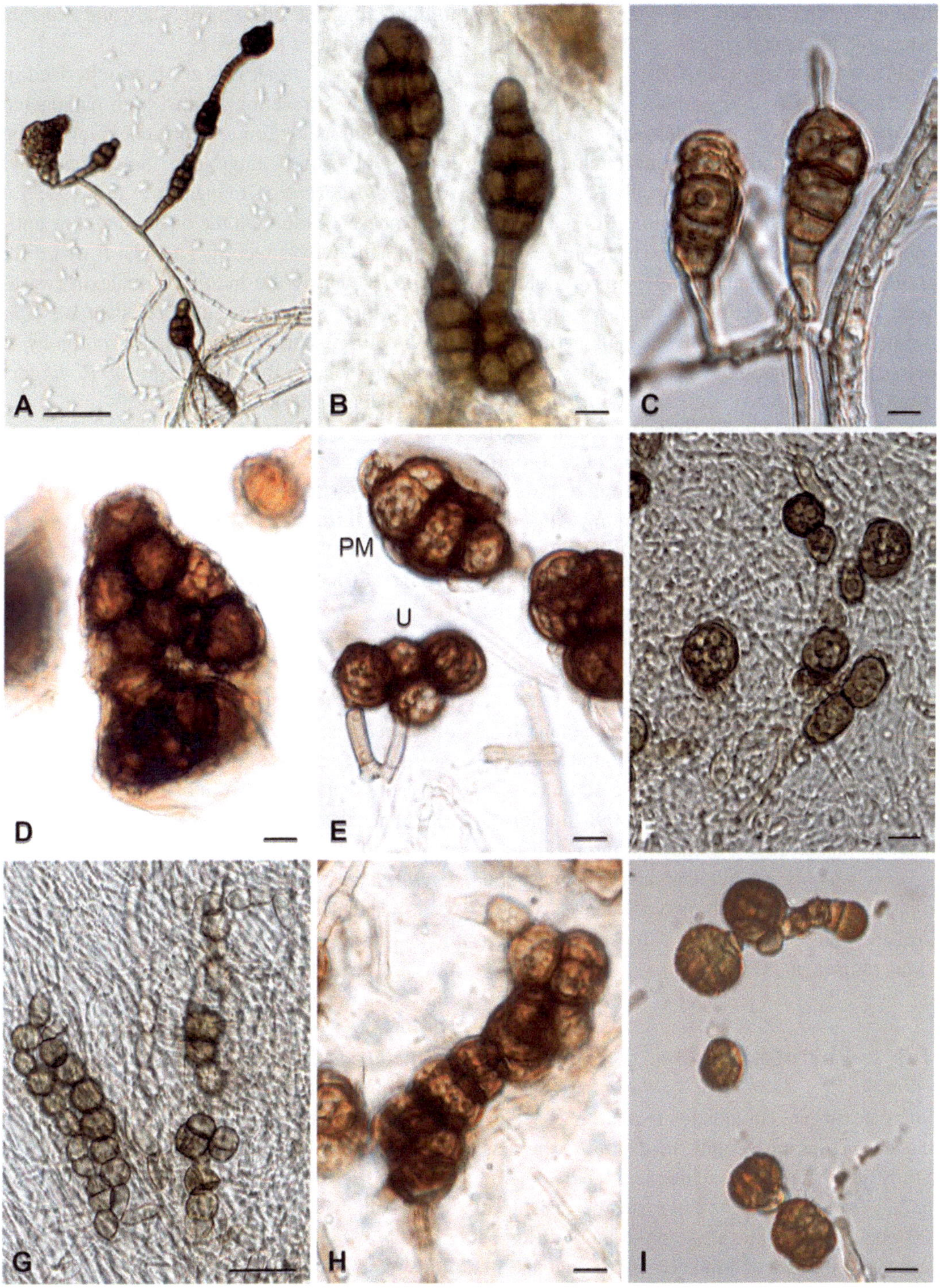

Fig. 7.2 (**a–i**) Chlamydospore morphology in *Phoma*. (**a**) Chain of chlamydospores of *P. glomerata*. (**b**) Alternaroid chlamydospores of *P. glomerata*. (**c**) Alternaroid chlamydospores of *P. jolyana*. (**d**) Botryoid chlamydospore as in *P. sorghina*. (**e**) Unicellular (u) and pseudosclerotoid multicellular chlamydospores (pm) in *P. chrysanthemicola*. (**f**) Unicellular chlamydospores of *P. pinodella*. (**g**) Chain of unicellular chlamydospores in *P. clematidina*. (**h**) Chain of unicellular chlamydospores in *P. chrysanthemicola*. (**i**) Epicoccoid chlamydospores in *P. epicoccina*. Scale bars: A = 100 μm, *G* = 50 μm, B–F and H–I = 10 μm. Reproduced from Aveskamp et al. (2008) under Creative Commons Attribution 4.0 International licence

during the rain periods can eventually help to determine the appropriate time for the fungicide application (Bordoy et al. 2018).

High humidity and periods of leaf wetness are the main drivers of enhancing most of the plant diseases caused by fungi, which affect many pathosystems at diverse phases of fungal growth. As a result, relative humidity measures could be employed as a forecaster for disease prediction (Davari and Hajieghrari 2008; Li et al. 2014). Cool weather is the favorable environmental condition for development of sugar beet seedling damping off disease which is resulted from seeds infected with *P. betae* (Leach and MacDonald 1976; Jacobsen 2006). *P. sclerotioides* favors low temperature to optimally grow on alfalfa, and a temperature of 0–27 °C can fit with the pathogen growth (Gray et al. 2004; Khani et al. 2017). The optimal temperature for *P. tracheiphila* growth is 25 °C, and for effective expansion of the symptom, the fungus needs a temperature degree of 20–22 °C (Smith et al. 1998; Liberato et al. 2006; Migheli et al. 2009).

Survival of soil-borne pathogens and their capability to infect the host crops in the next season depend mainly on temperature and moisture of the soil (Raaijmakers et al. 2009; Smolińska and Kowalska 2018). Potato gangrene disease severity was highest in areas having highest soil moisture content which is very necessary for growth and sustainability of the pathogen (Malcolmson and Gray 1968).

7.5 Phytotoxins

Phytoalexins are compounds that are released by the host crops in reaction to the fungal attack. Study on the biochemical interaction between the pathogenicity causes (toxins) and the resistance factors (phytoalexins) is providing different visions of the interaction between *Phoma* and its host (Rouxel and Balesdent 2005). Some *Phoma* species produce many phytotoxins such as phomalide, ascosonchine, deoxaphomin, phomalirazine, and herbarumin which are related to three chemical classes (El-Kady and Mostafa 1995; Evidente et al. 1996).

Coumarin phytoalexins are biosynthesized in sunflower as a reaction against both pathogenic and nonpathogenic fungi, which could inhibit the growth of some fungi like *Sclerotinia* and *Helminthosporium* (Urdangarin et al. 1999; Tal and Robeson 1986). On the other hand, low antifungal activity was recorded against *P. wasabiae* and *P. lingam* by the indole phytoalexin (wasalexin) which was produced by resistant *Wasabia japonica* in response to initial colonization by *L. maculans* where the anamorph is *P. lingam* (Pedras et al. 1999).

7.6 Management of *Phoma* Diseases

7.6.1 Cultural Practices

Many important methods of cultural practices can be conducted to reduce or eliminate the initial inoculum of *Phoma* pathogens and decrease the disease incidence and severity:

Phoma-free seed cultivation: Planting certified seeds that are free from black leg pathogen is a key management approach to prevent the outbreaks of disease since *Phoma lingam* and *Leptosphaeria biglobosa* can survive for many years on the infected seeds. It is recommended that untested seeds should not be planted and must be accompanied by official test indicating that the untreated seeds are free from pathogen (*P. lingam; L. maculans; L. biglobosa*) (West et al. 2001; Deb et al. 2020).

Crop rotation: A suitable crop rotation (with non-host plants for more than 4 years) should be implemented in areas that recorded the spread of *Phoma* diseases to prevent the pathogen from survival in the soil (Deb et al. 2020). *Phoma macdonaldii*, which causes *Phoma* black stem, overwinters in infected debris and spreads by splashing rain or insects. While no management strategy is completely effective, crop rotation will decrease the density of *Phoma* in the soil (High Plains Sunflower Handbook, Kansas State University 2005).

Debris removal: Burying of the crop stubble and eradication of decayed roots from the production fields can reduce occurrence of the disease (Jacobsen 2006; Davidson et al. 2011). Stubble management and good crop rotation minimize the likelihood of infection by ascospores that are released from colonized residue (West et al. 2001) and will prevent the rapid transformation to new healthy plants in areas where stubble residue still exists. Ideally, destroying the stem base debris would get rid of the continuing source of the pathogen inoculum (Johnson and Lewis 1994; West et al. 1999). In the recent studies, destroying the pathogen-infected remains is highly recommended for the control of the debris (Barbetti and Khangura 1999; Barbetti and Khangura 2000; West et al. 2001).

Removal of alternative host: Remains of the alternative volunteer plants should be removed, and crops rotation for more than 3 years without cultivations of cucurbit crops should be applied (Deb et al. 2020). In addition, avoid the rind injuries to cucurbit fruits since *P. cucurbitacearum* penetrates the host through wounds (Mathews et al. 2018). In order to control the black leg disease in *Brassica* vegetable crops, Aveskamp et al. (2008) advised not to cultivate the brassica, raphanus, or sinapis plants to be as cover crops or fodder fields near the cruciferous vegetables.

Cropping systems: Stem canker disease caused by *P. lingam* was controlled through the early sowing date and low application of the nitrogen fertilizers (West et al. 2001). The agriculture practices also are valid in controlling *P. macdonaldii* of sunflower, represented by planting the crop in low number of plants per square meter and low nitrogen fertilization under situation of lack of water (Seassau et al. 2012). To ensure disease escape and tolerance for the crown canker, early planting and low densities will help to avoid the infections at the most sensitive stage of the plant life cycle (Penaud et al. 1999; West et al. 2001). Aveskamp et al. (2008) reported many recommendations to decrease the *Phoma* disease severity through late summer planting to avoid the released spores from the infected debris and avoid storing crops in the fields that have recently recorded infections until infected residues have decayed and are no longer infectious. It is also important to provide enough space between plants to keep the leaves dry which will reduce the required time for infections.

Soil solarization: Soil solarization is a technique to kill pathogens via high temperatures. Soil solarization reduces the *P. lycopersici* infection that was recorded on carrot and tomato and was significantly effective in controlling *P. terrestris* (Katan et al. 1980; Mihajlović et al. 2017). The results of study by Lo et al. (2002) suggested that the management of the black rot disease of wasabi was effective by using pathogen-free plants and preventing secondary infection by covering with polyethylene sheets. As integrated control approach, when soil solarization was combined with dazomet (soil fumigant pesticide that inhibits development of fungal mycelium), the severity of pink rot was reduced and the yield increased by 100% (Molina-Vargas 2013).

Legislation: Legislation is very useful approach to protect fields and control for new entrance of infection of epidemiology of one of the severe diseases. Legislation was used in Canada in 1984 to control the *Phoma* stem canker disease when *L. maculans* species was identified as a pest officially, and the procedure for no sowing or transporting of the infected seeds was forbidden; thus, the farmers were not able to plant the oilseed rape for a period of 4 years in a land where the disease was recorded before (Stace-Smith et al. 1993). Even though the disease was reported in large scale in Canada, the percentage of spread was significantly decreased by these regulations (West et al. 2001).

Applying a fertilizer coated with fungicide: This approach was used through applying certain fertilizer that is coated with suitable fungicides, where this treated fertilizer was located near the seed in order to protect the young grown seedlings from being infected with the stem canker disease (Barbetti and Khangura 1999). This technique also could be applied in another location where severe stem canker disease happens (West et al. 2001).

Forecasting: Studies are conducted in France and the UK to decrease unnecessary fungicide use through optimizing the spraying to the time when *Phoma* pathogen is existing and the plant is susceptible (Fitt et al. 1997; Penaud et al. 1999; West et al. 1999).

7.7 Resistant Genotypes

Continuous efforts should be made to screen high number of host genotypes to determine the potential resistant species against *Phoma* diseases (Deb et al. 2020) and determine the level of resistance on the developed seedling and adult plants (Marcroft et al. 1999); recording the pathogenicity of *Phoma* populations would improve the choice of cultivars to be grown in a certain region (Balesdent and Rouxel 1998; Balesdent et al. 1998). In a field screening for the pigeon pea, MAL 11 and S.No.103 × MAL 23 were observed minimum *Phoma* stem canker disease incidence; these lines can be used in resistance breeding program against this disease caused by *Phoma cajani* (Pal et al. 2011). While the susceptibility or resistance to *Phoma* infection was different among the cultivars, scientists successfully identify the level of resistance (full or partial) among the varieties. The partial resistance in sunflower plants against seven *P. macdonaldii* isolates was carefully determined, and the quantitative trait loci (QTLs) which are responsible for the resistance were determined accordingly (Darvishzadeh et al. 2007; Kamphuis et al. 2008). Moreover, by studying the pea plants, appropriate sources of resistance against four isolates of *P. medicaginis* var. *pinodella* were also recognized (Ali et al. 1978). In other researches, clematis species were reported as highly resistant plants against infection by *P. clematidina*, and a method explained in this study can help the researchers to check the level of pathogen susceptibility towards new tested varieties (van de Graaf et al. 2001).

7.8 Chemical Control

Controlling of *Phoma* diseases generally based on the use of chemical fungicides, many fungicides are recommended and mostly benzimidazoles and thiophanate-methyl (Schmitz et al. 2006; Deb et al. 2020). Benzimidazole fungicide was recorded systemically as effective against the infection by the fungus *P. koolunga* and *P. exigua* var. *foveata*, while both thiabendazole and thiram fungicides reduced significantly the severity of Ascochyta blight disease on the plant seedlings (Khani et al. 2016). Foot rot disease was controlled effectively after the foliar spray of difenoconazole, prochloraz, and flusilazol plus tridemorph fungicides (Decognet et al. 1994). Chlorothalonil (protectant) and mancozeb (multisite protectant) fungicides were recorded as an effective fungicide against the fungus *P. cucurbitacearum* when they were applied in the early stages of the plant life cycle (Krieger 2001; Gullino et al. 2010). Furthermore, Dhyani et al. (1990) reported that thiram fungicide provided the best effective management against *P. destructiva* which is a seed-borne pathogen causing diseases of chili.

Treating the seeds with fungicides, soil or foliar fungicides were used to manage the *phoma* stem canker disease in many countries, where the methodology was selected based on the epidemiology of the disease and the importance of the crops (West et al. 2001). Spraying of foliar fungicides did not give significant control for the *Phoma* stem canker in Australia (Brown et al. 1976). In another study in Canada, the application of the foliar fungicide "propiconazole" did not display enough control of the *phoma* stem canker disease (Kharbanda et al. 1999), while in Europe, application of difenoconazole fungicide on the plant leaves was reported as significant controller of the crown canker diseases (Gladders et al. 1998).

Applying a mixture of more than two fungicides increased the antifungal activity and postponed the expression of resistance in the pathogen population (Cohen and Levy 1990; Gisi 1996). Periodical spraying of mixture of fungicides (captafol, prochloraz, dichlofluanid, chlorothalonil, and propiconazole) reduced the severity of the wilt disease caused by *P. clematidina* on the clematis plants (Smith and Cole 1991). Repeated application of copper and ziram fungicides during the rainy season – almost every three weeks – may decrease *phoma* disease severity, but these treatments are significantly costly (Solel et al. 1972; Smith et al. 1998).

It is important at the beginning of the growing season to protect the crops from the infection by *phoma* crown cankers since it is usually developed from the leaf spots (McGee and Petrie 1979; Hammond and Lewis 1986; Poisson and Peres 1999). Poisson and Peres (1999) exhibited that spraying of fungicides to control the leaf spot diseases may be late if applied on the young plants, since the fungus most probably will touch the stem before appearance of the leaf symptom. As a result, seed treatments could protect the young plants during this critical period, and there is no need for farmers to do foliar application of the fungicides (West et al. 2001).

Phoma stem canker of pigeon pea caused by *Phoma cajani* is one of the emerging diseases of the crop. Vitavax was found most effective fungicide to inhibit the mycelium growth in vitro (Pal et al. 2011).

7.9 Induced Systemic Resistance (ISR)

The induced systemic resistance (ISR) in plants is defined as signal-based defense mechanism that promotes the genetic expression of disease resistance as a reaction against bacteria, fungi, and viruses' infections. The resistance of *Phoma* diseases can be induced systemically by treatment with the rhizobacteria that promote the plant growth or pretreatment with weakened pathogen strains to reduce development of a virulent strain (Wei et al. 1991; Ramamoorthy et al. 2001). A previous study showed that seedlings of citrus that were inoculated with weak isolates of *P. tracheiphila* could considerably minimize the further infection by a virulent strain of this organism (Migheli et al. 2009).

7.10 Biological Control

Plant diseases are caused by bacterial and fungal pathogenic microorganisms. On the other hand, these organisms sometimes have many useful effects on crop productivity and as biocontrol agents. *Bacillus* as a biocontrol agent can produce antimicrobial compounds. Recently, the bacterium *Bacillus subtilis* (strain ALBA01) acted as a biocontrol agent against the fungus *P. terrestris* under both greenhouse and field conditions (Sayago et al. 2019), where the findings of this study noticeably indicated important biological protection expressed by *B. subtilis* against the soil-borne *P. terrestris* fungus. *P. terrestris* diseased plants inoculated with the bacterium showed usual development, particularly the photosynthetic activity; thus the strain ALBA01 of the *B. subtilis* bacteria may be employed to decrease the damage on the pink root (Sayago et al. 2019).

The development of the leaf spot disease caused by *P. medicaginis* on alfalfa was reduced significantly by the *Rhizophagus intraradices* fungus, decreasing the leaf spot incidence and disease index. The fungus colonization raises the alfalfa growth and induces the buildup of molecules that are involved in the plant defense mechanism (Li et al. 2019). Quinones, alkaloids, and saponins are important examples of secondary metabolites (endophytes) that function as powerful antimicrobial compounds (Istifadah et al. 2006; Ginting et al. 2013; Mousa et al. 2015; Gouda et al. 2016; Yuan et al. 2017; Baccari et al. 2018). *Pseudomonas* inhibited mal secco disease that is caused by *P. tracheiphila* (Coco et al. 2004). Antagonistic fungi are also employed in many cases to get rid of a plant pathogen, and several fungal parasites directly attack pathogens. *Beauveria bassiana* and *Metarhizium anisopliae* fungi were effective against *P. betae* in both greenhouse and field conditions (Heydari and Pessarakli 2010).

7.11 Conclusion

Phoma species are prevalent pathogens of different important plants worldwide. Classification and identification of *Phoma* spp., at the species level, are difficult depending on the morphological characterization in vitro. The recent identification is based on molecular markers.

Phoma spp. could cause infection in plants, animals, and humans, and they are widely reported as phytopathogens in varying conditions of temperature and humidity and geographical areas. The overwintering infected stubble, seeds, and pycnidia play important role in primary infection and spread of the disease, so debris removal is one of the most effective procedures to minimize the pathogenic inoculum of the fungi and reduce transferring the diseases to other non-infected plants and new clean areas.

Integrated disease management for the *Phoma* diseases should be applied in order to decrease the incidence and severity of the disease on the economically host crops. The package of controlling the diseases includes applying the best agricultural practices, cultivating medium- to high-resistant varieties, and planting disease-free seeds, while chemical control by spraying fungicides is considered mandatory in some severe cases.

References

Ali, S., Nitschke, L., Dube, A., Krause, M. and Cameron, B. (1978) Selection of pea lines for resistance to pathotypes of *Ascochyta pinodes, A. pisi* and *Phoma medicaginis* var. *pinodella, Australian Journal of Agricultural Research*, 29, 841–849.

Aveskamp, M., De Gruyter, J. and Crous, P. (2008) Biology and recent developments in the systematics of *Phoma*, a complex genus of major quarantine significance. *Fungal Diversity*, 31, 1–18.

Aveskamp, M.M., Verkley, G.J.M., de Gruyter, J., Murace, M.A., Perelló, A., Woudenberg, J.H.C. *et al.* (2009) DNA phylogeny reveals polyphyly of *Phoma* section Peyronellaea and multiple taxonomic novelties. *Mycologia*, 101, 363–382.

Aveskamp, M., De Gruyter, J., Woudenberg, J., Verkley, G. and Crous, P.W. (2010) Highlights of the Didymellaceae: a polyphasic approach to characterise *Phoma* and related pleosporalean genera. *Studies in Mycology*, 65, 1–60.

Baccari, C., Antonova, E. and Lindow, S. (2018) Biological control of Pierce's disease of grape by an endophytic bacterium. *Phytopathology*, 109, 248–256.

Balesdent M-H, Rouxel T, 1998. IMASCORE: une project de recherche europeÂen sur le phoma du colza. Les Rencontres Annuelles du CETIOM ± Colza 3 December, 30-1.

Balesdent M-H, Jedryczka M, Jain L, Mendes-Pereira E, Bertrandy J, Rouxel T, 1998. Conidia as a substrate for internal transcribed spacer-based PCR identification of members of the Leptosphaeria maculans species complex. Phytopathology 88, 1210-7.

Barbetti MJ, Khangura RK, 1999. Managing blackleg in the disease-prone environment of Western Australia. Proceedings of the 10th International Rapeseed Congress, 1999. Canberra, Australia. http://www.regional.org.au/papers/index.htm

Barbetti MJ, Khangura RK, 2000. Fungal Diseases of Canola in Western Australia. Perth, WA: Agriculture Western Australia, Bulletin no. 4406.

Boerema, G. (1973) The *Phoma* and *Ascochyta* species described by Wollenweber and Hochapfel in their study on fruit-rotting. *Studies in Mycology*, 3, 1–50.

Boerema, G. (1997) Contributions towards a monograph of *Phoma* (Coelomycetes). V. Subdivision of the genus in sections. *Mycotaxon*, 64, 321–333.

Boerema, G.H. (2004) *Phoma Identification Manual: Differentiation of Specific and Infra-specific Taxa in Culture*. Wallingford: CABI.

Bordoy, I. N., Quiroz, F. J., & Dosio, G. A. A. (2018). Increases in the source to sink ratio related to a higher carbohydrate concentration reduce phoma black stem in sunflower. European Journal of Plant Pathology, 150(3), 623-637.

Bridge, P.D., Roberts, P.J., Spooner, B.M. and Panchal, G. (2003). On the unreliability of published DNA sequences. New Phytologist 160: 43-48.

Bridge, P.D., Spooner, B.M. and Roberts, P.J. (2004). Reliability and use of published sequence data. New Phytologist 161: 15-17.

Broscious, S.C. and Kirby, H.W. (1988). Economic evaluation of fungicides for control of alfalfa foliar diseases. Phytopathology 78: 934-939.

Brown AGP, Barbetti MJ, Wood P, McR, 1976. Effect of benomyl on `blackleg' disease of rape in Western Australia. Australian Journal of Experimental Agriculture and Animal Husbandry 16, 276-9.

Castell-Miller, C., Zeyen, R. and Samac, D. (2007) Infection and development of *Phoma medicaginis* on moderately resistant and susceptible alfalfa genotypes. *Canadian Journal of Plant Pathology*, 29, 290–298.

Chen, C.A., Yu, J.K., Wei, N.W. (2000) Strategies for amplification by polymerase chain reaction of the complete sequence of nuclear large subunit ribosomal RNA-encoding gene in corals. Mar. Biotechnol. 6: 558-570

Chen, C.A., Wallace, C.C., Wolstenholme, J. (2002): Analysis of mitochondrial 12S RNA gene supports a two-clade hypothesis of the evolutionary history of scleractinian corals. Mol. Phylogenet. Evol. 23: 137-149.

Chen, Q., Jiang, J., Zhang, G., Cai, L. and Crous, P.W. (2015) Resolving the *Phoma* enigma. *Studies in Mycology*, 82, 137–217.

Coco, V., Grimałdi, V., Cirvilleri, G., Grasso, S., Catara, A. and Licciardello G. et al. Inhibition of *Phoma tracheiphila* by pseudomonads in citrus seedlings. In: El-Otmani, M. and Ait-Oubahou, A. (Eds.), *Proceedings of the International Citrus Congress* 2004. Riverside, CA: International Society of Citriculture, pp. 729–732.

Cohen, Y. and Levy, Y. (1990) Joint action of fungicides in mixtures: theory and practice. *Phytoparasitica*, 18, 159.

Darvishzadeh, R., Kiani, S.P., Dechamp-Guillaume, G., Gentzbittel, L. and Sarrafi, A. (2007) Quantitative trait loci associated with isolate specific and isolate non-specific partial resistance to *Phoma macdonaldii* in sunflower. *Plant Pathology*, 56, 855–8561.

Davari, M. and Hajieghrari, B. (2008) *Phoma negriana*, a new invasive pathogen for Moghan's vineyards in Iran. *African Journal of Biotechnology*, 7, 788–791.

Davidson, J.A., Krysinska-Kaczmarek, M., Wilmshurst, C.J., McKay, A. and Scott, E.S. (2011) Distribution and survival of ascochyta blight pathogens in field-pea-cropping soils of Australia. *Plant Disease*, 95, 1217–1223.

Deb D., Khan A., Dey N. *Phoma* diseases: Epidemiology and control. Plant Pathology 2020; 00:1–15.

Decognet, V., Cerceau, V. and Jouan, B. (1994) Control of *Phoma exigua* var. *linicola* on flax by seed and foliar spray treatments with fungicides. *Crop Protection*, 13, 105–108.

Descorps, C., Herbrard, C., Rakotonindraina, T., Dechamp-Guillaume, G., Mestries, E. and Aubertot, F. (2012) Advances in *Phoma macdonaldii* (*Leptosphaeria lindquistii*) epidemiology. In: *Proceedings of the 18th Sunflower Conference, 2012*. Paris: International Sunflower Association, pp. 199–204.

Dhyani, A., Sati, M. and Khulbe, R. (1990) Seed-borne fungi of chilli with special reference to pathogenicity and control of *Phoma destructiva. Indian Phytopathology*, 43, 110–113.

El-Kady, I. and Mostafa, M.E. (1995) Production of cytochalasins C, D and E from dematiaceous hyphomycetes. *Folia Microbiologica*, 40, 301–303.

Evidente, A., Capasso, R., Vurro, M. and Bottalico, A. (1996) Cytochalasin W, a new 24-oxa [14] cytochalasan from *Phoma exigua* var. *heteromorpha. Natural Toxins*, 4, 53–57.

Fitt BDL, Gladders P, Turner JA, Sutherland KG, Welham SJ, Davies JML, 1997. Prospects for developing a forecasting scheme to optimise use of fungicides for disease control on winter oilseed rape in the UK. Aspects of Applied Biology 48, 135–42.

Fitt, B.D., Brun, H., Barbetti, M. and Rimmer, S. (2006) World-wide importance of phoma stem canker (*Leptosphaeria maculans* and *L. biglobosa*) on oilseed rape (*Brassica napus*). In: Fitt, B.D.L., Evans, *Frontiers in Plant Science*, 6, 805.

Ginting, R.C.B., Sukarno, N., Widyastuti, U., Darusman, L.K. and Kanaya, S. (2013) Diversity of endophytic fungi from red ginger (*Zingiber officinale* Rosc.) plant and their inhibitory effect to *Fusarium oxysporum* plant pathogenic fungi. *HAYATI Journal of Biosciences*, 20, 127–137.

Gisi, U. (1996) Synergistic interaction of fungicides in mixtures. *Phytopathology*, 86, 1273–1279.

Gladders P, Symonds BV, Hardwick NV, Sansford CE, 1998. Opportunities to control canker (Leptosphaeria maculans) in winter oilseed rape by improved spray timing. International Organization for Biological Control Bulletin 21, 111–20.

Gouda, S., Das, G., Sen, S.K., Shin, H.-S. and Patra, J.K. (2016) Endophytes: a treasure house of bioactive compounds of medicinal importance. *Frontiers in Microbiology*, 7, 1538.

Gray, F., Hollingsworth, C., Groose, R., Reedy, C. and Larsen, R. (2004). Brown root rot resistance. *Phoma sclerotioides*. Available at: http://www.naaic.org/stdtests/brroo trot.pdf [Accessed 18 May 2020].

Gugel, R.K. Petrie, G.A. 1992. History, occurrence, impact, and control of blackleg of rape seed. Journal Canadian Volume 14, 1992 – Issue-1

Gullino, M.L., Tinivella, F., Garibaldi, A., Kemmitt, G.M., Bacci, L. and Sheppard, B. (2010) Mancozeb: past, present, and future. *Plant Disease*, 94, 1076–1087.

Hammond KE, Lewis BG, 1986. The timing and sequence of events leading to stem canker disease in populations of *Brassica napus var. oleifera* in the field. Plant Pathology 35, 551-64.

Harrington, T.C., and Rizzo, D.M. (1999): Defining species in the fungi. pp. 43-70. In: Structure and Dynamics of Fungal Populations. Worral, J.J. (Ed.) Kluwer Academic, Dordrecht

Heydari, A. and Pessarakli, M. (2010) A review on biological control of fungal plant pathogens using microbial antagonists. *Journal of Biological Sciences*, 10, 273–290.

High plains sunflower handbook, Kansas state university, 2005.

Hutchison, L.J., Chakravarty, P., Kawchuk, L.M. and Hiratsuka, Y. (1994). *Phoma etheridgei* sp. nov. from black galls and cankers of trembling aspen (*Populus tremuloides*) and its potential role as a bioprotectant against the aspen decay pathogen *Phellinus tremulae. Canadian Journal of Botany*, 72, 1424–1431.

Irinyi L, Kövics GJ, Rai MK, Sàndor E (2006) Studies of evolutionary relationships of Phoma species based on phylogenetic markers. pp. 99–113 In: 4th International Plant Protection Symposium at Debrecen University, 18–19 October 2006.

Istifadah, N., Saleeba, J.A. and McGee, P.A. (2006) Isolates of endophytic *Chaetomium* spp. inhibit the fungal pathogen *Pyrenophora tritici-repentis in vitro. Canadian Journal of Botany*, 84, 1148–1155.

Jacobsen, B.J. (2006) Root rot diseases of sugar beet. *Zbornik Matice Srpske za Prirodne Nauke*, 110, 9–19.

Johnson RD, Lewis BG, 1994. Variation in host range, systemic infection and epidemiology of Leptosphaeria maculans. Plant Pathology 43, 269-77.

Kaczmarek, J. and Jedryczka, M. (2011) Characterization of two coexisting pathogen populations of *Leptosphaeria* spp., the cause of stem canker of brassicas. *Acta Agrobotanica*, 64, 3–14.

Kamphuis, L.G., Lichtenzveig, J., Oliver, R.P. and Ellwood, S.R. (2008). Two alternative recessive quantitative trait loci influence resistance to spring black stem and leaf spot in *Medicago truncatula. BMC Plant Biology*, 8, 30.

Katan, J., Rotem, I., Finkel, Y. and Daniel, J. (1980) Solar heating of the soil for the control of pink root and other soilborne diseases in onions. *Phytoparasitica*, 8, 39–50.

Khani, M. (2014) *Aspects of Epidemiology of* Phoma koolunga *(Ascochyta Blight of Field Pea)*. PhD Thesis, Adelaide, University of Adelaide.

Khani, M., Davidson, J., Sosnowski, M. and Scott, E. (2016) Survival, transmission and control of *Phoma koolunga* in field pea seed and reaction of field pea genotypes to the pathogen. *Australasian Plant Pathology*, 45, 91–102.

Khani, M., Davidson, J. A., Sosnowski, M. R., & Scott, E. S. (2017). Atypical growth of Phoma koolunga on cultural media and on plants artificially inoculated in environmentally controlled conditions. Australasian Plant Pathology, 46(5), 511-514.

Kharbanda PD, Yang J, Beatty P, Jensen S, Tewari JP, 1999. Biocontrol of Leptosphaeria maculans and other pathogens Epidemiology of L. maculans on rapeseed 25 of canola with *Paenibacillus polymyxa* PKB1. Proceedings of the 10th International Rapeseed Congress, 1999. Canberra, Australia. http://www.regional.org.au/papers/index.htm

Khon, L.M. (1995): The clonal dynamic in wild and agricultural plant pathogen populations. Canadian Journal of Botany 73. 1231-1240.

Kövics, G.J. and Gruyter, J. de (1995): Comparable esterase isozyme analysis of some *Phoma* species occur on soybean. (A szóján el_forduló néhány *Phoma* faj észteráz izoenzim mintázatainak összehasonlító vizsgálata.) Proceedings of Debrecen Agricultural University (DATE Tudományos Közleményei) 31: 191-207.

Krieger, R. (2001) *Handbook of Pesticide Toxicology: Principles and Agents*. San Diego, CA: Academic Press.

Leach, L. and MacDonald, J. (1976) Seed-borne *Phoma betae* as influenced by area of sugar beet production, seed processing and fungicidal seed treatments. *Journal of the American Society of Sugar Beet Technologists*, 19, 4–15.

Li, Y., Uddin, W. and Kaminski, J. (2014) Effects of relative humidity on infection, colonization and conidiation of *Magnaporthe orzyae* on perennial ryegrass. *Plant Pathology*, 63, 590–597.

Li, Y., Duan, T., Nan, Z. and Li, Y. (2019) Arbuscular mycorrhizal fungus alleviates alfalfa leaf spots caused by *Phoma medicaginis* revealed by RNA-seq analysis. *Journal of Applied Microbiology*. https://doi.org/10.1111/jam.14387

Liberato, J., Cacciola, S., Di San, M. and Lio, G. (2006) Mal secco disease of citrus *Phoma tracheiphila*. Available at: http://www.padil.gov.au/pests -and-diseases/pest/main/136626 [Accessed 29 April 2020].

Liu, F., Bonthond, G., Groenewald, J., Cai, L. and Crous, P. (2019). Sporocadaceae, a family of coelomycetous fungi with appendage- bearing conidia. *Studies in Mycology*, 92, 287–415.

Lo, C.-T., Wang, K.-M., Hu, M.-F., Wang, C.-H. (2002) Integrated control of black rot disease of wasabi, caused by *Phoma wasabiae*. *Botanical Bulletin of Academia Sinica*, 43, 219–225.

Machowicz-Stefaniak, Z., Zalewska, E.D., Zimowska, B. and Król, E.D. (2014) Pathogenicity of *Phoma complanata* (Tode) Desm. Towards angelica (*Archangelica officinalis* Hoffm.). *Acta Scientiarum Polonorum Hortorum Cultus*, 13, 45–58.

Malcolmson, J.F. and Gray, E.G. (1968) Factors affecting the occurrence of gangrene (*Phoma exigua*) in potatoes. *Annals of Applied Biology*, 62, 77–87.

Marcroft SJ, Potter TD, Wratten N, Barbetti MJ, Khangura R, Salisbury PA, Burton WA, 1999. Alternative sources of resistance to Australian blackleg. Proceedings of the 10th International Rapeseed Congress, 1999. Canberra, Australia. http://www.regional.org.au/papers/index.htm

Maric, A.D., Masiveric, S. and Fayzalla, S. (1981) Presence of *Leptosphaeria lindquistii* Frezzi, last stade of *Phoma macdonaldii* Boerema, causal agent of black spot of sunflowers in Yugoslavia. *Zastija Bilja*, 32, 329–344.

Mathews, L., Paret, N.S.D., Freeman, J.H. and Olson, S.M. (2018). *Integrated Management of White Mold on Vegetables in Florida*. Gainesville, FL: IFAS Extension, University of Florida.

McGee DC, Petrie GA, 1979. Seasonal patterns of ascospore discharge by L. maculans in relation to blackleg of oilseed rape. Phytopathology 69, 586-9.

Migheli, Q., Cacciola, S.O., Balmas, V., Pane, A., Ezra, D. and Di San Lio, G.M. (2009) Mal secco disease caused by *Phoma tracheiphila*: a potential threat to lemon production worldwide. *Plant Disease*, 93, 852–867.

Mihajlović, M., Rekanović, E., Hrustić, J., Grahovac, M. and Tanović, B. (2017) Methods for management of soilborne plant pathogens. *Pesticidi i Fitomedicina*, 32, 9–24.

Miric, E., Aitken, E.A.B. and Goulter, K.C. (1999). Identification in Australia of the quarantine pathogen of sunflower *Phoma macdonaldii* (Teleomorph: *Leptosphaeria lindquistii*). Australian Journal of Agricultural Research 50: 325- 332.

Molina-Vargas, L.F. (2013) Mechanism of action of isothiocyanates. A review. *Agronomía Colombiana*, 31, 68–75.

Monte E, Bridge PD, Sutton BC (1991) An integrated approach to Phoma systematics. Mycopathologia 115:89–103

Mostert, L., Abeln, E.C.A., Halleen, F. and Crous, P.W. (2006). Genetic diversity among isolates of *Phaeomoniella chlamydospora* on grapevines. Australasian Plant Pathology 35: 453-460.

Mousa, W.K., Shearer, C.R., Limay-Rios, V., Zhou, T. and Raizada, M.N. (2015) Bacterial endophytes from wild maize suppress *Fusarium graminearum* in modern maize and inhibit mycotoxin accumulation. Front. Plant Sci. 6:805.

Nigro, F., Ippolito, A. and Salerno, M. (2011) Mal secco disease of citrus: a journey through a century of research. *Journal of Plant Pathology*, 93, 523–560.

Oliverio, M., Cervelli, M., Mariottini, P. (2002): ITS2 rRNA evolution and its congruence with the phylogeny of muricid neogastropods (Caenogastropoda, Muricoidea). Mol. Phylogenet. Evol. 25: 63-69.

Pal, D., P. Mishra, N. Sachan and A.K. Ghosh. 2011. Biological activities and medicinal properties of Cajanus cajan (L) Millsp. J. Adv. Pharm. Tech. Res. 2:207-214.

Pedras, M.S.C., Sorensen, J.L., Okanga, F.I. and Zaharia, I.L. (1999). Wasalexins A and B, new phytoalexins from wasabi: isolation, synthesis, and antifungal activity. *Bioorganic and Medicinal Chemistry Letters*, 9, 3015–3020.

Penaud A, Bernard C, Maisonneuve C, Peâreâs A, Pilorgeâ E, 1999. Decision rules for a chemical control of Leptosphaeria maculans. Proceedings of the 10th International Rapeseed Congress, 1999. Canberra, Australia. http://www.regional.org.au/papers/index.htm

Pethybridge, S. J., Esker, P., Hay, F., Wilson, C., and Nutter, F. W., Jr. 2005. Spatiotemporal description of epidemics caused by Phoma ligulicola in Tasmanian pyrethrum fields. Phytopathology 95:648-658.

Poisson B, Peres A, 1999. Study of rapeseed susceptibility to primary contamination of *Leptosphaeria maculans* in relation to plant vegetative stage. Proceedings of the 10th International Rapeseed Congress, 1999. Canberra, Australia. http://www.regional.org.au/papers/index.htm

Raaijmakers, J.M., Paulitz, T.C., Steinberg, C., Alabouvette, C. and Moënne-Loccoz, Y. (2009) The rhizosphere: a playground and battlefield for soilborne pathogens and beneficial microorganisms. *Plant and Soil*, 321, 341–361.

Rai, M.K., Tiwari, V.V., Irinyi, L. and Kövics, G.J. (2014) Advances in taxonomy of genus *Phoma*: polyphyletic nature and role of phenotypic traits and molecular systematics. *Indian Journal of Microbiology*, 54, 123–128.

Ramamoorthy, V., Viswanathan, R., Raguchander, T., Prakasam, V., & Samiyappan, R. (2001). Induction of systemic resistance by plant growth promoting rhizobacteria in crop plants against pests and diseases. Crop Protection, 20(1), 1-11.

Rodriguez, R. and Leath, K.T. (1992) Pathogenicity of *Phoma medicaginis* var. *medicaginis* to crowns of alfalfa. *Plant Disease*, 76, 1237–1240.

Roustaee, A., Dechamp-Guillaume, G., Gelie, B., Savy, C., Dargent, R. and Barrault, G. (2000) Ultrastructural studies of the mode of penetration by *Phoma macdonaldii* in sunflower seedlings. *Phytopathology*, 90, 915–920.

Rouxel, T. and Balesdent, M. (2005) The stem canker (blackleg) fungus, *Leptosphaeria maculans*, enters the genomic era. *Molecular Plant Pathology*, 6, 225–241.

Sayago, P., Juncosa, F., Orio, A.A., Luna, D.F., Molina, G., Lafi, J. and Ducasse, D.A.. (2019) *Bacillus subtilis* ALBA01 can mitigate onion pink root symptoms caused by *Setophoma terrestris*. *bioRxiv*, 601633

Schmitz, S., Zini, J., Etienne, M., Moreau, J.-M., Chandelier, A. and Cavelier, M. (2006) Effectiveness of thiophanate-methyl, trifloxystrobin and vinclozolin on canker caused by *Phoma exigua* Desm. on ash tree seedlings. *Biotechnologie, Agronomie, Société et Environnement*, 10, 25–31.

Schwanck, A.A., Savary, S., Debaeke, P., Vincourt, P. and Willocquet, L. (2016) Effects of plant morphological traits on phoma black stem in sunflower. *European Journal of Plant Pathology*, 145, 345–361.

Seassau, C., Dechamp-Guillaume, G., Mestries, E. and Debaeke, P. (2012). Low plant density can reduce sunflower premature ripening caused by *Phoma macdonaldii*. *European Journal of Agronomy*, 43, 185–193.

Shoemaker, R.A. and Brun, H. (2001). The teleomorph of the weakly aggressive segregate of *Leptosphaeria maculans*. Canadian Journal of Botany 79: 412-419.

Smith, G. and Cole, A. (1991) *Phoma clematidina*, causal agent of leafspot and wilt of clematis in New Zealand. *Australasian Plant Pathology*, 20, 67–72.

Smith, I.M., McNamara, D.G., Scott, P.R. and Harris, K.M. (1992). *Quarantine Pests for Europe. Data Sheets on Quarantine Pests for the European Communities and for the European and Mediterranean Plant Protection Organization*. CABI Publishing, United Kingdom & OEPP/EPPO, France.

Smith, I.M., Dunez, J., Phillips, D., Lelliott, R. and Archer, S. (1998). *European Handbook of Plant Diseases*. London: John Wiley & Sons.

Smolińska, U. and Kowalska, B. (2018) Biological control of the soil-borne fungal pathogen *Sclerotinia sclerotiorum* – a review. *Journal of Plant Pathology*, 100, 1–12.

Solel, Z., Pinkas, J. and Loebenstein, G. (1972) Evaluation of systemic fungicides and mineral oil adjuvants for the control of mal secco disease of lemon plants. *Phytopathology*, 62, 1007–1013.

Stace-Smith R, Bowler G, MacKenzie DJ, Ellis P, 1993. Monoclonal antibodies differentiate the weakly virulent from highly virulent strain of Leptosphaeria maculans, the organism causing blackleg of canola. Canadian Journal of Plant Pathology 15, 127-33.

Sullivan, R.F. and White, J.F. (2000) *Phoma glomerata* as a mycoparasite of powdery mildew. *Applied and Environmental Microbiology*, 66, 425–427.

Tal, B. and Robeson, D.J. (1986) The metabolism of sunflower phytoalexins ayapin and scopoletin: plant–fungus interactions. *Plant Physiology*, 82, 167–172.

Talhinhas, P., Sreenivasaprasad, S., Neves-Martins, J. Oliveira, H. (2002): Genetic and morphological characterization of *Colletotrichum acutatum* causing anthracnose of lupins. Phytopathology 92: 986-996.

Taylor, J.W., Jacobson, D.J., Kroken, S., Kasuga, T., Geiser, D.M., Hibbett, D.S., Fisher, M.C. (2000): Phylogenetic species recognition and species concepts in fungi. Fungal Genet. Biol. 31: 21-32.

Torres, M.S., White Jr, J.F., Cazares, G., Bergen, M., Bischoff, J.F. and Sullivan, R.F. (2005). A new species and its phylogenetic placement in the *Didymella/Phoma* complex (*Phaeosphaeriaceae, Pleosporales*). Mycotaxon 93: 297-308.

Urdangarin, C., Regente, M., Jorrin, J. and De La Canal, L. (1999) Sunflower coumarin phytoalexins inhibit the growth of the virulent pathogen *Sclerotinia sclerotiorum. Journal of Phytopathology*, 147, 441–443.

van de Graaf, P., O'Neill, T.M., Chartier-Hollis, J.M. and Joseph, M.E. (2001) Susceptibility of clematis varieties and species to stem infection by *Phoma clematidina* as an indicator for resistance to wilt. *European Journal of Plant Pathology*, 107, 607–614.

van de Graaf P, Joseph ME, Chartier-Hollis JM, O'Neill TM (2002) Prepenetration stages in infection of Clematis by Phoma clematidina. Plant Pathol 51:331–337

Voigt, K., Cozijnsen, A.J., Kroymann, J., Pöggeler, S. and Howlett, B.J. (2005). Phylogenetic relationships between members of the crucifer pathogenic *Leptosphaeria maculans* species complex as shown by their mating type (*MAT1-2*), actin, and β-tubulin sequences. Molecular Phylogenetics and Evolution 37: 541-557.

Weekers, P.H.H., Jonckheere, F.J. de, Dumont, H.J. (2001): Phylogenetic relationships inferred from ribosomal ITS sequences and biogeographic patterns in representative of the genus *Calopteryx* (Insecta: Odonata) of the West Mediterranean and adjacent west European zone. Mol. Phylogenet. Evol. 20**:** 89-99.

Wei, G., Kloepper, J.W. and Tuzun, S. (1991) Induction of systemic resistance of cucumber to *Colletotrichum orbiculare* by select strains of plant growth-promoting rhizobacteria. *Phytopathology*, 81, 1508–1512.

West JS, Biddulph JE, Fitt BDL, Gladders P, 1999. Epidemiology of Leptosphaeria maculans in relation to forecasting stem canker severity on winter oilseed rape in the UK. Annals of Applied Biology 135, 535-46.

West, J.S., Kharbanda, P., Barbetti, M. and Fitt, B.D. (2001) Epidemiology and management of *Leptosphaeria maculans* (phoma stem canker) on oilseed rape in Australia, Canada and Europe. *Plant Pathology*, 50, 10–27.

Wostemeyer J., 2002. Disease Management of Phoma Infections. Agricultural Applications, 2nd Edition. The Mycota XI. F. Kempken (Ed.) © Springer-Verlag Berlin Heidelberg 2013

Ningning Yan, Ren Na, Ruifang Jia, Jian Zhang and Jun Zhao. 2020. Occurrence of black stem on Helianthus annuus caused by Phoma macdonaldii and resistant evaluation of different sunflower varieties. OCL 2020, 27, 12.

Yuan, Y., Feng, H., Wang, L., Li, Z., Shi, Y., Zhao, L.H. *et al.* (2017) Potential of endophytic fungi isolated from cotton roots for biological control against verticillium wilt disease. *PLoS ONE*, 12, e0170557.

Zeilinger, S., Gupta, V.K., Dahms, T.E.S., Silva, R.N., Singh, H.B., Upadhyay, R.S. *et al.* (2015) Friends or foes? Emerging insights from fungal interactions with plants. *FEMS Microbiology Reviews*, 40, 182–207.

8 Major Plant Diseases Caused by *Phoma sensu lato* Species and Their Integrated Management Strategies

Fatemeh Salmaninezhad, Reza Mostowfizadeh-Ghalamfarsa, and Santa Olga Maria Cacciola

8.1 Introduction

The genus *Phoma* is considered as one of the most widely distributed genera within the order *Pleosporales* (Boerema 2004). *Phoma* is a polyphyletic genus, and its unclear species boundaries have arisen several controversies for taxonomists (Chen et al. 2015). This genus is reported to be the largest fungal genus comprising more than 3000 taxa known so far. Of these taxa, more than 100 are well-known for being pathogenic and occupy diverse ecological niches (Chen et al. 2015).

Phoma species are cosmopolitan and ubiquitous. They are found in plant tissues, water resources, and soil. The species of this genus can be saprobe, antagonistic, human and animal pathogens, as well as severe plant pathogens (Deb et al. 2020). *Phoma* plant pathogenic species cause devastating disorders and diseases in economically important crops (Table 8.1) resulting in annually huge yield losses (Aveskamp et al. 2008). They have been also reported as pathogens of ornamental and native plant species (Deb et al. 2020). The symptoms caused by *Phoma* species may vary; however, the most common symptoms are leaf and stem spots, vascular disorders, wilting, blight, canker, root rot, and even fruit rot (Aveskamp et al. 2008; Deb et al. 2020). Stem diseases are the most destructive diseases caused by *Phoma* species resulting in extensive economic losses in both crop and ornamental plants (Aveskamp et al. 2008). Certain *Phoma* species may induce various sorts of symptoms on different organs, such as leaves, stem, etc.

Pycnidia, pseudothecia, sclerotia, or pseudosclerotia can serve as the primary inoculum of this genus. They survive on plant residues. The conidia forming in the pycnidia or freely on hyphae are also considered as the secondary inoculum (Deb et al. 2020). The genus is known to produce a wide range of phytotoxins inducing symptoms and, in severe infections, plant death (Furukawa et al. 2007). Recognizing the main disease symptoms caused by *Phoma* species and the appropriate management strategies to control the diseases they cause are of great significance to plant pathologists (Furukawa et al. 2007; Aveskamp et al. 2008; Deb et al. 2020). While various kinds of management strategies have been reported, controlling the diseases caused by *Phoma* species is still a great challenge (Sun et al. 2020).

In this chapter, the most important diseases caused by *Phoma sensu lato* species in economically important crops or plants of environmental importance are discussed. These diseases were

F. Salmaninezhad · R. Mostowfizadeh-Ghalamfarsa (✉)
Department of Plant Protection, School of Agriculture, Shiraz University, Shiraz, Iran
e-mail: mostofi@shirazu.ac.ir

S. O. M. Cacciola
Department of Agriculture, Food and Environment (Di3A), University of Catania, Catania, Italy

M. Rai et al. (eds.), *Phoma: Diversity, Taxonomy, Bioactivities, and Nanotechnology*,
https://doi.org/10.1007/978-3-030-81218-8_8

Table 8.1 Major plant disease caused by *Phoma sensu lato* species

Disease	Pathogenic *Phoma* spp.	Host	Distribution
Branch diseases			
Branch blight	*Phoma fungicola*	Alfalfa; almond; *Chaenomeles sinensis*; *Malus × micromalus*; olive; peach; pistachio; tea	Spain, USA, Tunisia
Shoot Necrosis	*Phoma fungicola; Phoma glomerata; Phoma herbarum; Phoma incompta*	Peach; olive	Croatia, Greece, Italy, Spain
Fruit diseases			
Black rot	*Phoma vitis; Phoma herbarum*; *Phoma negriana; Phoma pomorum; Phoma macdonaldii*; *Phoma destructiva*; *Phoma strasseri; Phoma cucurbitacearum*	*Cucurbitaceae;* grapevine; peppermint; *Physalis pubescens;* sunflower; tomato	Brazil, Canada, China, India, Iran, Italy, Poland, USA
Gangrene	*Phoma foveata; Phoma eupyrena; Phoma exigua; Phoma exigua* var. *exigua; Phoma exigua* var. *foveata; Phoma matteuccicola*	Ostrich fern; potato	Australia, Canada, China, Europe (especially Italy, Sweden, Netherlands, and UK), Iran, Japan, New Zealand, Northern Africa, South American Andes, Ukraine
Ripening	*Phoma cucurbitacearum; Phoma destructiva; Phoma exigua; Phoma lycopersici; Phoma olea*	Balsam pear; olive; sunflower; tomato	Worldwide
Leaf diseases			
Leaf blight	*Phoma adonidicola; Phoma exigua* var. *exigua; Phoma glomerata; P. koolunga; Phoma matteucciicola; Phoma medicaginis* var. *pinodella; Phoma multirostrata*	Common bean; lavender; peas; pistachio; sugar beet	Australia, India, Iran, Italy, Turkey
Leaf spot	*Phoma betae; Phoma clematidina; Phoma destructiva; Phoma dictamnicola; P. glomerata; Phoma hedericola; Phoma herbarum; P. medicaginis; Phoma sorghina*	Wide range	Worldwide
Web blotch	*Phoma arachidicola*	Alsike clover; berseem clover; hairy vetch; Israel sweet clover; peanut	Africa, Argentina, China, Russia, Spain, USA
Root and crown diseases			
Brown root rot	*Phoma sclerotioides*	Alfalfa; alsike clover; bird's foot trefoil; common clover; Kentucky bluegrass; pigweed; red root amaranth; red fescue; wheat; winter rye	Canada, USA
Pink root rot	*Phoma terrestris*	Corn; *Cucurbitaceae* (squash, watermelon, etc.); onion	Australia, Brazil, Egypt, France, Iran, Israel, Italy, Japan, Mauritius, South Korea, Spain, USA
Root and crown rot	*Phoma betae; Phoma chrysanthemicola; Phoma exigua; Phoma macrostoma; Phoma medicaginis* var. *medicaginis*	Alfalfa; *Asteraceae* (chrysanthemum, pyrethrum, etc.); lettuce; strawberry; sunflower; whitetop	Canada, India, USA

(continued)

Table 8.1 (continued)

Disease	Pathogenic *Phoma* spp.	Host	Distribution
Root canker	*Phoma complanata*	Carrot; cow-parsnip; parsnip; Parsley	Poland, UK, USA
Seedling diseases			
Pre-emergence seedling damping-off	*Phoma betae; Phoma medicaginis; Phoma megasperma* f. sp. *medicaginis*	Alfalfa; barrel clover; sugar beet	Canada, eastern Europe, New Zealand, UK, and USA
Premature Ripening	*Phoma macdonaldii*	Sunflower	France
Stem diseases			
Blackleg	*Phoma lingam; Phoma multirostrata; Phoma wasabiae*	*Ajuga multiflora; Brassicaceae* (oilseed rape, turnips, wasabi, etc.); cabbage	Argentina, Australia, Canada, China, Europe, Iran, North America, Russia, South Korea
Black stem	*Phoma medicaginis* var. *medicaginis*; *Phoma macdonaldii; Phoma pinodella*	Alfalfa; fenugreek; sunflower	Africa, Australia, Canada, China, France, India, Italy, USA
Stem blight	*Phoma cucurbitacearum; Phoma destructiva; Phoma lingulicola*	*Cucurbitaceae*; papaya; pyrethrum; tomato	Australia, Brazil, India, South Africa, South Korea, Tasmania, USA, Turkey
Vascular diseases			
Mal secco	*Phoma tracheiphila*	*Citrus* spp.	Greece, Mediterranean region

categorized into sections based on their major symptoms on different plant organs or tissues. In each section, the pathogenic species, their hosts, the symptoms and the epidemiology of diseases they cause, as well as disease management strategies are reviewed.

8.2 Leaf Diseases

Leaf diseases are the most common diseases caused by *Phoma* species. The species infecting leaves also may express symptoms on other parts of the host. Leaf blight, leaf spot, and web blotch are the most important Phoma leaf diseases.

8.2.1 Leaf Blight

Leaf blight is reported to be one of the diseases caused by *Phoma* species worldwide. Although it is reported as leaf blight, other parts of the infected plants are usually involved. The disease has also been reported as spot blight disease and Ascochyta blight (Deb et al. 2020). Leaf blight has been mostly reported from Australia, India, Iran, Italy, and Turkey (Moral et al. 2018; Deb et al. 2020).

Symptoms: tiny black spots could be observed on the leaf surface and young shoots (Fig. 8.1b). Whole leaves and the host plant would become necrotic, with black necrotic lesions on the leaves. The symptoms could be especially observed in the leaves touching the forest ground as golden-brown color lesions which would later become gray and dry and finally fall down (Li et al. 2013). In severe infections, photosynthesis would be impaired resulting in plant death. In general, both leaves and stem would be infected (Li et al. 2013).

Phoma-related species and disease development: a number of *Phoma* species are reported to be the main causal agents of leaf blight, including *P. exigua* var. *exigua* Desm (syn. *Boeremia exigua* var. *exigua*); *P. glomerata* (Corda) Wollenw. & Hochapfel (syn. *Didymella glomerata* (Corda) Qian Chen & L. Cai); *P. matteucciicola* Aderkas, Gruyter, Noordel. & Strongman; *P. adonidicola* Yan Li, Q. Wang & G.Z. Lu; *P. multirostrata* (P.N. Mathur, S.K. Menon & Thirum.) Dorenb. & Boerema (syn. *Ectophoma multirostrata* (P.N. Mathur, S.K. Menon & Thirum.) Valenz.-Lopez, Cano, Crous, Guarro & Stchigel); *P. medicaginis* var. *pinodella* (L.K. Jones) Boerema (syn. *D. pinodella* (L.K. Jones) Qian Chen & L. Cai); and *P. koolunga* J.A. Davidson, D. Hartley, Priest, S. Kaczm., Herdina, A. McKay

Fig. 8.1 Types of Phoma symptoms on various hosts: (**a**) **Black stem** (*Phoma macdonaldii*) on sunflower (Photograph by Haward F. Shwartz [Available at] https://www.forestryimages.org/browse/detail.cfm?imgnum=5361538). (**b**) **Leaf spot** (*P. destructiva*) on Kava (*Piper methysticum* G. Forst.) (Photograph by S. Nelson [Available at] https://www.flickr.com/photos/scotnelson/24286094601/in/photostream/). (**c**) **Mal secco** (*P. tracheiphila*) on mature lemon (Photograph by S. O. Cacciola). (**d**) **Blackleg (stem canker)** (*P. lingam*) of Kava (Photograph by S. Nelson [Available at] https://www.flickr.com/photos/scotnelson/5694863874/). (**e**) **Pink root rot** (*P. terrestris*) on onion (Photograph by D. B. Langston [Available at] https://www.forestryimages.org/browse/detail.cfm?imgnum=5077074). (**f**) **Gummy stem blight** (*P. cucurbitacearum*) on watermelon (Photograph by R. A. Melanson [Available at] https://www.forestryimages.org/browse/detail.cfm?imgnum=5551694). (**g**) **Leaf blight** (*P. medicaginis* var. *pinodella*) on pea (Photograph by B. Watt [Available at] https://www.forestryimages.org/browse/detail.cfm?imgnum=5530358). (**h**) **Fruit rot** (*Phoma* sp.) of papaya (Photograph by S. Nelson [Available at] https://www.flickr.com/photos/scotnelson/37322627960/in/photostream/). (**i**) **Branch blight (and blackleg)** (*P. lingam*) of cabbage (Photograph by R. Zheng [Available at] https://www.forestryimages.org/browse/detail.cfm?imgnum=5562120). (**j**) **Seedling damping-off and blight** (*Phoma* sp.) on periwinkle (*Vinca* spp. L.) (Photograph by B. Watt [Available at] https://www.forestryimages.org/browse/detail.cfm?imgnum=5507692)

& E.S. Scott (Deb et al. 2020). Moisture and low temperature (12–18 °C) play the most significant role in disease development. Moreover, calcium and iron content of the soil is also another important factor conditioning the disease severity. Soils with an excessive amount of calcium and iron have been reported to favor Phoma blight (Li et al. 2013). The disease initiates from the buds by mycelial germination. In favorable conditions, such as heavy rain, defoliation and bud death could be also observed (Moral et al. 2018). The pathogen could also overwinter as pseudothecia or pseudosclerotia on plant debris or in soil.

Hosts: leaf blight has been reported from legumes, especially common bean (*Phaseolus vulgaris* L.) and peas (*Pisum sativum* L.), pistachio (*Pistacia vera* L.), lavender (*Lavandula spica* L.), and sugar beet (*B. vulgaris* L.) (Aveskamp et al. 2008; Li et al. 2013; Moral et al. 2018; Deb et al. 2020).

Integrated management strategies: removal of plant residues could be the most effective approach to control the disease, because the causal agent survives on plant debris. Removal of weeds could also effectively control the disease, since weeds may be a source of inoculum (Deb et al. 2020). The application of fungicides belonging to the groups of demethylation inhibitors, succinate dehydrogenase inhibitors, and quinone outside inhibitors has been reported to be the most effective chemical approach to control the disease (Moral et al. 2018).

8.2.2 Leaf Spot

Leaf spot is considered as the most common diseases caused by *Phoma* species. Most *Phoma* species may cause leaf spot symptoms. This disease has also been reported as one of the most devastating diseases and leads to huge economic loss both in crops and ornamental plants worldwide (Deb et al. 2020).

Symptoms: depending on the infected host, various sorts of symptoms have been reported. Numerous dark brown lesions could be observed on leaves which might kill the entire lower to middle leaves in severe infections (Fan et al. 2018). Circular to angular or irregular spots may also appear on leaves' margins and are separated or aggregated. The color of these spots is brown at the margin and pale brown to gray-white in the center (Aveskamp et al. 2008). Symptoms can be observed initially at the tips of the leaves as oval to circular light brown water-soaked sunken lesions which could progressively turn into dark brown to black with concentric ring patterns. In severe infections, leaves might become twisted, dry, and broken (Fig. 8.1g) or just show burning on which dark brown pycnidia differentiate (Fan et al. 2018). Chlorosis and defoliation might also be noticed in severe infections (Chand et al. 2019).

Phoma-related species and disease development: even though most *Phoma* species cause leaf spot on a wide host range of plants, *P. betae* A.B. Frank (syn. *Neocamarosporium betae* (Berl.) Ariyaw. & K.D. Hyde; previously known as *Pleospora betae* (Berl.) Nevod.): *P. clematidina* (Thüm.) Boerema (syn. *Calophoma clematidina* (Thüm.) Qian Chen & L. Cai); *P. destructiva* Plowr. (syn. *Boeremia exigua* (Desm.) Aveskamp, Gruyter & Verkley; and syn. *Remotididymella destructiva* (Plowr.) Valenz.-Lopez, Cano, Crous, Guarro & Stchigel); *P. dictamnicola* Boerema, Gruyter & Noordel. (syn. *Heterophoma nobilis* (Kabát & Bubák) Qian Chen & L. Cai); *P. glomerata* (Corda) Wollenw. & Hochapfel (syn. *Didymella glomerata* (Corda) Qian Chen & L. Cai); *P. hedericola* (Durieu & Mont.) Boerema (syn. *B. hedericola* (Durieu & Mont.) Aveskamp, Gruyter & Verkley); *P. herbarum* Westland; *P. medicaginis* Malbr. & Roum. (syn. *Ascochyta medicaginicola* Qian Chen & L. Cai); *P. multirostrata* (syn. *Ectophoma multirostrata*); and *P. sorghina* (Sacc.) Boerema, Dorenb. & Kesteren (syn. *Epicoccum sorghinum* (Sacc.) Aveskamp, Gruyter & Verkley) are reported as the main cause of leaf spot disease (Aveskamp et al. 2008; Deb et al. 2020). Overwintering occurs on plant debris as pseudothecia, chlamydospores, sclerotia, or pycnidia. Pycnidiospores cause secondary infections (Gao et al. 2018; Deb et al. 2020). The disease can cause severe damage in humid conditions (Deb et al. 2020).

Hosts: numerous hosts have been reported worldwide, including among others *Ficus elastica* Roxb. Ex Hornem, alfalfa (*Medicago sativa* L.), peas, *A. vera* (L.) Burm.f., ruffled fan palm (*Licuala grandis* (hort. ex W. Bull) H. Wendl.), sugarcane (*Saccharum officinarum* L.), coffee (*Coffea arabica* L.), oregano (*Origanum vulgare* L.), fenugreek (*Trigonella foenum-graecum* L.), *Medicago truncatula* Gaertn., Indian bean (*Lablab purpureus* (L.) Sweet), common bean, and white chrysanthemum (*Chrysanthemum morifolium* (Ramat.) Hemsl.) (Aveskamp et al. 2008; Chand et al. 2019; Deb et al. 2020).

Integrated management strategies: removing plant remains could be the most appropriate way to control the disease. This is mainly because of the ability of the pathogen to survive on plant debris (Aveskamp et al. 2008). The use of resistant cultivars for some crops, such as alfalfa (Fan et al. 2018), gives uncertain results as more than one *Phoma* species can infect the same crop and most *Phoma* species do not have host specificity and could infect more than one plant species (Chand et al. 2019). The application of benzimidazoles, as well as chlorothalonil, prochloraz, dichlofluanid, captafol, benomyl, mancozeb, and propiconazole, has been reported as an effective chemical approach to control leaf spot disease (Fan et al. 2018; Deb et al. 2020).

8.2.3 Web Blotch

Web blotch is also known as muddy spot or net blotch (Liu et al. 2020). This disease is one of the major diseases of peanut. The disease has been reported from various parts of Africa, Argentina, China, Russia, Spain, and the USA (Smith and Littrell 1980; McDonald and Beckerman 1985; Pettit et al. 1986) and is estimated to cause 10–30% yield loss annually. Severe infections may lead to more than 50% yield loss (Zhang et al. 2019; Liu et al. 2020).

Symptoms: symptoms may vary in different hosts, yet tan net-like bronzing and large light-brown blotches are the main symptoms of the disease (Anco 2019; Zhang et al. 2019). However, small, irregular, brown to reddish-brown lesions along the midrib of young and old leaves may be observed in early infections. Gradually, large, approximately circular tan or dark brown blotches might be recognized on the adaxial surface of leaves (Pettit et al. 1986). The lesions may cover the whole lamina. These lesions can be pale brown and less noticed. Older lesions may become darker, have a rough and dull surface, and become dried and cracked. Intensive defoliation and brittle leaves might occur in severe infections (Pettit et al. 1986; Anco 2019; Zhang et al. 2019).

Phoma-related species and disease development: although there has been confusion over the generic distinctions between *Phoma* and *Ascochyta* in the literature, it seems clear that *P. arachidicola* Marasas, Pauer & Boerema is the main pathogen of web blotch disease (Hong-hai et al. 2013; Zhang et al. 2019; Liu et al. 2020). The optimum temperature for mycelial growth and spore production is 25 °C. Disease severity is closely related to the inoculum availability, environmental conditions, and the susceptibility of the host cultivar (Mikunthan 1997). The fungus causes severe damages under cool and moist conditions (15–30 °C), especially under irrigation (Smith and Littrell 1980; McDonald and Beckerman 1985; Anco 2019). *Phoma arachidicola* overwinters in crop residues (McDonald and Beckerman 1985). Ascospores are reported to cause primary infections in early spring. Both pseudothecia and chlamydospores are reported to be effective sources of inoculum. Periods of extended thundershowers could result in severe damages (Liu et al. 2020). This is mainly due to the reduced light intensity and temperature which favor fungus growth (Zhang et al. 2019).

Hosts: legumes are reported to be the main host of the web blotch. Peanut (*A. hypogaea* L.), alfalfa, Israel sweet clover (*Trifolium* spp.), berseem clover (*T. alexandrinum* L.), alsike clover (*T. hybridum* L.), and hairy vetch (*Vicia hirsuta* (L.) Gray) are reported to be very susceptible to web blotch disease (Pettit et al. 1986).

Integrated management strategies: various management strategies have been pursued worldwide to control web blotch disease, among which crop rotation has been reported to be the most effective one (McDonald and Beckerman 1985; Subrahmanyam and Hildebrand 1994; Anco 2019). Due to the survival of the fungus in crop residues, removal of infected crop residues or deep plowing to bury the infected residues is highly recommended (Subrahmanyam and Hildebrand 1994; Anco 2019). Moreover, the removal of volunteer groundnut plants could also be a matter of importance. This is mainly because of the susceptibility of some peanut cultivars (especially Spinach cultivars) to the disease (Smith and Littrell 1980; Subrahmanyam and Hildebrand 1994). Furthermore, the application of resistant genotypes could also reduce the disease. Azoxystrobin, cyproconazole, propiconazole, boscalid, pyraclostrobin-metiram, kresoxim-methyl, and difenoconazole are reported as effective fungicides to control web blotch disease (He et al. 2014; Jing-jing et al. 2015; Anco 2019).

8.3 Stem Diseases

It has been claimed that the stem could be the most susceptible part of the host for *Phoma* species. A combination of a wide range of symptoms would appear on the host in response to stem infection by *Phoma* species. Blackleg, black stem, and stem blight have been reported as the most destructive Phoma diseases worldwide.

8.3.1 Blackleg

Blackleg is also known as blackleg stem canker due to the simultaneity occurrence of both disease symptoms (Hwang et al. 2016). Nevertheless, stem canker could be different from blackleg which can vary in different environmental conditions (Hwang et al. 2016). This disease is considered as a complex disease, causing a huge loss in various types of plants, especially crucifers (Ferdous et al. 2019). The disease has been reported several times from Argentina, Australia, Canada, China, Europe, Iran, North America, Russia, and South Korea (Hwang et al. 2016; Robin et al. 2017; Vakili et al. 2017; Barua et al. 2018; Burketová et al. 2018).

Symptoms: various types of symptoms, including leaf spots, stem lesions, seedling blight, and stem cankers, can be observed (Barua et al. 2018). Based on the age of the plants, the symptoms may vary. For instance, on young plants, black to dark brown lesions form on the stem between the first leaves and the ground; as a result, the disease is called blackleg (Fig. 8.1d) (Hwang et al. 2016). Progressively, the lesions may compress the stem and cause severe seedling blight (Sprague et al. 2018). Simultaneously, light green- to beige-colored spots may appear on the leaves which may turn brown to gray or white with a chlorotic halo. The lesions comprise little black pycnidia. Leaf lesions may be different among various hosts according to their resistance, the stage of lesion development, and the group (A or B) designation (Claassen 2016; Fernando et al. 2016). Group A forms pale green lesions that turn light brown to white with pycnidia, while group B forms smaller darker and greener spots with fewer pycnidia (Hwang et al. 2016). Furthermore, gray, oval, and superficial lesions can appear on stem and roots (Claassen 2016). In severe infections, stem cankers may be observed on the base of the stem. These cankers infect inner parts of stem and form lesions with dark brown, black to purple edges which might become gray in the center as disease evolves (Claassen 2016). On ornamental plants, brown water-soaked spots may be observed initially on at the base of petioles. The lesions merge and result in plant death (Lou et al. 2008).

Phoma-related species and disease development: *Phoma lingam* (Tode) Desm. (syn. *Leptosphaeria maculans* Ces. & De Not.), *P. multirostrata* (P.N. Mathur, S.K. Menon & Thirum.) Dorenb. & Boerema (syn. *Ectophoma multirostrata* (P.N. Mathur, S.K. Menon & Thirum.) Valenz.-Lopez, Cano, Crous, Guarro & Stchigel), and *Phoma wasabiae* Yokogi (syn. *L. biglobosa* Shoemaker & H. Brun) have been reported as the main pathogens causing blackleg (Moon et al. 2003; Lou et al. 2008; Hwang et al. 2016; Hossain et al. 2020; Vakili et al. 2017; Barua et al. 2018; Deb et al. 2020). Cankers at the stem base are usually caused by *L. maculans*. *Leptosphaeria biglobosa* has been considered as a less aggressive species causing upper stem lesions (Hossard et al. 2013). The primary source of inoculum has been reported to be the wind-dispersed ascospores that are released from the sexual fruiting bodies, i.e., the pseudothecia (Barua et al. 2018). The ascospores can also be dispersed by rainfalls. Pseudothecia can survive from 2 to 4 years in wet conditions and dry seasons, respectively (Hwang et al. 2016; Barua et al. 2018). The primary source of inoculum is the infected plant residues. The disease occurs between 8 and 24 °C (Fernando et al. 2016; Barua et al. 2018). Pycnidiospores form in the pycnidia on leaf spots and act as the secondary inoculum source (Claassen 2016). The disease can also be seed-borne (Wang et al. 2020). Stem cankers develop in high-temperature conditions (Vakili et al. 2017). Fungal spores can remain viable for 1 year on death tissues. In addition, agricultural materials play an important role in disease distribution (Barua et al. 2018). On ornamental plants, the disease is most severe during the flowering stage (Lou et al. 2008).

Hosts: Brassicaceae family members, especially *Brassica* spp. including oilseed rape or canola (*B. napus* L.), cabbage (*B. oleracea* L.), and turnips (*B. rapa* L.), as well as wasabi (*Wasabia japonica* Matsum.), and the ornamental plant *Ajuga multiflora* Bunge (Labiatae) are the main hosts of the blackleg disease (Moon et al. 2003; Lou et al. 2008; Claassen 2016; Hwang et al. 2016; Hossain et al. 2020; Sprague et al. 2018; Deb et al. 2020).

Integrated management strategies: the disease is difficult to control. A wide range of management strategies has been applied to control blackleg, including fungicide applications, use of biological control agents, and cultural practices (McCredden et al. 2018). Due to the survival of sexual structures on plant debris, removal of plants residues is a matter of importance (Claassen 2016). Moreover, the disease can also be dispersed via human activities; as a result, sanitary restrictions should also be considered (Wang et al. 2020). Certified seeds could effectively reduce the disease incidence. The application of benomyl solution for seed disinfection has also been proposed (Moon et al. 2003). Different microbial antagonists have been claimed to be effective in controlling blackleg disease, among which *Bacillus*, *Pseudomonas*, and *Serratia* bacterial species strains, as well as fungal species strains of *Gliocladium* and *Trichoderma*, are the most well-known ones. *Serratia plymuthica* HRO-C48 and *Gliocladium catenulatum* J1446 have proven to be the most effective biological control agents to control blackleg disease (Claassen 2016). The application of resistant cultivars of rapeseed to blackleg has been reported to be the best management strategy to control the disease. However, due to the broad host range of the disease, the application of resistant cultivars is still a challenge (Hwang et al. 2016; Robin et al. 2017). Crop rotation, stubble management, and tillage aim at reducing the amount of inoculum from previous crop (Hossard et al. 2018). In addition, changing the sowing date may avoid exposure to airborne ascospores when the crop is most vulnerable (Aubertot et al. 2006; Hossard et al. 2018). Separation distances between the current and previous season crops could also decrease the disease incidence. It has been reported that higher plant densities could reduce the risk of infections on the stem (Aubertot et al. 2006; Zamani-Noor and Knüfer 2018). High autumn nitrogen availability could also promote Phoma stem canker severity (Javaid et al. 2017). Fluquinconazole, flutriafol, prothioconazole, penthiopyrad, picoxystrobin, and glyphosate-ammonium could be used as seed treatments, coated fertilizer granules, or foliage sprays for chemical control (West et al. 2002; Aubertot et al. 2006; Zamani-Noor and Knüfer 2018). *Erwinia herbicola* (Löhnis) Dye, *Paenibacillus polymyxa* (Prazmowski) Macé strain PKB1, *Cyathus striatus* (Huds.) Willd., and *Trichoderma harzianum* Rifai have been proposed as effective biological control agents of stem canker disease (Aubertot et al. 2006). Even though it is still considered a challenging approach, resistant cultivars could be the best long-term strategy to control Phoma stem canker (Mitrousia et al. 2018).

8.3.2 Black Stem

The black stem is considered as one of the most destructive diseases caused by some species of *Phoma*. The disease has been reported approximately from all countries, especially from African countries, such as Tunisia, Australia, Canada, China, France, India, Italy, and the USA (Deb et al. 2020; Sun et al. 2020). The black stem is also known as spring black stem on alfalfa (Fan et al. 2018; Deb et al. 2020).

Symptoms: the petiole base is the first entry point of the pathogen. The symptoms, which at the beginning are dark lesions on the petiole base, later appear on stem and leaves as stunting and weaken stem (Fig. 8.1a) (Fan et al. 2018). In sunflower (*Helianthus annuus* L.), the reduction in seed production and weight loss are the two common consequences of black stem disease (Sun et al. 2020). In alfalfa, seedling blight, stem canker, leaf defoliation, leaf spot, and chlorosis, as well as root and crown rot, are the major symptoms of the disease (Fan et al. 2018; Deb et al. 2020).

***Phoma*-related species and disease development:** *Phoma medicaginis* var. *medicaginis*, *P. pinodella* (L.K. Jones) Morgan-Jones & K.B. Burch (syn. *Didymella pinodella* (L.K. Jones) Qian Chen & L. Cai), and *P. macdonaldii* Boerema (syn. *Leptosphaeria lindquistii* Frezzi) are the causal agents of black stem disease (Deb et al. 2020). On alfalfa, the disease is also called spring black stem because of disease spread in cool and wet environmental conditions. The conidia within the pycnidia could rapidly disperse via rainfalls (Gao et al. 2018). Pseudothecia, pycnidia, and mycelium are reported to be the overwintering structures on plant debris. These structures could be carried out by insects or wind which severely damages the next growing season. *Phoma macdonaldii* is a seed-borne fungus which also causes premature ripening (Deb et al. 2020). Both fungi infect hosts by direct penetration or through wounds and stomata (Sun et al. 2020).

Hosts: alfalfa, barrel clover, fenugreek (*Trigonella foenum-graecum* L.), and sunflower are the most susceptible hosts of black stem disease (Deb et al. 2020).

Integrated management strategies: due to the importance of seeds in spreading the disease, certified seeds should be required (Sun et al. 2020). Removal of plant debris could also significantly reduce the disease in the next growing season (Deb et al. 2020). The application of benomyl and propiconazole could effectively reduce disease incidence (Gao et al. 2018). Utilization of resistant cultivars has been proposed several times, for both alfalfa and sunflower; yet it is still a challenging approach (Deb et al. 2020). For instance, only seven varieties of sunflowers, including Xinshikui6, CY101, Longshikui3, 118, KJ003, 7 K512, and T33 in China, are reported to be resistant to sunflower black stem (Sun et al. 2020). However, it is not known whether they could also be used in other regions.

8.3.3 Stem Blight

Stem blight disease is caused by various *Phoma* species worldwide. Due to the wide host range of the pathogen, the disease is known by different names, such as ray blight, stem blight, tomato rot, and gummy stem blight. Stem blight causes a huge loss universally, especially in Australia, Brazil, India, South Africa, South Korea, Tasmania, the USA, and Turkey (Aveskamp et al. 2008; Pethybridge et al. 2011; Basım et al. 2016; Colmán et al. 2018; Deb et al. 2020).

Symptoms: according to the host, different symptoms may be observed. For instance, on chrysanthemum (*Chrysanthemum indicum* L.) and pyrethrum (*Tanacetum cinerariifolium* (Trevir.) Schultz-Bip.), the disease is called ray blight, causing necrotic spots and lesions on stem, flower, and buds, leading to blight stem both in greenhouse and field conditions. In severe infections, the entire leaf surface would be covered with necrotic lesions, which would also appear on the stem, causing stem twist and dieback. The appearance of a shepherd's crook symptoms on the infected buds and flowers has also been reported several times (Pethybridge et al. 2011; Deb et al. 2020). On tomato, blight symptoms would be noticeable after 5 days on leaves, flowers, and stems which is also known as tomato rot (Colmán et al. 2018). On cucurbits, the disease is known as gummy stem blight. In other words, the pathogen can cause a wide range of symptoms, from seedlings damping-off to stem, leaf, and fruit blight (Basım et al. 2016). Leaf spot and black rot have been reported as other important symptoms of the disease in cucurbits. Dark brown to brown, ellipsoid to oblong-shaped lesions appear on the stem which would later be observed as stem cankers. Waxy water-soaked brown gummy exudates are produced superficially (Fig. 8.1f). The cankers progress in the cortical tissue. In severe infections, dry and rough cankers and the cracking surface may appear. Circular beige-colored leaf spots with yellow borders could be present on leaves; the spots can eventually merge causing leaf blight (Basım et al. 2016; Deb et al. 2020).

***Phoma*-related species and disease development:** different *Phoma* species have been reported as the causal agent of stem blight, including *P. destructiva*; *P. cucurbitacearum* (Fr.) Sacc. (syn. *Stagonosporopsis cucurbitacearum* (Fr.) Aveskamp, Gruyter & Verkley; syn.

Didymella bryoniae (Fuckel) Rehm.); *P. lingulicola* Boerema (syn. *S. chrysanthemi* (F. Stevens) Crous, Vaghefi & P.W.J. Taylor; syn. *D. ligulicola* Baker, Dimock & Davis); *S. citrulli* M.T. Brewer & J.E. Stewart; and *S. caricae* (Syd. & P. Syd.) Aveskamp, Gruyter & Verkley (Pethybridge et al. 2011; Basım et al. 2016; Colmán et al. 2018; Deb et al. 2020). Ascospores are considered as primary inoculum spreading by wind and rain. Overwintering occurs on plant debris and infected seeds. Pycnidia could serve as secondary inoculum. Symptoms appear from the late of May (Basım et al. 2016).

Hosts: tomato (*Solanum lycopersicum* L.) (Colmán et al. 2018), pyrethrum (Pethybridge et al. 2011), *Cucurbitaceae* family members (Basım et al. 2016; Deb et al. 2020), and papaya (*Carica papaya* L.) (Basım et al. 2016) have been reported as the main hosts of the pathogen.

Integrated management strategies: due to the overwintering of the pathogen on plant debris, removing the infected remains could be a matter of importance (Deb et al. 2020). Certified seeds should also be used. Moreover, it is of great significance to control moisture and temperature in seed storage to reduce disease incidence (Aveskamp et al. 2008). Chemical control could be effective with the application of difenoconazole, pyraclostrobin, and carboxin (Kaewkham et al. 2016). *Gliocladium catenulatum* J1446, *Pseudomonas aeruginosa* 23, and *Bacillus subtilis* QST713 as putative biological control agents have been proved to effectively control the disease (Kaewkham et al. 2016; Deb et al. 2020). Although difficult, the application of resistant cultivar could also be a promising way to control stem blight. However, due to the broad host range of the disease, it is a challenging option (Kaewkham et al. 2016).

8.4 Branch Diseases

Branch diseases are caused by a few species of *Phoma*. This disease has been reported mainly from stone fruits worldwide. Branch blight and shoot necrosis are the major branch diseases.

8.4.1 Branch Blight

Branch or shoot blight is a fatal foliage disease causing a huge loss in woody plants and has been reported from Arizona (USA), Spain, and Tunisia (Chen et al. 2013; Triki et al. 2019; Deb et al. 2020).

Symptoms: the most common symptoms are blighted shoots and dieback (Aveskamp et al. 2008). Wilted leaves remain attached to the blighted shoots (Fig. 8.1i). Twigs might become dead, and in severe infections, canker, bud morality, leaf chlorosis, and fruit rot would be observed (Chen et al. 2013). Buds with severe gummosis and barks with abundant pycnidia have also been reported as the consequences of the branch and shoot blight (Triki et al. 2019). Twisted twigs have also been recorded in some crops (Deb et al. 2020).

Phoma-related species and disease development: *Phoma fungicola* (syn. *Nothophoma quercina*) has been reported as the main pathogen causing branch blight. The fungus overwinters on dead shoots attached to the tree or on plant debris (Chen et al. 2013; Moral et al. 2017). The fungus produces abundant pycnidia on plant residues which would later become the primary source of infection. Under favorable conditions, such as cloudy and cool weather (moisture above 70% and temperature between 20 and 25 °C), the fungus can immediately reproduce asexually and kill the host (Deb et al. 2020).

Hosts: pistachio, olive (*Olea europaea* L.), peach (*Prunus persica* (L.) Batsch), *Malus* × *micromalus* Makino, *Chaenomeles sinensis* Schneid, tea (*Camellia sinensis* (L.) Kuntze), and almond (*Prunus dulcis* (Mill). D. A. Webb) have been recognized as the main hosts for branch blight (Chen et al. 2013; Moral et al. 2017; Triki et al. 2019).

Integrated management strategies: because of the importance of plant debris and death shoots as primary inoculum, collection and eradication of these materials might effectively control the disease (Deb et al. 2020). The application of carbendazim along with crop rotation has been claimed to be an appropriate way to control the disease (Chen et al. 2013).

8.4.2 Shoot Necrosis

Shoot necrosis is a destructive disease caused by *Phoma* species in both crop and woody plants (Deb et al. 2020). This disease could result in huge economic damages and has been reported from Croatia, Greece, Italy, and Spain (Ivic et al. 2010; Thomidis et al. 2011).

Symptoms: young shoots are more vulnerable to the disease, expressing red to brown lesions. Leaf wilting and decaying, shoot necrosis with some leaves attached, and, in severe infections, shoot canker might also be observed (Ivic et al. 2010). Under the canker, brown necrosis lesions on the bark could be observed (Thomidis et al. 2011). Blossoms and young fruits are also susceptible to wilting and necrosis (Ivic et al. 2010).

Phoma-related species and disease development: various *Phoma* species have been reported as the shoot necrosis causal agent, including *P. glomerata*, *P. incompta* Sacc. & Martelli (syn. *Comoclathris incompta* (Sacc. & Martelli) Ariyaw. & K.D. Hyde), *P. herbarum*, and *P. fungicola* (syn. *Nothophoma quercina*) (Aveskamp et al. 2008; Ivic et al. 2010; Thomidis et al. 2011; Deb et al. 2020). Overwintering occurs on the blighted shoot on trees as pycnidia (Ivic et al. 2010). As a consequence, pycnidia could cause the primary infection on new shoots. The optimal growth temperature for the fungus is 25 °C (Thomidis et al. 2011).

Hosts: peach and nectarine (*Prunus persica* L.), olive, and alfalfa are the main hosts of the pathogen (Fan et al. 2018; Deb et al. 2020).

Integrated management strategies: removal of plant debris and the use of resistant cultivars have been claimed as the best approaches to control the disease (Ivic et al. 2010). However, cultivation of resistant cultivars could be a challenging approach for peaches and nectarines because all cultivars have been reported to be susceptible (Thomidis et al. 2011). Chemical control has also been successfully conducted with thiophanate methyl, carbendazim, and tebuconazole (Thomidis et al. 2011; Deb et al. 2020).

8.5 Vascular Diseases

Phoma species are not specialized vascular plant pathogens like *Fusarium* or *Verticillium* species; however, in some cases, they may infect xylem which results in various symptoms such as wilting, shoot chlorosis, and trunk diseases of *Citrus* spp. Even though some researchers claimed that seedling diseases would also result in wilting, several reports are of the opinion that mal secco disease is the only vascular disease caused by *Phoma* species.

8.5.1 Mal secco

Mal secco, an Italian name meaning "dry disease," is a vascular wilt disease. It is the most destructive and economically relevant disease of lemon (*Citrus limon* (L.) Osbeck) in the Mediterranean region where it is regarded as a limiting factor to the expansion of the industry of this citrus species (Khanchouch et al. 2017; Deb et al. 2020). The disease was firstly identified in Greece (Nachmias et al. 1979). However, currently, all citrus-growing regions of the Mediterranean Basin, except for Morocco, Portugal, and Spain, are reported to show the disease symptoms. An initial outbreak of the disease was recently identified in Spain, the first lemon producer country in the Mediterranean region, but was promptly eradicated. Although, presently, mal secco disease is restricted to the Mediterranean and Black Sea regions, it is regarded as a threat to lemon industry worldwide (Migheli et al. 2009; Ben-Hamo et al. 2020).

Symptoms: This disease shows three distinct forms based on the symptoms. In the most common form, leaf vein chlorosis and shoot wilting followed by leaf dropping and shedding as well as dieback of apical twigs appear in spring. Dieback symptoms progress rapidly downward and extend to entire branches. Typically, desiccation affects only a sector of the tree canopy, reflecting the distribution of the fungus in the vascular system (Fig. 8.1c). Vigorous plants react with the emergence of suckers from the trunk below the branches that are drying up. A salmon-pink or brown-reddish wood discoloration under the bark of withering twigs, infected branches, and trunk is the most typical symptom of the disease, which is due to the gum production within the xylem vessels (Migheli et al. 2009). Scattered, black pin-point spots (the pycnidia of the pathogen), barely visible with naked eyes, appear within lead- or ash-gray areas on 2–3-year old withered twigs. The disease progresses basipetally, more or less rapidly, and finally the tree dies (Migheli et al. 2009; Deb et al. 2020).

If the causal agent remains latent in the tissue, the supposedly healthy plant might suddenly develop the disease and collapse. Mal nero is the second form of the disease; it is a chronic infection, which is characterized by a dark brown and orange discoloration of the heartwood, without any external symptoms. Nevertheless, as the disease progresses, the pathogen infects the outer functional xylem, and the tree might finally collapse abruptly (Hajlaoui et al. 2008; Migheli et al. 2009). In the third form, i.e., "Mal fulminante," a rapid wilting and collapse of the whole tree might result from a systemic invasion of the functional xylem by the pathogen due to root or stem infection (Migheli et al. 2009).

Phoma-related species and disease development: The fungus of mal secco was originally described as *Deuterophoma tracheiphila* Petri. Subsequently, the name of the species was changed in *Bakerophoma tracheiphila* (Petri) Cif. and then in *Phoma tracheiphila* (Petri) L.A. Kantsch. & Gikaschvili. The current name, after the taxonomical revision of the genus *Phoma*, is *Plenodomus tracheiphilus* (Petri) Gruyter, Aveskamp & Verkley, in the *Leptosphaeriaceae* family (de Gruyter et al. 2013).

Disease progress depends on both the susceptibility of citrus species or variety and the tree's age. The pathogen may kill the plant within a few months or years (Migheli et al. 2009). Moreover, root infections result in a more rapid occurrence of symptoms than foliar or twig infections, due to the more efficient upward passive inoculum transport in the tree xylem. Fungal conidia are reported to be the primary source of inoculum (Migheli et al. 2009). Optimum temperatures for xylem colonization by the pathogen are between 15 and 22 °C, whereas at or below 10 and above 28 °C, colonization of the xylem is inhibited (Migheli et al. 2009). Inoculum dispersal takes place mainly by rainfall, while long-distance transport occurs prevalently through infected propagation material and plants. The entrance of the pathogen is through wounds and leaf scars. As a consequence, plants injured by hail or damaged by frost are more prone to infections. Length of the incubation period varies according to season, and in young trees it ranges between 2 and 7 months, whereas it can last several years in the mal nero form of the disease (Migheli et al. 2009). The fungus can remain viable in the soil within infected plant residues for months and even years, depending on the characteristics of the soil; it survives longer in clayey soils than in sandy soils (Nigro et al. 2011).

Hosts: Even though lemon (*Citrus* × *lemon*) has been reported as the main host for the pathogen, the causal agent of mal secco disease can also infect other *Citrus* species such as citron (*C. medica* L.), bergamot (*C.* × *bergamia* Risso), chinotto (*C.* × *myrtifolia* Raf.), lime (*C.* × *aurantifolia* Christ.), sour orange (*C.* × *aurantium* L.), rough lemon (*C.* × *jambhiri* Lush), citrus alemow (*C.* × *macrophylla* Wester), Volkamer lemon (*C.* × *volkameriana* Ten. & Pasq), rangpur lime (*C.* × *limonia* Osbeck), and mandarin (*C. deliciosa* Tenore) as well as several genera of the Rutaceae family, including *Fortunella*, *Poncirus*, and *Severina* (Migheli et al. 2009; Abbate et al. 2019). Moreover, some intergeneric and interspecific hybrids from the Rutaceae family have recently been proved to be susceptible to the infections by this pathogen (Tusa et al. 2000; Migheli et al. 2009). At present, in the field, mal secco disease is a problem for lemon and for other minor citrus, such as lime, citron, and bergamot. However, almost all citrus species proved to be susceptible to *Pl. tracheiphilus*, at least when artificially inoculated (Migheli et al. 2009). Mandarin has reported showing relative tolerance to mal secco disease, yet it is still considered as one of the major hosts of the pathogen. Among rootstocks, alemow, a likely hybrid between citron and biasong (*C. micrantha*), and Volkamer lemon are extremely susceptible. Sour orange, widely used as a lemon rootstock in the Mediterranean region, is considered to be very susceptible in Italy, but only moderately affected in Israel (Migheli et al. 2009). Actually, the susceptibility of sour orange decreases with the age of the plant. The rootstock influences the susceptibility of the scion to mal secco. In a comparative test with other citrus rootstocks, including, among others, sour orange, alemow, and Volkamer lemon, the tangelo Siamelo CRC 2486 (*C.* × *tangelo* J.W. Ingram & H.E. Moore), a hybrid between 'King' tangor × unknown grapefruit and Ichang lemon CRC 1215 (*C.* × *wilsonii* Tanaka), proved to be tolerant (Nigro et al. 2011). However, these two rootstocks are not widely used in commercial lemon plantings.

Integrated management strategies: strict quarantine measures are applied worldwide to avoid the expansion of the geographical distribution range of mal secco, which presently is restricted to countries around the Mediterranean and Black Sea, and to prevent its introduction into new areas. *Plenodomus tracheiphilus* is a quarantine pathogen in the A2 list of the European and Mediterranean Plant Protection Organization (EPPO), Asia and Pacific Plant Protection Commission, Caribbean Plant Protection Commission, North American Plant Protection Organization, and Inter-African Phytosanitary Council (Migheli et al. 2009).

In countries where the disease has already been established, an integrated approach is the pursued strategy, including regulations to prevent the spread of the fungus responsible of the disease and reduce the amount of inoculum, phytosanitary certification schemes of nursery plants, cultural practices and other agronomic means, use of host genetic resistance, and chemical and biological control. Despite this, the disease is difficult to manage, and only limited success has been achieved. In Italy it is compulsory to prune or uproot symptomatic trees and to burn the pruning residues (Migheli et al. 2009). Yearly pruning of symptomatic twigs and removal of suckers are recommended. Pruning is better performed in spring when symptoms are more easily recognized. Soil tillage in rainy periods and no tillage combined with chemical control of weeds enhance the risk of root infections. Windbreaks and hail nets reduce injuries and, consequently, the risk of infections on the canopy. They are particularly recommended in nurseries. Chemical treatments with copper compounds and mancozeb, in autumn and winter or immediately after hailstorms or frosts, might prevent infections and are especially recommended in nurseries. Promising biocontrol agents, such as endophytic bacteria with a strong antagonistic activity against *Pl. tracheiphilus*, were identified (Kalai-Grami et al. 2014, 2016). Moreover, induction of resistance by pre-inoculation with other non-pathogenic fungal species or non-pathogenic mutants of *Pl. tracheiphilus* was explored as an alternative to control mal secco (Migheli et al. 2009). Presently, however, biological control cannot be considered as a practical option to manage this disease in commercial lemon orchards. Early diagnosis of the pathogen in nurseries through molecular approaches (Balmas et al. 2005; Licciardello et al. 2006; Demontis et al. 2008; Migheli et al. 2009; Kalai et al. 2010) and the use of tolerant lemon cultivars (Solel and Spiegel-Roy 1978; Tuzcu et al. 1989; Thanassoulopoulos 1991; Gentile et al. 1992; Deng et al. 1995; Baş and Koç 2006; Migheli et al. 2009) appear as the most effective management options. Since the first appearance of the disease, clonal selection has been pursued in several countries, especially in Italy, and tolerant lemon cultivars, such as 'Monachello' and 'Interdonato,' or moderately susceptible, such as 'Femminello S. Teresa Riva,' 'Femminello Zagara Bianca,' and 'Femminello Continella,' have been obtained. However, most of these cultivars do not perform well outside the area of origin, and some have a reduced yield or produce fruits of poor quality. Alternative breeding strategies and new technologies that could be applied for obtaining mal secco-tolerant lemon cultivars have been recently reviewed (Abbate et al. 2019).

Presently, in lemon-growing areas where it has established, mal secco is an unresolved problem and represents the most serious limiting factor for the expansion of lemon industry. A resurgence of the disease, e.g., has been recently observed in eastern Sicily, the first lemon producer region in Italy, due to the concurrence of several factors, including hailstorms, the widespread use of high-yield producing 'Femminello Siracusano 2Kr' lemon variety, and the use of susceptible rootstocks, such as alemow and Volkamer lemon, due to their better performance than sour orange in inducing more vigorous tree growth as well as a more abundant and earlier production (Cacciola and Gullino 2019).

8.6 Root and Crown Diseases

Root and crown diseases can be divided into two main groups: rot diseases and canker diseases. The symptoms might also occur in other parts of plants such as leaves and petioles. The management of root rot is a challenge, and the disease could cause severe damages in both ornamental and crops.

8.6.1 Brown Root Rot

Brown root rot has been claimed as the major disease of alfalfa in Canada and the USA (Larsen et al. 2007) which causes a huge loss. It can also affect a wide range of other plants (Hollingsworth et al. 2005).

Symptoms: disease detection based only on symptoms could be challenging. This is mainly due to the similarity of the symptoms with those caused by *Aphanomyces*, *Phytophthora*, *Pythium*, and *Rhizoctonia* species (Larsen et al. 2002). The causal agent affects secondary roots and taproots, causing dark brown to black lesions that could be sunken or rotten. These brown lesions may be surrounded by a darker black border. Commonly, the upper secondary roots and taproots of dead and living plants are infected (Hollingsworth and Gray 1999). Pycnidia and pycnosclerotia form superficially and are embedded in the vicinity of root lesions. Pycnidia could also be found on the lower stems and petioles (Larsen et al. 2007). Stunting and decline have also been proved to be the symptoms of brown root rot. Due to the root infection, crown and shoots may also be affected and die. The infected plants could not feasibly break the dormancy in early spring (Wunsch et al. 2007).

Phoma-related species and disease development: brown root rot is caused by *Phoma sclerotioides* G. Preuss ex Sacc. (syn. *Leptosphaeria sclerotioides* (Preuss ex Sacc.) Gruyter, Aveskamp & Verkley; previously called *Plenodomus meliloti* Dearn. & G. B. Sanfornd) (Wunsch et al. 2008). This fungus overwinters at low temperatures on legumes and cereals (Larsen et al. 2007). Lesions can be observed on both roots and crowns and may cause yield loss. The disease is associated with winterkill. The optimum growth rate of the pathogen is 15–16 °C. The fungus is not able to survive at high temperatures and, therefore, in summer (Hollingsworth et al. 2003). *Phoma sclerotioides* has been reported to be both pathogenic and saprophytic. In late winter and early spring, it causes yield loss in alfalfa and sweet clover. In contrast, it has been reported as a saprophyte on canola and annual cereal debris (Larsen et al. 2002).

Hosts: alfalfa and sweet clover are the main hosts of the pathogen. However, *P. sclerotioides* has been reported from other legumes and cereal crops, such as winter wheat (*Triticum aestivum* L.), common clover (*Trifolium* spp.), alsike clover (*T. hybridum* L.), bird's foot-trefoil (*Lotus corniculatus* L.), redroot amaranth (*A. retroflexus* L.), red fescue (*Festuca rubra* L.), pigweed (*A. amaranthoides* L.), Kentucky bluegrass (*Poa pratensis* L.), and winter rye (*Secale cereale* L.) (Larsen et al. 2007; Wunsch et al. 2008).

Integrated management strategies: due to the disease association with winterkill, it is suggested a proper management of alfalfa plantation in the field to minimize winter stress (Larsen et al. 2002). Moreover, crop rotation with non-host plants and those with spring-sown small grain could effectively reduce inoculum density (Wunsch and Bergstrom 2011). Some alfalfa cultivars, such as Heinrichs and Peace cultivars, have been reported as resistant cultivars to brown root rot (Larsen et al. 2007).

8.6.2 Pink Root Rot

Pink root rot also known as red root rot is considered as one of the most devastating diseases of onions and recently of cucurbits in Australia, Brazil, Egypt, France, Iran, Israel, Italy, Japan, Mauritius, South Korea, Spain, and the USA (Ikeda et al. 2012).

Symptoms: the disease symptoms initiate from the roots, which turn into dark-red to pink (Fig. 8.1e) (Koenning et al. 2007). Water-soaked, crispy or shrivel roots and eventually plant death may be observed. Wilting, etiolation, decline, and stunting have also been reported as disease symptoms. Onion bulbs would remain small and have low storage quality (Esfahani and Pour 2008). The decline has been reported as a major symptom in cucurbits (Ikeda et al. 2012).

Phoma-related species and disease development: pink root rot is a complex disease caused not only by a *Phoma* species but also by *Fusarium*, *Pythium*, and *Rhizoctonia* species. *Phoma terrestris* H. N. Hansen (syn. *Setophoma terrestris* (H.N. Hansen) Gruyter, Aveskamp & Verkley; previously known as *Pyrenochaeta terrestris* (E. M. Hans.) Gorrenz, J.C. Walker, & R. H. Larson) causes pink root rot (Ikeda et al. 2012). The pathogen is known to be a cosmopolitan saprophytic microorganism which could also be a weak parasite of the roots of various hosts. The pathogen dispersal occurs through infected plant debris, movement of water and infected soil, as well as mechanical cultivation. The fungus can survive in the soil for years. The optimal growth condition for the fungus is a relatively cool temperature with adequate soil moisture (Esfahani and Pour 2008).

Hosts: onion (*A. cepa* L.), *Cucurbitaceae* family members, including watermelon (*Citrullus lanatus* (Thunb.) Matsum. & Nakai) and squash (*Cucurbita moschata* Duchesne ex Poir), and corn (*Zea mays* L.) have been reported as the main hosts of *P. terrestris* (Newby et al. 1997; Ikeda et al. 2012). However, the most susceptible host is onion.

Integrated management strategies: pink root rot is difficult to control because of the survival of fungus in the soil for several years. Moreover, the causal agent has a wide host range (Koenning et al. 2007). Hence, crop rotation could not be appropriate. Even though soil solarization and the application of onion-resistant cultivars have been proposed as effective methods to control the disease, pink root rot has not been controlled ideally, as yet (Ikeda et al. 2012). Recently, the application of leaf powders of some plants including basil (*Ocimum basilicum* L.), neem (*A. indica* A. Juss.), argel (*Solenostemma argel* L.), and lantana (*Lantana camara* L.) together with the fungicide Premis could effectively reduce the pink root rot disease in onion fields (Mohammed and Eskov 2020).

8.6.3 Root and Crown Rot

Root and crown rot has been reported on certain plants from approximately all parts of the world, especially Canada, India, and the USA, causing huge economic losses (Menzies and Colhoun 1976). In some literature, the disease is also called root black rot (Vaghefi et al. 2019).

Symptoms: the symptoms could be different based on the infected host plant. In the chrysanthemums, the roots could be severely damaged. Initially, brown to dark brown or reddish-brown necrotic lesions appear on roots which results in root rot and pre-matured plant death. Some parts of the roots might also be separated from the main root system leading to dwarfing (Menzies and Colhoun 1976). In pyrethrum, severe necrosis and chlorosis of the plant tissue may be observed. Both root and crown rot symptoms have been reported on alfalfa. Dry necrosis and black lesions could be observed on alfalfa tip and lateral roots resulting in severe stunting and root discoloration (Chand et al. 2019). In severe infections, foliage wilting and chlorosis might be observed in response to root destruction. Furthermore, the crown diameter is also reduced in diseased alfalfa. On lettuce (*Lactuca sativa* L.), dark brown to black immersed and sunken cavities are the symptoms on crown and taproot. No fungal fruiting bodies can be found on these dry and firm cavities. As the plant grows, the cavities remain on one side of the weakened crown leading to plant malformation (Koike et al. 2006). On sugar beet, root rot initiates from the core tissue of roots and crown and then progresses into the lower parts and outer parts of the taproot and surrounding crown tissues.

Phoma-related species and disease development: various *Phoma* species have been reported as root and crown rot disease causal agent, including *P. betae*, *P. exigua* (syn. *Boeremia exigua*), *P. medicaginis* var. *medicaginis*, *P. macrostoma* Mont. (syn. *Didymella macrostoma* (Mont.) Qian Chen & L. Cai), and *P. chrysanthemicola* Hollós (syn. *Paraphoma chrysanthemicola* (Hollós) Gruyter, Aveskamp & Verkley) (Caesar et al. 2012; Deb et al. 2020). On alfalfa, *P. medicaginis* var. *medicaginis* has been claimed to be the primary pathogen of root rot and causes wilting. It is also hypothesized that root-feeding insects might be vectors for *P. medicaginis* var. *medicaginis* (Chand et al. 2019). In contrast to alfalfa, no symptom of wilting has been recorded for *P. chrysanthemicola* (Menzies and Colhoun 1976). In sugar beet, *P. betae* could be both soil-borne and seed-borne, and the disease is also called black rot due to the formation of dark black lesions on roots (Vaghefi et al. 2019). *Phoma exigua* could affect both roots and crowns of lettuce and strawberry (Koike et al. 2006; Grantina-Ievina and Kalniņa 2016). Root and crown wounds are the main points of entry for the pathogen. Conidia are considered the primary inoculum. Soil moisture is another important factor for disease dispersal (Deb et al. 2020).

Hosts: nearly all *Asteraceae* family members are susceptible to root and crown rot, including chrysanthemum and pyrethrum. Besides, legumes such as alfalfa, lettuce, strawberry (*Fragaria* × *ananassa* Duchesne), sugar beet, and weeds, such as whitetop (*Lepidium draba* L.), have been recorded as the main hosts of the pathogens (Koike et al. 2006; Caesar et al. 2012; Mahish and Ghritlahare 2017).

Integrated management strategies: numerous management strategies have been proposed to control the disease due to its negative effects on a wide range of host plants. Because of the importance of seeds in disease dispersal, certified seeds should be used (Chand et al. 2019). Carbendazim, mancozeb, zineb, pyraclostrobin + boscalid, cyprodinil + fludioxonil, and azoxystrobin have been reported to be effective in controlling the disease on chrysanthemums and lettuce (Mahish and Ghritlahare 2017). Moreover, the application of plants' extracts, such as neem, has been claimed to be a reliable control method (Mahish and Ghritlahare 2017). Removing the surrounding weeds is also a matter of importance because of the survival of the pathogen on weeds (Caesar et al. 2012). The application of phosphate, nitrogen, and potassium fertilizers has also been suggested to reduce the susceptibility to the disease (Menzies and Colhoun 1976). Avoiding crown wounds could also be an effective measure to avoid infections. Despite being challenging, using resistant cultivar could be a promising approach to reduce disease incidence. However, it cannot be necessarily applicable due to the wide host range of the pathogen and its soilborne nature (Deb et al. 2020).

8.6.4 Root Canker

Root canker is claimed to be one of the major diseases of parsnip in Poland, the UK, and the USA (Cerkauskas 1987; Deb et al. 2020).

Symptoms: symptoms may appear on leaves, petioles, and roots, respectively. Pale brown to dark brown, varied in size and shape, necrotic lesions with yellowish-green surroundings are observed on the leaves. Pycnidia formation occurs on leaves. Curled, twisted, and wilted leaves may also be observed in severe infections (Cerkauskas 1987). Petioles may express tan and oval lesion symptoms. With the disease development, these lesions may become darker and merge which results in canker formation. Twisted petioles might be apparent where the cankers form. If the cankers are formed on the upper part of the petioles, a phenomenon called "shepherd's crook" takes place (Cerkauskas 1987). Root cankers appear as black to dark brown lesions with pycnidia which could be produced superficially or embedded in root tissues. Root canker can be observed on the upper or lateral parts of roots. Due to the root infections, seed germination and seedling damping-off might also be observed (Deb et al. 2020).

***Phoma*-related species and disease development:** *Phoma complanata* (Tode) Desm. (syn. *Calophoma complanata* (Tode) Qian Chen & L. Cai) is the causal agent of root canker (Deb et al. 2020). The fungus has been reported from the upper northern hemisphere and could be an alternative pathogen of weeds. The disease has been reported as seed-borne. Pycnidia could survive on plant debris at low temperatures (0–1 °C) for more than 100 days. No chlamydospore production has been reported for *P. complanata* (Aveskamp et al. 2008).

Hosts: parsnip (*Pastinaca sativa* L.), parsley (*Petroselinum crispum* (Mill.) Fuss), carrot (*Daucus carota* subsp. *sativus* (Hoffm.) Schübl. & G. Martens), and cow-parsnip (*Heracleum maximum* W. Bartram) have been reported to be susceptible to *P. complanata* (Cerkauskas 1987; Deb et al. 2020).

Integrated management strategies: the application of certified seeds or seed treatment with thiram could effectively reduce disease incidence due to the seed-borne nature of the disease (Cerkauskas 1987). Moreover, removing and destroying plant debris and roots with canker symptoms could reduce the incidence of root canker. This is mainly because the fungus overwinters on plant debris (Aveskamp et al. 2008). Appropriate tillage practices as well as crop rotation may effectively reduce the inoculum, because no chlamydospores are formed (Deb et al. 2020). Some American cultivars of parsnip, including All American and Hollow Crown cultivars, have shown significant resistance to root canker (Cerkauskas 1987).

8.7 Fruit Diseases

Fruit diseases are caused by various species assigned to *Phoma*. These *Phoma* species cause huge economic losses especially as postharvest diseases. Black rot, gangrene, and ripening are the most important fruit diseases worldwide.

8.7.1 Black Rot

Black rot is considered a destructive disease of fruits; in particular, it has been reported as a postharvest fruit rot worldwide (Aveskamp et al. 2008). The disease has been reported from Brazil, Canada, China, India, Iran, Italy, Poland, and the USA (Pethybridge et al. 2011; Basım et al. 2016; Colmán et al. 2018).

Symptoms: irregular light-colored spots (usually white spots) could be observed on fruits. These rot spots are leathery and sunken (Davari and Hajieghrari 2008). Pycnidia could be observed in dark areas on the fruits. In severe infections, the fruits may become mummified and dried out (Deb et al. 2020). The disease could also affect leaves and stem causing small brownish spots on leaves and canker on the stem (Davari and Hajieghrari 2008).

***Phoma*-related species and disease development:** *Phoma vitis* Schulzer, *P. herbarum*, *P. negriana* Thüm. (syn. *Didymella negriana* (Thüm.) Qian Chen & L. Cai), *P. pomorum* Thüm. (syn. *D. pomorum* (Thüm.) Qian Chen & L. Cai), *P. macdonaldii*, *P. destructiva*, *P. strasseri* Moesz (syn. *Boeremia strasseri* (Moesz) Aveskamp, Gruyter & Verkley), and *P. cucurbitacearum* (syn. *Stagonosporopsis cucurbitacearum*; syn. *D. bryoniae*) are reported the main causal agents of black rot (Basım et al. 2016; Colmán et al. 2018; Deb et al. 2020). The maximum infection rate occurs in humid conditions (more than 90% humidity and temperatures between 16 and 24 °C or even warmer). The fungus can be easily dispersed by wind and rain. The conidia could successfully infect leaves, fruit, petioles, and shoots during the season (Davari and Hajieghrari 2008). The fungus overwinters as mycelia and pycnidia in plant debris. *Phoma cucurbitacearum* has also been reported as a seed-borne pathogen (Colmán et al. 2018).

Hosts: grapevine (*Vitis vinifera* L.), sunflower, peppermint (*Mentha* × *piperita* L.), tomato, *Physalis pubescens* L., and cucurbits are the main hosts of black rot disease (Colmán et al. 2018; Gao et al. 2018; Deb et al. 2020).

Integrated management strategies: eradication of plant debris is the most practical approach to control the disease (Basım et al. 2016). The application of *Rhizobia* bacterium as plant growth-promoting Rhizobacteria (PGPR) agents could effectively control black rot disease (Aveskamp et al. 2008). In the case of seed-borne species, the application of certified seeds could be a matter of importance (Deb et al. 2020). Furthermore, the seeds should be collected from healthy plants. Due to the wide host range of the pathogen, crop rotation and the application of resistant cultivar could be challenging. However, a 2- to 3-year crop rotation with no cucurbit crops would also be practical (Deb et al. 2020).

8.7.2 Gangrene

Gangrene is a potato (*Solanum tuberosum* L.) disease, causing huge economic losses worldwide in stored potatoes. The disease has been reported several times from Australia, Canada, China, Europe (especially from Italy, Sweden, the Netherlands, and the UK), Iran, Japan, New Zealand, Northern Africa, South American Andes, and Ukraine (Jellis 1982; Hide and Read 1990; MacDonald et al. 2000; Sharifi et al. 2002; Boerema 2004; Yang et al. 2012).

Symptoms: various types of symptoms have been reported worldwide (Abbas et al. 2013). Sunken, round, oval, or irregular lesions might be observed on umbilicus or buds of the infected potato tubers (Cullen et al. 2007; Yang et al. 2012). The symptoms may appear on tubers after 30 days of storage at 4 °C. The lesions might become khaki to lilac (Jiang et al. 2013). Sunken lesions are 1–3 cm in size and may expand progressively, becoming irregular in shape, and usually with a recognized wavy edge externally (Yang et al. 2012; Jiang et al. 2013). Some other researchers reported that the symptoms may first become recognizable as small dark depressions or disorders related to previous wounds (Cullen et al. 2007). These depressions may become larger to produce irregularly formed areas known as thumb marks. In severe infections, internal rot and formation of cavities beneath these depressions may also take place. The symptoms may sometimes be confused with dry rot disease caused by *Fusarium* species (Cullen et al. 2007). In less severe infections, the lesions are limited to the surface and the disease is called pit rot, which might appear as immature gangrene lesions (A'Hara 2015).

Phoma-related species and disease development: *Phoma foveata* Foister (syn. *Boeremia foveata* (Foister) Aveskamp, Gruyter & Verkley), *P. exigua* (syn. *B. exigua*), *P. exigua* Desm. var. *foveata* (Foister) Boerema, *P. exigua* var. *exigua* Desm., *P. eupyrena* Sacc. (syn. *Juxtiphoma eupyrena* (Sacc.) Valenz.-Lopez, Crous, Stchigel, Guarro & Cano), and *P. matteuccicola* Aderkas, Gruyter, Noordel & Strongman are reported as the causal agents for gangrene (Abbas et al. 2013; A'Hara 2015). *Phoma exigua* is regarded as a species complex (Aveskamp et al. 2008). *Phoma exigua* var. *exigua* has a wide host range and wide geographical distribution (Europe and Asia) compared to *P. foveata* and *P. exigua* var. *foveata* (Cimmino et al. 2008; Machowicz-Stefaniak et al. 2008; A'Hara et al. 2012). Besides, *P. eupyrena* is the less aggressive pathogen compared to the other mentioned species which causes pit rot on tubers (A'Hara 2015). On the other hand, *P. matteuccicola* has been reported from Canada and the USA on a spring vegetable (Von Aderkas et al. 1992; Boerema 2004). *Phoma exigua* var. *foveata* can appear on potato tubers 1 month after storage and can overwinter in soil and field. It is once reported that *P. exigua* var. *foveata* might be able to aerially spread, and the local rain splash could be effective on its dispersal (Deb et al. 2020). It has been stated that pathogen growth is temperature related. In other words, low temperatures (between 2 and 6 °C) enhance the disease severity, whereas high temperatures (between 7 and 31 °C) have a negative impact on pathogen growth. It is also noted that the age of tubers has an impact on their susceptibility to gangrene disease. The longer storage period, the more susceptible tubers to lesion extension (A'Hara 2015). Furthermore, the Mg:Ca ratio is considered as an important factor in tuber resistance to gangrene because of the effects of these cations on plant cell wall against enzymatic degradation of gangrene causal agents (Olsson 1988). Hence, tubers with high levels of Mg:Ca are less susceptible to gangrene (Olsson 1988). Although this disease leads to huge economic losses in storage potato, the infected seed tubers may become the sources of inoculum for the next crop season. As a result, favorable environmental conditions for the pathogen may lead to significant yield loss (Giebel and Dopierała 2004).

Hosts: this disease has been reported on potato tubers, bulbs, and stem base. Moreover, it has been also observed on ostrich fern (*Matteuccia struthiopteris* (L.) Todero) (Von Aderkas et al. 1992).

Integrated management strategies: various management strategies have been proposed to control the disease, including certified seeds, resistant potato varieties, appropriate handling, and storage at 15 °C as well as the application of chemical fungicides, such as benomyl thiabendazole, benomyl (1%), captofol (1%), and chitosan (0.5%) (Abbas et al. 2013; Mohammed and Eskov 2020). Moreover, *P. exigua* var. *foveata* is included in the lists of quarantine pathogens of the European Plant Protection Organization (Giebel and Dopierała 2004; Polozhenets and Nemerytska 2019). The application of molecular techniques, such as PCR-RFLP and multiplex quantitative real-time PCR, may assist in early pathogen identification which could be helpful for plant pathologists to apply an appropriate management strategy to control the disease (Cullen et al. 2007; Aveskamp et al. 2008; A'Hara et al. 2012).

8.7.3 Ripening

Ripening or simply fruit rot is one of the *Phoma* spp.-inducing diseases both in the field and storage (Colmán et al. 2018; Deb et al. 2020).

Symptoms: brown to black spots have been reported on tomato fruits, leading to yield loss (Colmán et al. 2018). Moreover, irregular necrotic lesions could also be observed on olive fruits (Moral et al. 2017). Leaves might also be infected. Take balsam pear (*Momordica charantia* L.) as an example in which orange to brown lesions might appear on the leaves (Furukawa et al. 2007). These lesions are irregular in shape and may contain black pycnidia (Fig. 8.1h). The infected leaves may become scorched and cracked. In cucurbits, the symptoms may appear with gummy stem blight symptoms (Furukawa et al. 2007).

Phoma-related species and disease development: *Phoma exigua*, *P. destructiva*, *P. lycopersici* Cooke (syn. *Boeremia lycopersici* (Cooke) Aveskamp, Gruyter & Verkley; syn. *Didymella lycopersici* Kleb), *P. cucurbitacearum* (*D. bryoniae*), and *P. olea* (DC) Sacc. (syn. *Dothiora oleae* (DC.) Crous) have been reported as the main pathogens, causing ripening and fruit rot (Furukawa et al. 2007; Aveskamp et al. 2008; Deb et al. 2020). High relative humidity and temperatures between 25 and 30 °C are reported to be the best environmental conditions for the fungal attack (Furukawa et al. 2007; Colmán et al. 2018). The fungus could be easily spread via infected seeds. Furthermore, plant debris are considered as the main primary inoculum, containing fungal resistant structures, such as perithecia, pycnidia, and mycelia. Ascospores and conidia can infect the plants in rainy seasons (Deb et al. 2020).

Hosts: sunflower, tomato, balsam pear, and olive are the main hosts of the disease (Furukawa et al. 2007; Colmán et al. 2018; Deb et al. 2020; Sun et al. 2020).

Integrated management strategies: the application of certified seeds, removal of plant residues, controlling the fruit storage humidity and temperature, as well as the application of mancozeb, thiram, and captofol fungicides can effectively control the ripening caused by *Phoma* species (Deb et al. 2020).

8.8 Seedling Diseases

The most common seedling disease caused by *Phoma* species is damping-off. Certain *Phoma* species infections lead to pre-emergence damping-off. The disease symptoms are different from those caused by *Pythium* species. Moreover, premature ripening, which causes a crisis in all sunflower cultivation areas, is another devastating seedling disease.

8.8.1 Damping-Off

Seedling damping-off is one of the reported diseases of some plants from Canada, eastern Europe, New Zealand, the UK, and the USA (Slimene et al. 2012; Chand et al. 2019; Koenick et al. 2019; Deb et al. 2020).

Symptoms: young seedlings are susceptible to the disease and would be infected first (Chand et al. 2019). The symptoms have been observed initially as brown to black lesions on young seedlings which results in damping-off (Fig. 8.1j) (Slimene et al. 2012). Brown, circular to oval-shaped lesions with dark concentric rings may also appear on leaves (as leaf spot) (Chand et al. 2019). The spots may quickly merge leading to leaves' defoliation and death (Deb et al. 2020). Then, the disease may spread throughout the crown, causing crown rot (Chand et al. 2019). Moreover, taproot has also been reported as a susceptible organ to the disease (Koenick et al. 2019). The pathogen can also cause wounds on roots (Pethybridge et al. 2018; Chand et al. 2019). Root rot with dark and sunken spots, which later become watery and soft, has also been reported as one of the major symptoms of seedlings' damping-off in severe infections (Slimene et al. 2012).

Phoma-related species and disease development: *Phoma medicaginis* Malbr. & Roum., *P. megasperma* f. sp. *medicaginis* Drechs., and *P. betae* (syn. *Neocamarosporium betae*) have been reported to cause seedling damping-off (Handelsman et al. 1990; Slimene et al. 2012; Pethybridge et al. 2018; Chand et al. 2019). Asexual reproduction of the fungi results in the rapid spread of the disease (Chand et al. 2019). In favorable environmental conditions, i.e., cool rainy seasons, a large number of conidia would be formed in pycnidia which could be promptly distributed by heavy rains (Koenick et al. 2019). The fungi produce pseudothecia as the sexual structure which in turn produces aerially dispersed ascospores over long distances (Koenick et al. 2019). *Phoma betae* has also been recorded as seed-borne pathogen (Chand et al. 2019). Hence, the physical contact of conidia and ascospores with the plant's reproduction structures (such as floral parts) may result in disease spread via seed (Chand et al. 2019).

Hosts: sugar beet (*B. vulgaris* and *B. vulgaris* subsp. *vulgaris*), alfalfa, and barrel clover (*M. truncatula* L.) have been reported numerous times as a major host pathogen (Handelsman et al. 1990; Garibaldi et al. 2007; Slimene et al. 2012; Deb et al. 2020). Even though other members of *Chenopodiaceae* family have also been recorded as major hosts, *B. vulgaris* is the most susceptible host (Chand et al. 2019; Koenick et al. 2019; Deb et al. 2020).

Integrated management strategies: crop rotation with non-*Chenopodiaceae* family members and the application of certified seeds could appropriately reduce disease damages (Chand et al. 2019). Seed treatments with thiram and hot water could significantly reduce disease initiation (Koenick et al. 2019; Deb et al. 2020). Due to the survival of the pathogen on plant debris and surrounding weeds between growing season, removal of plant debris and infected weeds could be useful as a management strategy (Chand et al. 2019). The effective application of biological control agents, such as *Bacillus subtilis* (Ehrenberg) Cohn L194 strain and *B. cereus* Frankland & Frankland UW85 strain, has been claimed in the literature (Slimene et al. 2012).

8.8.2 Premature Ripening

Premature ripening is a lethal seedling disease of sunflower caused by *Phoma* species in all sunflower cultivation regions, especially France, causing huge economic losses (Bordat et al. 2017; Deb et al. 2020; Sun et al. 2020).

Symptoms: wilt and stock rot are reported as the main symptoms of premature ripening. The leaves have been claimed to be wilted and necrotic (Aveskamp et al. 2008). Plant stalk would be brownish-black, and in severe infections, the plant might die before maturity (Bordat et al. 2017). Plant head would become smaller. Seed yield reduction and reduced weight and oil content are other responses to premature ripening disease (Bordat et al. 2017). The causal agent can also attack the stem, causing canker (Sun et al. 2020).

Phoma-related species and disease development: *Phoma macdonaldii* has been reported as the causal agent of premature ripening (Aveskamp et al. 2008). The fungus could feasibly attack the plant via collar infection at the flowering stage (Bordat et al. 2017). The disease severity could be increased with a high nitrogen content of the fertilizer applied, resulting in collar infection (Seassau et al. 2010). The fungus overwinters as ascospores or pycnidia in plant debris, which represent the primary inoculum (Deb et al. 2020). High precipitations would result in severe infections, due to the importance of humidity in releasing the ascospores (Deb et al. 2020; Sun et al. 2020).

Hosts: sunflower is the only host reported for the premature ripening disease (Bordat et al. 2017).

Integrated management strategies: the application of certified seeds, eradication of plant residues, as well as chemical control with benomyl and propiconazole are the most appropriate management strategies for ripening control (Seassau et al. 2010; Deb et al. 2020). Moreover, long tillage and crop rotation can also reduce disease incidence (Seassau et al. 2010). The use of resistant cultivars, such as FU, XRQ, PAZ2, and PSC8, has also been proposed and claimed to be a promising approach to effectively control the disease (Bordat et al. 2017).

8.9 Conclusion

The correct and early detection of the symptoms caused by *Phoma* species on various hosts is crucial for applying preventive and effective management strategies based on the integrated pest management to control plant diseases. Hence, crop sanitation, certified seeds, appropriate crop practices, chemical and biological control, as well as resistant cultivars may effectively control the diseases caused by *Phoma* species. Even though nowadays researchers believe that the use of resistant cultivars would be the best approach to manage *Phoma* diseases, it is still not always feasible because of the wide host range of the pathogens and it is widespread in all types of soils and plant materials. Moreover, a single disease could be caused by more than one *Phoma* species. Thus, the application of resistant cultivars is still a challenge.

References

A'Hara, D. (2015). Detection and identification of *Phoma* pathogens of potato. In *Plant Pathology* (pp. 17–27). Humana Press, New York, NY.

Abbas, M. F., Naz, F., & Irshad, G. (2013). Important fungal diseases of potato and their management – a brief review. *Mycopathology, 11*(1), 45–50.

Abbate, L., Mercati, F., & Fatta del Bosco, S. (2019). (An overview on citrus mal secco disease: approaches and strategies to select tolerant genotypes in *C. limon. Crop Breeding, Genetics and Genomics, 1*(2), 1–29.

A'Hara, D., Mulholland, V., & Ventisei, H. M. (2012). Development of a multiplex real-time quantitative PCR assay to detect *Phoma* pathogens of potato. In *Proceedings Crop Protection in Northern Britain* (pp. 279–284).

Anco, D. (2019). Peanut disease management. *Peanut Money-Maker, 42,* 184–196.

Aubertot, J. N., West, J. S., Bousset-Vaslin, L., Salam, M. U., Barbetti, M. J., & Diggle, A. J. (2006). Improved resistance management for durable disease control: a case study of Phoma stem canker of oilseed rape (*Brassica napus*). *European Journal of Plant Pathology, 114,* 91–106.

Aveskamp, M. M., De Gruyter, J., & Crous, P. W. (2008). Biology and recent developments in the systematics of *Phoma*, a complex genus of major quarantine significance. *Fungal Diversity, 31,* 1–18.

Balmas, V., Scherm, B., Ghignone, S., Salem, A. O. M., Cacciola, S. O., & Migheli, Q. (2005). Characterisation of Phoma tracheiphila by RAPD-PCR, microsatellite-primed PCR and ITS rDNA sequencing and development of specific primers for in planta PCR detection. *European Journal of Plant Pathology, 111*(3), 235–247.

Barua, P., You, M. P., Bayliss, K. L., Lanoiselet, V., & Barbetti, M. J. (2018). Inert materials as long-term carriers and disseminators of viable *Leptosphaeria maculans* ascospores and wider implications for ascomycete pathogens. *Plant Disease, 102*(4), 720–726.

Baş, B., & Koç, N. K. (2006). In vitro selection of Kütdiken lemon 20b to canditate for resistance to Phoma tracheiphila. *Plant Pathology Journal (Faisalabad), 5*(1), 35–40.

Basım, E., Basım, H., Abdulai, M., Baki, D., & Öztürk, N. (2016). Identification and characterization of *Didymella bryoniae* causing gummy stem blight disease of watermelon (*Citrullus lanatus*) in Turkey. *Crop Protection, 90,* 150–156.

Ben-Hamo, M., Ezra, D., Krasnov, H., & Blank, L. (2020). Spatial and temporal dynamics of mal secco disease

spread in lemon orchards in Israel. *Phytopathology, 110*(4), 863–872.
Boerema, G. H. (Ed.). (2004). *Phoma identification manual: differentiation of specific and infra-specific taxa in culture*. CABI. The Netherlands.
Bordat, A., Marchand, G., Langlade, N. B., Pouilly, N., Muños, S., Dechamp-Guillaume, G., & Bret-Mestries, E. (2017). Different genetic architectures underlie crop responses to the same pathogen: the {*Helianthus annuus* Phoma macdonaldii*} interaction case for black stem disease and premature ripening. *BMC Plant Biology, 17*(1), 167–182.
Burketová, L., Jindřichová, B., Stehlík, D., Trdá, L., Mazáková, J., Fajemisin, O., & Ryšánek, P. (2018). *Leptosphaeria* species causing blackleg disease in oilseed rape and *L. maculans* race incidence in the Czech Republic. *Welcome at the Faculty of Agriculture of the University of Zagreb, 136*, 189–190.
Cacciola, S.O. & Gullino, M.L. (2019). Emerging and re-emerging fungus and oomycete soil-borne plant diseases in Italy. *Phytopathologia Mediterranea, 58*, 451–472.
Caesar, A. J., Lartey, R. T., & Caesar-Ton-That, T. C. (2012). First report of a root and crown disease of the invasive weed *Lepidium draba* caused by *Phoma macrostoma*. *Plant Disease, 96*(1), 145.
Cerkauskas, R. F. (1987). Pathogenicity and survival of *Phoma complanata*. *Canadian Journal of Plant Pathology, 9*(1), 63–67.
Chand, N., Jones, E. E., & Casonato, S. (2019). Pathogenicity of *Phoma betae* isolates from red beet (*Beta vulgaris*) at seed farms in Canterbury, New Zealand. *New Zealand Plant Protection, 72*, 21–26.
Chen, S. F., Morgan, D. P., & Michailides, T. J. (2013). First report of *Phoma fungicola* associated with stem canker and fruit blight of pistachio in Arizona. *Journal of Plant Pathology, 95*(2), 447–452.
Chen, Q., Jiang, J. R., Zhang, G. Z., Cai, L., & Crous, P. W. (2015). Resolving the *Phoma* enigma. *Studies in Mycology, 82*, 137–217.
Cimmino, A., Andolfi, A., Berestetskiy, A., & Evidente, A. (2008). Production of phytotoxins by *Phoma exigua* var. *exigua*, a potential mycoherbicide against perennial thistles. *Journal of Agricultural and Food Chemistry, 56*(15), 6304–6309.
Claassen, B. J. (2016). Investigations of black leg and light leaf spot on *Brassicaceae* hosts in Oregon. MSc. Thesis, Oregon University, USA.
Colmán, A. A., Alves, J. L., da Silva, M., & Barreto, R. W. (2018). *Phoma destructiva* causing blight of tomato plants: a new fungal threat for tomato plantations in Brazil?. *Tropical Plant Pathology, 43*(3), 257–262.
Cullen, D. W., Toth, I. K., Boonham, N., Walsh, K., Barker, I., & Lees, A. K. (2007). Development and validation of conventional and quantitative polymerase chain reaction assays for the detection of storage rot potato pathogens, *Phytophthora erythroseptica, Pythium ultimum* and *Phoma foveata*. *Journal of Phytopathology, 155*(5), 309–315.
Davari, M., & Hajieghrari, B. (2008). *Phoma negriana*, a new invasive pathogen for Moghan's vineyards in Iran. *African Journal of Biotechnology, 7*, 788–791.
de Gruyter, J., Woudenberg, J.H.C., Aveskamp, M.M., Verkley, G.J.M., Groenewald J.Z., & Crous P.W. (2013). Redisposition of *Phoma*-like anamorphs in *Pleosporales*. *Studies in Mycology 75*: 1–36.
Deb, D., Khan, A., & Dey, N. (2020). Phoma diseases: Epidemiology and Control. *Plant Pathology, 1*, 1–15.
Demontis, M. A., Cacciola, S. O., Orrù, M., Balmas, V., Chessa, V., Maserti, B. E., & Migheli, Q. (2008). Development of real-time PCR systems based on SYBR® Green I and TaqMan® technologies for specific quantitative detection of *Phoma tracheiphila* in infected Citrus. *European Journal of Plant Pathology, 120*(4), 339–351.
Deng, Z. N., Gentile, A., Domina, F., Nicolosi, E., & Tribulato, E. (1995). Selecting lemon protoplasts for insensitivity to *Phoma tracheiphila* toxin and regenerating tolerant plants. *Journal of the American Society for Horticultural Science, 120*(6), 902–905.
Esfahani, M. N., & Pour, B. A. (2008). Differences in resistance in onion cultivars to pink root rot disease in Iran. *Journal of General Plant Pathology, 74*(1), 46–52.
Fan, Q., Creamer, R., & Li, Y. (2018). Time-course metabolic profiling in alfalfa leaves under *Phoma medicaginis* infection. *PloS one, 13*(10), e0206641.
Fernando, W. G., Zhang, X., & Amarasinghe, C. C. (2016). Detection of *Leptosphaeria maculans* and *Leptosphaeria biglobosa* causing blackleg disease in canola from Canadian canola seed lots and dockage. *Plants, 5*(1), 12–23.
Ferdous, M. J., Hossain, M. R., Park, J. I., Robin, A. H. K., Jesse, D. M. I., Jung, H. J., ... & Nou, I. S. (2019). Inheritance pattern and molecular markers for resistance to blackleg disease in cabbage. *Plants, 8*(12), 583–600.
Furukawa, T., Ono, Y., & Kishi, K. (2007). Gummy stem blight of balsam pear caused by *Didymella bryoniae* and its anamorph *Phoma cucurbitacearum*. *Journal of General Plant Pathology, 73*(2), 125–128.
Gao, P., Li, Y., Guo, Y., & Duan, T. (2018). Co-inoculation of lucerne (*Medicago sativa*) with an AM fungus and a rhizobium reduces occurrence of spring black stem and leaf spot caused by *Phoma medicaginis*. *Crop and Pasture Science, 69*(9), 933–943.
Garibaldi, A., Gilardi, G., Bertetti, D., & Gullino, M. L. (2007). First report of leaf spot and root rot caused by *Phoma betae* on *Beta vulgaris* subsp. *vulgaris* (Garden beet Group) in Italy. *Plant Disease, 91*(11), 1515.
Gentile, A., Tribulato, E., Continella, G., & Vardi, A. (1992). Differential responses of citrus calli and protoplasts to culture filtrate and toxin of *Phoma tracheiphila*. *Theoretical and Applied Genetics, 83*(6–7), 759–764.
Giebel, J., & Dopierała, U. (2004). Pathogenesis of potato gangrene caused by *Phoma exigua* var. *foveata*: II. Activities of some hydrolases and dehydrogenases. *Journal of Phytopathology, 152*(7), 399–403.

Grantina-Ievina, L., & Kalniņa, I. (2016). Strawberry crown rot–a common problem in 2015. *Environmental and Experimental Biology, 14*, 51–52.

Hajlaoui, M. R., Kalai, L., Mnari-Hattab, M., Guermech, A., & Abdelaal, N. B. (2008). Occurrence of 'mal nero' disease on mandarin and orange trees in Tunisia. *Plant Pathology, 57*(4), 784.

Handelsman, J., Raffel, S., Mester, E. H., Wunderlich, L., & Grau, C. R. (1990). Biological control of damping-off of alfalfa seedlings with *Bacillus cereus* UW85. *Applied and Environmental Microbiology, 56*(3), 713–718.

He, J., Zhou, R., Xu, Z., Cui, J., & Fu, J. (2014). Toxicity measurement of different fungicides on *Phoma arachidicola. The Proceedings of Chinese Society of Plant Protection in 2014, Fujian, China, 5–7 December, 2014*, 522–526.

Hide, G. A., & Read, P. J. (1990). Effect of neighboring plants on the yield of potatoes from seed tubers affected with gangrene (*Phoma foveata*) or from plants affected with stem canker (*Rhizoctonia solani*). *Annals of Applied Biology, 116*(2), 233–243.

Hollingsworth, C. R., & Gray, F. A. (1999). First report of brown root rot on alfalfa caused by *Phoma sclerotioides* in the continental United States. *Plant Disease, 83*(11), 1071.

Hollingsworth, C. R., Gray, F. A., Koch, D. W., Groose, R. W., & Heald, T. E. (2003). Distribution of *Phoma sclerotioides* and incidence of brown root rot of alfalfa in Wyoming, USA. *Canadian Journal of Plant Pathology, 25*(2), 215–217.

Hollingsworth, C. R., Gray, F. A., & Groose, R. W. (2005). Evidence for the heritability of resistance to brown root rot of alfalfa, caused by *Phoma sclerotioides. Canadian Journal of Plant Pathology, 27*(1), 64–70.

Hong-hai, Y. A. N., Xue-wu, W. A. N. G., Ru-qin, Z. H. A. N. G., Yu-cheng, C. H. I., & Shu-chun, X. I. A. (2013). Endogenous salicylic acid expression and its resistance to peanut web blotch. *Chinese Journal of Oil Crop Sciences, 35*(6), 723–727.

Hossard, L., Jeuffroy, M. H., Pelzer, E., Pinochet, X., & Souchere, V. (2013). A participatory approach to design spatial scenarios of cropping systems and assess their effects on Phoma stem canker management at a regional scale. *Environmental Modelling & Software, 48*, 17–26.

Hossard, L., Souchere, V., & Jeuffroy, M. H. (2018). Effectiveness of field isolation distance, tillage practice, cultivar type and crop rotations in controlling Phoma stem canker on oilseed rape. *Agriculture, Ecosystems & Environment, 252*, 30–41.

Hossain, M. R., Ferdous, M. J., Park, J. I., Robin, A. H. K., Natarajan, S., Jung, H. J., ... & Nou, I. S. (2020). In-silico identification and differential expression of putative disease resistance-related genes within the collinear region of Brassica napus blackleg resistance locus LepR2'in Brassica oleracea. Horticulture, *Environment, and Biotechnology, 61*(5), 879–890.

Hwang, S. F., Strelkov, S. E., Peng, G., Ahmed, H., Zhou, Q., & Turnbull, G. (2016). Blackleg (*Leptosphaeria maculans*) severity and yield loss in canola in Alberta, Canada. *Plants, 5*(3), 31–42.

Ikeda, K., Kuwabara, K., Urushibara, T., Soyai, P., Miki, S., & Shibata, S. (2012). Pink root rot of squash caused by *Setophoma terrestris* in Japan. *Journal of General Plant Pathology, 78*(5), 372–375.

Ivic, D., Ivanovic, A., Milicevic, T., & Cvjetkovic, B. (2010). Shoot necrosis of olive caused by *Phoma incompta*, a new disease of olive in Croatia. *Phytopathologia Mediterranea, 49*(3), 414–416.

Javaid, A., Gajula, H., Fitt, B. D. L., & Huang, Y. (2017). Investigating the risk of severe Phoma stem canker caused by *Leptosphaeria biglobosa* on winter oilseed rape in UK. *Aspects of Applied Biology*, (134), 53–57.

Jellis, G. J. (1982). Laboratory assessments of the susceptibility of potato tubers to gangrene (*Phoma exigua* var. *foveata*). *Plant Pathology, 31*(2), 171–177.

Jiang, H., Yang, C., Xue, L., Pu, C., Chen, X., Shang, X., & Li, C. (2013). Identification and biological characteristics of the pathogen causing the potato gangrene in Gansu Province. *Acta Prataculturae Sinica, 22*(2), 123–131.

Jing-jing, H. E., Ru-jun, Z. H. O. U., Jian-chao, C. U. I., Zhe, X. U., Jun-fan, F. U., & Nan, W. U. (2015). Indoor toxicity measurement and field control effects of different fungicides against *Phoma arachidicola. Chinese Journal of Oil Crop Sciences, 37*(4), 525–531.

Kaewkham, T., Hynes, R. K., & Siri, B. (2016). The effect of accelerated seed ageing on cucumber germination following seed treatment with fungicides and microbial biocontrol agents for managing gummy stem blight by *Didymella bryoniae. Biocontrol Science and Technology, 26*(8), 1048–1061.

Kalai, L., Mnari-Hattab, M., & Hajlaoui, M. R. (2010). Molecular diagnostics to assess the progression of *Phoma tracheiphila* in *Citrus aurantium* seedlings and analysis of genetic diversity of isolates recovered from different *Citrus* species in Tunisia. *Journal of Plant Pathology, 92*, 629–636.

Kalai-Grami, L., Slimane, I. B., Mnari-Hattab, M., Rezgui, S., Aouani, M. A., Hajlaoui, M. R., & Limam, F. (2014). Protective effect of *Bacillus amyloliquefaciens* against infections of *Citrus aurantium* seedlings by *Phoma tracheiphila. World Journal of Microbiology and Biotechnology, 30*(2), 529–538.

Kalai-Grami, L., Karkouch, I., Naili, O., Slimene, I. B., Elkahoui, S., Zekri, R. B., & Limam, F. (2016). Production and identification of iturin A lipopeptide from Bacillus methyltrophicus TEB1 for control of Phoma tracheiphila. *Journal of Basic Microbiology, 56*(8), 864–871.

Khanchouch, K., Pane, A., Ali Chriki, A., & Cacciola, S. O. (2017). Major and emerging fungal diseases of citrus in the Mediterranean region. In Citrus Pathology Gill, H., & Garg, H. (Eds.) (pp. 3–30), IntechOpen, London. UK.

Koenick, L. B., Vaghefi, N., Knight, N. L., du Toit, L. J., & Pethybridge, S. J. (2019). Genetic diversity and differentiation in *Phoma betae* populations on table beet in New York and Washington States. *Plant Disease, 103*(7), 1487–1497.

Koenning, S. R., Frye, J. W., Pataky, J. K., Gibbs, M., & Cotton, D. (2007). First report of *Phoma terrestris* causing red root rot on sweet corn (Zea mays) in North Carolina. *Plant Disease, 91*(8), 1054.

Koike, S. T., Subbarao, K. V., Verkley, G. J., Fogle, D., & O'Neill, T. M. (2006). Phoma basal rot of romaine lettuce in California caused by *Phoma exigua*: occurrence, characterization, and control. *Plant Disease, 90*(10), 1268–1275.

Larsen, R. C., Hollingsworth, C. R., Vandemark, G. J., Gritsenko, M. A., & Gray, F. A. (2002). A rapid method using PCR-based SCAR markers for the detection and identification of *Phoma sclerotioides*: the cause of brown root rot disease of alfalfa. *Plant Disease, 86*(9), 928–932.

Larsen, J. E., Hollingsworth, C. R., Flor, J., Dornbusch, M. R., Simpson, N. L., & Samac, D. A. (2007). Distribution of *Phoma sclerotioides* on alfalfa and winter wheat crops in the North Central United States. *Plant Disease, 91*(5), 551–558.

Li, Y., Zheng, C., Zhao, Z., Lu, G., Fu, X., Mei, R., & Wang, Q. (2013). Characterization of *Phoma adonidicola* causing a spot blight on Adonis palaestina. *European Journal of Plant Pathology, 137*(1), 127–136.

Licciardello, G., Grasso, F. M., Bella, P., Cirvilleri, G., Grimaldi, V., & Catara, V. (2006). Identification and detection of *Phoma tracheiphila*, causal agent of citrus mal secco disease, by real-time polymerase chain reaction. *Plant Disease, 90*(12), 1523–1530.

Liu, H., Sun, Z., Zhang, X., Qin, L., Qi, F., Wang, Z., & Li, S. (2020). QTL mapping of web blotch resistance in peanut by high-throughput genome-wide sequencing. *BMC Plant Biology, 20*, 1–11.

Lou, B. G., Hu, X. Q., & Wu, L. (2008). First report of blackleg of *Ajuga multiflora* caused by *Phoma multirostrata* in China. *Plant Disease, 92*(6), 981.

Macdonald, J. E., White, G. P., & Côté, M. J. (2000). Differentiation of *Phoma foveata* from *P. exigua* using a RAPD generated PCR-RFLP marker. *European Journal of Plant Pathology, 106*(1), 67–75.

Machowicz-Stefaniak, Z., Zimowska, B., & Zalewska, E. (2008). The occurrence and pathogenicity of *Phoma exigua* Desm. var. *exigua* for selected species of herbs. *Acta Agrobotanica, 61*(2), 157–166.

Mahish, P. K., & Ghritlahare, A. (2017). Pathogenicity of *Phoma chrysanthemicola* to chrysanthemum plants (*Asteraceae* Family) and control of pathogen by chemical and biological approach. *Biosciences Biotechnology Research Asia, 14*(3), 1191–1200.

McCredden, J., Cowley, R. B., Marcroft, S. J., & Van de Wouw, A. P. (2018). Changes in farming practices impact on spore release patterns of the blackleg pathogen, *Leptosphaeria maculans*. *Crop and Pasture Science, 69*(1), 1–8.

McDonald, D., & Beckerman, S. R. (1985). *Proceedings of the Regional Groundnut Workshop for Southern Africa 26–29 March 1984 Lilongwe, Malawi*. International Crops Research Institute for the Semi-Arid Tropics.

Menzies, S. A., & Colhoun, J. (1976). Control of Phoma root rot of chrysanthemums by the use of fertilizers. *Transactions of the British Mycological Society, 67*(3), 455–462.

Migheli, Q., Cacciola, S. O., Balmas, V., Pane, A., Ezra, D., & di San Lio, G. M. (2009). Mal secco disease caused by *Phoma tracheiphila*: a potential threat to lemon production worldwide. *Plant Disease, 93*(9), 852–867.

Mikunthan, G. (1997). First report of web blotch of peanut caused by *Phoma arachidicola* in the dry zone of Sri Lanka. *Plant Disease, 81*(7), 832–832.

Mitrousia, G. K., Huang, Y. J., Qi, A., Sidique, S. N. M., & Fitt, B. D. L. (2018). Effectiveness of Rlm7 resistance against *Leptosphaeria maculans* (Phoma stem canker) in UK winter oilseed rape cultivars. *Plant Pathology, 67*(6), 1339–1353.

Mohammed, S. R., & Eskov, I. D. (2020). The use of chitosan against potato gangrene during storage. *The Agrarian Scientific Journal*, (1), 17–21.

Moon, J. S., Kim, H. M., Choi, D. C., Hong, Y. K., Sung, M. H., Jang, Y. J., & Choi, Y. G. (2003). Disinfection of seed borne black leg disease (*Phoma wasabiae*) in Wasabi (*Wasabia japonica* Matsum.). *Protected Horticulture and Plant Factory, 12*(4), 180–183.

Moral, J., Agustí-Brisach, C., Pérez-Rodríguez, M., Xaviér, C., Raya, M. C., Rhouma, A., & Trapero, A. (2017). Identification of fungal species associated with branch dieback of olive and resistance of table cultivars to *Neofusicoccum mediterraneum* and *Botryosphaeria dothidea*. *Plant Disease, 101*(2), 306–316.

Moral, J., Lichtemberg, P. S. F., Papagelis, A., Sherman, J., & Michailides, T. J. (2018). *Didymella glomerata* causing leaf blight on pistachio. *European Journal of Plant Pathology, 151*(4), 1095–1099.

Nachmias, A., Barash, I., Buchner, V., Solel, Z., & Strobel, G. A. (1979). A phytotoxic glycopeptide from lemon leaves infected with *Phoma tracheiphila*. *Physiological Plant Pathology, 14*(2), 135–140.

Newby, D. L., Carroll, R. B., & Whittington, D. P. (1997). Incidence of *Phoma terrestris*, causal agent of red root rot, in corn roots and soil in Delaware and on the eastern shore of Maryland. *Plant Disease, 81*(2), 232.

Nigro, F., Ippolito, A. & Salerno, M. (2011). Mal secco disease of citrus: a journey through a century of research. *Journal of Plant Pathology, 93* (3), 523–560.

Olsson, K. (1988). Resistance to gangrene (*Phoma exigua* var. *foveata*) and dry rot (*Fusarium solani* var. *coeruleum*) in potato tubers. 1. The influence of pectin-bound magnesium and calcium. *Potato Research, 31*(3), 413–422.

Pethybridge, S. J., Gent, D. H., & Hay, F. S. (2011). Epidemics of ray blight on pyrethrum are linked to seed contamination and overwintering inoculum of *Phoma ligulicola* var. *inoxydabilis*. *Phytopathology, 101*(9), 1112–1121.

Pethybridge, S. J., Kikkert, J. R., Hanson, L. E., & Nelson, S. C. (2018). Challenges and prospects for building resilient disease management strategies and tactics for the New York table beet industry. *Agronomy*, *8*(7), 112–129.

Pettit, R. E., Philley, G. L., Smith, D. H., & Taber, R. A. (1986). Peanut Web Blotch: II Symptoms and Host Range of Pathogen. *Peanut Science*, *13*(1), 27–30.

Polozhenets, V., & Nemerytska, L. (2019). Phomous rot is a dangerous disease of potato tubers. *Quarantine and Plant Protection*, 12, 28–32.

Robin, A. H. K., Larkan, N. J., Laila, R., Park, J. I., Ahmed, N. U., Borhan, H., & Nou, I. S. (2017). Korean *Brassica oleracea* germplasm offers a novel source of qualitative resistance to blackleg disease. *European Journal of Plant Pathology*, *149*(3), 611–623.

Seassau, C., Dechamp-Guillaume, G., Mestries, E., & Debaeke, P. (2010). Nitrogen and water management can limit premature ripening of sunflower induced by *Phoma macdonaldii*. *Field Crops Research*, *115*(1), 99–106.

Sharifi, K., Ershad, D., & Sahihi, Z. (2002). Identification of gangrene disease on imported seed potato tubers. *Iranian Journal of Plant Pathology*, *38*(3/4), 277–285.

Slimene, I. B., Tabbene, O., Djebali, N., Cosette, P., Schmitter, J. M., Jouenne, T., & Limam, F. (2012). Putative use of a *Bacillus subtilis* L194 strain for biocontrol of *Phoma medicaginis* in *Medicago truncatula* seedlings. *Research in Microbiology*, *163*(5), 388–397.

Smith, D. H. & Littrell, R. H. (1980). Management of Peanut Foliar Diseases. *Plant Disease*, *64*, 356–361.

Solel, Z., & Spiegel-Roy, P. (1978). Methodology of selection of lemon clones for tolerance to mal secco (Phoma tracheiphila). *Phytoparasitica, 6*(3), 129–134.

Sprague, S. J., Graham, J. M., Brill, R., & McMaster, C. (2018). Infection of *Brassica napus* after stem elongation by *Leptosphaeria maculans* (blackleg): disease progression and yield loss. *AusCanola 2018 Co-hosts*, 179.

Subrahmanyam, P., & Hildebrand, G. L. (1994). Integrated disease management: an important component in sustaining groundnut production in the SADC Region. In *Sustainable groundnut production in southern and eastern Africa: proceedings of a Workshop, 5–7 Jul 1994, Mbabane, Swaziland* (Vol. 502, p. 45).

Sun, H., Kav, N. N., Liang, Y., Sun, L., & Chen, W. (2020). Proteome of the fungus *Phoma macdonaldii*, the causal agent of black stem of sunflower. *Journal of Proteomics*, 103878.

Thanassoulopoulos, C. C. (1991). "Ermioni" a new lemon cultivar resistant to mal secco disease (*Phoma tracheiphila*). *Journal of Phytopathology*, *131*(3), 234–242.

Thomidis, T., Michailides, T. J., & Exadaktylou, E. (2011). *Phoma glomerata* (Corda) Wollenw. & Hochapfel a new threat causing cankers on shoots of peach trees in Greece. *European Journal of Plant Pathology*, *131*(2), 171–178.

Triki, M. A., Gharbi, Y., Bouazizi, E., Cheffi, M., Krid, S., Feki, F. A., & Bouhamed, J. (2019). First report of branch blight of almond trees caused by *Nothophoma quercina* in Tunisia. *Journal of Plant Pathology*, *101*(4), 1277.

Tusa, N., Fatta del Bosco, S., Nigro, F., & Ippolito, A. (2000). Response of cybrids and a somatic hybrid of lemon to *Phoma tracheiphila* infections. *Horticultural Science*, *35*(1), 125–127.

Tuzcu, O., Cinar, A., Kaplankiran, M., Erkilic, A., & Yesiloglu, T. (1989). Resistance of some Citrus species and hybrids to mal secco (Phoma tracheiphila Kanc. et Ghik.) disease. *Fruits, 44*(3), 139–148.

Vaghefi, N., Silva, A., Koenick, L. B., & Pethybridge, S. J. (2019). Genome resource for *Neocamarosporium betae* (syn. *Pleospora betae*), the cause of Phoma leaf spot and root rot on *Beta vulgaris*. *Molecular Plant-Microbe Interactions*, *32*(7), 787–789.

Vakili, Z. Z., Rahnama, K., Narollah, N. S., & Yamchi, A. (2017). Molecular identification of *Leptosphaeria maculans* and determination of aggressive new pathotypes canola Phoma stem canker in north Iran. *Journal of Plant Protection (Agricultural Science and Technology), 31,* 296–311.

Von Aderkas, P., De Gruyter, J., Noordeloos, M. E., & Strongman, D. B. (1992). *Phoma matteuccicola* sp. nov., the causal agent of gangrene disease of ostrich fern. *Canadian Journal of Plant Pathology*, *14*(3), 227–228.

Wang, Y., Strelkov, S. E., & Hwang, S. F. (2020). Yield losses in canola in response to blackleg disease. *Canadian Journal of Plant Science, 1,* 1–7.

West, J. S., Fitt, B. D., Leech, P. K., Biddulph, J. E., Huang, Y. J., & Balesdent, M. H. (2002). Effects of timing of *Leptosphaeria maculans* ascospore release and fungicide regime on Phoma leaf spot and Phoma stem canker development on winter oilseed rape (*Brassica napus*) in southern England. *Plant Pathology*, *51*(4), 454–463.

Wunsch, M. J., & Bergstrom, G. C. (2011). Genetic and morphological evidence that *Phoma sclerotioides*, causal agent of brown root rot of alfalfa, is composed of a species complex. *Phytopathology*, *101*(5), 594–610.

Wunsch, M. J., Schindelbeck, R. R., Van Es, H. M., & Bergstrom, G. C. (2007). Distribution, impact, and soil environment of *Phoma sclerotioides* in northeastern US alfalfa fields. *Plant Disease*, *91*(10), 1293–1304.

Wunsch, M. J., Dillon, M. A., Torres, R., Schwartz, H. F., & Bergstrom, G. C. (2008). First report of brown root rot of alfalfa caused by *Phoma sclerotioides* in Colorado and New Mexico. *Plant Disease*, *92*(4), 653.

Yang, C. D., Chen, X. R., Jiang, H. X., & Pu, C. J. (2012). First report of potato gangrene caused by *Phoma foveata* in China. *Plant Disease*, *96*(11), 1698–1698.

Zamani-Noor, N., & Knüfer, J. (2018). Effects of host plant resistance and fungicide application on Phoma stem canker, growth parameters and yield of winter oilseed rape. *Crop Protection*, *112*, 313–321.

Zhang, X., Xu, M., Wu, J., Dong, W., Chen, D., Wang, L., & Chi, Y. (2019). Draft genome sequence of *Phoma arachidicola* Wb2 causing peanut web blotch in China. *Current Microbiology*, *76*(2), 200–206.

Phoma as an Infectious Pathogen in Medical Practice

9

Yelena V. Kukhar and Vladimir S. Kiyan

Abbreviations

AgNPs	Silver nanoparticles
AR	Agglutination reaction
AuNPs	Gold nanoparticles
BSL	Biosafety level
CAR	Coagglutination reaction
CFR	Complement fixation reaction
CID	Counter immunodiffusion reaction
DNA	Deoxyribonucleic acid
DPR	Diffusion precipitation reaction
ELISA	Enzyme-linked immunosorbent assay
IDR	Immunodiffusion reaction
IEP	Immunoelectrophoresis
IFR	Immunofluorescence reaction
ITS	Internal transcribed spacer regions
KOH	Potassium hydroxide
LA	Latex agglutination
MIC	Minimum inhibitory concentration
NaOH	Sodium hydroxide
PCR	Polymerase chain reaction
PHAR	Passive hemagglutination reaction
RNA	Ribonucleic acid

Y. V. Kukhar (✉) · V. S. Kiyan (✉)
Research Platform of Agricultural Biotechnology, Saken Seifullin Kazakh Agrotechnical University, Nur-Sultan, Kazakhstan

9.1 Introduction

Phoma is an anamorphic complex consisting of *Dematiaceae* fungi belonging to the phylum *Ascomycota*, class *Dothideomycetes*, order *Pleosporales*, and family *Didymellaceae* (Jones et al. 2011; Parr et al. 2014; Valenzuela-Lopez et al. 2018). These organisms inhabit soil, organic debris, and water and include species that parasitize other fungi, insects, and vertebrates. Of the more than 200 currently known *Phoma* varieties, about 10 species are considered opportunistic or primarily pathogenic (De Hoog et al. 2000; Boerema et al. 2004; Aveskamp et al. 2010). Among them are *Phoma oculo-hominis* (Valenzuela-Lopez et al. 2018), *Phoma eupyrena* (Bakerspigel et al. 1981), *Phoma minutella* (Baker et al. 1987), *Phoma hibernica* (Roehm et al. 2012), *Phoma sorghina* (Rai 1989), *Phoma herbarum*, *Didymella heteroderae* (Tullio et al. 2010; Valenzuela-Lopez et al. 2017), *Phoma cava* (Zaitz et al. 1997; Valenzuela-Lopez et al. 2018), *Phoma minutispora* (Shukla et al. 1984), *Phoma insulana* (Hernández-Hernández et al. 2018), *Phoma macrostoma* (Kukhar et al. 2020), *Phoma exigua* (Balis et al. 2006), *Phoma glomerata* (Errera et al. 2008), and other fungi of the *Phoma* genus (Bennett et al. 2018).

The first human case of *Phoma* (*P. hibernica*) infection in 1970 was reported by Bakerspigel (1970). Lesions of human fungi *Phoma* spp. is a rather rare occurrence. However, as the population

M. Rai et al. (eds.), *Phoma: Diversity, Taxonomy, Bioactivities, and Nanotechnology*,
https://doi.org/10.1007/978-3-030-81218-8_9

of the population with a weakened immune system increases, there are growing reports of these infections. There are currently over 30 reports of human infections caused by *Phoma* spp. (Bennett et al. 2018).

Fungi of the genus *Phoma* can be found in foods. It is known that *Phoma* spp. infect seeds, nuts, soybeans, potatoes, bananas, sorghum, corn, kiwi berries, lemons, tomatoes, eggplants, and pomegranates (Bennett et al. 2018). It has been argued that food sources contaminated with *Phoma* spp. may pose a much more significant threat to immunosuppressed people than simply defeat to agricultural crops. Given the constant increase in the number of opportunistic fungal infections, correlated with an increase in the number of immunosuppressed people (Zaitz et al. 1997), the consumption of foods containing *Phoma* spp. can become dangerous for them. Contaminants can serve as fomites in invasive fungal infections.

Also, it should be noted that medical advances have made it possible and helped to increase the number of transplants of parenchymal organs, the use of chemotherapy for the treatment of malignant tumors, and the use of other immunosuppressive agents, resulting in an increased risk of infection of the population affected by various types of fungi of the genus *Phoma*. A number of publications related to the mention of fungi of the genus *Phoma* as an etiological factor in systemic and superficial mycoses, especially frequent pathologies in patients with immunosuppression, are reported (Zaitz et al. 1997). This is evidenced by the data of cultural-morphological and molecular-genetic studies provided by various authors (Zaitz et al. 1997; Balis et al. 2006; Bennett et al. 2018; Kukhar et al. 2020).

There is another potential danger of fungi of the genus *Phoma*. *Phoma* spp. can produce metabolites such as cytochalasin A and B, deoxaphomine, proxyfomin, and tenuazonic acid, which can be cytotoxic (Weidenbörner 2001; Visconti et al. 1987; Cole et al. 2003; Lugauskas et al. 2006). For example, *P. sorghina* is known to produce tenuazonic acid. This mycotoxin causes acute toxic effects in animals, such as precancerous changes in the esophageal mucosa in mice treated with tenuazonic acid for 10 months (Yekeler et al. 2001). Oliveira et al. (2017) report that the production of tenuazonic acid after ingestion of grains infected with *P. sorghina* may correlate with a hemorrhagic disorder in humans known in Brazil as onyalay.

Currently, human infection with *Phoma* spp. becomes more and more frequent. The spectrum of diseases caused by *Phoma* spp. is extremely wide, from lesions of the skin and its appendages to systemic invasive diseases. Opportunistic infections caused by *Phoma* spp. are manifested in the form of subcutaneous mycoses, onychomycosis, systemic phaeohyphomycosis lung damage, damage to the paranasal sinuses, sinusitis, and disseminated infection with the formation of brain abscesses in immunodeficiencies after inhalation of conidia (Kiran Kumar and Kalleker 2008; Mishra et al. 2011). At the same time, the clinical and histopathological picture of diseases caused by *Phoma* spp. does not have its own peculiarities; the pathogen is detected by chance during complex diagnostic measures (Young et al. 1973).

Phoma spp. are isolated from various biological materials—skin, subcutaneous tissue, phlegm, purulent discharge from the eyes, and paranasal sinuses. In the diagnosis of mycosis caused by *Phoma* spp. in skin scrapings or sputum or tissue biopsies treated with 10% KOH, septate hyphae were detected from light gray to brown (Zaitz et al. 1997; Balis et al. 2006; Bennett et al. 2018; Kukhar et al. 2020).

For the diagnosis of *Phoma* spp., inoculations are made on nutrient media such as Sabouraud dextrose agar, where typical colonies grow. The genus *Phoma* is characterized by spherical, ellipsoidal colonies immersed in the substrate, rarely protruding pycnidia, and unicellular, colorless conidia (Boerema et al. 2004).

Colonies of *Phoma* species are greenish-gray to brown; some species may develop red or pinkish pigmentation. Growth rate varies widely between species. Dark pycnidia of various shapes (from globular to subglobular or pear-shaped) with one or more stomata can form individually or in the form of aggregates. The hyaline conidiogenic cells lining the inner walls of the pycnidia produce their conidia by percurrent proliferation, which are then squeezed out into the mucous masses (Crous et al. 2014). Conidia of various shapes are unicellular and transparent. *Alternaria*-like chlamydospores are found in some species like *P. glomerata*. The main parameters of identification determined by morphological characters are the formation of conidia (asexual spores), pycnidia (asexual fruiting bodies), and chlamydospores (enlarged, thick-walled vegetative cells in hyphae or at the apex of hyphae) (Kukhar et al. 2020).

Thus, while there is the potential for infection with *Phoma* spp. through the consumption of contaminated food, there is not enough data on the body's state that can lead to infection. There is no professional medical information about clinical signs and main symptom complexes in human diseases caused by *Phoma* spp. Also, there is no legislatively approved methodological instruction (approved standard, detection, and identification protocol) for the detection of this organism, although there are guidelines for the identification of *Phoma* in culture. But the correct laboratory identification of the etiological agent is crucial for the correct strategy and tactics of treating patients (Sutton et al. 2009).

There are no standardized treatments, but voriconazole, posaconazole, and itraconazole show the most consistent in vitro activity against this group of fungi (Bennett et al. 2018).

This section contains a discussion of the epidemiology, clinical manifestations of diseases caused by fungi of the *Phoma* genus, and its biological characteristics, which can be essential for diagnosing various manifestations of phaeohyphomycosis.

9.2 Spectrum of Diseases Caused by *Phoma* spp.

Phoma spp. can cause subcutaneous, deep, superficial mycoses. The infection is called phaeohyphomycosis. The spectrum of diseases caused by *Phoma* spp. is extremely wide. Opportunistic infections caused by *Phoma* spp. are manifested in the form of subcutaneous mycoses, onychomycosis, lung damage, damage to the paranasal sinuses, sinusitis, and eye infections. In contrast to the genus *Pyrenochaeta*, no cases of eumycetoma caused by *Phoma* have been reported.

A review of the available literature revealed 33 cases of phaeohyphomycosis caused by *Phoma* spp. At the same time, in 66.7% of the total number of cases, patients were diagnosed with "superficial or deep skin lesions" (22 patients) (Bakerspigel 1970; Young et al. 1973; Gordon et al. 1975; Punithalingam 1979; Bakerspigel et al. 1981; Shukla et al. 1984; Baker et al. 1987; Hirsh and Schiff 1996; Rosen et al. 1996; Zaitz et al. 1997; Arrese et al. 1997; Oh et al. 1999; Everett et al. 2003; Suh 2005; Faisal et al. 2007; Vasoo et al. 2011; Schieffelin et al. 2014; Crous et al. 2014); 6% of cases were diagnosed as onychomycosis (2 patients) (Tullio et al. 2010; Kukhar et al. 2020). In 9% of cases (3 patients), the diseases concerned lung lesions (Janke 1956; Morris et al. 1995; Balis et al. 2006); in 3% of cases (1 patient), invasive rhinosinusitis was revealed (Balis et al. 2006). The remaining 15% of cases of phaeohyphomycosis (5 patients) were associated with *Phoma* spp. as a result of trauma or wearing contact lenses (trauma of the eyeball, keratitis, ulcers, and corneal abscess) (Punithalingam 1976; Rishi and Font 2003; Errera et al. 2008; Kumar et al. 2015; McElnea et al. 2015).

In this case, the course of phaeohyphomycosis can proceed both in a superficial form, affecting the skin of the legs, hands, fingers, and ears or mucous membranes of the eyes or nasal sinus, and in the form of disseminated infection with the formation of brain abscesses in immunodeficiency after inhalation of conidia (Kiran Kumar and Kalleker 2008; Mishra et al. 2011) or in the form of systemic phaeohyphomycosis (Everett et al. 2003; Chowdhary et al. 2014).

The age range of patients diagnosed with *Phoma* spp.-related phaeohyphomycosis ranged from 1 month to 77 years, including three cases of infection in newborns (9%) (Punithalingam 1979; Bakerspigel et al. 1981; Kalyani et al. 2006).

A retrospective analysis of the data on the incidence of phaeohyphomycosis carried out by Bennett et al. (2018) showed that the first recorded case of human infection with an infection caused by *Phoma* spp. was recorded in 1956 (Janke 1956). The next case of *Phoma* spp. was described in a young Canadian farmer who complained of skin lesions of the lower limb that disappeared after treatment (Bakerspigel 1970).

Systemic lesions of *Phoma* spp. cause lung disease and brain abscesses. Systemic phaeohyphomycosis associated with *Phoma* spp. is typically seen in severely immunocompromised patients and usually has a poor outcome. Risk factors or underlying diseases associated with *Phoma* infections may include diabetes mellitus, corticosteroid therapy, and cancer chemotherapy (Zaitz et al. 1997; Everett et al. 2003; Balis et al. 2006; Kalyani et al. 2006; Kiran Kumar and Kalleker 2008; Mishra et al. 2011; Chowdhary et al. 2014). The diagnosis of *Phoma* spp. phaeohyphomycosis in immunocompromised patients following kidney transplant surgery was reported by Young et al. in 1973 and Everett et al. in 2003. There are known cases of detection of phaeohyphomycosis caused by *Phoma* spp. in patients with strong immunosuppression due to chemotherapy associated with a blood disease: acute myeloid leukemia and acute lymphoblastic leukemia (Balis et al. 2006; Kalyani et al. 2006).

Mixed infections of phaeohyphomycosis have also been reported, when the disease is caused by several rare fungal organisms. A famous case reported in the literature is that of *Phoma* spp. and *Acremonium* in an infant with invasive rhinosinusitis that resulted in the death of the newborn (Kalyani et al. 2006). At the same time, according to official statistics, the overall mortality rate from diseases caused by *Phoma* spp. is quite low compared to mortality from other common fungal infections and currently accounts for only 6% of deaths (2 patients).

Until recently, the identification of *Phoma* spp. was quite difficult, therefore, in early publications; the pathogen in most cases (17 out of 33) was noted as a representative of *Phoma* spp. without specifying the species, which is 51.5%. Advances in modern diagnostics allow more accurate identification of the pathogen using molecular genetics methods with an indication of the species, Table 9.1.

Phaeohyphomycosis. Phaeohyphomycosis (phaeomycotic cyst) is caused by a variety of dematized (brown-pigmented) fungi (genus *Exophiala, Phialophora, Wangiella, Bipolaris, Exserohilum, Cladophialophora, Phaeoannellomyces, Aureobasidium, Cladosporium, Curvularia, Alternaria, Phoma*) usually living on decaying plants and wood (Revankar (2006). Compared to the rest of the pathogens mentioned above, fungi of the genus *Phoma* are rare pathogens of phaeohyphomycosis (Silveira and Nucci 2001).

Some authors call *Phoma cava* the primary causative agent of phaeohyphomycosis (Sudduth et al. 1992; Padhye et al. 1998), although cases of infection with other species of *Phoma* spp. are described. Most infections occur sporadically, more often in older men who work with plant materials as farmers, carpenters, or in other professions (Fader and McGinnis 1998).

Most infections of phaeohyphomycosis are observed in immunocompetent individuals. Patients are more often immunologically compromised by such major diseases as diabetes mellitus, tuberculosis, leprosy, and leukemia. Some patients have local immunosuppression due to long-term use of topical corticosteroids. Also, with an increase in the number of people with immunodeficiency, infections are more and more often recorded during stem cell transplantation (Mesa et al. 1999).

Phaeohyphomycosis is registered everywhere. Human-to-human transmission has not been established. Lesions can occur in almost any part of the body but usually at sites that are exposed to external influences, especially on the arms and legs (Everett et al. 2003; Suh 2005).

Infection with *Phoma* spp. occurs as a result of traumatic implantation of plants, woody material, or soil particles contaminated with *Phoma* spp. (Young et al. 1973; Sutton et al. 2009). The most common cause is the implantation of contaminated plant material.

The most common and typical lesions are skin or subcutaneous cysts or abscesses. The primary lesion begins as a single, discrete, asymptomatic small nodule. It is palpable under smooth and slightly raised skin. In the affected tissue, a pain-

Table 9.1 Described cases of infections caused by *Phoma* spp. in humans

Phoma spp.	Source	Immunosuppression	Treatment	Reference
Phoma exigua	Pulmonary	Acute leukemia, diabetes mellitus, chemotherapy	Amphotericin B	Janke (1956); Morris et al. (1995); Balis et al. (2006)
Phoma spp.	Sinus (invasive rhinosinusitis) [mixed infection]	Acute lymphoblastic leukemia; s/p chemotherapy	Amphotericin B, posaconazole, voriconazole, debridement	Roehm et al. (2012)
Phoma insulana	Chromoblastomycosis (deep foot)	Chronic alcoholism, smoker	None/lost to follow-up	Hernández-Hernández et al. (2018)
Phoma spp., incl. *Phoma oculo-hominis*, *Phoma glomerata*	Eye (corneal ulcer, keratitis, endophthalmitis, corneal ulcer and abscess); globe trauma; retinal detachment surgery after penetrating globe injury (risk factor, used contact lenses)	Otherwise healthy	Amphotericin eye (intravitreal), amphotericin B (intravitreal), voriconazole (intravitreal), oral itraconazole, debridement, keratectomy	Punithalingam (1976); Rishi and Font (2003); Errera et al. (2008); Kumar et al. (2015); McElnea et al. (2015)
Phoma herbarum, *Phoma macrostoma (syn. Didymella macrostoma)*	Nail, toe (mixed infection, non-dermatophytic onychomycosis)	Otherwise healthy, chronic alcoholism	Allylamine, sertaconazole	Tullio et al. (2010); Kukhar et al. (2020)
Phoma spp., incl. *Phoma cruris-hominis*	Skin (deep: subcutaneous, subcutaneous phaeohyphomycosis)	Otherwise healthy	Itraconazole	Punithalingam (1979); Oh et al. (1999)
Phoma spp., incl. *Phoma cava*, *Phoma eupyrena*, *Phoma minutispora*, *Phoma sorghina*	Skin (face, superficial ear, perioral lesions, deep face, neck, hands)	Otherwise healthy, oral steroids, typhoid fever, topical steroids	Oral griseofulvin, clotrimazole, topical clotrimazole, topical miconazole, ketoconazole, amphotericin B, dimethicone corticosteroid, 15% zinc oxide paste	Gordon et al. (1975); Bakerspigel et al. (1981); Shukla et al. (1984); Rai (1989); Rosen et al. (1996); Suh (2005)
Phoma spp., incl. *Phoma cava*	Skin (deep/hands, ganglion cysts on wrist, forearm)	Otherwise healthy, pulmonary sarcoidosis, oral steroids s/p renal transplant, diabetes mellitus	Itraconazole, oral itraconazole, ketoconazole, amphotericin B, surgical debridement, surgical excision	Hirsh and Schiff (1996); Zaitz et al. (1997); Everett et al. (2003); Vasoo et al. (2011)

(continued)

Table 9.1 (continued)

Phoma spp.	Source	Immunosuppression	Treatment	Reference
Phoma spp., incl., *Phoma hibernica*, *Phoma minutella*	Skin (deep leg, deep foot, deep heel, plantar foot, deep knee) [mixed infection]	Topical steroids, steroid therapy, diabetes mellitus, azathioprine; prednisone, s/p renal transplant, s/p liver transplant, s/p pancreas transplant, atopic dermatitis	Oral griseofulvin, oral ketoconazole, oral itraconazole, debridement, surgical excision, amputation for secondary gangrene, topical bifonazole and ketoconazole	Bakerspigel (1970); Young et al. (1973); Baker et al. (1987); Arrese et al. (1997); Schieffelin et al. (2014)

less, enclosed mass is formed, which is necrotic. The nodule gradually develops into an encapsulated, fluctuating abscess with fluid content: a sinus tract or ulcer forms on the surface of the epidermis. If the main focus of infection is in the epidermis and dermis, then sometimes a granulomatous, slightly raised plaque may appear.

Cases of detection of *Phoma* spp. are described in deep lesions of the skin of the face and neck with localization of foci on the ears, around the mouth, and eyes; with lesions of the skin of the trunk and extremities (arms and legs) with localization on the knees, feet, heels, soles of the feet, and toenails; and with foot injuries.

Lesions often appear on the hands (English 1980) or legs, but it happens on other areas of the skin (Dawson and Lepper 1970). As a rule, there is only one lesion. A dense subcutaneous nodule appears that is not welded to the underlying tissues and is easily palpable and painless. The skin over the nodule is not changed. The nodule enlarges and turns into a fluctuating but also painless abscess with purulent contents. The skin above it changes if an abscess breaks open or ulcerates in this place (Hirsh and Schiff 1996).

The existence of an abscess usually does not affect the general condition of the patient. No typical symptoms (fever, weight loss); the hemogram does not change.

Typically, *Phoma* spp. are described in connection with superficial or deep skin infections except in the case described by Young et al. (1973). During diagnostic measures, *P. minutella* was isolated from the tissues and contents of the abscess to identify the pathogen from the subcutaneous inflammatory process on the farmer's leg undergoing corticosteroid therapy for myasthenia gravis. It was possible to diagnose and identify the pathogen using complex cultural, histological, and microscopic measures. The authors identified the fungus *P. minutella* by direct microscopic examination of tissue homogenates and aspirates from the affected area, as well as stained biopsies. In addition, mold was recovered on multiple isolation media inoculated with portions of the same clinical specimens (Baker et al. 1987).

When the nails of the fingers and toes are affected by *Phoma* spp. fungi, patients develop onychomycotic lesions of the nails and skin of the periungual space (Tullio et al. 2010; Kukhar et al. 2020).

Phoma spp. can also cause systemic opportunistic infections, including sinusitis in patients with chronic allergic rhinitis or immunosuppression, or brain abscess in immunodeficiencies after inhalation of conidia (Errera et al. 2008; Kumar et al. 2015).

Lymph node involvement and dissemination are rare. However, in some cases, infection with phaeohyphomycosis can involve the central nervous system or other internal organs, such as the liver, lungs, or pancreas. It can manifest itself as hematogenous metastasis from foci on the skin, under the skin, or without visible damage (Bakerspigel 1970; Young et al. 1973; Gordon et al. 1975; Bakerspigel et al. 1981; Shukla et al. 1984; Baker et al. 1987; Rai 1989; Rosen et al. 1996; Zaitz et al. 1997; Arrese et al. 1997; Oh et al. 1999; Suh 2005; Al-Abdely et al. 2005; Vasoo et al. 2011; Schieffelin et al. 2014).

Chromoblastomycosis. Chromoblastomycosis (chromomycosis) is a chronic infection caused by fungal pigments affecting the skin and subcutaneous tissues, characterized by warty nodules or plaques, caused by *Cladophialophora carrionii*, *Fonsecaea compacta*, *Fonsecaea pedrosoi*, and *Phialophora verrucosa*. In endemic areas, *Fonsecaea pedrosoi* and *Cladophialophora carrionii* are mainly found (Hospenthal 2014).

Chromoblastomycosis is a chronic granulomatous fungal skin disease that develops at the site of a previous skin injury (Schell and Esterre 2005; Lopez Martinez and Mendez Tovar 2007). Chromomycosis is considered an occupational disease of agricultural workers and loggers, who often walk barefoot (Queiroz-Telles et al. 2009). The disease is characterized by a relatively benign course with a tendency to progression and the formation of granulomatous-verrucous eruptions (Hay 2008; Salgado et al. 2009). The disease is characterized by damage mainly to the lower extremities. The onset of the disease is characterized by the presence of several warty bluish-red bumps at the site of injury. Their number increases over time; the disease progresses and spreads to neighboring areas of the skin and deep into the soft tissues; dense, painless nodes appear in the dermis and subcutaneous fat. After their opening, an ulcer with a papillomatous bottom is formed (papillomatous-ulcerative stage). Without treatment, it grows into a cauliflower-like mass, leading to lymphatic blockage and elephantiasis-like edema of the lower limb. The process is most often localized in the buttocks, lower back, back of the thighs, knee joints, legs, feet, hands, shoulders, on the face and neck, and the anterior abdominal wall. In some cases, chromoblastomycosis spreads from primary foci to other parts of the body through lymphatic or blood vessels, as a result of which, the liver, lungs, and brain tissues may be damaged (Milam and Fenske 1989).

Phoma is an unusual and very rare causative agent of human chromoblastomycosis (Esterre et al. 1996). In 2018, a case was described of detecting chromoblastomycosis caused by *P. insulana* in a patient with a history of a lacerated foot wound with subsequent warty growth and scaly lesions that developed over 27 years. On visual examination of the patient, chromoblastomycosis was clinically confirmed; microscopy of the scales showed the presence of fumagoid cells. During culture studies, a dematiaceous fungus was grown, which, based on morphology and PCR sequencing, was identified as *P. insulana* (Hernández-Hernández et al. 2018).

Onychomycosis. Onychomycosis affects about 1/10 of the world's population and is ubiquitous (Baran and Kaoukhov 2005). Up to 1/4 of nail infections can be caused by molds and yeasts that are resistant to a number of antifungal agents (Markinson and Elewski 2007; Veer et al. 2007; Kaur et al. 2008). The lesion begins as whitish, whitish-yellow longitudinal stripes on the lateral surface of the nail plate. Usually, fungi penetrate the nail plate from foci of fungal infection of the interdigital folds of the feet and soles (Araújo et al. 2003). Fungi can also infect the nail from the free edge in the form of subungual hyperkeratosis.

There are three types of nail damage (Arenas 2011). Normotrophic type is when only a discoloration of the nail appears or a slight loosening from the free edge. Trophism of the nail plate does not suffer. The hypertrophic type of lesion is characterized by a thickening of the nail due to subungual hyperkeratosis. Nails are dull, uneven, yellow-gray, crumbling. A feature of this mycosis is partial damage to the nail plate; the area of the hole often remains intact. The atrophic type is characterized by thinning of the nail, its destruction, or detachment from the nail bed in the form of onycholysis or koilonychia (Kaur et al. 2008).

Keratitis. Fungal keratitis is the inflammation of the cornea of the eyes, resulting from a fungal infection of the body. It primarily affects the corneal epithelium and stroma, although the endothelium and anterior chamber of the eye may get involved in more severe disease (Tuli 2011). *Phoma* spp. can affect the eyes during operations on retinal detachment after a penetrating injury of the eyeball, with endophthalmitis, keratitis, corneal ulcers, and eye injuries.

Plant/organic trauma is a predisposing event for fungal keratitis. A typical example would be a puncture wound to the cornea of a farmer with awn, chaff, or durum wheat pieces during harvest. The wound promotes the implantation of the fungus directly into the cornea. The fungus grows slowly and multiplies with penetration into the anterior and posterior stromal layers of the cornea. There are cases when the fungus grew through the Descemet's membrane and entered the anterior chamber of the cornea of the eye (Tuft and Tullo 2009; Gower et al. 2010).

Symptoms of fungal keratitis include blurred vision, red and painful eyes that do not improve with contact lens removal or antibiotic treatment, increased sensitivity to light (photophobia), watery eyes, or uncharacteristic discharge. Symptoms are noticeably less pronounced compared to those of a bacterial ulcer. Signs of such a lesion of the eyelids and their appendages are edema, hyperemia, and irritation of the conjunctiva. Ulceration around the cornea may appear in the form of a dry ulcer, predominantly white, fluffy in appearance. Often such damage to the cornea is accompanied by bacterial suppuration of the fungal ulcer.

This disease is quite common in the tropics and in countries with large rural populations. Keratitis caused by molds is mainly reported in India, where injuries account for up to 20% of cases (Gopinathan et al. 2002; Thomas 2003; Thomas and Kaliamurthy 2013). Cases of fungal keratitis are regularly reported in Florida in the United States (Keay et al. 2011). Widespread in tropical and temperate regions, *Sphaeropsis subglobosa*, coelomycete associated with *Phoma* spp. has been identified as the causative agent of keratomycosis in Scotland after corneal injury with a bamboo cane (Kirkness et al. 1991). The lesion was accompanied by a corneal ulcer without any pigmentation.

Pulmonary infections. Fungal infections of the lung are potentially life-threatening. These infections are observed mainly in patients with weakened immunity or in patients with underlying lung disease and may occur in immunocompetent patients (Chowdhary et al. 2014). A wide variety of molds can be involved in the infection process, including *Phoma* spp. Clinical manifestations of fungal lung infections include pneumonia, pulmonary nodules, and endobronchial lesions. Treatment of pulmonary mycoses consists of intravenous administration of liposomal amphotericin B or azoles active against mold, with the exception of ketoconazole for a long period. However, the mortality rate in immunocompromised patients is high if the main defects of the host defense are not eliminated (Hollingsworth et al. 2007; Mullane et al. 2007; Al-Aidaroos et al. 2007; Woo et al. 2008; Elinav et al. 2009; Lastoria et al. 2009; Revankar and Sutton 2010). A solitary pulmonary nodule in immunocompetent patients can be treated surgically (Borges Jr et al. 1991; Greig et al. 2001; Revankar and Sutton 2010).

Cerebral infection. Fungal infections of the brain are more often caused by endemic neurotropic fungi. Cerebral abscesses have been described that can cause *Rhinocladiella mackenziei* (Middle East), *Cladophialophora bantiana* (India), and *Exophiala dermatitidis* (East Asia).

Cerebral abscess caused by *Phoma* spp. is extremely rare but often fatal. A high pattern of occurrence of these infections in individuals with obvious immunocompetence was revealed (Flanagan and Bryceson 1997; Ebright et al. 1999; Filizzola et al. 2003; Carter and Boudreaux 2004; Delfino et al. 2006; Chang et al. 2009; Al-Tawfiq and Boukhamseen 2011; Revankar 2011). Typically, cerebral abscesses caused by *Phoma* spp. are haematogenous. The initial source of infection is probably a subclinical pulmonary focus. Spores may spread from the lesion of the nasal sinuses or after surgery. Cerebral abscesses are often geographically limited.

9.3 Diagnosis of Mycoses Caused by Species of the Genus *Phoma*

For the diagnosis of mycoses caused by *Phoma* spp., the standard measures used to identify the causative agent of any mycosis are carried out.

Microscopy. Microscopic examination of the pathological material is mandatory and reliable when spores and mycelium of the fungus are found in the pathological material. A positive microscopy result is a basis for a positive diagnosis (Lease and Alexander 2011).

For microscopy, the material is taken from the affected areas of the skin, hair and nails, parenchymal organs, and fresh but already fully developed lesions (Aly 1996).

Microscopic examination of pathological material for fungi is carried out in native and stained preparations by the method of hanging or crushed drops.

In the express diagnosis of mycoses, complex methods of staining smears are often used: according to Gram, treatment with 10% KOH (NaOH), staining with nigrosin or ink, staining according to Romanovsky-Giemsa or Wright, staining with methenamine silver according to Gomori, staining according to Gridley, staining with periodic acid and Schiff's reagent (according to McManus), and direct fluorescence method. Mold fungi are stained with hematotoxylin in blue-gray, gentian violet (in violet), azure-eosin (in blue), and Schiff's reagent—in a deep pink or red color (Zaias and Rebell 1996).

Histological examinations. In a histological study, the tissue reaction is assessed only after the pathogen is detected: the same fungus, depending on the virulence and state of the body, causes tissue reactions of a different nature, or a similar tissue reaction is formed in various mycoses (Zaias and Rebell 1996; Gurta et al. 2000). When stained with methenamine silver, according to Gomori, fungal cells are dark gray or black. When treated with 10% caustic potassium, the structures of pathogens in the fragments of the skin and its appendages are visualized due to the destruction of keratin. At the same time, fungal cells remain unchanged (Aly 1996). When cytological preparations are stained with Calcofluor white, a blue or green glow appears, depending on the filter (Lease and Alexander 2011).

Immunofluorescent staining methods are used to confirm the diagnosis in the absence of fungi in the infiltrate. For these purposes, fluorescein-labeled antiserum or antibodies are used (Guarner and Brandt 2011). The reaction of sulfation and impregnation of histological sections with silver (Lever and Schaumburg-Lever 1983) and staining according to Gram-Wegert (fungal elements—dark violet) and Grocott (dirty blue-green and black staining of fungi) (Guarner and Brandt 2011) is also used.

The culture method. Isolation of pure culture is a highly sensitive and specific method for laboratory diagnosis of mycoses. The cultural method allows you to identify pathogens and determine their genus and species. The cultural method is especially indispensable for the diagnosis of latent forms of mycoses, carriage by healthy people (Sutton et al. 2009).

The culture is determined by its appearance and microscopic features. To identify the isolate isolated from the clinical material, the forms of the sporulation organs and mycelium are identified, and the morphology is studied at the macro- and microscopic levels. The reason is that many unrelated fungi can be identical in morphological characteristics of colonies and are characterized by unusual pleomorphism depending on the growth conditions (Zaitz et al. 1997; Balis et al. 2006; Sutton et al. 2009; Bennett et al. 2018; Kukhar et al. 2020).

Usually, fungal cultures are investigated in dynamics, determining the color, size, nature of the surface, pattern, consistency, growth rate, enzymatic activity, and pathogenicity.

To identify the type of fungus, the nature of the growth of the colonies is taken into account, and then preparations are made. Material for research is prepared from the periphery of the colony and its central part. Usually, unstained preparations are considered, but staining is also possible. Microscopy of the primary culture should be performed several times as the culture grows (Kozel and Wickes 2014).

Immunological methods. Immunological research methods are used for serological diagnosis of fungal diseases (Virella 1997).

For the diagnosis of fungal diseases, a number of serological methods have been developed based on the detection in the blood serum of animal antibodies specific to the causative agent of the disease: immunodiffusion reaction (IDR), complement fixation reaction (CFR), immunoelectrophoresis (IEP), agglutination reaction (AR), immunofluorescence reaction (IFR), passive hemagglutination reaction (PHAR), counter immunodiffusion reaction (CID) and the method of rosette formation, diffusion precipitation reaction (DPR), and coagglutination reaction (CAR) (Bloomfield et al. 1963; Bauman and Smith 1975; Sweet et al. 1979; Marier et al. 1979; Williams et al. 1981; Greenberger and Patterson 1982; de Repentigny 1989; Kozel and Wickes 2014).

Enzyme-linked immunosorbent assay (ELISA) and latex agglutination (LA) are becoming standard methods for diagnosing diseases (Scott et al. 1980; George et al. 1981; Davies and Sarosi 1987; Lambert and George 1987; Richardson and Page 2018). To diagnose various mycoses, latex agglutinating tests for cryptococcosis, maduromycosis, blastomycosis, histoplasmosis, paracoccidioidomycosis, sporotrichosis, and candidiasis have been developed (Marier et al. 1979; Greenberger and Patterson 1982).

Allergic diagnostics. Mycoses are often accompanied by changes in the immunobiological state of the patient's body, manifested in the form of allergic reactions (Simon-Nobbe et al. 2008). The allergic state of the body with mycoses is detected by allergic skin tests: cutaneous, epidermal, and intradermal. Mycoallergens have a more or less pronounced antigenic cross section, but the percentage of positive skin tests is higher with homologous infections (histoplasmin, with histoplasmosis; candidine, with candidiasis; etc.) (Pfavayi et al. 2020).

Molecular diagnostics and molecular identification of the causative agent. Molecular diagnostics make it possible to detect pathogen cell fragments or products of its metabolism in various environments of the organism (Shukla et al. 1984; Bennett et al. 2018). A PCR assay was developed to detect *Phoma exigua* DNA in dewaxed lung biopsy material (Balis et al. 2006).

Molecular analyses that have been used to determine the genus *Phoma* include nuclear rDNA (*ITS*, internal transcribed spacer regions), *28 Snr* RNA gene fragments (*LSU*), RNA polymerase II gene (*rpb2*), and beta-tubulin gene (*tub2*) (Faggi et al. 2001; Kardjeva 2006; Rai et al. 2014).

Despite the number of methods and a relatively high percentage of detection of patients, there are still problems in the diagnosis of mycoses. Difficulties in early diagnosis are often associated with the omission of the initial period of the disease and the delay in mycological examination (Ghannoum et al. 2000; Liu et al. 2002).

Morphology of species of the genus *Phoma*. Human damage caused by species of the genus *Phoma* can be superficial or deep, affecting the skin, cornea, subcutaneous tissue, and lungs. The selection of material for the detection of *Phoma* spp. depends on the location of the lesion, the type of pathology, the duration of the process, and the time of treatment.

The material from which *Phoma* spp. is isolated is the skin, subcutaneous tissue, phlegm, purulent discharge from the eyes, and paranasal sinuses.

In the diagnosis of mycosis caused by *Phoma* spp., in skin scrapings, sputum, or tissue biopsies treated with 10% KOH, septate hyphae were detected from light gray to brown. The mycelium separated before examination should be centrifuged. Detection of hyphae of the pathogen in pathological or biopsy material is a prerequisite for a positive diagnosis.

Tullio V. et al. collected samples of the skin and nails, suspended in a 20% KOH solution, and examined under a light microscope. This made it possible to identify the mycelium of the pathogen (Tullio et al. 2010).

The *Phoma* spp. grow the maximum at a pH close to 5.5.

To carry out cultural and morphological diagnostics, inoculations are usually made on nutrient media such as Sabouraud dextrose agar and its modifications, where typical colonies of *Phoma* spp. appear. When identifying, it should be remembered that the type of nutrient medium contributes to achieving the best growth characteristics in *Phoma* spp.; oat agar supports abundant pycnidia production, while malt extract agar stimulates pigment production and crystal formation (De Hoog et al. 2000).

The main parameters of identification, determined by morphological features, are the formation of conidia (asexual spores), pycnidia (asexual fruiting bodies), and chlamydospores (enlarged, thick-walled vegetative cells in hyphae or at the top of hyphae) (Bennett et al. 2018).

The genus *Phoma* is characterized by spherical, ellipsoidal colonies immersed in the substrate, rarely protruding pycnidia, and unicellular, colorless conidia.

Distinctive features of *Phoma* spp. include the ability to form asexual fruiting bodies lined with conidiogenic fungi, which are spore-producing hyphae. Colonies tend to have a powdery or velvety texture and overgrowth, and pigmentation can range from greenish-gray to brown (Punithalingam 1979).

To detect the growth of fungi in the cultural diagnosis, Tullio et al. (2010) cultivated the pathogen in agar with mycobiotics and Sabouraud dextrose agar containing chloramphenicol. *Phoma herbarum*, *Chaetomium globosum*, and *Microascus cinereus* have been isolated and identified using these methods.

Phoma spp. in vitro exhibit morphological features such as chlamydospores, conidia, and pycnidial conidiomas (fruit structures that act as a means of dispersing conidia), which are unique to this genus, helping to distinguish them from other dematiaceous fungi (Baker et al. 1987; Boerema et al. 2004).

The fungus forms ostiolate fruiting bodies known as pycnidia and numerous small asexual conidia. Pycnidia are grayish dark-brown to black, globular, subglobose, or pear-shaped. Pycnidia with one or more stomata are either immersed in the substrate or on the surface of the agar. Pycnids can be formed individually or as aggregates.

The hyaline conidiogenic cells lining the inner walls of pycnidia produce their conidia by percurrent proliferation, which is then squeezed out as a mucous mass.

Conidia of various shapes are unicellular and transparent. In some *Phoma* species, *Alternaria* chlamydoconidia are found.

Conidia (pycnidiospores) are formed from the phialids lining the inner wall of pycnidia and are hyaline, unicellular but sometimes 2–3 septate, elliptical, rod-shaped, or curved (De Hoog et al. 2000; Roehm et al. 2012).

Thus, *Phoma* species typically form a slow-growing colony of greenish-gray to dark olive-gray to brown or dark brown. Some species may develop red or pinkish colony pigmentation (Kukhar et al. 2020). Growth rate varies widely between species.

The high variability of microscopic morphology leads to ambiguity in the classification of the genus; therefore, phenotypic characters are not always distinguishable between *Phoma* spp.

A case of isolation of a microscopic filamentous fungus of the *Didymellaceae* family of the genus *P. macrostoma*, strain H11, from the distal edge of the affected nail plate of a patient in a narcological dispensary, showed a morphology that was fundamentally different from the classical descriptions of *Phoma* spp.

With the primary isolation of the strain, the formation of a colony on the agar surface appeared on the second day. Colonies are fast-growing, the growth of a white colony was noted from the third day, and the formation of colonies was completed by 10–18 days. As they matured, flat colonies turned pale pink; with age, the colony acquired a pinkish-orange tint in the center with a yellowish edge. The center of the mature colony rose, forming voids; the reverse side of fungi turned to yellow-brown. Embossed radial stripes were formed (Fig. 9.1).

On the second day of growth, there is a massive growth of hyphae from germinated pycnidiospores, the first stages of branching of hyaline septic hyphae of multicellular mycelium *Phoma* spp. On the third day of cultivation of *P. macrostoma* H11, the formation of pycnidia and chlamydospores was observed.

The pycnidia of *Phoma* spp. are on short or barely noticeable special outgrowths of multicellular mycelium, which does not differ in color, thickness, or density from other hyphae. The shell of pycnidia is well expressed, thin, dense, and leathery with well-visible exit holes. The walls of the pycnidia are noticeably darker than the rest of the mycelium (Fig. 9.2).

The culture of *P. macrostoma* isolated by the authors, which caused damage to the patient's nail plate, was distinguished by characteristic micromorphological-specific features (Yekeler et al. 2001).

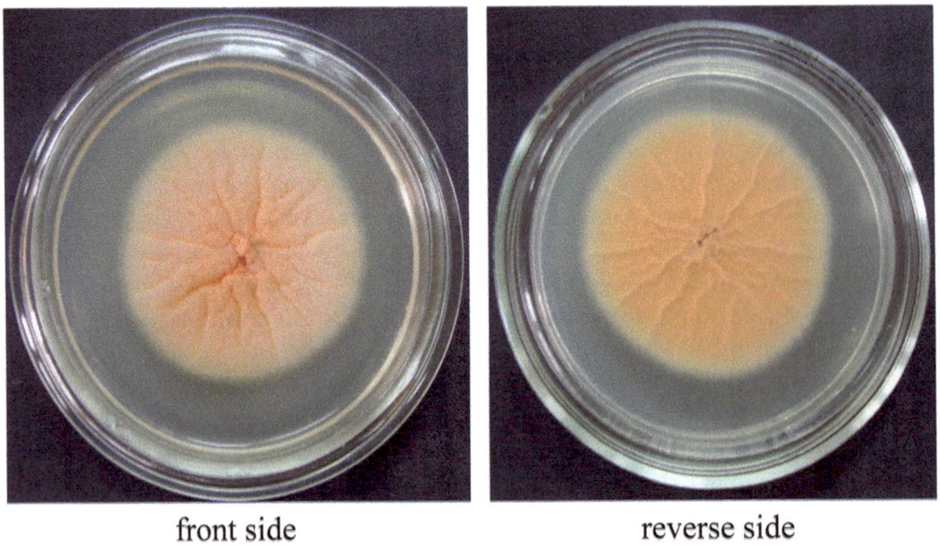

Fig. 9.1 Mature colonies of *P. macrostoma* H11

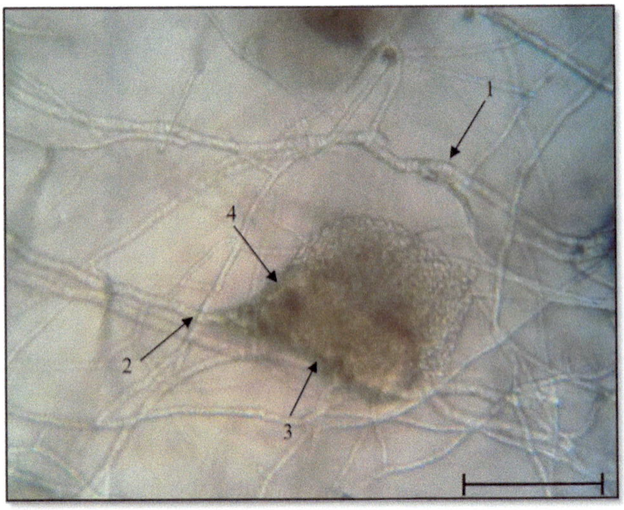

Fig. 9.2 Microscopic structures of *P. macrostoma* like the mycelium are septate, branching, and colorless (1); conidiophores are long, thick, and multicellular (2); pyriform pycnidia (3); ostiole (4), scale bar = 50 μm

9.4 Treatment, Prevention, and Control of Mycoses Caused by *Phoma* spp.

Pathogenesis of diseases associated with *Phoma* spp. The causative agent *Phoma* spp. penetrates into tissues along with a foreign body and causes an inflammatory granulomatous reaction there. Initially, a small abscess is formed with purulent contents and a pathogen inside. Then a capsule is formed around the abscess, and the so-called phaeogyphomycosis cyst is formed.

There are three stages of cyst development:

1. Tuberculoid
2. Stellate abscess
3. A fluctuating abscess

The lesion is usually located in the dermis or subcutaneous tissue. Sometimes the cyst opens on its own, forming a fistulous course, especially in patients with severe immunosuppression. In such patients, the capsule is formed worse. Sometimes, instead of an abscess, a focus of granulomatous inflammation may develop. The primary focus turns into a dense subcutaneous infiltrate, which can ulcerate or capture the underlying tissue. Against the background of severe immunodeficiency, necrotic foci appear.

Lesions can occur in almost any part of the body but usually in areas that are exposed to external influences, especially on the arms and legs (Mirhendi et al. 2013). The infectious agent can enter through wounds made with contaminated plant materials. The most common and typical lesions are skin or subcutaneous cysts or abscesses. The primary lesion begins as a single, discrete, asymptomatic small nodule. It is palpable under smooth and slightly raised skin. The nodule gradually develops into an encapsulated, fluctuating abscess with fluid content. A sinus tract or ulcer forms on the surface of the epidermis (Ono et al. 2004). Sometimes a granulomatous, slightly raised plaque can appear when the main focus of infection is in the epidermis and dermis. Lymph node involvement and dissemination are rare. However, in some cases, infection with phaeogyphomycosis can involve the central nervous system or other internal organs, such as the liver, lungs, or pancreas. It can manifest as hematogenous metastasis from lesions on the skin, under the skin, or without visible damage.

In case of damage to the skin and its appendages on the hands and feet, onychomycotic damage to the nails and mycosis of the feet are recorded (Tullio et al. 2010; Kukhar et al. 2020). Extensive lesions of the nails on the fingers or toes appear in the form of clouding and thickening. Nails have lesions occupying one to two thirds of the surface, have a dirty gray color, are dull, are thickened due to subungual hyperkeratosis, and crumble at the free end. The skin of the feet is thickened due to hyperkeratosis, peeling of the skin is noticeable in the interdigital spaces, and there may be bubbles with serous-purulent contents, an unpleasant odor.

The main treatments for a phaeohyphomycotic cyst include surgical removal of the cyst or resection of infected tissue whenever possible.

Removal of cysts, nodules, and other skin lesions without antifungal treatment is usually sufficient for an immunocompetent host. For immunocompromised patients, concurrent systemic antifungal treatment is usually required (Restrepo et al. 1988).

Of interest to medicine and veterinary medicine is the fact that some antifungal drugs act in vitro on *Phoma* "strains." Rosen et al. (1996), studying the minimum inhibitory concentration (MIC, μg/ml) of some fungistatic agents on the *Phoma* (*Pleurophoma*) strain isolated from skin lesions, obtained the following results: itraconazole, £ 0.0018; ketoconazole, 0.2; fluconazole, 5.0; and griseofulvin, 0.5 (Gordon et al. 1975). Baker et al. (1987) in a study of the *P. minutella* strain isolated from a case of subcutaneous phaeohyphomycosis, MIC values of 0.625 μg/mL for amphotericin B and 0.06 μg/mL for 5-FC were obtained (Baker et al. 1987). Gordon et al. (1975), in a study of *P. cava* isolated from deerskin lesions, referred to the authors which recorded the sensitivity of this species to nystatin and griseofulvin, without citing the results obtained for the MIC of these agents (Restrepo et al. 1988; Al-Aidaroos et al. 2007; Hollingsworth et al. 2007; Errera et al. 2008; Queiroz-Telles et al. 2009).

In the treatment of 33 cases of *Phoma* species infections reported in the literature, antifungal agents have been used topically and orally. Typically, topical medications such as clotrimazole and miconazole are used for most skin infections. For infections accompanied by damage to the eyeball, cornea, and other, intravitreal amphotericin B or voriconazole is used.

Systemic therapy is used in patients with infections other than skin diseases and associated with sinusitis, with damage to the lungs and brain. Also, systemic antimycotics are prescribed in case of violation of the immune status of patients.

In the described treatment regimens, antifungal agents used included amphotericin B, ketoconazole, itraconazole, voriconazole, and posaconazole. Earlier, before the approval of azoles for the treatment of mycoses caused by *Phoma* spp., oral griseofulvin was used.

Thus, triazoles (itraconazole and voriconazole) (Hirsh and Schiff 1996) and amphotericin B (Everett et al. 2003) are recommended for the treatment of skin lesions. In view of the fact that the use of amphotericin B, flunol, and ketoconazole was not very effective for pathologies associated with *Phoma* spp., the appointment of these drugs should be considered inappropriate.

More effective drugs during the first few months of treatment of a phaeogyphomycosis cyst, especially in combination with surgical methods, are Sporagal or pound. A good effect was found from the alternation of amphotericin and pound or Sporagal.

The in vitro sensitivity of pathogenic fungi to antifungal agents is important information for the physician when choosing a suitable antifungal agent and choosing a route of administration. For filamentous fungi, species-specific fungistatics have been proposed for only a limited number of fungal species, and clinically proven fungicides are not available for most detected pathogenic molds (Valenzuela-Lopez et al. 2017). The authors reported this in one study aimed at identifying the frequency of occurrence of coelomycetes in clinical samples, studying their phenotypic and molecular characteristics, and analyzing antifungal sensitivity in vitro of nine antifungal drugs (Attapattu 1997; Pereiro Jr et al. 2004).

The in vitro susceptibility data for *Phoma* species are based on isolated case reports with minimum inhibitory concentrations of itraconazole and voriconazole ranging from 0.25 to 8 mg/L and MIC of amphotericin B from 0.5 to 1 mg/L (Errera et al. 2008).

Antifungal trials have shown that terbinafine, echinocandins (caspofungin, micafungin, and anidulafungin), and amphotericin B are most active against *Phoma* spp. (Bakerspigel et al. 1981) and *P. herbarum* (Rai 1989).

In another report by Sutton et al. (2009), there was similar results with amphotericin B. The most active were itraconazole, and the MIC of fluconazole was 32 (Mishra et al. 2011). Everett et al. (2003) reported the resistance of *Phoma* spp. to fluconazole and flucytosine in case of infection in a patient with a kidney transplant (Hospenthal 2014).

The approved treatment protocol for infections caused by *Phoma* spp. does not exist. This is due to the lack of fungicidal drugs that cause clinical cure and the difficulties in conducting antifungal susceptibility testing.

Prevention and control measures. It is believed that infection with *Phoma* spp. can pose a much greater threat to people than it currently appears.

According to the results of reports from the Working Group on the Study of Dangerous Fungi at the European Confederation of Medical Mycology, according to the degree of risk (BSL), all known pathogens belong to 1–3 BSL groups. Conditionally pathogenic fungi, which include *Phoma* spp., causing deep mycoses by chance, as a rule, only against the background of immunodeficiency or a critical condition of the patient, are assigned to group 1 BSL. This also includes the occasional causative agent of non-dermatophytic onychomycosis. *P. cruris-hominis* and *P. oculo-hominis* are included in the second BSL group, since they are better adapted to living in the environment of the macroorganism. Moreover, cases of deep mycoses caused by them occur quite regularly.

Human infection with celiac fungi is becoming more frequent and ranges from superficial to systemic distribution. The number of infections caused by *Phoma* spp. is increasing with the development of medicine, primarily due to an increase in the number of patients at risk from immunosuppression. In this case, the identification of the pathogen in medical institutions occurs, as a rule, by chance, against the background of a worsening of the patient's condition.

9.5 Conclusion

Summing up the above, it should be noted that *Phoma* spp. are currently becoming important etiological agents of mycoses in humans and animals. At the same time, the absence of a methodological approach and standardized protocols for the diagnosis and treatment of opportunistic mycoses caused by *Phoma* spp. are the major challenges.

The variety of *Phoma* species, a high level of pathogen polymorphism, and the absence of this pathogen in the list of pathogenic fungi do not allow to carry out a quick and unambiguous diagnosis. The authors made an attempt to generalize the experience of detecting and identifying pathogens of the genus *Phoma* spp. and to identify the risks of this infection.

General recommendations include the following:

- When a patient enters a medical clinic, take into account the nature of the profession, gender, climate, and the presence or likelihood of traumatic injuries.
- Include a history of chronic diseases (diabetes mellitus, tuberculosis, leprosy, leukemia, and others), the use of local corticosteroids, and organ or stem cell transplantation operations.
- In case of superficial and subcutaneous mycoses, include *Phoma* spp. as an etiological factor, with the obligatory classical laboratory manipulations: microscopy with KOH (NaOH), and cultural and morphological studies of a pure culture of the pathogen to identify morphological features such as chlamydospores, conidia, and pycnidial conidiomas.
- It is imperative to carry out a histological and microscopic examination of tissue homogenates and aspirates from the affected area and stained biopsies using appropriate dyes.
- To identify fungi of the genus *Phoma* to a species, use molecular markers for typing the pathogen.
- Voriconazole, posaconazole, itraconazole, and amphotericin B are recommended for the treatment of *Phoma* spp. To increase the effectiveness of antifungal treatment, select antimycotics after testing the pathogen for sensitivity.

References

Al-Abdely H, Alkhunaizi A, Al-Tawfiq J (2005) Successful therapy of cerebral phaeohyphomycosis due to Ramichloridium mackenziei with new triazole, posaconazole. Med. Mycol 43:91–95.

Al-Aidaroos A, Bin-Hussain I, El Solh H (2007) Invasive *Chaetomium* infection in two immunocompromised pediatric patients. Pediatr Infect Dis J 26:456–458.

Al-Tawfiq JA, Boukhamseen A (2011) Cerebral phaeohyphomycosis due to *Rhinocladiella mackenziei* (formerly Ramichloridium mackenziei): case presentation and literature review. J Infect Public Health 4:96–102.

Aly R (1996) Microbial Infections of Skin and Nails. In: Baron S, editor. Medical Microbiology. Galveston (TX): University of Texas Medical Branch at Galveston 4th edition.

Araújo AJG, Bastos OM., Souza MAJ, Oliveira JC (2003) Onychomycosis caused by emergent fungi: clinical analysis, diagnosis and revision. An. Bras. Dermatol 78:445–455.

Arenas R (2011) Micología Médica Ilustrada. México City: McGraw-Hill-Interamericana 4th edition.

Arrese JE, Piérard-Franchimont C, Piérard GE (1997) Unusual mould infection of the human stratum corneum. J. Med. Vet. Mycol 35:225–227.

Attapattu MC (1997) Chromoblastomycosis – a clinical and mycological study of 71 cases from Sri Lanka. Mycopathologia 137:145–151.

Aveskamp MM, de Gruyter J, Woudenberg JHC, Verkley GJM, Crous PW (2010) Highlights of the *Didymellaceae*: a polyphasic approach to characterise *Phoma* and related pleosporalean genera. Stud Mycol 65:1–60.

Baker JG, Salkin IF, Forgacs P, Haines JH, Kemna ME (1987) First Report of Subcutaneous *Phaeohyphomycosis* of the Foot Caused by *Phoma Minutella.* Journal of Clinical Microbiology 25:2395–2397.

Bakerspigel A (1970) The isolation of *Phoma hibernica* from a lesion on a leg. Sabouraudia 7:261–264.

Bakerspigel A, Lowe D, Rostas A (1981) The isolation of *Phoma eupyrena* from a human lesion. Arch Dermatol 117:362–363.

Balis E, Velegraki A, Fragou A, Pefanis A, Kalabokas T, Mountokalakis T (2006) Lung mass caused by *Phoma exigua.* Scand. J. Infect. Dis 38:552–555.

Baran R, Kaoukhov A (2005) Topical antifungal drugs for the treatment of onychomycosis: an overview of current strategies for monotherapy and combination therapy. J. Eur. Acad. Dermatol. Venereol 19:21–29.

Bauman DS, Smith CD (1975) Comparison of immunodiffusion and complement fixation tests in the diagnosis of histoplasmosis. Journal of Clinical Microbiology 2:77–80.

Bennett A, Ponder MM, Garcia-Diaz J (2018) *Phoma* Infections: Classification, Potential Food Sources, and Its Clinical Impact. Microorganisms 6(3):58.

Bloomfield N, Gordon MA, Elmendorf DF (1963) Detection of *Cryptococcus neoformans* antigen in body fluid by latex particle agglutination. Proceedings of the Society of Experimental Biology and Medicine 114:64–67.

Boerema GH, J de Gruytere, Noordeloos ME, Hamers MEC (2004) *Phoma* identification manual. Differentiation of specific and infra-specific taxa in culture. Cabi Publishing 470.

Borges MC Jr, Warren S, White W, Pellettiere EV (1991) Pulmonary phaeohyphomycosis due to Xylohypha bantiana. Arch Pathol Lab Med 115:627–629.

Carter E, Boudreaux C (2004) Fatal cerebral phaeohyphomycosis due to *Curvularia lunata* in an immunocompetent patient. J Clin Microbiol 42:5419–5423.

Chang X, Li R, Yu J, Bao X, Qin J (2009) Phaeohyphomycosis of the central nervous system caused by *Exophiala dermatitidis* in a 3-year-old immunocompetent host. J Child Neurol 24:342–345.

Chowdhary A, Meis JF, Guarro J, de Hoog GS, Kathuria S, Arendrup MC, Arikan-Akdagli S, Akova M, Boekhout T, Caira M (2014) ESCMID and ECMM joint clinical guidelines for the diagnosis and management of systemic phaeohyphomycosis: Diseases caused by black fungi. Clin. Microbiol. Infect 20:47–75.

Cole RJ, Jarvis BB, Schweikert MA (2003) Handbook of Secondary Fungal Metabolites. Academic Press: Amsterdam 3rd edition.

Crous PW, Shivas RG, Quaedvlieg W, van der Bank M, Zhang Y, Summerell BA, Guarro J, Wingfield MJ, Wood AR, Alfenas AC, Braun U, Cano-Lira JF, García D, Marin-Felix Y, Alvarado P, Andrade JP, Armengol J, Assefa A, den Breeÿen A, Camele I, Groenewald JZ (2014) Fungal Planet description sheets: 214-280. *Persoonia 32*:184–306.

Davies SF, Sarosi GA (1987) Serodiagnosis of histoplasmosis and blastomycosis. American Review of Respiratory Disease 136: 254–255.

Dawson CO, Lepper AWD (1970) *Peyronellaea glomerata* infection of the ear pinna in gots. Sabouraudia 8:145–148.

De Hoog GS, Guarro J, Gene J, Figueras MJ (2000) Atlas of clinical fungi. Centraalbureau voor Schimmelcultures. Virgili 2nd edition 645–668.

de Repentigny L (1989) Serological techniques for diagnosis of fungal infection. Eur J Clin Microbiol Infect Dis 8(4):362–75.

Delfino D, De Hoog S, Polonelli L (2006) Survival of a neglected case of brain abscess caused by *Cladophialophora bantiana*. Med Mycol 44:651–654.

Ebright JR, Chandrasekar PH, Marks S, Fairfax MR, Aneziokoro A, McGinnins MR (1999) Invasive sinusitis and cerebritis due to *Curvularia clavata* in an immunocompetent adult. Clin Infect Dis 28:687–689.

Elinav H, Izhar U, Benenson S (2009) Invasive Scytalidium dimidiatum infection in an immunocompetent adult. J Clin Microbiol 47:1259–1263.

English MP (1980) Infection of the finger-mail by *Pyrenochaeta* unguis-hominis. J. Derm 103:91–93.

Errera MH, Barale PO, Nourry H (2008) Usefulness of voriconazole in treatment of *Phoma glomerata* after penetrating injury. J Fr Ophtalmol 31:62–66.

Esterre P, Andriantsimahavandy A, Ramarcel ER, Pecarrere JL (1996) Forty years of chromoblastomycosis in Madagascar: a review. Am J Trop Med Hyg 55:45–7.

Everett JE, Busick NP, Sielaff T, Wahoff DC, Dunn DL (2003) A deeply invasive *Phoma* species infection in a renal transplant recipient. Transplant Proc 35:1387–1389.

Fader RC, McGinnis MR (1998) Infections caused by dematiaceous fungi: Chromoblastomycosis and phaeohyphomycosis. Infect. Dis. Clin. North. Am 2:925–938.

Faggi E, Pini G, Campisi E, Bertellini C, Difonzo E, Mancianti F (2001) Application of PCR to distinguish common species of dermatophytes. J. Clin. Microbiol 39(9):3382–3385.

Faisal M, Elsayed E, Fitzgerald SD, Silva V, Mendoza L (2007) Outbreaks of phaeohyphomycosis in the chinook salmon (oncorhynchus tshawytscha) caused by *Phoma herbarum*. Mycopathologia 163(1):41–48.

Filizzola MJ, Martinez F, Rauf SJ (2003) Phaeohyphomycosis of the central nervous system in immunocompetent hosts: report of a case and review of the literature. Int J Infect Dis 7:282–286.

Flanagan KL, Bryceson AD (1997) Disseminated infection due to *Bipolaris australiensis* in a young immunocompetent man: case report and review. Clin Infect Dis 25:311–313.

George DB, Lambert RS, Bruce MJ, Pickering TW, Wolcott RM (1981) Radioimmunoassay: a sensitive screening test for histoplasmosis and blastomycosis. American Review of Respiratory Disease 124:407–410.

Ghannoum MA, Hajjeh RA, Scher R (2000) A large-scale North American study of fungal isolates from nails: the freguency of onychomycosis, fungal distribution, and antifungal susceptibility patterns. J. Am. Acad. Dermatol 43(4):641–648.

Gopinathan U, Garg P, Fernandes M, Sharma S, Athmanathan S, Rao GN (2002) The epidemiological features and laboratory results of fungal keratitis: a 10-year review at a referral eye care center in South India. Cornea 21:555–559.

Gordon MA, Salkin IF, Stone WB (1975) *Phoma (Peyronellaea)* as zoopathogen. Sabouraudia 13:329–333.

Gower EW, Keay LJ, Oechsler RA, Iovieno A, Alfonso EC, Jones DB, Colby K, Tuli SS, Patel SR, Lee SM, Irvine J, Stulting RD, Mauger TF, Schein OD (2010) Trends in fungal keratitis in the United States, 2001 to 2007. Ophthalmology 117(12):2263–7.

Greenberger P A, Patterson R (1982) Application of enzyme-linked immunosorbent assay (ELISA) in diagnosis of allergic bronchopulmonary aspergillosis. Journal of Laboratory and Clinical Medicine 99: 288–293.

Greig JR, Khan MA, Hopkinson NS, Marshall BG, Wilson PO, Rahman SU (2001) Pulmonary infection with *Scedosporium prolificans* in an immunocompetent individual. J Infect 43:15–17.

Guarner J, Brandt ME (2011) Histopathologic diagnosis of fungal infections in the 21st century. *Clin Microbiol Rev* 24(2):247–280.

Gurta AK, Gurta MA, Summerbell RC (2000) The epidemiology of onychomycosis: possible role of smoking and peripheral arterial disease. J. Eur. Acad. Dermatol. Venereol 14(6)466–469.

Hay RJ (2008) Deep fungal infections. In: Wolff K, Goldsmith LA, Katz SI, Gilchrest BA, Paller AS, Leffell DJ, editors. Fitzpatrick's Dermatology in General Medicine 7th edition. New York: Mc Graw Hill 1833–1835.

Hernández-Hernández F, Vargas-Arzola J, Ríos-Cruz OP, Córdova-Martínez E, Manzano-Gayosso P, Segura-Salvador A (2018) First case of chromoblastomycosis due to *Phoma insulana*. Enferm. Infecc. Microbiol. Clin 36:95–99.

Hirsh AH, Schiff TA (1996) Subcutaneous phaeohyphomycosis caused by an unusual pathogen: *Phoma* species. J Am Acad Dermatol 34:679–680

Hollingsworth JW, Shofer S, Zaas A (2007) Successful treatment of Ochroconis gallopavum infection in an immunocompetent host. Infection 35:367–369.

Hospenthal DR (2014) Agents of chromoblastomycosis.

Janke D (1956) About a human-pathogenic new species of *Peyronellaea* bred from lung changes. Mycopathologia 7:229–240.

Jones MDM, Forn I, Gadelha C, Egan MJ, Bass D, Massana R, Richards TA (2011) Discovery of novel intermediate forms redefines the fungal tree of life. Nature 474:200–203.

Kalyani M, Ranjitham M, Sekar UP (2006) Subcutaneous infection by *Phoma* species in a renal allograft, recipient a case report. Indian J Practising Doctor 3.

Kardjeva V (2006) Forti-eight-hour diagnosis of onychomycosis with subtyping of Trichophyton rubrum strains. J. Clin. Microbiol 44(4):1419–27.

Kaur R, Kashyap B, Bhalla P (2008) Onychomycosis – epidemiology, diagnosis and management. Indian Journal of Medical Microbiology 26(2):108–116.

Keay LJ, Gower EW, Iovieno A, Oechsler RA, Alfonso EC, Matoba A, Colby K, Tuli SS, Hammersmith K, Cavanagh D, Lee SM, Irvine J, Stulting RD, Mauger TF, Schein OD (2011) Clinical and microbiological characteristics of fungal keratitis in the United States, 2001-2007: a multicenter study. Ophthalmology 118(5):920–926.

Kiran Kumar K, Kaveri Kalleker (2008) Phaeohyphomycosis. Indian J. Pathol. Microbiol 51:556–558.

Kirkness CM, Seal DV, Clayton YM, Punithalingam E (1991) Sphaeropsis subglobosa keratomycosis-first reported case. Cornea 10(1):85–9.

Kozel TR, Wickes B (2014) Fungal diagnostics. Cold Spring Harb Perspect Med 4(4).

Kukhar E, Smagulova A, Kiyan V (2020) Biological properties of Phoma macrostoma related to non-dermatophyte onychomycosis. Medical Mycology Case Reports 27:55–58.

Kumar P, Thomas S, Papagiannuli E, Hardman SC, Jenkins D, Prydal J (2015) A case of *Phoma* fungal keratitis in a contact lens user. JRSM Open 6.

Lambert RS, George RB (1987) Evaluation of enzyme immuno-assay as a rapid screening test for histoplasmosis and blastomycosis. American Review of Respiratory Disease 136:316–319.

Lastoria C, Cascina A, Bini F (2009) Pulmonary Cladophialophora boppii infection in a lung transplant recipient: case report and literature review. J Heart Lung Transplant 28:635–637.

Lease ED, Alexander BD (2011) Fungal diagnostics in pneumonia. Semin Respir Crit Care Med 32(6):663–672.

Lever WF, Schaumburg-Lever L (1983) Hystopathology of the Skin. Lippincott compani. Philadelphia 6 edition.

Liu D, Pearce L, Lilleu G, Coloe S, Baird R, Pedersen J (2002) Identification of dermatophytes fungi *Trichophyton rubrum*, *T. soudanense* and *T. gourvilii*. J. Med. Microbiol 51:117–122.

Lopez Martinez R, Mendez Tovar LJ (2007) Chromoblastomycosis. Clin. Dermatol 25:188–194.

Lugauskas A, Raila A, Railiene M, Raudoniene V (2006) Toxic micromycetes in grain raw material during its processing. Ann. Agric. Environ 13:147–161.

Marier R, Smith W, Jansen M, Andriole VT (1979) A solid-phase radioimmunoassay for the measurement of antibody to *Aspergillus* in invasive aspergillosis. Journal of Infectious Diseases 140:771–779.

Markinson BC, Elewski B (2007) Onychomycosis: diagnosis and definition of cure. J. Am. Acad. Dermatol 56(6):939–944.

McElnea E, Farrell S, Lynch B, Bishop K, Mullen D, Borman A, Higgins G (2015) A rare case of fungal keratitis: Diagnosis and management. JMM Case Rep.

Mesa A, Henao J, Gil M, Durango G (1999) Phaeohyphomycosis in kidney transplant patients. Clinical Transplantation 13(3):273–276.

Milam CP, Fenske NA (1989) Chromoblastomycosis. Derm. Clin 7:219–225.

Mirhendi H, Fatemi MJ, Bateni H (2013) First case of disseminated phaeohyphomycosis in an immunocompetent individual due to Alternaria malorum. Med Mycol 51:196–202.

Mishra D, Singal M, Rodha MS, Subramanian AJ (2011) Subcutaneous phaeohyphomycosis of foot in an immunocompetent host. A. J. Lab. Physicians 3(2):122–124.

Morris JT, Beckius ML, Jeffery S, Longfeld RN, Heaven RF, Baker WJ (1995) Lung mass caused by *Phoma* species. Infect. Dis. Clin. Pract 4:58–59.

Mullane K, Toor AA, Kalnicky C, Rodriguez T, Klein J, Stiff P (2007) Posaconazole salvage therapy allows successful allogeneic hematopoietic stem cell transplantation in patients with refractory invasive mold infections. Transpl Infect Dis 9:89–96.

Oh CK, Kwon KS, Lee JB, Jang HS, Chung TA, Suh SB (1999) *Subcutaneous phaeohyphomycosis* caused by *Phoma* species. Int. J. Dermatol 38:874–876.

Oliveira RC, Goncalves SS, Oliveira MS, Dilkin P, Mallmann CA, Freitas RS, Bianchi P, Correa B (2017) Natural occurrence of tenuazonic acid and *Phoma sorghina* in Brazilian sorghum grains at different maturity stages. Food Chem 230:491–496.

Ono M, Nishigori C, Tanaka C, Tanaka S, Tsuda M, Miyachi Y (2004) Cutaneous alternariosis in an immunocompetent patient: analysis ofthe internal transcribed spacer region of rDNA and Brm2 of isolated Alternaria alternata. Br J Dermatol 150:773–775.

Padhye AA, Davis MS, Baer D (1998) Phaeohyphomycosis caused by *Phaeoacremonium* inflatipes. J. Clin. Microbiol 36:2763–2765.

Parr CS, Wilson N, Leary P, Schulz KS, Lans K, Walley L, Hammock JA, Goddard A, Rice J, Studer M et al. (2014) The Encyclopedia of Life. Providing Global Access to Knowledge about Life on Earth. Biodivers. Data J. 2:e1079.

Pereiro M Jr, Pereiro Ferreiros MM, De Hoog GS, Toribio J (2004) Cutaneous infection caused by *Alternaria* in patients receiving tacrolimus. Med Mycol 42:277–282.,

Pfavayi LT, Sibanda EN, Mutapi F (2020) The Pathogenesis of Fungal-Related Diseases and Allergies in the African Population: The State of the Evidence and Knowledge Gaps. Int Arch Allergy Immunol 181:257–269.

Punithalingam E (1976) *Phoma oculohominis* sp. nov. from corneal ulcer. Trans. Br. Mycol. Soc 67:142–143.

Punithalingam E (1979) Sphaeropsidales in culture from humans. Nova Hedwig 31:119–158.

Queiroz-Telles F, Esterre P, Perez-Blanco M (2009) Chromoblastomycosis: an overview of clinical manifestations, diagnosis and treatment. Med. Mycol 47:3–15.

Rai MK (1989) *Phoma sorghina* infection in human being. Mycopathologia (Den Haag) 105:167–170.

Rai M K, Tiwari VV, Irinyi L, Kövics GJ (2014). Advances in taxonomy of genus phoma: polyphyletic nature and role of phenotypic traits and molecular systematics. Indian journal of microbiology 54(2):123-128.

Restrepo A, Gonzalez A, Gomez I, Arango M, de Bedout C (1988) Treatment of chromoblastomycosis with itraconazole. Ann N Y Acad Sci 544:504–516.

Revankar SG (2006) Phaeohyphomycosis. Infect. Dis. Clin. North. Am 20:609–620.

Revankar SG (2011) *Cladophialophora bantiana* brain abscess in an immunocompetent patient. Can J Infect Dis Med Microbiol 22:149–150.

Revankar SG, Sutton DA (2010) Melanized fungi in human disease. Clin Microbiol Rev 23:884–928.

Richardson M, Page I (2018) Role of Serological Tests in the Diagnosis of Mold Infections. Curr Fungal Infect Rep 12:127–136.

Rishi K, Font RL (2003) Keratitis caused by an unusual fungus, Phoma species. Cornea 22:166–168.

Roehm CE, Salazar JC, Hagstrom N, Valdez TA (2012) Phoma and Acremonium invasive fungal rhinosinusitis in congenital acute lymphocytic leukemia and literature review. Int J Pediatr Otorhinolaryngol 76(10):1387–1391.

Rosen T, Rinaldi MJ, Tschen JA, Stern JK, Cernoch P (1996) Cutaneous lesions due to *Pleurophoma* (*Phoma*) complex. South Med J 89:431–433.

Salgado CG, da Silva MB, Yamano SS, Salgado UI, Diniz JA, da Silva JP (2009) Cutaneous localized annular chromoblastomycosis. J. Cutan. Pathol 36(2):257–261.

Schell WA, Esterre P (2005) Chromoblastomycosis. In: Hay RJ, editor. Topley and Wilson's microbiology and microbial infections 10th edition.

Schieffelin JS, Garcia-Diaz JB, Loss GE, Jr, Beckman EN, Keller RA, Staffeld-Coit C, Garces JC, Pankey GA (2014) Phaeohyphomycosis fungal infections in solid organ transplant recipients: Clinical presentation, pathology, and treatment. Transpl. Infect. Dis 16:270–278.

Scott EN, Muchmore HG, Felton FG (1980) Comparison of enzyme immunoassay and latex agglutination methods for detection of *Cryptococcus neoformans* antigen. American Journal of Clinical Pathology 73:790–794.

Shukla NP, Rajak RK, Agarwasl GP, Gupta D (1984) *Phoma minutispora* as a human pathogen. Mykosen 27:255–258.

Silveira F, Nucci M (2001) Emergence of black moulds in fungal disease: epidemiology and therapy. Curr. Opin. Infect. Dis 14:679–684.

Simon-Nobbe B, Denk U, Pöll V, Rid R, Breitenbach M (2008) The Spectrum of Fungal Allergy. Int Arch Allergy Immunol 145:58–86.

Sudduth EJ, Crumbley AJ, Farrar WE (1992) Phaeohyphomycosis due to *Exophiala* species: clinical spectrum of disease in humans. Clin. Infect. Dis 15(247):639–644.

Suh MK (2005) Phaeohyphomycosis in Korea. Nihon Ishinkin Gakkai Zasshi 46:67–70.

Sutton DA, Rinaldi MG, Sanche SE (2009) Dematiaceous fungi. In Clinical Mycology. Churchill Livingstone (Elsevier) 2nd edition.

Sweet GH, Cimprich RS, Cook AC, Sweet DE (1979) Antibodies in histoplasmosis detected by use of yeast and mycelial antigens in immunodiffusion and electroimmunodiffusion. American Review of Respiratory Disease 120:441–449.

Thomas PA (2003) Fungal infections of the cornea. Eye 17:852–862.

Thomas PA, Kaliamurthy J (2013) Mycotic keratitis: epidemiology, diagnosis and management. Clin Microbiol Infect 19:210–220.

Tuft SJ, Tullo AB (2009) Fungal keratitis in the United Kingdom 2003-2005. Eye (Lond) 23(6):1308–13.

Tuli SS (2011) Fungal keratitis. Clin Ophthalmol 5:275–9.

Tullio V, Banche G, Allizond V, Roana J, Mandras N, Scalas D, Panzone M, Cervetti O, Valle S, Carlone N, Cuffini AM (2010) Non-dermatophyte moulds as skin and nail foot mycosis agents: *Phoma herbarum*, *Chaetomium globosum* and *Microascus cinereus*. Fungal Biol 114(4):345–9.

Valenzuela-Lopez N, Sutton DA, José F. Cano-Lira, Katihuska P, Nathan W, Josep G, Alberto MS (2017) Coelomycetous Fungi in the Clinical Setting: Morphological Convergence and Cryptic Diversity. Journal of Clinical Microbiology 55(2):552–567.

Valenzuela-Lopez N, Cano-Lira JF, Guarro J, Sutton DA, Wiederhold N, Crous PW, Stchigel AM (2018) Coelomycetous Dothideomycetes with emphasis on the families *Cucurbitariaceae* and *Didymellaceae*. Stud. Mycol. 90:1–69.

Vasoo S, Yong LK, Sultania-Dudani P, Scorza ML, Sekosan M, Beavis KG, Huhn GD (2011) Phaeomycotic cysts caused by *Phoma* species. Diagn. Microbiol. Infect. Dis 70:531–533.

Veer P, Patwardhan NS, Damle AS (2007) Study of onychomycosis: prevailing fungi and pattern of infection. Indian J. Med. Microbiol 25:53–56.

Virella G (1997) Microbiology and infectious Diseases. Baltimore: Williams and Wilkins 3rd edition.

Visconti A, Logrieco A, Vurro M, Bottalico A (1987) Tenuazonic acid in blackmold tomatoes: Occurrence, production by associated Alternaria species, and phytotoxic properties. Phytopathol. Mediterran 26:125–128.

Weidenbörner M (2001) Pine nutes: The mycobiota and potential mycotoxins. Can. J. Microbiol 47:460–463.

Williams JE, Murphy R, Standard PG, Phair JP (1981) Serologic response in blastomycosis: diagnostic value of double immunodiffusion assay. American Review of Respiratory Disease 123:209–212.

Woo PC, Lau SK, Ngan AH, Tse H, Tung ET, Yuen KY (2008) Lasiodiplodia theobromae pneumonia in a liver transplant recipient. J Clin Microbiol 46:380–384.

Yekeler H, Bitmis K, Özcelik N, Doymaz MZ, Metin C (2001) Analysis of toxic effects of Alternaria toxins on esophagus of mice by light and electron microscopy. Toxicol. Pathol 29:492–497.

Young NA, Kwon-chung KJ, Freeman J (1973) Subcutaneous abscess caused by *Phoma* sp. resembling *Pyrenochaeta romeroi*: Unique fungal infection occurring in immunosuppressed recipient of renal allograft. Am. J. Clin. Pathol 59:810–816.

Zaias N, Rebell G (1996) Chronic dermatophytosis syndrome due to Trichophyton rubrum. Int. J. Dermatol 35(9):614–617.

Zaitz C, Heins-Vaccari EM, de Freitas RS, Arriagada GL, Ruiz L, Totoli SA (1997) Subcutaneous phaeohyphomycosis caused by *Phoma cava*. Report of a case and review of the literature. Rev Inst Med Trop Sao Paolo 39: 43–8.

Part III

Secondary Metabolites and Bioactivities

10 Fruitful Decade of *Phoma* Secondary Metabolites from 2011 to 2020: Chemistry, Chemical Diversity, and Biological Activities

Hidayat Hussain, Elizbit, Iftikhar Ali, Nilufar Z. Mamadalieva, Ghulam Abbas, Maroof Ali, Gohar Zaman, Amjad Khan, Uzma Hassan, and Ivan R. Green

10.1 Introduction

Natural products (NPs) are a crucial and invaluable source of innovation in the modern drug design process. Notably, chemical structures of NPs occupy various interesting biologically related chemical spaces, and their properties cannot be proficiently explored by investigating synthetic molecules only (Lachance et al. 2012; Gerry and Schreiber 2018). Moreover, NPs illustrated a more dominant structural foundation over synthetic compounds in the drug development process. The NP scaffolds are considered as "privileged" because NP derivatives demonstrated a wide range of biological activities (Davisona and Brimble 2019). In addition, these NP privileged scaffolds

H. Hussain (✉)
Department of Bioorganic Chemistry, Leibniz Institute of Plant Biochemistry, Halle (Saale), Germany
e-mail: Hidayat.Hussain@ipb-halle.de

Elizbit
Department Materials Engineering, National University of Sciences and Technology (NUST), Islamabad, Pakistan

I. Ali
Shandong Key Laboratory of TCM Quality Control Technology, Shandong Analysis and Test Center, Jinan, Shandong Province, People's Republic of China

Department of Chemistry, Karakoram International University, Gilgit, Pakistan

N. Z. Mamadalieva
Department of Bioorganic Chemistry, Leibniz Institute of Plant Biochemistry , Halle (Saale), Germany

Institute of the Chemistry of Plant Substances of the Academy Sciences of Uzbekistan, Tashkent, Uzbekistan

G. Abbas
Department of Biological Sciences and Chemistry, College of Arts and Sciences, University of Nizwa, Nizwa, Sultanate of Oman

M. Ali
College of life Sciences, Anhui Normal University, Wuhu, China

G. Zaman
Department of Chemistry, Quaid-i-Azam University, Islamabad, Pakistan

A. Khan
Department of Pharmacy, Quaid-i-Azam University, Islamabad, Pakistan

U. Hassan
Institute of Chemical Sciences, University of Peshawar, Peshawar, Pakistan

I. R. Green
Department of Chemistry and Polymer Science, University of Stellenbosch, Stellenbosch, South Africa

M. Rai et al. (eds.), *Phoma: Diversity, Taxonomy, Bioactivities, and Nanotechnology*,
https://doi.org/10.1007/978-3-030-81218-8_10

act as crucial, biologically important platforms to create bioactive compound libraries for the drug development (Davisona and Brimble 2019).

Fungal-derived natural products have proved to be a tremendous source of numerous clinically approved drugs, and notably, these drugs possess a diverse mode of action (González-Medina et al. 2017). Moreover, fungi are widely present in nature and produce novel secondary metabolites with intriguing chemical diversity. Fungal chemical diversity is generated from simple starting materials, viz., sugars, organic acids, terpenes, amino acids, and bases (González-Medina et al. 2017). The genus *Phoma* was introduced over 180 years ago, and over 2000 species have been studied and reported from various hosts/substrates (Rai et al. 2009). Moreover, *Phoma* spp. have been chemically investigated in the last 10 years and produce intriguing chemical diversity. This chapter gives a comprehensive review about *Phoma* sp., and 208 secondary metabolites with diverse range of biological effects.

10.2 Secondary Metabolites of *Phoma* spp. and Their Biological Efficacy

10.2.1 α-Pyrone Derivatives

Sang et al. (2017) isolated two pairs of enantiomeric α-pyrone dimers, (±) phomones A (**1**) and B (**2**) along with rosellisin (**3**) (Fig. 10.1) from *Phoma* sp. Metabolite **1** illustrated a novel 6/4/5/6 tetracyclic ring system. Compounds **1** and **2** did not exhibit cytotoxic effects against the three human cancer cell lines, viz., HL-60, PC-3, and HCT-116. Four new α-pyrone derivatives, viz., phomones C–F (**4–7**) together with the known rosellisin diacetate (**8**), rosellisin (**9**), 2-propenoic acid (**10**), and macrophic acid (**11**), were isolated from the *Phoma* spp. Compounds **5** and **8** were the most active compounds on three cancer cells, viz., HL-60, PC-3, and HCT-116, with IC_{50} values in the range of 0.52 to 2.31 μM. On the other hand, **9** was active against HL-60 and PC-3 with IC_{50} of 1.04 and 5.90 μM, respectively (Sang et al. 2017a).

In another study, rosellisin (**9**) and macrophin (**12**) were reported from *Phoma macrostoma*. Macrophin (**12**) illustrated cytotoxic effects against MDA-MB-231 (IC_{50}, 14.8 μM), T47D (IC_{50}, 8.1 μM), and MCF-7 (IC_{50}, 13.0 μM). Notably, this compound possesses potent effects toward MIAPaCa-2 with an IC_{50} of 0.9 μM (Nalli et al. 2019). Phomanones B (**13**) and **14** were reported from *Phoma* sp. but were inactive toward K562, HL-60, HCT-116, BEL-7402, A549, HeLa, L-02, SH-SY5Y, HO8910, PC-3, U87, MGC-803, and MDA-MB-231 cancer cells (Lu et al. 2019).

10.2.2 Isocoumarins

(3*R*)-5-Hydroxymellein (**15**) (Fig. 10.2) was obtained from *Phoma* sp. and illustrated good antibacterial, antialgal, and antifungal effects (Hussain et al. 2014). Another isocoumarin analog named chenisocoumarin (**16**) was obtained from *Phoma chenopodiicola* and unfortunately demonstrated no phytotoxic effects (Evidente et al. 2015). In another investigation, diaportinol (**17**), desmethyldiaportinol (**18**), citreoisocoumarinol (**19**), and citreoisocoumarin (**20**) were reported from *Phoma* sp. (Shi et al. 2017). It was found that (–)-6-methoxymellein (**21**) and 7-hydroxy-3,5-dimethyl-isochromen-1-one (**22**) were produced by the *Phoma* sp., and the latter compound illustrated antiviral activity against the influenza A virus with an IC_{50} of 20.9 μg/mL (Liu et al. 2019). In another study (-)-(3*R*)-6-methoxy-7-chloromellein (**23**) and (3*R*,4*S*)-4-hydroxy-6-methoxy-7-chloromellein (**24**) were obtained from *Phoma* sp., and the structure of compound **23** was confirmed via an X-ray crystallographic study (Elsebai and Ghabbour 2016).

10.2.3 Anthraquinones and Xanthones

The anthraquinone analog, emodin (**25**), was reported from the *Phoma* sp. (Hussain et al. 2014). In another investigation on the constituents of *Phoma* sp., 7-(γ,γ)-

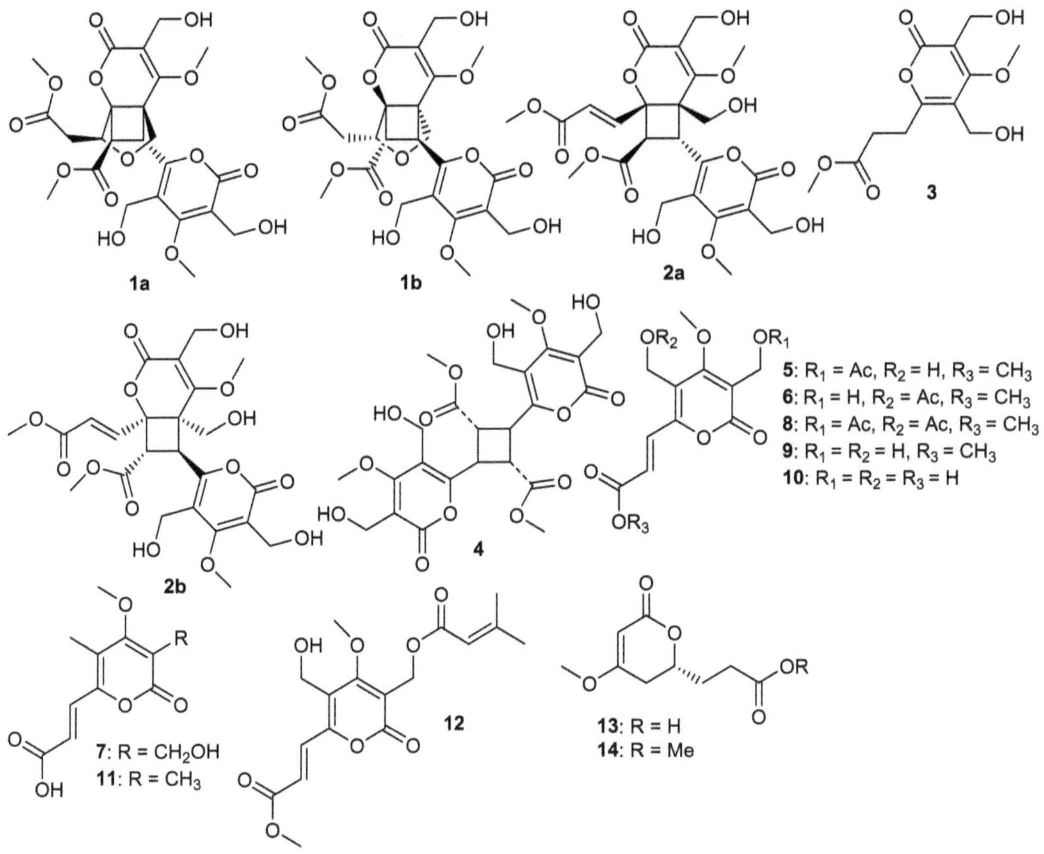

Fig. 10.1 Structures of α-pyrones **1–14**

dimethylallyloxymacrosporin (**26**), macrosporin (**27**), 7-methoxymacrosporin (**28**), tetrahydroaltersolanol B (**29**), altersolanol L (**30**), and ampelanol (**31**) were isolated (Huang et al. 2017) (Fig. 10.3). Antimicrobial evaluation of these isolates illustrated that macrosporin (**27**) possesses antifungal effects toward *Fusarium graminearum*, *F. oxysporum*, *Colletotrichum musae*, *C. gloeosporioides*, *Penicillium italicum*, and *Rhizoctonia solani* with MIC values ranging from 3.75 to 100 μg/mL. Interestingly, the inhibitory effects of compound **27** toward *F. oxysporum* (MIC, 3.75 μg/mL) was higher than the positive control (carbendazim, MIC, 6.25 μg/mL). On the other hand, hydroanthraquinone **30** illustrated moderate antifungal effects against *R. solani* and *P. italicum* with MICs of 50 and 35 μg/mL, respectively (Huang et al. 2017). In another investigation, the anthraquinone questin (**32**) and the xanthone norlichexanthone (**33**) were isolated from *Phoma* sp. obtained cultures of *Phoma* sp. (Xia et al. 2015; Liu et al. 2019).

10.2.4 Thiodiketopiperazines

The thiodiketopiperazines, phomazines A–C (**34–36**) (Fig. 10.4), along with epicorazine A (**37**), epicorazine B (**38**), epicoccin D (**39**), epicoccin A (**40**), epicoccin C (**41**), epicoccin E (**42**), epicoccin B (**43**), epicorazine C (**44**), exserohilone A (**45**), and rostratin A (**46**) were isolated from the *Phoma* sp. OUCMDZ-1847. Compounds **35** (MGC-803, IC_{50}, 8.5 μM), **37** (HL-60, IC_{50}, 0.05 μM; HCT-116, IC_{50}, 0.33 μM; A549, IC_{50}, 2.3 μM; K562, IC_{50}, 1.5 μM; MGC-803, IC_{50}, 2.7 μM), **38** (HL-60, IC_{50}, 1.6 μM; K562, IC_{50}, 4.6 μM; MGC-803, IC_{50}, 5.2 μM), **44**

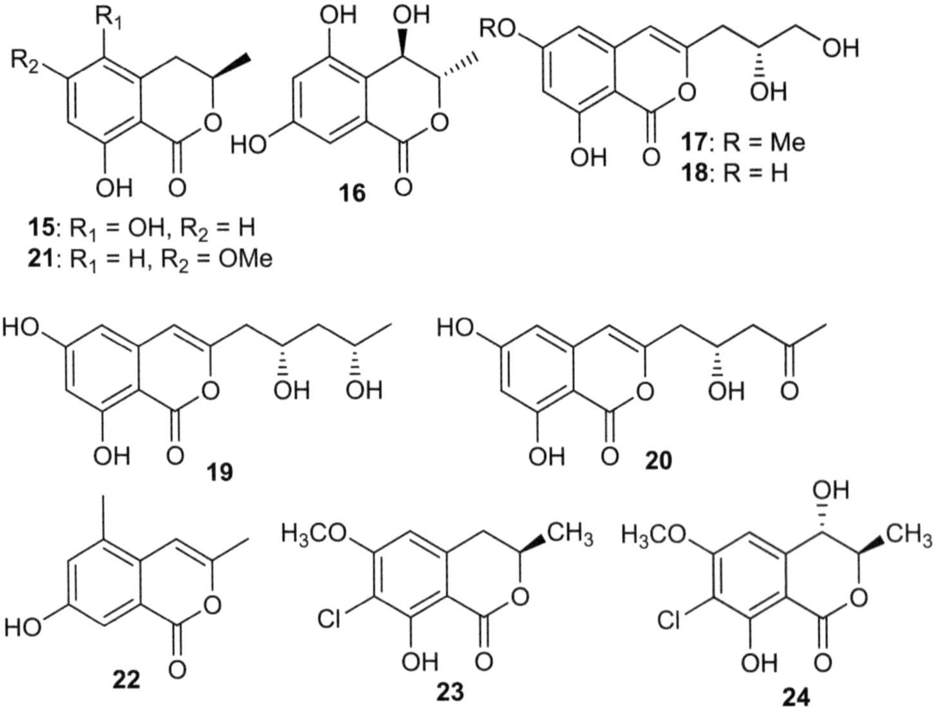

Fig. 10.2 Structures of isocoumarins **15–24**

(HL-60, IC_{50}, 3.5 μM; HCT-116, IC_{50}, 7.1 μM; MGC-803, IC_{50}, 3.4 μM), and **45** (HL-60, IC_{50}, 3.5 μM; HCT-116, IC_{50}, 4.0 μM; MGC-803, IC_{50}, 4.2 μM) all illustrated significant cytotoxic effects toward the various cancer cell lines cited (Kong et al. 2014). In another study, thiodiketopiperazines **47** and **48** were reported from *Phoma* sp., associated with *Glycyrrhiza glabra*. Antimicrobial results showed both these compounds were significantly active toward *Staphylococcus aureus*, *S. pyogenes*, and methicillin-resistant *S. aureus* with $IC_{50} < 10$ μM (Arora et al. 2016).

10.2.5 Cytochalasin Derivatives

The small library of cytochalasin derivatives, viz., cytochalasin B_2 (**49**), deoxaphomin B (**50**), deoxaphomin C (**51**), 20-deoxycytochalasin F (**52**), and cytochalasin B (**53**) (Fig. 10.5), was obtained from *Phoma* sp. Compounds **49** and **51–53** exhibited cytotoxicity toward SK-OV-3, A549, SK-MEL-2, and HCT15 cell lines with IC_{50} < 10 μM except for compound **52** whose activity toward A549 had an IC_{50} of 12.4 μM. In addition, cytochalasin B (**53**) illustrated cytotoxicity toward HeLa cells with IC_{50} of 7.9 μM (Kim et al. 2012). Ergocytochalasin A (**54**) was reported from the fungus *Phoma multirostrata*. Interestingly, this metabolite illustrated cytotoxic effect toward MCF-7, HT-29, HepG2, A549, HCT116, and CT26 cancer cells with IC_{50} values ranging from 6.9 to 26.6 μM. In addition, compound **54** possesses immunosuppressive effects toward ConA-induced T cell and LPS-induced B cell. Moreover, this metabolite was also active against the human influenza A virus (H1N1, EC_{50}, 40.9 μM), dengue virus type 3 (DV3, EC_{50}, 12.5 μM), and the respiratory syncytial virus (RSV, EC_{50}, 6.4 μM) and additionally with SI of 1.1, 1.8, and <1, respectively (Peng et al. 2020).

Phomachalasins A–D (**55–58**) (Fig. 10.6) were obtained from *Phoma exigua*, and none of these metabolites were found to possess antimicrobial effects. Moreover, these compounds were

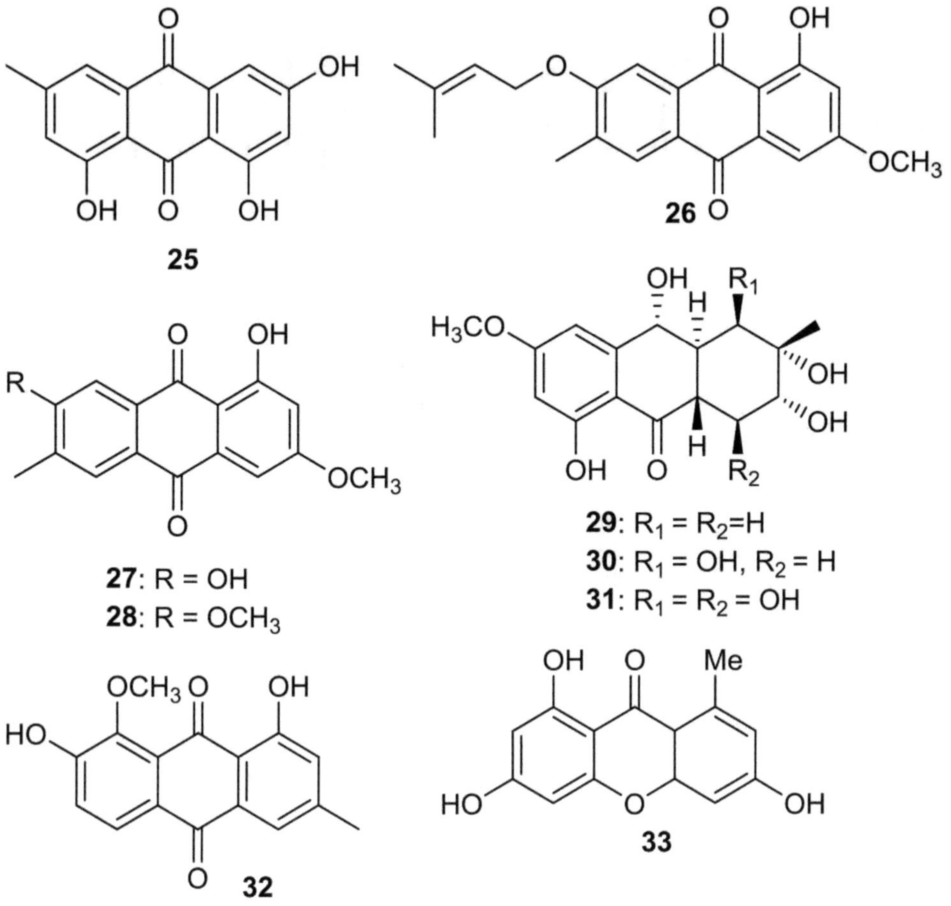

Fig. 10.3 Structures of isocoumarins **25–33**

found to bear the 1,2,3,4,6,7-hexasubstituted bicycle[3.2.0]heptene moieties coupled to the macrocyclic core system (Evidence et al. 2011). Cytochalasins B3–B6 (**59–62**), along with proxiphomin (**63**), were obtained from the jellyfish-derived fungus *Phoma* sp. These metabolites illustrated moderate cytotoxic effects toward KB, A549, and HCT116 cancer cells (Kim et al. 2015). The cytochalasins, multirostratin A (**64**) and 20-oxo-deoxaphomin (**65**), deoxaphomin (**66**), cytochalasin A (**67**), cytochalasin B (**68**), cytochalasin Z2 (**69**), and cytochalasin F (**70**), were isolated from the fungus *Phoma multirostrata*. Compounds **64** and **65** were active toward the five cancer cells, viz., HL-60, A-549, SW-480, SMMC-7721, and MCF-7 with IC_{50} values ranging from 7.7 to 15.8 μM (Chen et al. 2015).

10.2.6 Diphenyl Ether Derivatives

Phomaethers A–C (**71–73**) (Fig. 10.7) and 2,3′-dihydroxy-4-methoxy-5′,6-dimethyl diphenyl ether (**74**) were found to be produced by the fungus *Phoma* sp. (TA07-1). Compounds **71, 73,** and **74** possess antibacterial effects toward *Staphylococcus albus*, *S. aureus*, *Escherichia coli*, *Vibrio parahaemolyticus*, and *V. anguillarum* with MIC values ranging between 0.156 and 10.0 uM (Shi et al. 2017). Biphenyl ether derivatives **75–77** were isolated from the fungus *Phoma* sp. TPU1222. Biphenyl ether **76** possesses a 32% PTP1B inhibition. On the other hand, biphenyl ethers **75** and **77** illustrated PTP1B inhibitory effects with IC_{50} of 13 μM and 17 μM, respectively (Sumilat et al. 2017).

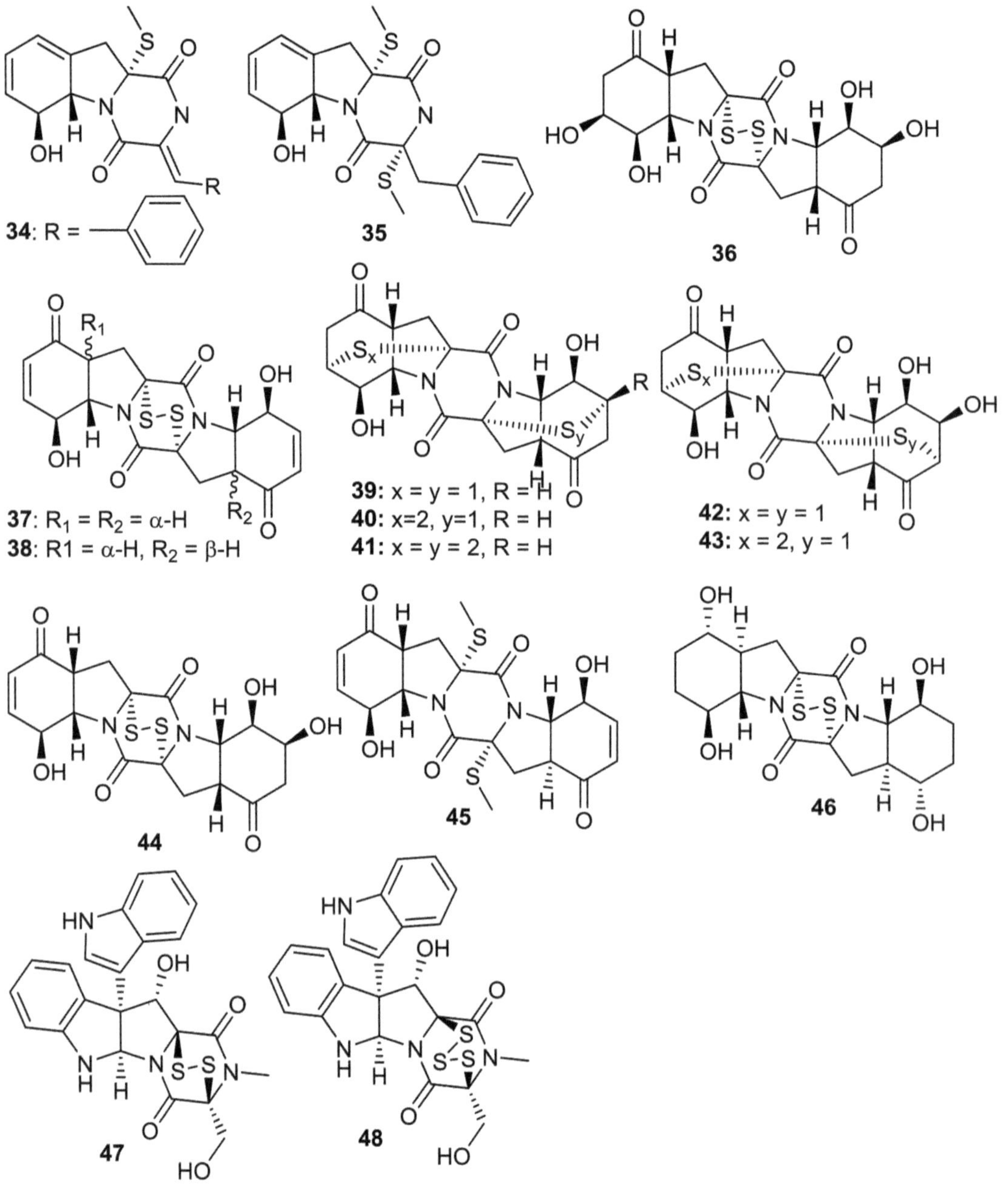

Fig. 10.4 Structures of thiodiketopiperazines **34–48**

Quercilolin (**78**) (Fig. 10.8) was obtained from the *Phoma* sp. CZD-F11, and this compound was found to be inactive on PC-3 and BRD4 cancer cells (Wu et al. 2018). In another investigation, Fang et al. (2012) reported on the three diphenyl ether derivatives: asterric acid (**79**), methyl asterrate (**80**), and methyl 2-(2-formyl-3-hydroxy-5-methylphenoxy)-5-hydroxy-3-methoxybenzoate (**81**) isolated from the *Phoma* sp. strain SHZK-2. Barceloneic acid A (**82**) was isolated from cultures of *Phoma* sp. JS752, and this compound was found to be inactive in an antibacterial assay (Xia et al. 2015).

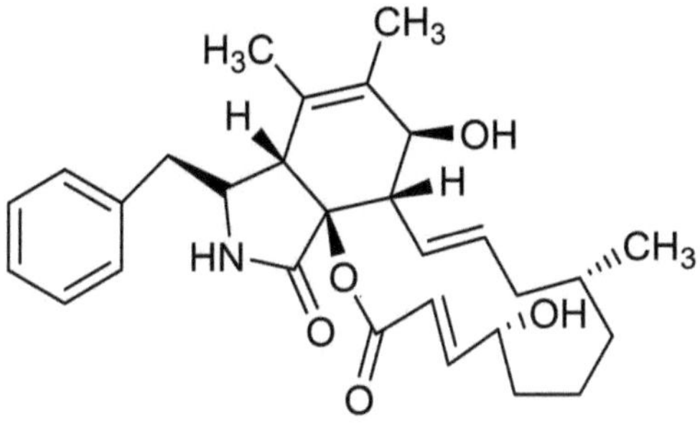

49

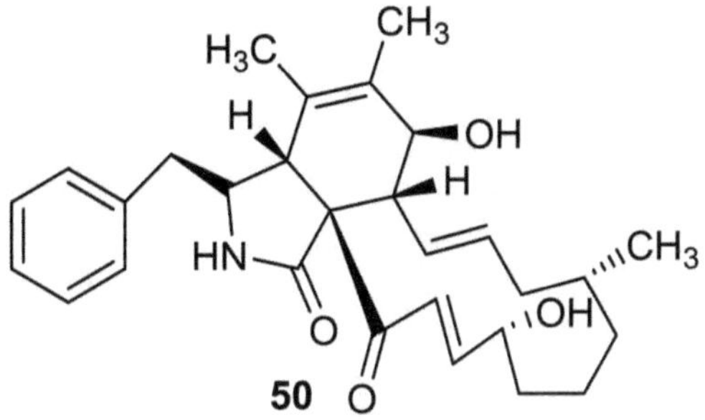

50

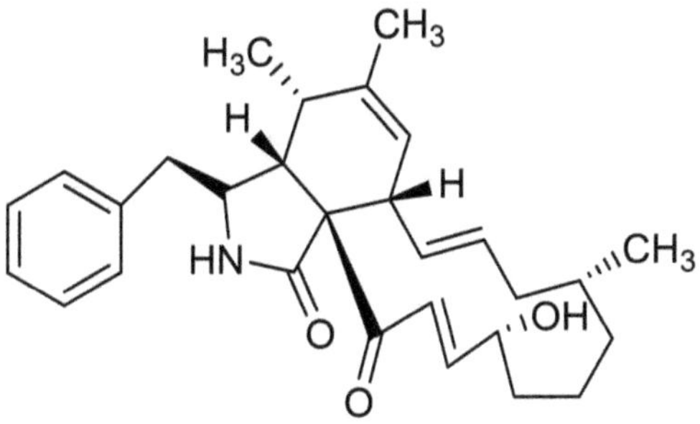

51

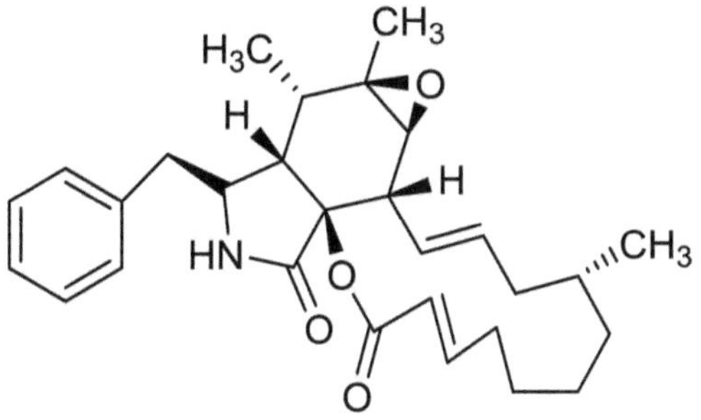

52

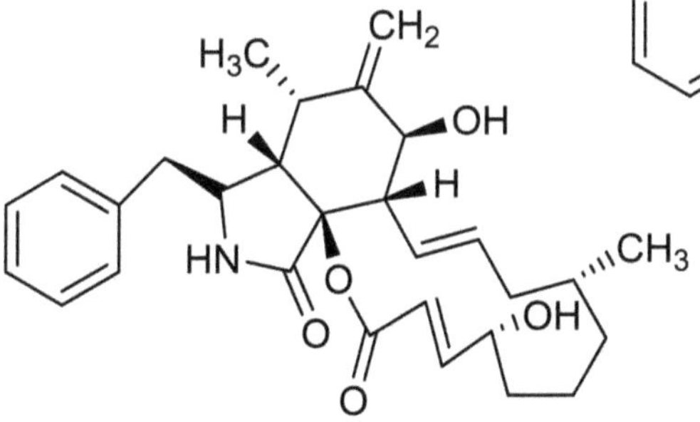

53

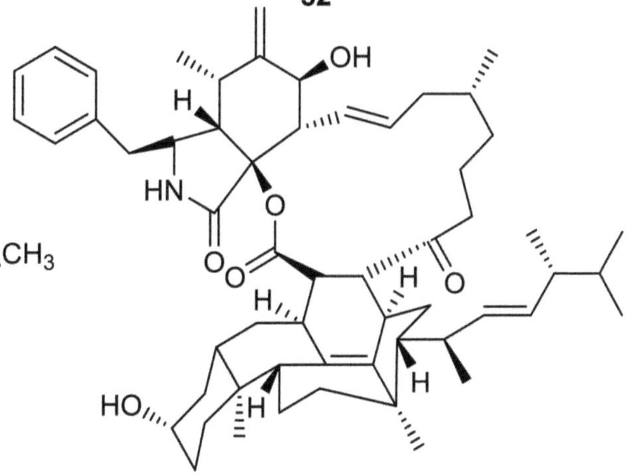

54

Fig. 10.5 Structures of cytochalasin derivatives **49–54**

10.2.7 Furopyrans

The tetrasubstituted furopyrans, chenopodolans E (**83**) and F (**84**) (Fig. 10.9) along with chenopodolan B (**85**) were reported from fungus *Phoma chenopodiicola.* In a zootoxic assay on *A. salina*, compound **83** was found to be active, causing 75% larval mortality, while compound **84** was inactive (Evidente et al. 2016). In another study of *Phoma chenopodicola*, chenopodolan D (**86**) and 11-O-acetylchenopodolan D (**87**) were iso-

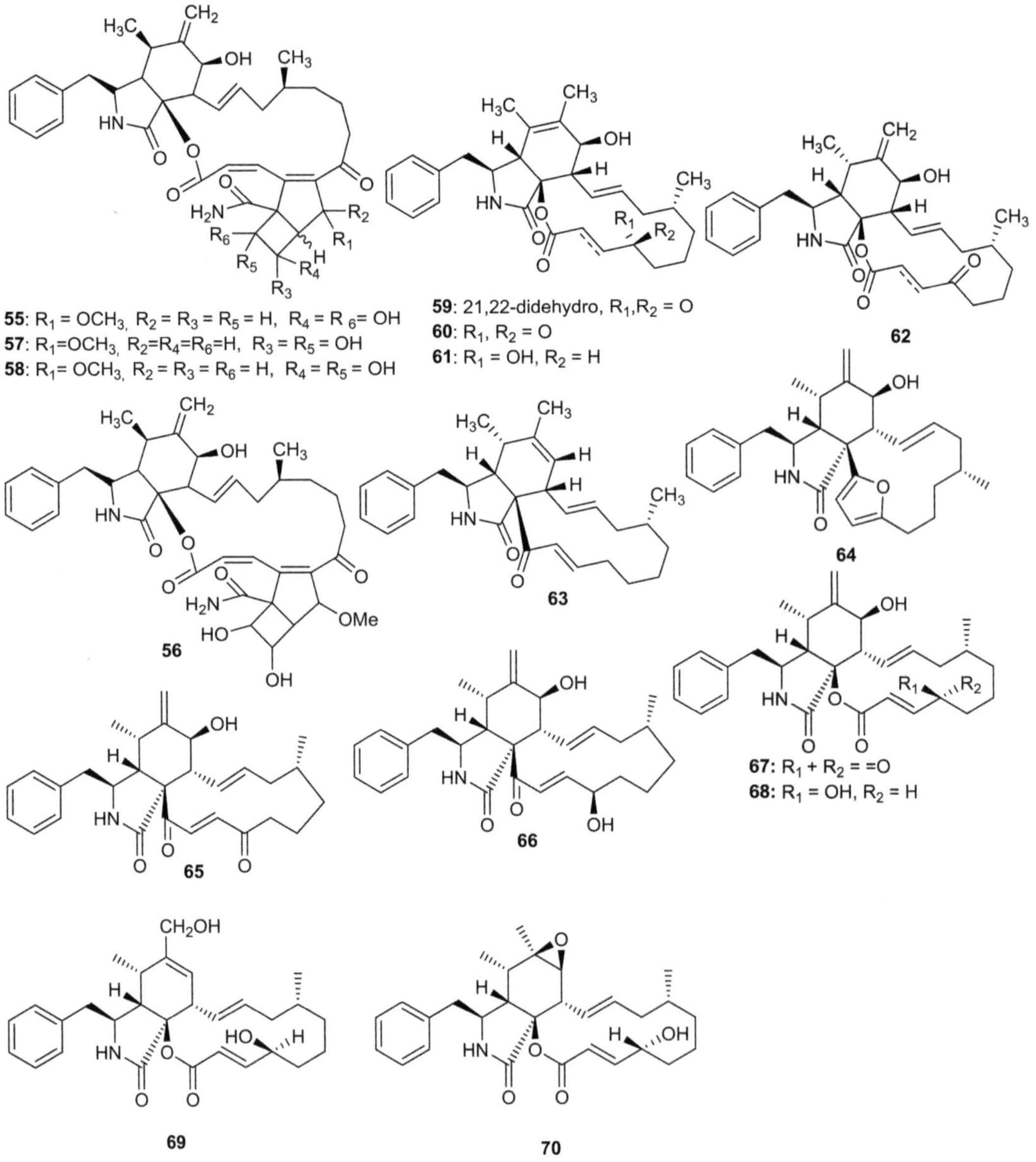

Fig. 10.6 Structures of cytochalasin derivatives **55–70**

lated, and the former compound was found to possess phytotoxic effects (Evidente et al. 2015). In another investigation, *Phoma chenopodiicola* produced chenopodolans A–C (**88–90**) and 11-O-acetyl-chenopodolan A (**91**). Chenopodolan A (**88**) and 11-O-acetylchenopodolan A (**91**) possess the same level of phytotoxic effects toward *Sonchus oleraceus* and *Mercurialis annua* (Cimmino et al. 2013).

10.2.8 Xyloketals and Chromones

The six xyloketals, viz., phomeketales A–F (**91–96**) (Fig. 10.10), were produced by the *Phoma* sp., which were isolated from *Sumbaviopsis* J. J. Smith. Compounds **91–96** were evaluated against PC-3, HL-60, and Molm 13 cell lines. Results of this evaluation demonstrated that metabolite **93** possesses cytotoxic activity on HL-60 cells (IC_{50},

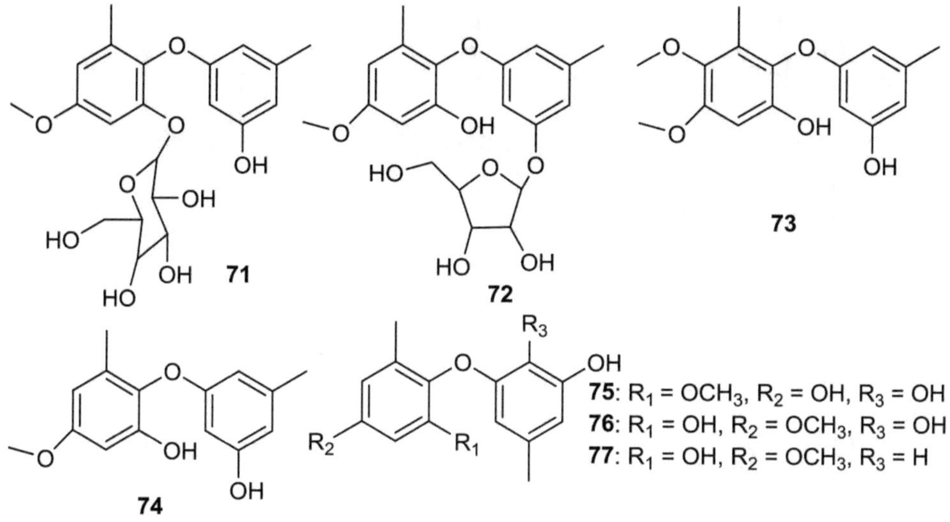

Fig. 10.7 Structures of diphenyl ether derivatives **71–77**

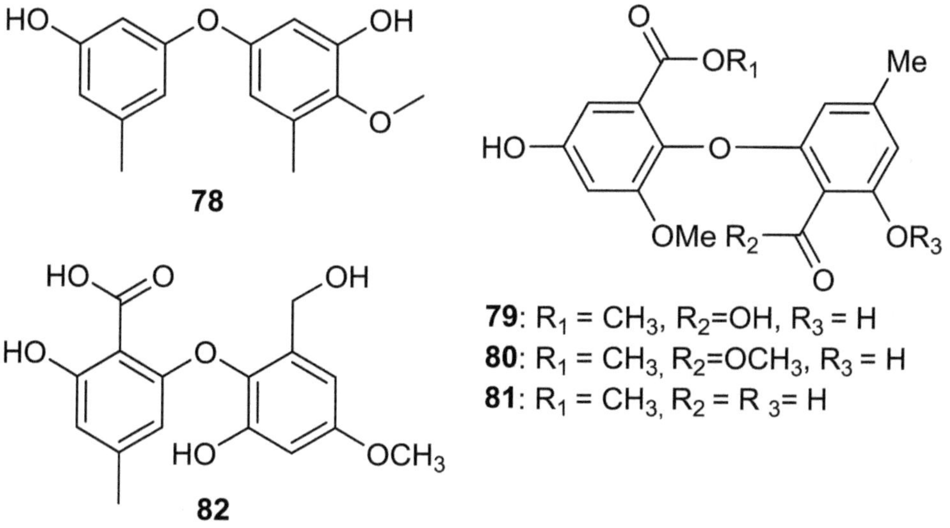

Fig. 10.8 Structures of diphenyl ether derivatives **78–82**

12.39 μM) and is active against the AChE enzyme with an IC_{50} of 40 μM (Sang et al. 2016). The chromone analog, chaetochromone D (**97**), was reported from the *Phoma* sp. SYSU-SK-7 and was found to possess radical scavenging effects with EC_{50} values of 68.1 μM (Chen et al. 2019). The chromone derivatives, spiciferone F (**98–100**), spiciferone A (**101**), spiciferol A (**102**), and compound **103,** were reported from the fungus *Phoma betae*. These metabolites were tested for their cytotoxic effects against HeLa, HCT 116, and MCF7 cells where it was found that only compound **98** illustrated cytotoxic effects on MCF7 with an IC_{50} of 7.73 μM (Tan et al. 2018). Phomalichenones C (**104**) and D (**105**), 8-ethyl-7-hydroxy-5-methoxy-2-methylchroman-4-one (**106**), and LL-D253 γ (**107**) were isolated from the *Phoma* sp. EL002650 (Kim et al. 2018).

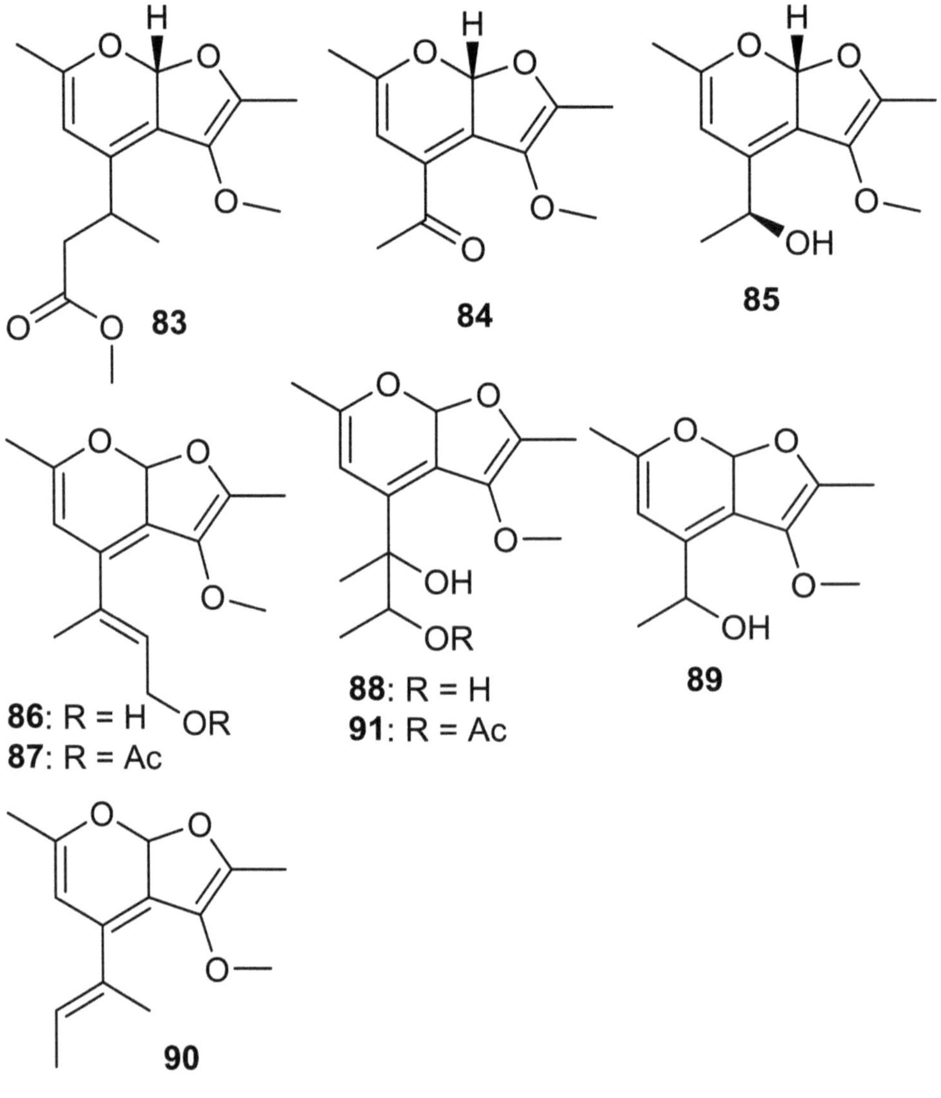

Fig. 10.9 Structures of furopyrans **83–90**

10.2.9 Meroterpenoids and Diterpenoids

The *ent*-pimaradiene diterpene chenopodolin (**108**) (Fig. 10.11) was isolated from the *Phoma chenopodiicola* and was found to possess phytotoxic effects (Cimmino et al. 2013a). Six meroterpenoids, phomanolides A–F (**109–114**), were reported from *Phoma* sp. Compounds **111–114** were screened for their cytotoxic effects toward the glioma cells (H4, SH-SY5Y, and U251), breast cancer (MCF-7), and the cervical cell (HeLa). Compounds **112** illustrated weak activity on SH-SY5Y (IC_{50}, 131.9 μM), U251 (IC_{50}, 38.8 μM), H4 (IC_{50}, 44.6 μM), and MCF7 (IC_{50}, 68.4 μM). On the other hand, compound **114** was weakly active toward SH-SY5Y (IC_{50}, 83.5 μM) and HeLa (IC_{50}, 50.0 μM) (Zhang et al. 2019). Phomanolides C–F (**111–114**) represent tropolonic meroterpenoids bearing unusual structural

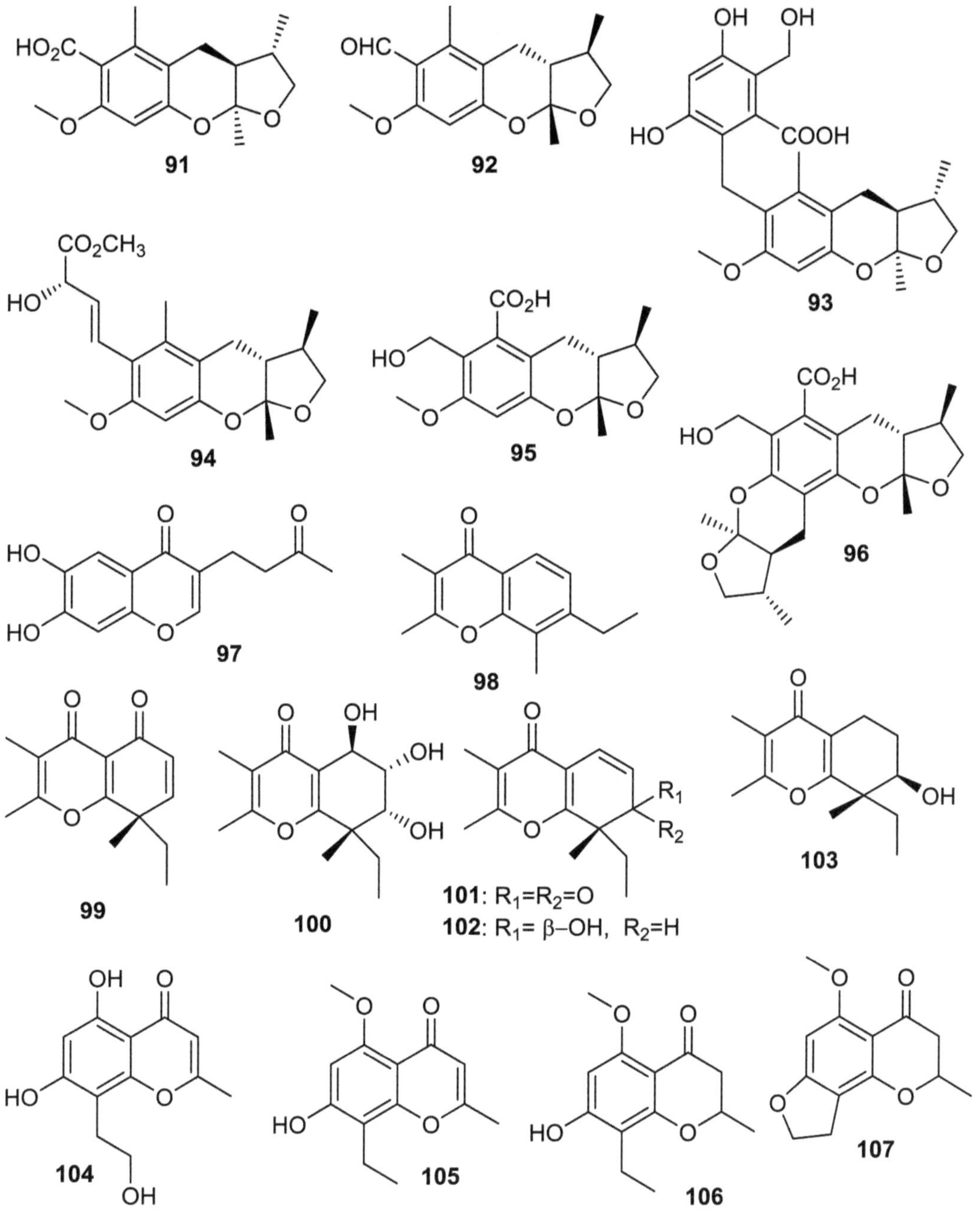

Fig. 10.10 Structures of xyloketals and chromones **91–107**

features. Meroterpenoids **111–113** have the rather unusual trioxa[4.4.3]propellane core, and oxygen-containing propellane cores are present in a few natural products, viz., neosetophomone A (El-Elimat et al. 2019) and cephalotanins A and B (Xu et al. 2016).

10.2.10 Alkaloids

Phomapyrrolidones A–C (**115–117**) (Fig. 10.12) having the cyclopenta[*b*]fluorene core system were reported to be isolated from the fungus *Phoma* sp. NRRL 46751. It

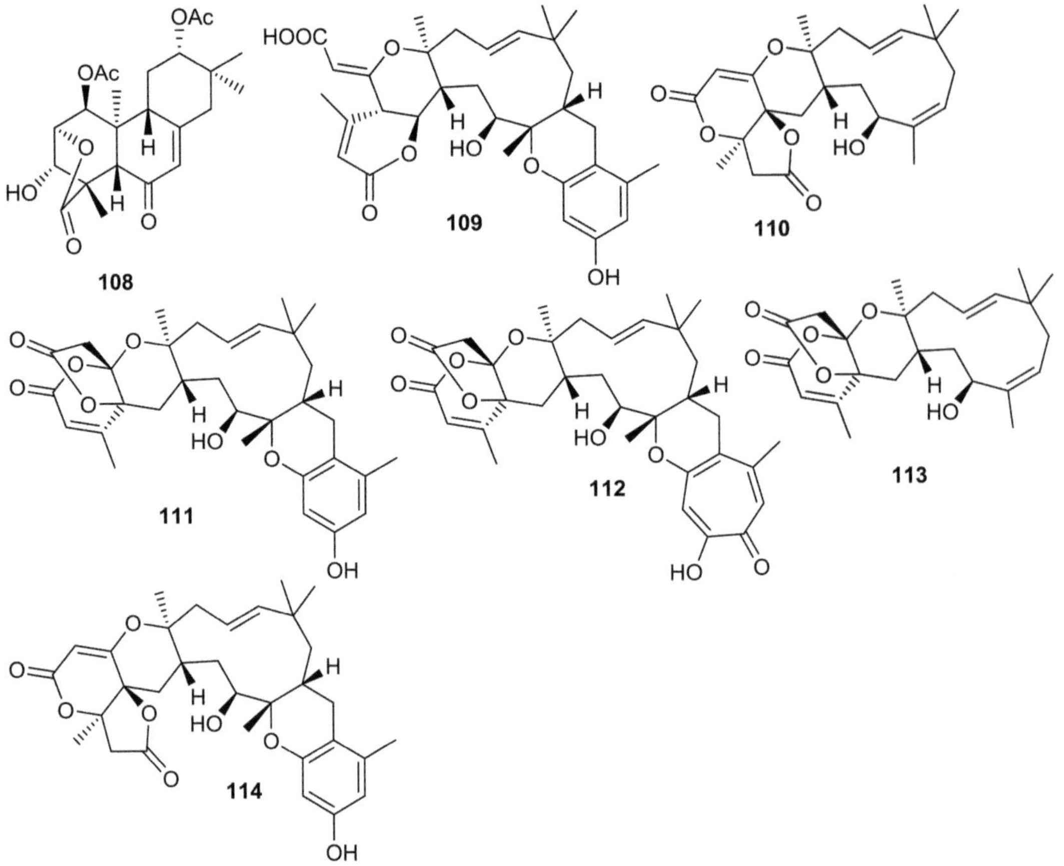

Fig. 10.11 Structures of meroterpenoids and diterpenoids **108–114**

was found that metabolite **116** and **117** illustrated weak antitubercular activity (Wijeratne et al. 2013). In another investigation on the fungus *Phoma herbarum*, it was found to produce eight pyrazinone alkaloids, viz., the terezines E–K (**118–124**) and one pyrazine alkaloid, terezine L (**125**). Terezines K (**124**) and L (**125**) were found to be inactive in screening for antimalarial, antibacterial, antimycobacterial, and cytotoxic activities (Maha et al. 2016). Maha et al. (2017) later reported four tyrosine-derived alkaloids, terezine M (**126**) and phomarosines A–C (**127–129**), together with the two hydantoin-derived alkaloids, (*S*)-5-isopropyl-3-methoxyimidazolidine-2,4-dione (**130**) and (*S*)-5-(4-hydroxybenzoyl)-3-isobutyrylimidazolidine-2,4-dione (**131**), to be isolated from the *Phoma herbarum*.

10.2.11 Nonaromatic Polyketides

The polyketides, phomaketides (132–136) and pseurotins A_3 (**137**) and G (**138**) (Fig. 10.13), were reported from the *Phoma* sp. NTOU4195. Among these metabolites, compound **132** illustrated significant antiangiogenic effects IC_{50} of 8.1 μM, while metabolite 134 exhibited potent inhibitory effects of LPS-induced NO production (IC_{50}, 8.8 μM) (Lee et al. 2016). Phomaketide A (**139**) was reported from *Phoma* sp. and illustrated anti-lymphangiogeneic potential (IC_{50} = 3.7 μM) and cytotoxic effects on lung cancer (A549, IC_{50}, 3.0 μM) (Tai et al. 2019). Phomretones A–F (**140–145**) were reported from the cocultivation of *Phoma* sp. and *Armillaria* sp. All the isolated metabolites (**140–145**) were found to be inactive in cytotoxicity, anti-acetylcholinesterase, and anti-PTP1B assays (Li et al. 2020).

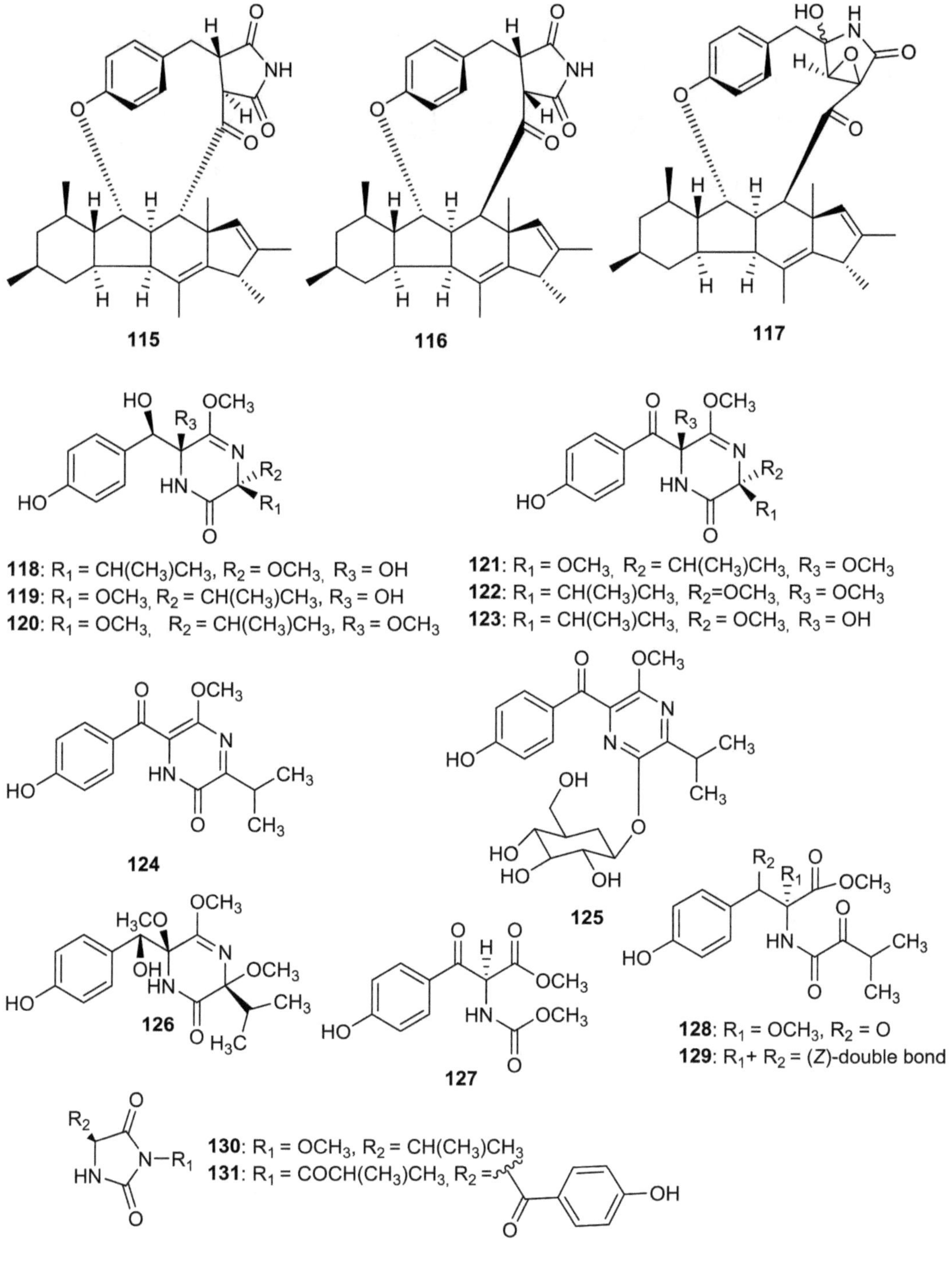

Fig. 10.12 Structures of alkaloids **115–131**

Polyketides, bellidisins A–D (**146–149**) (Fig. 10.14), pinolidoxin (**150**), 5,6-epoxypinolidoxin (**151**), and 2-epi-herbarumin II (**152**), were produced by *Phoma bellidis*. Metabolite **149** illustrated significant cytotoxic effects on A549, HL-60, MCF-7, SMMC-7721, and SW480 cell lines with IC_{50} ranged from 3.4 to 15.2

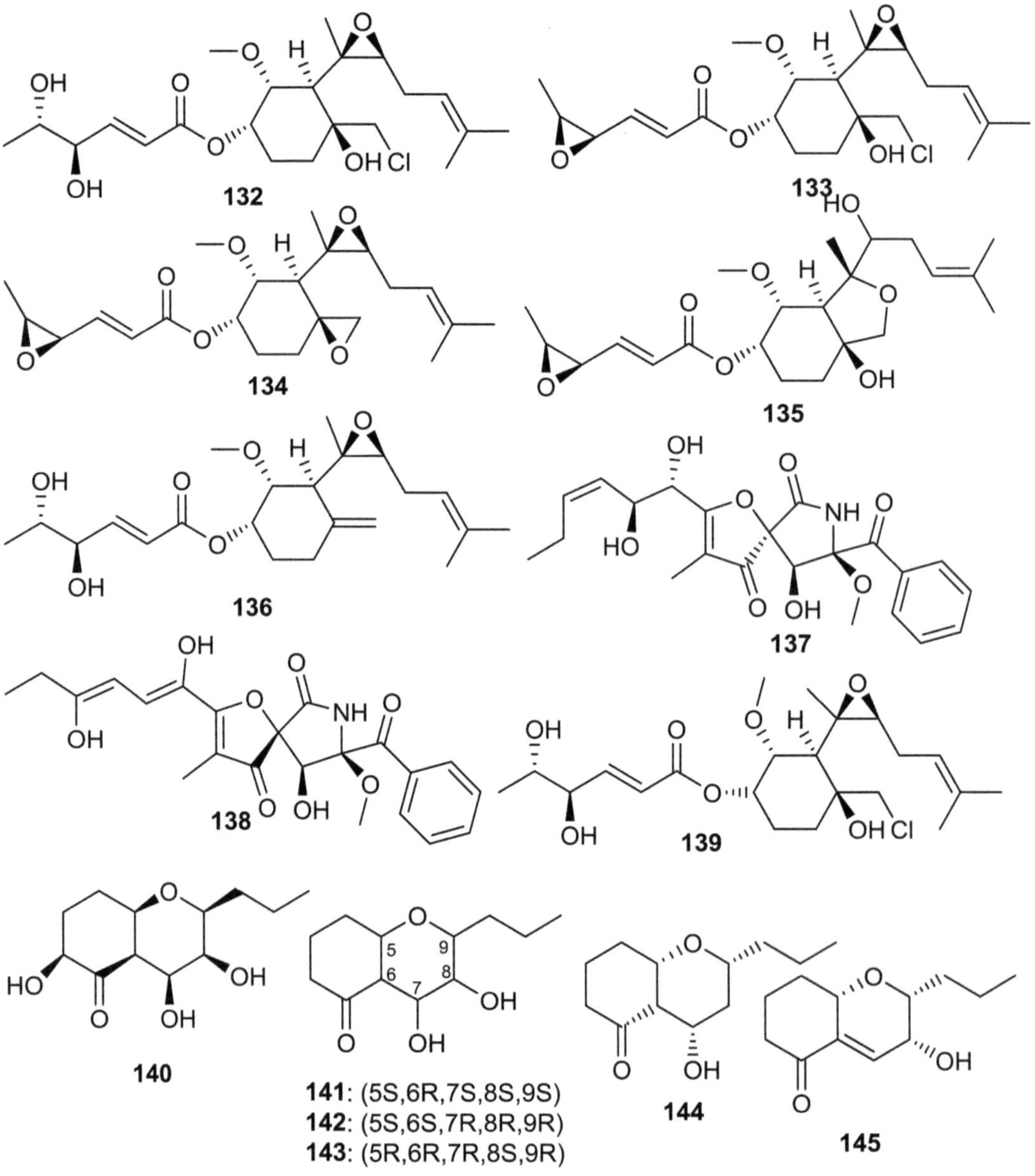

Fig. 10.13 Structures of nonaromatic polyketides **132–145**

μM. Notably compound **149** activities were better than cisplatin (IC_{50}, 4.8–27.7 μM). On the other hand compound **146** was only active toward HL-60 with IC_{50} of 20.6 μM (Wang et al. 2019). In another study arthropsadiol C (**153**) and massarilactone H (**154**), together with massarilactone B (**155**), massarilactone G (**156**), massarigenin C (**157**), massarigenin D (**158**), and massarilactone E (**159**), were produced by *Phoma herbarum*. Metabolites **154**, **156**, and **157** illustrated moderate neuraminidase inhibition with IC_{50} ranged from 4.15 to 9.16 μM (Zhang et al. 2012).

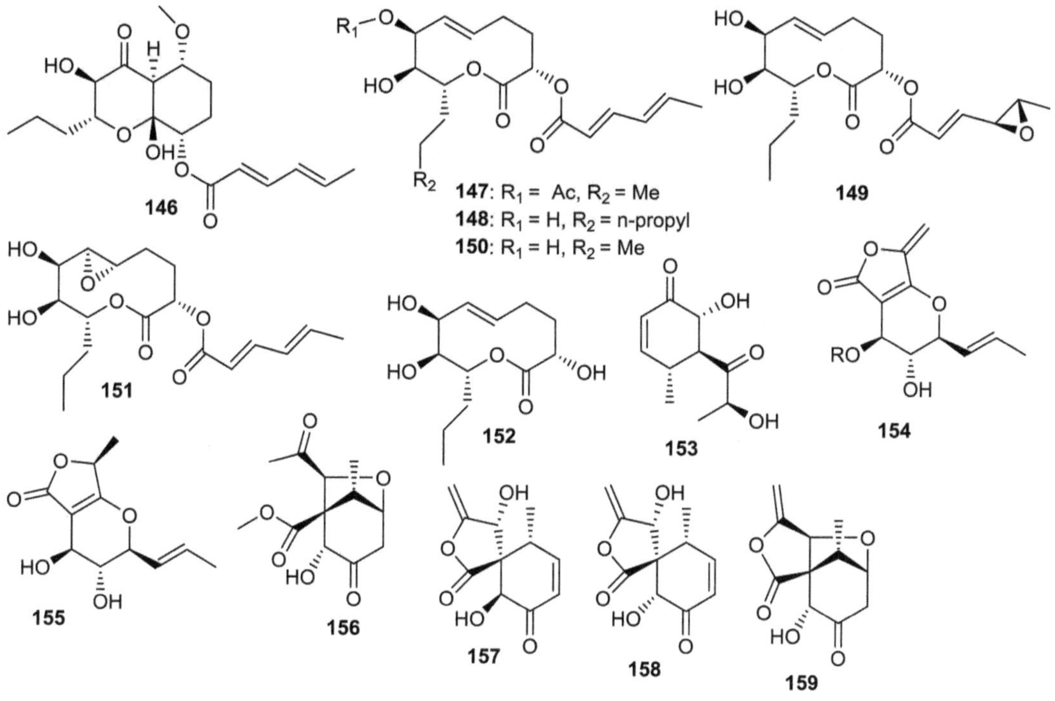

Fig. 10.14 Further structures of nonaromatic polyketides **146–159**

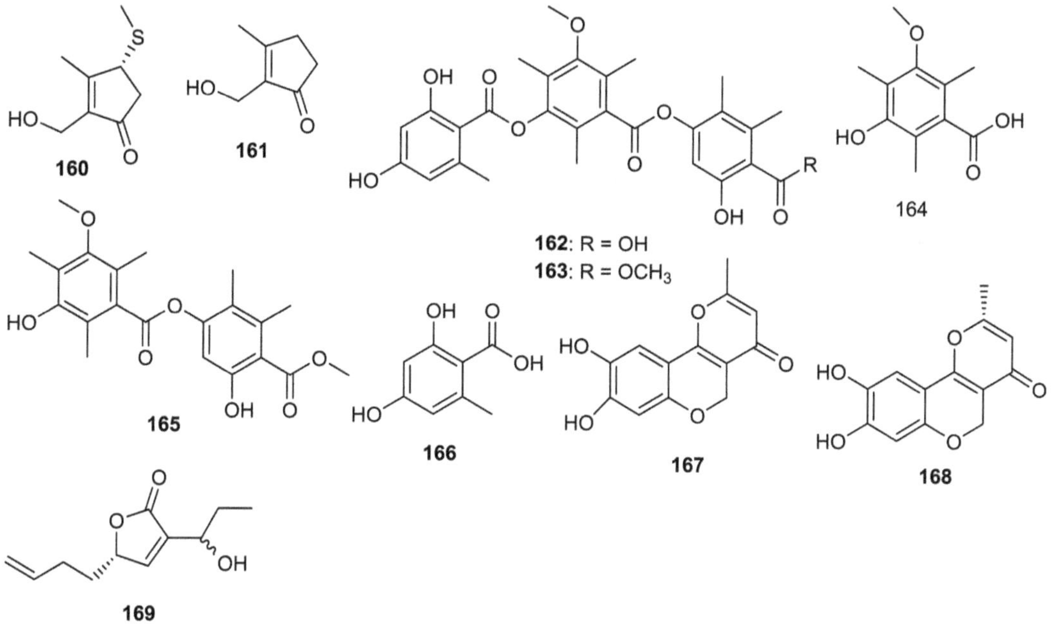

Fig. 10.15 Structures of compounds **160–169**

10.3 Miscellaneous

Phomanone A (**160**) (Fig. 10.15) and 2-hydroxymethyl-3-methylcyclopent-2-enone (**161**) were produced by *Phoma* sp., and these metabolites were not active on various cancer cell lines (Lu et al. 2019). Colletotric A (**162**) and B (**163**), 3-hydroxy-5-methoxy-2,4,6-trimethylbenzoic acid (**164**), colletotric C (**165**), orsellinic acid (**166**), citromycin (**167**),

(−)-2,3-dihydrocitromycin (**168**), and 8-hydroxy-pregaliellalactone B (**169**) were produced by *Phoma* sp. SYSU-SK-7. Compounds **162** and **163** illustrated antimicrobial effects against *Pseudomonas aeruginosa* and *C. albicans* with MIC ranging from 1.67 to 6.28 μg/mL. Moreover, metabolites **162–166** possess *α*-glucosidase effects in which the IC_{50} values ranged from 36.2 to 90.6 μM. On the other hand, metabolite **163** illustrated cytotoxicity against MDA-MB-435 (IC_{50}, 16.8 μM) and A549 cells (IC_{50}, 20.7 μM). In addition, compound **167** was active toward DPPH (EC_{50}, 11.8 μM) (Chen et al. 2019).

The authors of this chapter have isolated phomafuranol (**170**) and phomalacton (**171**) (Fig. 10.16) from the fungus *Phoma* spp., and these metabolites illustrated good antimicrobial effects (Hussain et al. 2014). The same authors have also reported sclerodin (**172**), 8,9-dihydro-3,5,7-trihydroxy-1,8,8,9-tetramethyl-5-(2-oxopropyl)-4*H*-phenaleno[1,2-*b*]furan-4,6(5*H*)-dione (**173**), atrovenetinone (**174**), and sclerodione (**175**) from the *Phoma* sp. Metabolite **172** illustrated moderate antialgal effects, while compounds **173–175** possess moderate antifungal properties (Hussain et al. 2015). In another investigation, the benzodioxin (**176**) and methoxyphenoxyacrylic acid (**177**) were isolated from the fungus *Phoma macrostoma* (Nalli et al. 2019), and botryosphaeridione (**178**) was obtained from *Phoma* sp. LN-16 (Zhang et al. 2012a).

(3*S*)-3,6,7-Trihydroxy-α-tetralone (**179**), cercosporamide (**180**), ß-sitosterol (**181**), and trichodermin (**182**) (Fig. 10.17) were isolated from a *Phoma* species. These compounds showed antimi-

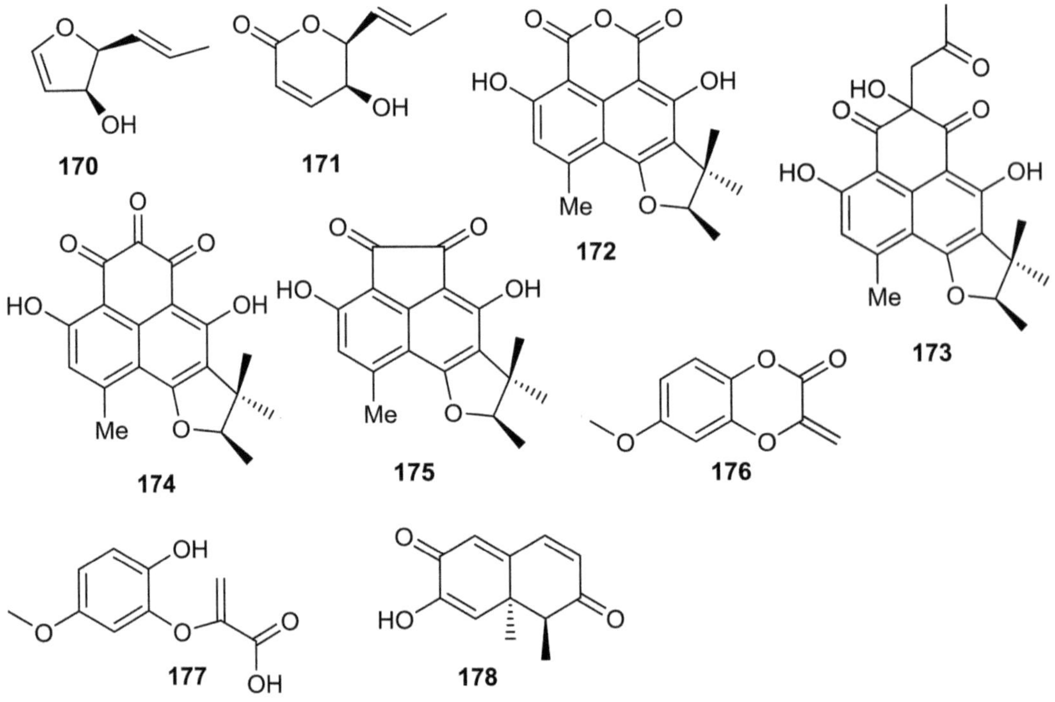

Fig. 10.16 Structures of compounds **170–178**

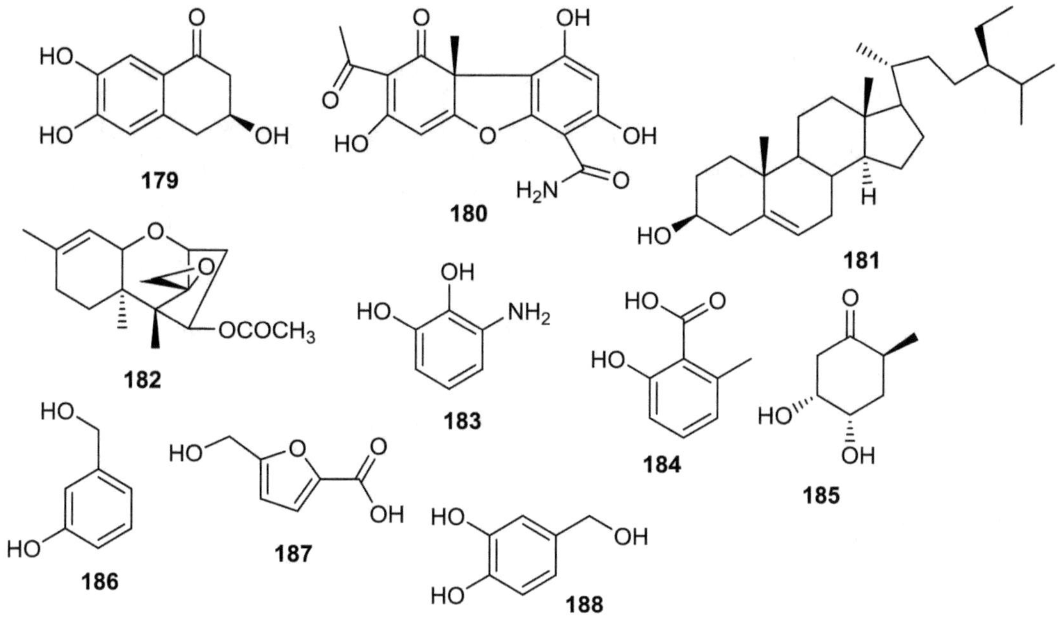

Fig. 10.17 Structures of compounds **179–188**

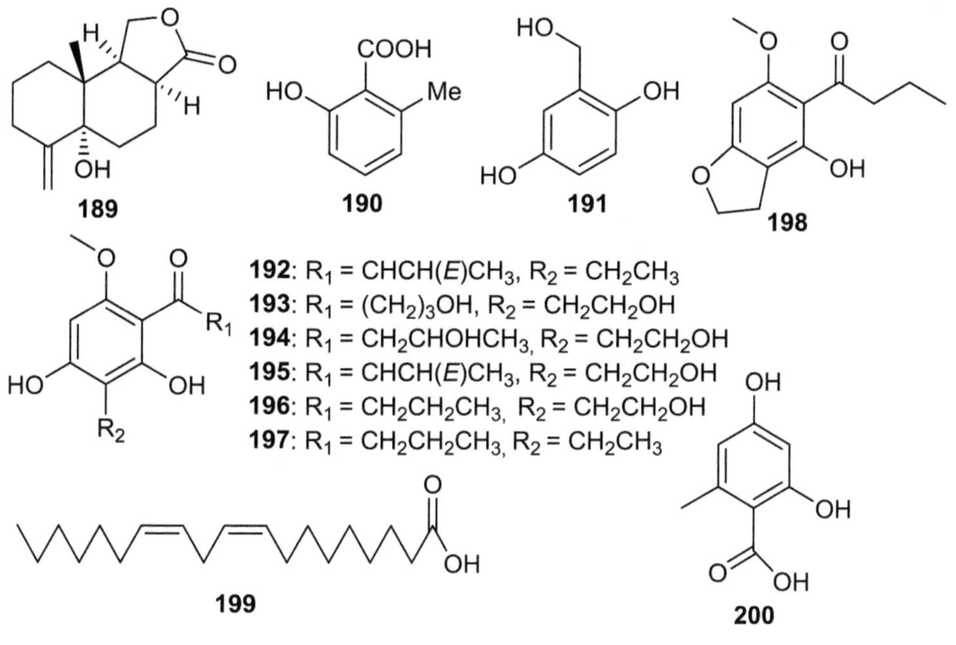

Fig. 10.18 Structures of compounds **189–200**

crobial activity against the four plant pathogenic fungi: *Magnaporthe oryzae*, *Fusarium oxysporum*, *Colletotrichum gloeosporioides*, and *Rhizoctonia solani*, and two further plant pathogenic bacteria (*Xanthomonas oryzae* and *X. campestris*). Biological studies demonstrated that compound **179** stopped the growth of *Rhizoctonia solani* and *Fusarium oxysporum* with EC_{50} of 48.5 and 413.2 μg/mL, respectively. On the other hand, compound **180** illustrated cytotoxic effects on

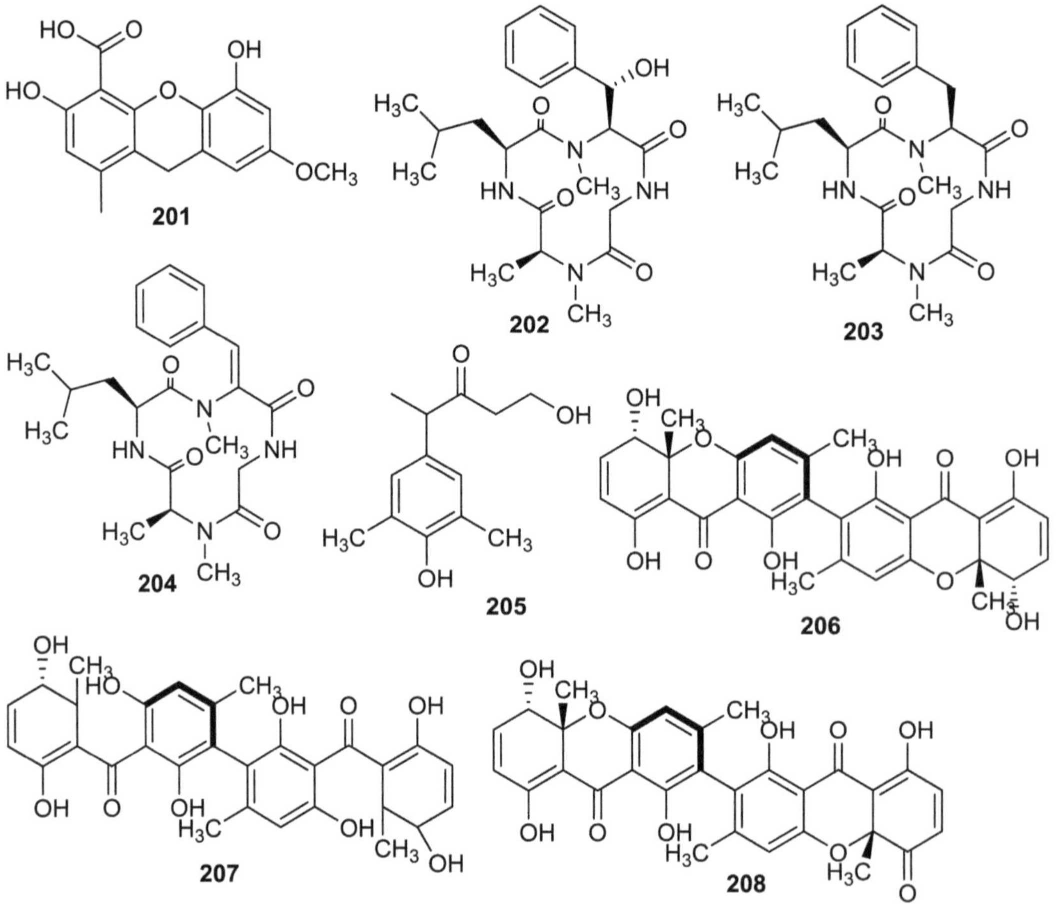

Fig. 10.19 Structures of compounds **201–208**

HT-29, MGC80-3, SMMC-772, MCF-7, HL-60, and P388 cell lines with the IC_{50} ranging from 9.3 to 48.7 μM, respectively (Wang et al. 2012). In another study, *Phoma* sp. CZD-F11 produced 1,2-dihydroxy-3-aminobenzene (**183**), 2-hydroxy-6-methylbenzoic acid (**184**), (2*S*, 4*S*, 5*R*)-4,5-dihydroxy-2-methylcyclohexanone (**185**), 3-hydroxybenzyl alcohol (**186**), 5-hydroxymethyl-furan-2-carboxylic acid (**187**), and 4-(hydroxy-methyl)catechol (**188**). Compound **187** was found to possess moderate cytotoxic effects on PC-3 cell lines (66.1% inhibition), while compound **187** (78.5% inhibition) and **188** (76.4% inhibition) demonstrated good bromodomain protein 4 (BRD4) inhibition (Wu et al. 2018).

Liu et al. (2019) isolated phomanolide (**189**) (Fig. 10.18) along with two aromatic analogs named 6-methylsalicylic acid (**190**) and gentisyl alcohol (**191**) from the endophytic fungus *Phoma* sp. In another study, *Phoma* sp. EL002650 was found to produce phomalichenones A (**192**) and B (**193**) together with compounds **194–198**. Biological studies revealed that compounds **192** and **195** stopped production of NO in LPS-stimulated RAW264.7 macrophages (Kim et al. 2018). 9(Z),12(Z)-Nonadecadienoic acid (**199**) and orsellinic acid (**200**) were produced by a *Phoma* sp. strain SHZK-2, and these metabolites illustrated moderate cytotoxic effects against HEPG2 and Raji cells (Fang et al. 2012).

Barceloneic acid C (**201**) was found to be produced by *Phoma* sp. JS752 and possesses moderate antibacterial effects toward *Listeria monocytogenes* and *Staphylococcus pseudintermedius* with an MIC of 1.02 μg/ml (Xia et al. 2015). In another investigation, three peptides **202–204** were isolated from the fungus *Phoma* sp. isolated from *Nemopilema nomurai* (Kim et al. 2012a). *Phoma herbarum* FGCC#54 was found to produce 1-hydroxy-4-(3′,5′-dimethyl-4′-hydroxyphenyl)-3-one (**205**) and illustrated herbicidal effects on test weeds (Kalam et al. 2014). In another study, the polyketides, phomalevones A–C (**206–208**), were obtained from *Phoma* sp. (MYC-1734 = NRRL 39060; Cucurbitaceae) (Shim et al. 2011) (Fig. 10.19).

10.4 Conclusion and Future Perspectives

Secondary metabolites, with their impressive diversity and complexity, remain an unlimited source and inspiration of numerous new potential pharmaceutical drugs for treatment of a wide range of diseases. Fungal secondary metabolism remains a vitally important topic as illustrated here in *Phoma.* This fungus is a privileged resource of numerous molecules belonging to a wide variety of classes of secondary metabolites with interesting bioactivities. The 208 *Phoma* fungal molecules presented in this chapter represent only a small subfraction of fungal metabolite chemistry. Various metabolites isolated from *Phoma* sp. possess important and potent biological effects. Moreover, isolated compounds with high in vitro biological effects should be developed further for in vivo testing. In addition, semisynthetic derivatives of active secondary metabolites should be prepared in order to evaluate them comprehensively and especially their structure activity relationship (SAR) studies.

Acknowledgments The authors Hidayat Hussain and Nilufar Z. Mamadalieva are thankful to the Alexander von Humboldt Foundation for its generous support in providing the research opportunity in Germany.

References

Arora P, Wani ZA, Nalli Y, Ali A, Riyaz-Ul-Hassan S (2016) Antimicrobial potential of thiodiketopiperazine derivatives produced by *Phoma* sp., an endophyte of *Glycyrrhiza glabra* Linn. Microb Ecol 72:802-812.

Chen ZM, Chen HP, Li Y, Feng T, Liu JK (2015) Cytochalasins from cultures of endophytic fungus *Phoma multirostrata* EA-12. J Antibiot 68(1):23-26.

Chen Y, Yang W, Zou G, Chen S, Pang J, She Z (2019) Bioactive polyketides from the mangrove endophytic fungi *Phoma* sp SYSU-SK-7. Fitoterapia 139:104369.

Cimmino A, Andolfi A, Zonno MC, Avolio F, Berestetskiy A, Vurro M, Evidente A (2013) Chenopodolans A–C: phytotoxic furopyrans produced by *Phoma chenopodiicola*, a fungal pathogen of *Chenopodium album.* Phytochemistry 96:208-213.

Cimmino A, Andolfi A, Zonno MC, Avolio F, Santini A, Tuzi A, Berestetskyi A, Vurro M, Evidente A (2013a) Chenopodolin: a phytotoxic unrearranged ent-pimaradiene diterpene produced by *Phoma chenopodiicola*, a fungal pathogen for *Chenopodium album* biocontrol. J Nat Prod 76:1291-1297.

Davisona EK, Brimble MA (2019) Natural product derived privileged scaffolds in drug discovery. Curr Opin Chem Biol 52:1–8.

El-Elimat T, Raja HA, Ayers S, Kurina SJ, Burdette JE, Mattes Z, Sabatelle R, Bacon JW, Colby AH, Grinstaff MW, Pearce CJ, Oberlies NH (2019) Meroterpenoids from Neosetophomasp: A dioxa[433]propellane ring system, potent cytotoxicity, and prolific expression. Org Lett 21:529-534.

Elsebai MF, Ghabbour HA (2016) Isocoumarin derivatives from the marine-derived fungus *Phoma* sp. 135. Tetrahedron Lett 57:354-356.

Evidence A, Cimmino A, Andolfi A, Berestetskiy A, Motta A (2011) Phomachalasins A–D, 26-oxa [16] and [15] cytochalasans produced by *Phoma exigua* var. *exigua*, a potential mycoherbicide for *Cirsium arvense* biocontrol. Tetrahedron 67:1557-1563.

Evidente M, Cimmino A, Zonno MC, Masi M, Berestetskyi A, Santoro E, Superchi S, Vurro M, Evidente A (2015) Phytotoxins produced by *Phoma chenopodiicola*, a fungal pathogen of *Chenopodium album.* Phytochemistry 117:482-488.

Evidente M, Cimmino A, Zonno MC, Masi M, Santoro E, Vergura S, Evidente A (2016) Chenopodolans E and F, two new furopyrans produced by *Phoma chenopodiicola* and absolute configuration determination of chenopodolan B. Tetrahedron 72(51):8502-8507.

Fang MJ, Fang H, Li WJ, Huang DM, Wu Z, Zhao YF (2012) A new diphenyl ether from *Phoma* sp. strain, SHZK-2. Nat Prod Res 26:1224-1228.

Gerry CJ, Schreiber SL (2018) Chemical probes and drug leads from advances in synthetic planning and methodology. Nat Rev Drug Discov 17:333-352.

González-Medina M, Owen JR, El-Elimat T, Pearce CJ, Oberlies NH, Figueroa M, Medina-Franco JL

Bioactive Secondary Metabolites from Endophytic *Phoma* spp.

11

Pramod Kumar Mahish, Shweta Singh, and Ravishankar Chauhan

Abbreviations

BEAS-2B	Human bronchial epithelial cells
COSY	Correlated spectroscopy
DV-3	Human dengue virus type 3
EtOAc	Ethyl acetate
FT-IR	Fourier-transform infrared spectroscopy
GC-MS	Gas chromatography-mass spectrometry
H1N1	Influenza A virus subtype
HeLa	Henrietta Lacks (immortal cell line)
HIV	Human immunodeficiency virus
HL	Human leukemia cell line
HMBC	Heteronuclear multiple bond correlation
HMQC	Heteronuclear multiple quantum coherence
IL	Interleukin
LC-MS	Liquid chromatography-mass spectrometry
MCF-7	Breast cancer cell line (MCF = Michigan Cancer Foundation)
NMR	Nuclear magnetic resonance
NOESY	Nuclear Overhauser effect spectroscopy
PAF	Platelet-activating factor
ROS	Reactive oxygen species
RS	Reciprocal syncytial virus
SMMC	Human hepatocarcinoma cell line
SW480	Human colorectal cancer cell lines
VOCs	Volatile organic compounds

11.1 Introduction

Microbes and plants are found in continuous interaction with each other. Among them, some microbes cause symptomatic diseases, while some are associated with beneficial significance to the host plant. Endophytes are described as the internal mycota of living plants. The word endophyte was first coined by de Bary in 1866. It consists of two Greek words endon = within and phyte = plant. It includes foliar to root symbionts of fungi, bacteria, actinomycetes, and mycoplasm (Arnold 2007). Endophytes are fungi that colonize living plant tissue without causing any immediate, overt negative effects (Khan et al. 2016).

The abundance of endophytic fungi is so vast. The endophytes were isolated from nearly 300,000 land plant species with one or more fungi (Arnold 2008). The term endophyte regu-

P. K. Mahish (✉) · S. Singh
Department of Biotechnology, Government Digvijay (Autonomous) P.G. College, Rajnandgaon, Chhattisgarh, India

R. Chauhan
National Centre for Natural Resources (NCNR), Pandit Ravishankar Shukla University, Raipur, Chhattisgarh, India

M. Rai et al. (eds.), *Phoma: Diversity, Taxonomy, Bioactivities, and Nanotechnology*,
https://doi.org/10.1007/978-3-030-81218-8_11

larly alludes to a particular host type, a taxonomic group of hosts, or the kind of tissue involved. Microfungi are predominant parts of those arrays, colonizing foliar and twig surfaces (epiphytes), inside tissues of foliage (foliar endophytes), young and adult bark (bark endophytes), as well as wood (xylem endophytes and wood decomposers) (Stone et al. 2012).

Endophytic fungi are classified into two groups clavicipitaceous endophytes (C-endophytes) and nonclavicipitaceous endophytes (NC-endophytes) which are based on the taxonomy, tissue specificity, host range, ecology function, etc. (Schaechter 2011). Clavicipitaceous belongs to the *Ascomycota* associated with grasses (Bacon and White 2000). These fungi are present in the intracellular space of rhizome, leaf sheath, and culms (Moy et al. 2000; Dugan et al. 2002). NC-endophytes found an association with the tropical tree, boreal, and arctic/Antarctic communities (Arnold et al. 2000; Gamboa and Bayman 2001). These fungi belong to ascomycetes and basidiomycetes.

Endophytes show host selection when the endophyte forms a relationship with two related flora but demonstrate a preference in one particular host (Cohen 2006). The endophyte also demonstrates host preference; it is used when endophytes show common occurrence on a particular host (Paulus et al. 2006). Fungal endophytes are the course of attention by scientific communities from the last few decades because of its promising benefits to the host plant in a different way.

The molecule produced by organisms that are not essential for their development, growth, and survival is known as secondary metabolites. The term "secondary metabolites" was coined by Albrecht Kossel in 1891. Bacteria, fungi, and plants are the fundamental living organisms responsible for the production of secondary metabolites. Microbial secondary metabolites are petite biomolecules delivered during the late development stage (Sanchez and Demain 2011). The production of secondary metabolites from the fungal source has little advantages like easy growth, manipulation, low nutrient requirement, and simple extraction process. Endophytic fungi associated with plants are a valuable source of phenolics, alkaloids, polyketides, steroids, enzymes, peptides, and quinines (Meena et al. 2019).

Since the discovery of anticancer Taxol from endophytic fungi, the attraction of scientific communities rises to investigate molecules from plant-associated microbes, which leads to the finding of antibacterial, antiviral, antioxidant, immunosuppressant, anti-inflammatory, antidiabetic, and anticancer metabolites (Uzma et al. 2019). Endophytic fungi are also the source of oxidative and hydrolytic extracellular enzymes which are useful in healthcare, textile, beverage, food, leather, and confectionery industries (Mishra et al. 2019).

Endophytes are vast hidden biodiversity of fungal flora as they are little explored in the developing countries having rich plant resources (Rodriguez et al. 2009). Therefore, there is a high probability of investigating potential endophytic fungi with various applications in therapeutics, agriculture, and biotechnology.

11.2 Endophytic *Phoma*

Phoma is a well-known genus of phylum *Ascomycota*, class *Dothideomycetes*, order *Pleosporales*, and family *Didymellaceae* (Bennett et al. 2018). There are about 2000 *Phoma* species that have been identified throughout the world (Boerema et al. 2004). *Phoma* was reported as soil born, saprophytic, plant pathogen, and endophytes. The identification of *Phoma* can be determined by its morphological characteristics (Rai and Rajak 1993). The colonies of *Phoma* give greenish-gray to brown in agar media. The formation of asexual spore is known as conidia with unicellular and hyaline as well as different in shape (Boerema et al. 2004). Pycnidia can be observed as asexual fruiting bodies which are globose to subglobose to pyriform with a dark appearance. The ostioles are found single or multiple (Aveskamp et al. 2008). Thick-walled vegetative cells within hyphae or at hypha are known as chlamydospores (Guégan et al. 2016; Choi and Kim 2017).

The presence of endophytic *Phoma* has been recorded from continents like Asia, Europe, Africa, North America, Australia, and South America. More endophytes are investigated from the Indian subcontinent, China, California, Mediterranean, Peru, Brazil, etc. (Gazis and Chaverri 2010; Bharathidasan and Panneerselvam 2011; Moricca et al. 2012; Maheswari and Rajagopal 2013; Tan et al. 2018a; Chen et al. 2020; Gama et al. 2020). *Phoma* is reported from host plants of almost all kinds of the climatic conditions like tropical, subtropical, arid, and mangroves (Loro et al. 2012; Humzah et al. 2018). Host plants of the mountain, desert, forest, mangrove, and river belt also contain *Phoma* as endophytic fungi (Guo et al. 2008; Sun et al. 2012; El-Nagerabi et al. 2013; Du et al. 2020). Tree, herb, shrub, grasses, and palm inhabits the endophytic *Phoma* which are perennial and evergreen (Gazis and Chaverri 2010; Ghimire et al. 2011; Rivera-Orduña et al. 2011; Khan et al. 2016; Selim et al. 2018). The fungi were obtained from various parts of plants like leaf, stem, root, petioles, barks, etc. (Fernandes et al. 2015; Szucs et al. 2018; Radiastuti et al. 2019).

The finding of the new natural and eco-friendly bioactive compound is always a focused field of researchers. The agrochemical and pharmaceutical areas are very good examples of such a field. In the recent year, phomodione, cercosporamide, phomenon, phomin, cytochalasins, and dehydrophomin were identified from *Phoma* as novel secondary metabolites. These metabolites showed a significant role of antiviral, antifungal, antibacterial, antiprotozoal, anticancer, and weedicide activities (Elkhateeb and Daba 2020).

11.3 Secondary Metabolites of Endophytic *Phoma*

Several thousands of fungus species are very diverse in their morphological characters. The word fungus on hearing it always clicks in our mind that is an infective agent that spreads so many kinds of infection. But, the fungus is a rich source of natural metabolite producer despite only being an infectious agent. Secondary metabolites produced by the endophytes may directly or indirectly influence the host plant in many ways like protecting from plant pathogen or inducing the production of secondary metabolites from the host plant, etc.

The genus *Phoma* is one of the fungal endophytes reported from many plant species (Bennett et al. 2018) and comes under plant pathogenic fungi that cause many kinds of diseases in plant, viz., lesions in leaves, stems, pods, and blossoms, discoloration in hypocotyl, cotyledons, etc. (Cimmino et al. 2008; Mahish and Ghritlahare 2017). Although *Phoma* spp. are parasites, they are tremendous sources of natural metabolites also. *Phoma* spp. are morphologically similar, but discrimination of the species is based on the culture characteristics, growth conditions, and the most impotent the production of secondary metabolites. Some metabolites are directly produced by them or some are induced. Secondary metabolites produced by *Phoma* have many pharmaceutical, clinical, or biotechnological applications (Uzma et al. 2019; Chen et al. 2020).

Different *Phoma* species produce distinct metabolites and have different biotechnological exploitations (Rai et al. 2020). Two new isochromanone derivatives (3S,4S)-3,8-dihydroxy-6-methoxy-3,4,5-trimethyl-lisochroman-1-one and methyl (S)-8 hydroxy-6-methoxy-5-methyl-4a-(3-oxobutan-2-yl)benzoate are produced by *Phoma* sp. PF2 is isolated from *Artemisia princeps* Pamps (Kim et al. 2018). Some *Phoma* species produce new secondary metabolites with symbiont such as *Phoma* sp. YUD17001 produces five new compounds promoter A, B, and C, and phexandiols A and B with co-culture of *Armillaria* sp. (Li et al. 2019). Along with terrestrial *Phoma* sp., marine *Phoma* spp. are also an enormous source of different secondary metabolite contents enolides, lactone, quinine, phthalate, diterpenes, etc. and are different from terrestrial *Phoma* sp. (Ostarhage et al. 2000; Rai et al. 2018). It is reported that they also synthesize amino acid derivative like *Phoma herbarum* PSU-H256 and can synthesize tyrosine derivative temezine M and phomarosine A, B, and C, and hydantoin derivatives S-5-isoprppyl-3-methoxyimidazolidine-2,4-dione and S-5-(4-hydroxybenzoyl)-3-isobutyrylimidazolidine-2,4-

dione show weak antibacterial activity (Maha et al. 2017). Fermentation broth of *Phoma* sp. YE 3135 obtained from *Aconitum vilmorinianum* contained rare 14-nordrimane-type sesquiterpenoid along with 6-methoxymellein; 7-hydroxy-3,5-dimethylisochromane-1-one; norlichxanthone; 6-methylsalicylic acid; and gentisyl alcohol with antiviral activity (Liu et al. 2019). Macrophin, rosellisin, 2-(2-hydroxy-5-6-methylene-1,4-benzodioxin-2(3H)-one, and methoxyphenoxy acrylic acid were isolated first time from *Phoma macrostoma* endophyte of the medicinal plant *Glycyrrhiza glabra* with cytotoxic potential (Nalli et al. 2019). *Phoma* sp. SYSU-SK-7 produces polyketides colletotric B, 3-hydroxy-5-methoxy-2,4,6-trimethylbenzoic acid, colletotric C, chaetochromone D, and 8-hydroxy-pregaliellalactone having an antimicrobial activity (Chen et al. 2018). Some *Phoma* spp. produce growth promoters like gibberellin and cytokinin (Kedar et al. 2014) and volatile compounds like ethanol, 1-butanol, 1-hexane, 2-(p-anisyl)-4-methyl, epi-bicyclosequiphellandrene, butanoic acid, propanoic acid, hexadecanoic acid, dodecanoic acid, etc. (Strobel et al. 2011). Along with the phomalactone, (3R)-5-hydroxymellein, and emodin, Hussain et al. (2014) isolated new dihydrofuran derivatives known as phomafuranol. Some secondary metabolites of endophytic *Phoma* spp. are summarized in Table 11.1.

11.4 Bioanalytical Techniques to Determine Secondary Metabolites of *Phoma* Species

It is very interesting to know that 6 of the 20 commonly prescribed medications are secondary metabolites of fungal origin (Rai et al. 2009). In this connection, the genus *Phoma* was introduced more than 180 years ago having numerous significant secondary metabolites which include antimicrobials, mycoherbicides, and phytotoxins. However, one of the chief challenges is the identification of such novel secondary metabolites. In metabolomics, eventually, all paths lead to the identification and quantification of certain key compounds. To be sure, without key compound distinguishing proof, the consequences of any metabolomic investigations are biochemically uninterpretable (Wishart 2011). Due to the huge chemical diversity and characters of most metabolites and metabolomic data, respectively, metabolite identification is intrinsically difficult. Thus, a lot of exertion in metabolomics over the previous decade has been centered around making metabolite identification accurate and quicker. In this context, nuclear magnetic resonance (NMR) spectroscopy is a perfect analytical tool for the characterization of metabolites (Wilson and Burlingame 1974). This analytical tool is one of the first techniques to be adopted in metabolomics as a nonbiased tool that allows the identification of novel compounds. NMR is particularly amenable to detecting compounds that are less traceable to LC-MS/GC-MS analyses, such as amines, sugars, or volatile liquids having molecular weight > 500 Da, or relatively nonreactive compounds (Reo 2002). Generally, the ^{1}H and/or ^{13}C NMR spectrum is sufficient for positive identification since NMR data offer two or three orthogonal measures concerning a compound's identity such as chemical shifts, peak intensities, and spin coupling patterns (Wishart 2008).

In line, using NMR spectroscopy very recently, four new polyketides, particularly bellidisin A (Fig. 11.1a) and bellidisins B-D, were isolated from *Phoma bellidis* (Wang et al. 2019). The absolute configurations and structures of these polyketides were determined by 1D and 2D NMR. In their investigation, they noticed that bellidisin D exhibited stronger cytotoxicity against human cancer cell lines HL-60, A549, SMMC-7721, MCF-7, and SW480 in comparison to cisplatin—a standard cytotoxic drug (Wang et al. 2019). In the same year, Liu et al. (2019) also reported phomanolide (Fig. 11.1e), a new rare 14-nordrimane-type sesquiterpenoid from an endophytic fungus *Phoma* sp. isolated from the roots of *Aconitum vilmorinianum*. The structure of this compound was determined by combined analyses of the ^{1}H-^{1}H COSY, HMBC, HMQC, and NOESY spectra. The coupling series from H-1 to H-3 and from H-6 to H-9 could be without a doubt set up by following connections

Table 11.1 Secondary metabolites from endophytic *Phoma*

S. no.	Fungi	Secondary metabolites	Bioactivity	References
1	*Phoma bellidis* Neergaard	Polyketides	Cytotoxic	Wang et al. (2019)
2	*Phoma betae* A.B. Frank	Diterpenes, aphidicolin	Antifungal	Oikawa et al. (2001)
		Spiciferones A, G–H, spiciferol A	Anticancer	Tan et al. (2018b)
		Taxol	Anticancer	Kumaran et al. (2012)
3	*Phoma cucurbitacearum* (E.M. Fries) P.A. Saccardo	Thiodiketopiperazine	Antimicrobial	Arora et al. (2016)
4	*Phoma etheridgei* Hutchison and Hiratsuka	Phomalone, dihydrobenzofuran	Antifungal	Ayer and Jemenez (1994)
5	*Phoma fungicola* Aveskamp, Gruyter, and Verkley	Polyketide spiciferone A	Herbicidal	Mohamed et al. (2017)
6	*Phoma glomerata* (A.C.J. Corda) Wollenw and Hochapfel	Salvianolic acid	Cholesterol control	Li et al. (2016)
7	*Phoma herbarum* G.D. Westendorp	Brefeldin A, nonenolides like herbarumin I and herbarumin II	Phytotoxin	Rai et al. (2009)
		3-Nitrophthalic acid	Phytotoxin	Vikrant et al. (2006)
		Terezine M, phomarosine A–C, S-5-isoprpyl-3-methoxyimidazolidine-2,4-dione	Antibacterial and cytotoxic	Maha et al. (2017)
		Gentisyl alcohol	Antagonistic to plant pathogen	Gupta et al. (2016)
		Gibberellin	Growth promoter	Hamayun et al. (2009)
		Tanshinone	Growth promoter	Chen et al. (2020)
8	*Phoma lingam* (Tode) Desmazières, J.B.H.J.	Sirodesmin PL, deacetylsirodesmin	Phytotoxin	Rai et al. (2009)
9	*Phoma macrostoma* J.P.F.C. Montagne	Macrophin	Anticancer	Nalli et al. (2019)
10	*Phoma medicaginis* Malbranche and Roumeguère	Brefeldin A	Antifungal	Weber et al. (2004)
		Glucosidases	Growth promoter	Khan et al. (2016)
		Paclitaxel	Anticancer	Zaiyou et al. (2017)
11	*Phoma multirostrata* (P.N. Mathur, S.K. Menon, and Thirum.) Dorenb and Boerema	Multirostratin A, 20-oxo-deoxaphomin, deoxaphomin, cytochalasin A, cytochalasin B, cytochalasin Z2, cytochalasin F	Anticancer	Chen et al. (2014)
		Ergocytochalasin A	Antiviral	Peng et al. (2020)
12	*Phoma radicina* section *Paraphoma radicina* (McAlpine) Morgan-Jones and J.F. White	Cryptotanshinone	Growth promoter	Teimoori-Boghsani et al. (2020)
13	*Phoma sorghina* (P.A. Saccardo) Boerema, Dorenb., and Kesteren	6-Methylsalicylic acid, epoxydon, desoxyepoxydon, phyllostine, 2,3′-dihydroxy-4-methoxy-5′,6-dimethyl diphenyl ether, and epoxydon 6-methylsalicylate ester	Phytotoxin	Venkatasubbaia et al. (1992)

(continued)

Table 11.1 (continued)

S. no.	Fungi	Secondary metabolites	Bioactivity	References
14	*Phoma* sp.	Cyperin-2-O-R-D-glucoside	Anticancer	Wu et al. (2018)
		Phomalichenones A–D	Anti-inflammatory	Kim et al. (2018)
		Thiodiketopiperrazine derivatives	Antimicrobial	Arora et al. (2016)
		Gibberellic acid	Growth promoter	Khan et al. (2014)
		Anthraquinone, 7-(γ,γ)-dimethylallyloxymacrosporin, macrosporin, 7-methoxymacrosporin, tetrahydroaltersolanol B, altersolanol L, and ampelanol	Antifungal and antibacterial	Huang et al. (2017)
		Phomadecalins A–D and phomapentenone A	Antibacterial	Che et al. (2002)
		Phomaketides A–E, pseurotin A3, pseurotin G	Antiangiogenesis, anti-inflammatory	Lee et al. (2016)
		Phomapyrrolidones A–C	Antituberculosis	Wijerante et al. (2013)
		YM-215343	Antifungal	Shibazaki et al. (2004)
		Pyrenopherol, dihydropyrenophorin	Antimicrobial	Zhang et al. (2008)
		Phomodione	Antifungal	Hoffman et al. (2007)
		Isochromanone derivatives like (3S,4S)-3,8-dihydroxy-6-methoxy-3,4,5-trimethyl-lisochroman-1-one, methyl (S)-8 hydroxy-6-methoxy-5-methyl-4a-(3-oxobutane-2-yl) benzoate	Inhibitory activity in nitric oxide	Kim et al. (2018)
		Phomaethers A–C, 2,30-dihydroxy-4-methoxy-50,6-dimethyl diphenyl ether, diaportinol, desmethyldiaportinol, citreoisocoumarin, citreoisocoumarinol	Antibacterial	Shi et al. (2017)
		2-Hydroxy-6-methylbenzoic acid	Antibacterial	Zhang et al. (2012)
		Squalestatin S1	Cholesterol control	Aldred et al. (2001)
		Epoxydines A and B	Antifungal, antibacterial	Qin et al. (2010)
		Barceloneic acid A, barceloneic acid C, questin	Antibacterial	Xia et al. (2014)
		Barceloneic acid A	Anticancer	Jayasuriya et al. (1995)
		Methyl 2-(2-formyl-3-hydroxy-5-methylphenoxy)-5-hydroxy-3-methoxybenzoate, asterric acid, methyl asterrate, 9(*Z*),12(*Z*)-nonadecadienoic acid and orsellinic acid	Cytotoxic	Fang et al. (2012)
		Cryptophomic acid, cryptodiol, cryptotriol, and dihydroisocoumarin derivative	Antibacterial	Elsebai et al. (2018)
		Phomanolide	Antiviral	Liu et al. (2019)
		3,16-Diketoaphidicolan	Antibacterial	Dai et al. (2010)
		Equisetin	Anti-HIV	Singh et al. (1998)
		Biopolymer	Herbicidal	Luft et al. (2019)
		(10′*S*)-verruculide B, vermistatin, dihydrovermistatin, (*S*,*Z*)-5-(3′,4′-dihydroxybutyldiene)-3-propylfuran-2(5H)-one, nafuredin	Inhibitor of protein tyrosine phosphatases	Gubiani et al. (2017)
		1-Methoxy-3,5′-dimethyl-2,3′-oxybiphenyl-5,1′,2′-triol	Antidiabetic	Sumilat et al. (2017)
		Thiodiketopiperazines	Anticancer	Kong et al. (2014)
		Pleofungins A, B, C, and D	Antifungal, antibiotics	Aoyagi et al. (2007)
		Alpha tetralone, (3S)-3,6,7-trihydroxy-alpha-tetralone, trichodermin, cercosporamides	Cytotoxic	Wang et al. (2011)
		Squalestatin S1	Cholesterol control	Aldred et al. (2001)
15	*Phoma terrestris* H.N. Hansen	Ergoflavin	Anti-inflammatory and anticancer	Deshmukh et al. (2009)
		N-amino-3-hydroxy-6-methoxyphthalimide and 5H-dibenz[B, F]azepine	Antimicrobial	Park et al. (2015)

in the ^{1}H-^{1}H COSY spectrum and was the absolute first case of 14-nordrimane-type sesquiterpene till the present (Liu et al. 2019). Five new metabolites, comprising phexandiols A (Fig. 11.1d) and B and phomesters A-C, were obtained by the co-culture of the endophytic fungus *Phoma* sp. YUD17001 from *Gastrodia elata* with *Armillaria* sp. (Li et al. 2019). Considering the biogenetic origins, the absolute configurations and structure of these compounds were determined by Mosher's method and electronic circular dichroism spectra, following 1D and 2D NMR spectroscopy (Li et al. 2019).

In 2018, five new polyketides, particularly colletotric B (Fig. 11.1c) and C, 3-hydroxy-5-methoxy-2,4,6-trimethylbenzoic acid, chaetochromone D (Fig. 11.1b), and 8-hydroxy-pregaliellalactone B, were obtained from the mangrove endophytic fungus *Phoma* sp. SYSU-SK-7 (Chen et al. 2018). Their structures were elucidated by analyses of extensive 1D and 2D NMR spectroscopic data following mass spectrometry. Moreover, four new phomones, phomone C (Fig. 11.1f) and phomones D–F, were isolated from the endophytic fungus *Phoma* sp. YN02-P-3 by Sang et al. (2017) which is the primary case in point of 6-α,β-unsaturated ester-2-pyrone dimers through intermolecular symmetrical cyclo-expansion. The chemical structures of these phomones were elucidated using FT-IR and 1D/2D NMR spectroscopic data following the MS data (Sang et al. 2017). Thus, an ample number of new secondary metabolites from *Phoma* sp. were determined by extensive NMR, FT-IR, and mass spectroscopic techniques for the establishment of a novel drug.

11.5 Bioactivity and Biotechnological Significance of Secondary Metabolites Obtained from Endophytic *Phoma*

In the constantly changing phase of today, we always need to discover new drugs to save our communities. The plants are like a boon for us. We have been using the plant as medicine for a very long time. Not only plants help us to produce a new drug, there are rather lots of medicinal metabolites produced by endophytes of host plants that could be utilizesd in pharmaceutical applications. The bioactivity of some metabolites produced by *Phoma* sp. endophytes is described below.

11.5.1 Antimicrobial Activity

In the past 30 years, 50% of drugs are natural products or their derivatives. Moreover, the adequacy of present-day drugs is significantly diminishing particularly because of the ever-increasing microorganism opposition that is right now an issue of extraordinary general well-being concern (Alpert 2016). Endophytes obtained from *Glycyrrhiza glabra* Linn., designated as *Phoma* GG1F1 closely related to *Phoma cucurbitacearum* and *Phoma* sp. URM 7221 from the medicinal plant *Schinus terebinthifolius* (Silva et al. 2017), showed significant antimicrobial activity. The chemical compounds from *Phoma* GG1F1 extract are two thiodiketopiperazine derivatives. Both the compounds inhibited the growth of several bacterial pathogens especially *Staphylococcus aureus*, *Bacillus subtilis*, and *Streptococcus pyogenes* (Arora et al. 2016). One new dihydrofuran derivative, named phomafuranol together with three known compounds, viz., phomalacton, (3R)-5-hydroxymellein, and emodin, produced by EtOAc extract of the culture of *Phoma* sp. isolated from *Fucus serratus*, showed good antifungal (*Microbotryum violaceum*) activity and significant antibacterial *Bacillus megaterium* and antialgal *Chlorella fusca* activities (Hussain et al. 2014).

Phomaether A and 2,30-dihydroxy-4-methoxy-50,6-dimethyl diphenyl present in EtOAc concentrate of *Phoma* (TA07-1) from *Dichotella gemmacea*, and many *Phoma* spp. show noteworthy antibacterial activity toward Gram-positive, *Staphylococcus albus* and *S. aureus*, and Gram-negative bacteria, *Escherichia coli*, *Vibrio parahaemolyticus*, *V. anguillarum*, *Staphylococcus albus* (Huang et al. 2017), and *Mycobacterium phlei* (Elsebai et al. 2018). Inositol phosphorylceramide (IPC) synthase an important factor for fungus viability is mostly targeted for fungal antibiotics (Figueiredo et al. 2005). One of the inhibitors of this metabolic enzyme is pleofungins (also called F-15078) produced by the endophytes *Phoma* sp. SANK 13899 and is active against *Saccharomyces cerevisiae*, *Aspergillus fumigatus*, *Candida albicans*, and *Cryptococcus neoformans* (Aoyagi et al. 2007). Phomafungin a cyclic lipodepsipeptide exhibits a broad spectrum of antifungal activity in very low concentration against *Candida* sp., *Aspergillus fumigatus*, and *Trichophyton mentagrophytes* (Herath et al. 2009). Likewise, barceloneic acid C a new compound exhibited antibacterial activities against *Listeria monocytogenes* and *Staphylococcus pseudintermedius*, *Bacillus cereus* 13,061, and Gram-negative bacteria *Salmonella typhimurium* produced by *Phoma* sp. JS752 (Xia et al. 2014). Several species of *Phoma* have demonstrated antiviral activity, viz., *Phoma* sp. YE3135 from *Aconitum vilmorinianum* possesses antiviral activity against the H1N1 virus (Liu et al. 2019); even ergocytochalasin A acts against DV-3 (human dengue virus type 3), H1N1 virus, and RS virus (reciprocal syncytial virus) (Peng et al. 2020).

11.5.2 Anticancer Activity

Cancer is a multistage process. Cancer cells are characterized by the aid of using the large replicative potential, apoptotic resistance, invasive ability, and one of the major causes of fatalities. Many studies have indicated the possible probability of healthful plants that host endophytic fungi capable of producing pharmacologically natural medicine. The healthful properties of those plants might be thanks to the endophytes residing inside them (Uzma et al. 2019). Taxol is a diterpene an extraordinarily anticancer drug widely used in disease related to tissue proliferation. It was isolated from the bark of the yew tree of the *Taxus* genus. These trees are uncommon and slow-developing, and a lot of bark may be handled to get a modest quantity of the medication (Gangadevi and Muthumary 2007). Taxadiene synthase (ts), a one of a kind quality in the gene arrangement of the taxane skeleton, was affirmed for Taxol biosynthesis. *Phoma betae* is an excellent candidate for Taxol supply and can serve as a potential species for genetic engineering to enhance the production of Taxol to a higher level (Kumaran et al. 2012). Multirostratin A and 20-oxo-deoxaphomin metabolites produced by the endophytic fungus *Phoma multirostrata* EA-12 show moderate cytotoxicity against five tumor cell lines (HL-60, A-549, SMMC-7721, MCF-7, and SW-480) (Chen et al. 2014). Similarly, ergosterol, ergosterol peroxide (EP), and 9,11-dehydroergosterol peroxide (DEP) produced by *Phoma* sp. show anticancer property against different kinds of cancer cell lines A549, J5, HeLa, and MCF-7 and normal lung cell line Beas-2b, as well as macrophages RAW 264.7, and can induce apoptosis, autophagy, and ROS generation (Wu et al. 2018). *Phoma* sp. from the medicinal plant *Pereskia bleo* produces hydrolyzer of L-asparagine, L-asparaginase enzyme (Chow and Ting 2014) that catalyzes the conversion of L-asparagine to L-aspartate and ammonia (Theantana et al. 2009); as asparagine helps to grow cancer cell, this enzyme removes it from the serum (Verma et al. 2007). Barceloneic acid A, B, and barceloneic lactone from *Phoma* sp. inhibit farnesyl protein transferase enzyme which is related to Ras mutation and a factor to induce tumor formation (Jayasuriya et al. 1995).

11.5.3 Anti-inflammatory Activity

Agents that reduce pain, fever, swelling (edema), and redness-like inflammatory responses are called anti-inflammatory substances. Inflammation may result in autoimmune or autoinflammatory disorders, neurodegenerative disease, or cancer. Using many anti-inflammatory substances is a new way to treat many of the diseases with the use of metabolites like anticytokine and inhibition of many enzymes (Dinarello 2010). Phomactin A derived from marine *Phoma* sp. exhibits antagonistic activity against the PAF platelet-activating factor which activates the platelet aggregations involved in many inflammatory reactions, as well as respiratory reactions

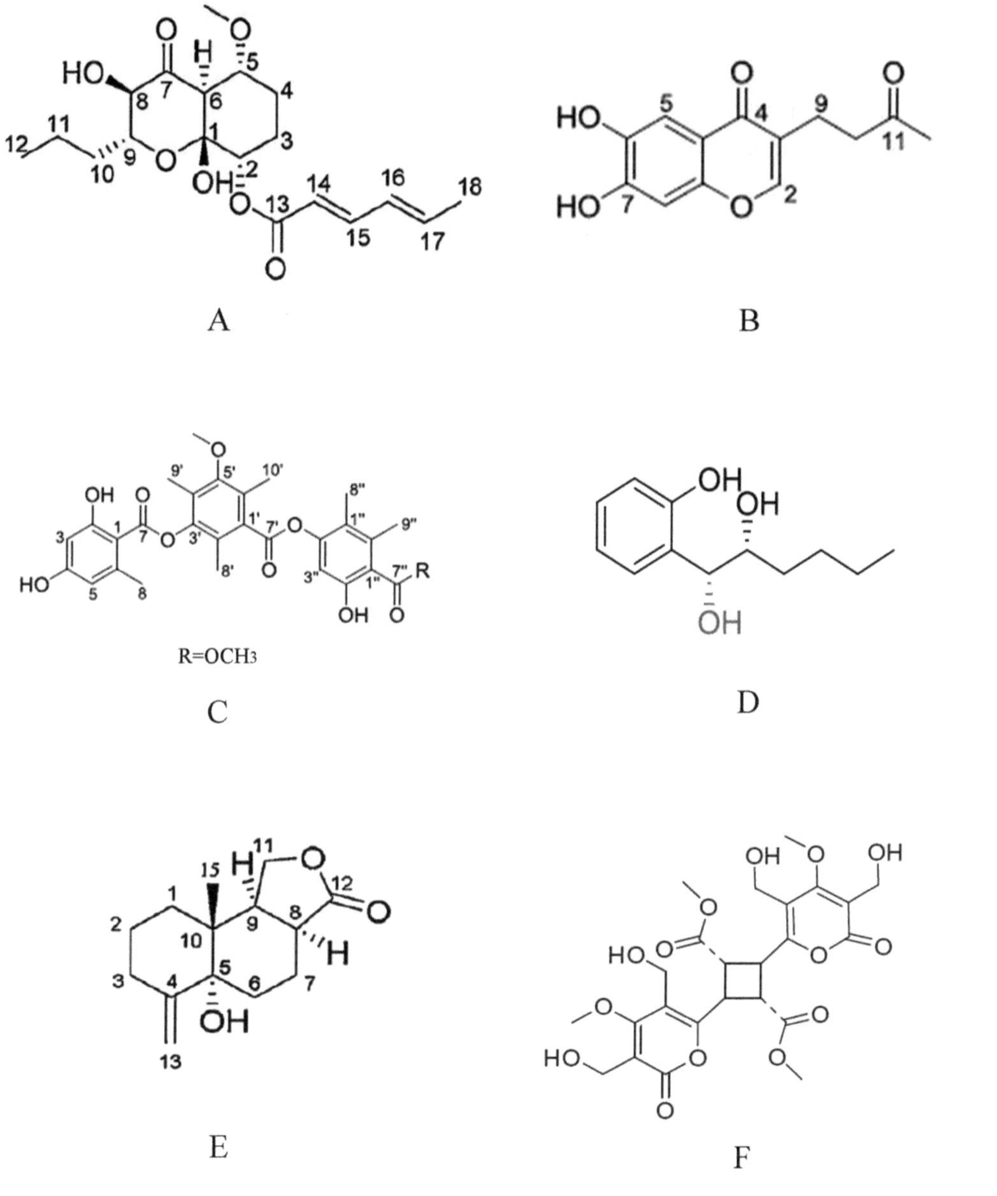

Fig. 11.1 Molecular structure of some secondary metabolites obtained from endophytic *Phoma*. (**a**) Bellidisin A (Wang et al. 2019); (**b**) Chaetochromone D (Chen et al. 2018); (**c**) Colletotric B (Chen et al. 2018); (**d**) Phexandiols A (Li et al. 2019); (**e**) Phomanolide (Liu et al. 2019); (**f**) Phomone C (Sang et al. 2017)

(Sugano et al. 1991). Phomalichenone A produced from *Phoma* sp. EL002650 inhibits the expression of the gene of inflammatory agents like tumor necrosis factor-α, nitric oxide synthase (iNOS), cyclooxygenase-2 (COX-2), and many other cytokines like IL-1 β and IL6 (Kim et al. 2018).

11.5.4 Anti-HIV Activity

Acquired immunodeficiency syndrome (AIDS) caused by HIV is a global health issue and has always been a challenge for scientists (Pommier et al. 2005). Many proteins are involved in the multiplication of HIV, one of those is integrase. Inhibition of integrase is antiretroviral therapy by blocking integration over the T cell (Hajimahdi and Zarghi 2016). The compounds equisetin and phomasetin as a first natural inhibitor isolated from *Phoma* sp. MF6070 are inhibitors of integrase that inhibit the integration reactions catalyzed by preintegration complexes by HIV-1 (Singh et al. 1998; Hazuda et al. 1999).

11.5.5 Angiogenesis Inhibition

Angiogenesis is the formation of new blood vessels and is diagnosed in numerous debilitating human diseases, together with cancer, age-associated macular degeneration (AMD), and numerous inflammatory diseases (Carmeliet 2003). Edible red algae endophyte *Phoma* sp. NTOU4195 produces polyketides phomaketide A showing potent antiangiogenesis activity by suppressing the tube formation of endothelial progenitor cell along with significant A549 cell growth inhibition and anti-inflammatory activity.

11.5.6 Plant Growth Promotion

Some *Phoma* sp. can promote the growth of host plants; this growth enhancement effect is at least in part due to endophyte-producing growth hormones. *Phoma* sp. isolated from *Tinospora cordifolia* and *Calotropis procera* induces the growth of *Zea mays* (Kedar et al. 2014), and in rice, it induces seed germination (Khan et al. 2016). It has also been found that gibberellic acid-deficient rice cultivar (a dwarf variety of rice) shows an increase in plant length when cultivated with culture filtrate of *Phoma* TK 2-4 a new strain of *Phoma herbarum* and *Phoma* LK-13 from *Moringa peregrina* (Hamayun et al. 2009). Recently, tanshinone from *Phoma herbarum* (Chen et al. 2020) and cryptotanshinone from *Phoma radicina* section *Paraphoma radicina* (Teimoori-Boghsani et al. 2020) were identified and found significant role in plant growth promotion.

11.5.7 Bio-herbicide

Weeds are one of the primary issues looked at by horticulturists, which are liable for huge misfortunes in numerous vegetable yields. The use of microbial bioproducts is an effective option in contrast to weed control (Toredo et al. 2018). We can reduce the problem of weeds in a more comfortable way through an integrated and coordinated system. The use of synthetic compounds couldn't viably control the weeds and causes to build up resistance in weeds as well as contaminates nature also (Radhakrishnan 2018). Metabolites of *Phoma* sp. are one of the great resources of bio-herbicide. *Phoma macrostoma* is found to be used for biological weed control and has great potential for the control of dandelion (Bailey et al. 2011) and now is registered as a bio-herbicide in North America to control broadleaved weed species in turfgrass (Bailey et al. 2013).

11.5.8 Biofuel Production

It has been observed that many endophytes can produce many volatile organic compounds (VOCs) that show a special relationship between plant, endophytes, and nature (Stadler and Schulz 2009). Endophytic fungi are now being recognized as an unexplained source of volatile low molecule hydrocarbon and can be used as a producer of biofuel of the next generation directly due to the low number of biosynthetic phases (Bhagobaty 2015). It has been found that *Phoma* sp. from *Larrea tridentata* (Creosote bush) produces volatile compounds such as sesquiterpenoids, some alcohol, and reduced form of naphthalene derivatives along with the trans-caryophyllene. All VOCs show the best antimicrobial property against *Verticillium*, *Ceratocystis*, and *Cercospora* and least sensitivity for *Sclerotinia*, *Trichoderma*, *Colletotrichum*, and *Aspergillus* (Strobel et al. 2011). *Phoma exigua* is an excellent source of producing cellulase and xylanase hydrolytic enzyme used in the saccharification of biomass (Tiwari et al. 2013), a process of breaking a complex carbohydrate into simple monosaccharide sugar, key steps of bio-ethanol production from biomass (Agwa et al. 2017).

11.6 Conclusion

Over the last few decades, exploring natural sources for potential biomolecules has been advanced incredibly. Microbial sources are significant due to their easy operation, growth, and manipulation. The biomolecules fulfill the call for pharmaceutical, food, beverage, agrochemical, textile, dye, and paint industries. *Phoma* spp. are exceptionally interesting among the organisms on account of their extensive abundance and novel characteristics. The nonpathogenic plant endophytes have been accounted for a significant source of many natural secondary metabolites of antimicrobial, cytotoxic, and anti-inflammatory properties that may reach the medicinal demand. The secondary metabolites of various *Phoma* endophytes have been found as growth promoters and also show antagonism against plant pathogens which are useful for agrochemical users. Accordingly, the present chapter highlights the diversity of endophytic *Phoma*, secondary metabolites of novel significance, identification tools of metabolites, and bioactivity of molecules. This compiled literature might encourage new expansion in the field and provide insight to the researchers.

References

Agwa OK, Nwosu IG, Abu GO (2017) Saccharification of bio-ethanol fermentation of carbohydrate extracted microalgal biomass by genetically identified organisms. Biotechnol Biomater 8:1-7.

Aldred D, Magan N, Lane BS (2001) Influence of water activity and nutrients on growth and production of squalestatin S1 by a *Phoma* sp. J Appl Microbiol 87:842–848.

Alpert PT (2016) Superbugs: antibiotic resistance is becoming a major public health concern. Home Health Care Manag Pract 29:1-4.

Aoyagi A, Yano T, Kozuma S, Takastu T (2007) Pleofungins, Novel Inositol Phosphorylceramide Synthase Inhibitors, from *Phoma* sp. SANK 13899. J Antibiot 60:143-152.

Arnold AE 2007 Understanding the diversity of foliar endophytic fungi: progress, challenges, and frontiers. Fungal Biol 21:51–66.

Arnold, AE 2008 Endophytic Fungi: Hidden Components of Tropical Community Ecology In Tropical Forest Community Ecology. Blackwell Scientific 1st edition 254:271

Arnold AE, Maynard Z, Gilbert G, Coley PD, Kursar TA (2000) Are tropical endophytic fungi hyper diverse. Ecol Lett 3:267–274.

Arora P, Wani ZA, Nalli Y, Ali A, Hassan SR (2016) Antimicrobial potential of Thiodiketopiperazine derivatives produced by *Phoma* sp., an endophytes of *Glycyrrhiza glabra* Linn. Microb Ecol 72:802-812.

Aveskamp MM, De Gruyter J, Crous PW (2008) Biology and recent developments in the systematics of *Phoma*, a complex genus of major quarantine significance. Fungal Divers 31:1-18.

Ayer W, Jemenez D (1994) Phomalone, an antifungal metabolite of *Phoma etheridgei*. Can J Chem 72:2326-2332.

Bacon CW, White JFJ (2000) Physiological Adaptations in the Evolution of Endophytism in the Clavicipitaceae In Microbial Endophytes; Marcel Dekker Inc 237:261.

Bailey KL, Pitt WM, Falk S, Derby J (2011) The effects of *Phoma macrostoma* on nontarget plant and target weed species. Biol Control 58:379–386.

Bailey KL, Falk S, Durby JA, Melzer, Bolend G (2013) The effect of fertilizers on the efficacy of the bioherbicide, *Phoma macrostoma*, to control dandelions in turfgrass. Biol Control 65:147-15.

Bennett A, Ponder MM, Diaz JG (2018) *Phoma* infections: classification, potential food sources and their clinical impact. Microorganisms 6:1-12.

Bhagobaty RK (2015) Endophytic fungi: prospects in Biofuel production. Proc Natl Acad Sci India Sect B Biol Sci 85:21-25.

Bharathidasan R, Panneerselvam A (2011) Isolation and identification of endophytic fungi from *Avicennia marina* in Ramanathpuram district, Karankadu, Tamilnadu, India. Eur J Exp Biol 1:31-36.

Boerema GH, de Gruyter J, Noordeloos ME, Hamers MEC (2004) *Phoma* identification manual. CABI Publishing 1st Edition.

Carmeliet P (2003) Angiogenesis in health and disease. Nat Med 9:653-60.

Che Y, Gloer JB, Wicklow DT (2002) Phomadecalins A, B, C, D and Phomapentenone A: New bioactive metabolites from *Phoma* sp. NRRL 25697, a fungal colonist of Hypoxylon stromata. J Nat Prod 65:399-402.

Chen ZM, Chen HP, Li Y, Feng T, Liu JK (2014) Cytochalasins from cultures of endophytic fungus *Phoma multirostrata* EA-12. J Antibiot 68:23–26.

Chen Y, Yang W, Zou G, Chen S, Pang J, She Z (2018) Bioactive polyketides from the mangrove endophytic fungi *Phoma* sp. SYSU-SK-7. Fitoterapia 140:1-17.

Chen HM, Wu HX, He XY, Zhang HH, Miao F, Liang ZS (2020) Promoting tanshinone synthesis of *Salvia miltiorrhiza* root by a seed endophytic fungus, *Phoma herbarum* D603. China journal of Chinese Materia Medica 45:65-71.

Choi JJ, Kim SH (2017) A genome tree of life for the fungi kingdom. Proc Natl Acad Sci USA 114: 9391–9396.

Chow YY and Ting ASY (2014) Endophytic L-asparaginase-producing fungi from plants associated with anticancer properties. J Adv Res 56:1-8.

Cimmino A, Andolfi A, Berestetskiy A, Evidente A (2008) Production of phytotoxin by *Phoma exigua* var exigua, a potential mycoherbicide against perennial thistles. J Agr Food Chem 56:630-634.

Cohen, SD (2006) Host selectivity and genetic variation of *Discula umbrinella* isolates from two oak species: Analyses of intergenic spacer region sequences of ribosomal DNA. Microbol Ecol 52:463–469.

Dai D, Hussaina H, Drägerb S, Schulzb B, Kurtánc T, Pescitellid G, Flörkea U, Krohn K (2010) Metabolites from the Fungus *Phoma* sp. 7210, associated with *Aizoon canariense*. Nat Prod Commun 5:1175-1180.

Deshmukh SK, Mishra PD, Kulkarni-Almeida A, Verekar S, Sahoo MR, Periyasamy G, Goswami H, Khanna A, Balakrishnan A, Vishwakarma R (2009) Anti-Inflammatory and Anticancer Activity of Ergoflavin Isolated from an Endophytic Fungus. Chem Biodivers 6(5): 784–789.

Dinarello C (2010) Anti-inflammatory Agents: Present and Future. Cell 140:935–950.

Du W, Yao Z, Li J, Sun C, Xia J, Wang B, Shi D, Ren L (2020) Diversity and antimicrobial activity of endophytic fungi isolated from *Securinega suffruticosa* in the Yellow River Delta. PLoS One 15:1–18.

Dugan F, Sitton, Sullivan R, James W (2002) The neotyphodium endophyte of wild barley (*Hordeum brevisubulatum* sub sp. *violaceum*) grows and sporulates on leaf surfaces of the host. Symbiosis 32:147–159.

Elkhateeb WA, Daba GM (2020) The Exceptional Endophytic Fungi, *Emericella* (Berk.) and *Phoma* (Sacc.) Genera. International Journal of Research in Pharmacy and Biosciences 7:10-15.

El-Nagerabi SAF, Elshafie AE, Alkhanjari SS (2013) Endophytic fungi associated with *Ziziphus* species and new records from mountainous area of Oman. Biodiversity 14:10-16.

Elsebai MF, Hazem A, Ghabbour HA, Legrave N, Vive FF, Mehiri M (2018) New bioactive chlorinated cyclopentene derivatives from the marine-derived Fungus *Phoma* sp. Med Chem Res 27:1885–1892.

Fang MJ, Fang H, Li WJ, Huang DM, Wu Z, Zhao YF (2012) A new diphenyl ether from *Phoma* sp. strain, SHZK-2. Nat Prod Res 26:1224-1228.

Fernandes EG, Pereira OL, da Silva CC, Bento CBP, de Queiroz MV (2015) Diversity of endophytic fungi in *Glycine max*. Microbiol Res 181:84–92.

Figueiredo JM, Dias WB, Previato LM, Previato JO, Heise N (2005) Characterization of the inositol phosphorylceramide synthase activity from *Trypanosoma cruzi*. Biochem J 387: 519–529.

Gama DS, Santos IAFM, Abreu LM, Medeiros FHV, Duarte WF, Cardoso PG (2020) Endophytic fungi from Brachiaria grasses in Brazil and preliminary screening of *Sclerotinia sclerotorium* antagonists. Sci Agr 77.

Gamboa, MA, Bayman, P (2001) Communities of endophytic fungi in leaves of a tropical timber tree (*Guarea guidonia*: Meliaceae). Biotropica 33:352–360.

Gangadevi V, Muthumary EJ (2007) Taxol, an anticancer drug produced by an endophytic fungus *Bartalinia robillardoides* Tassi, isolated from a medicinal plant, *Aegle marmelos* Correa ex Roxb. World J Microbiol Biotechnol 24:717–724.

Gazis R, Chaverri P (2010) Diversity of fungal endophytes in leaves and stems of wild rubber trees *Hevea brasiliensis* in Peru. Fungal Ecol. 3:240–254.

Ghimire SR, Charlton ND, Bell JD, Krishnamurthy YL, Craven KD (2011) Biodiversity of fungal endophyte communities inhabiting switchgrass *Panicum virgatum* L growing in the native tallgrass prairie of northern Oklahoma. Fungal Div 471:19-27.

Gubiani JR, Wijerante EMK, Shi T, Araujo AR, Arnold AE, Chapman E, Gunatalika AAL (2017) An epigenetic modifier induces production of (10′*S*)-verruculide B, an inhibitor of protein tyrosine phosphatases by *Phoma* sp. nov. LG0217, a fungal endophyte of *Parkinsonia microphylla*. Bioorg Med Chem *25:1860-1866*.

Guégan S, Garcia-Hermoso D, Sitbon K, Ahmed S, Moguelet P, Dromer F, Lortholary O (2016) French Mycosis Study Group. Ten-Year Experience of Cutaneous and/or Subcutaneous Infections Due to *Coelomycetes* in France. Open Forum Infect Dis X 3:1-13.

Guo LD, Huang GR, Wang Y (2008) Seasonal and Tissue age influences on endophytic fungi of *Pinus tabulaeformis* (Pinaceae) in the Dongling mountains, Beijing. J Integr Plant Biol 50:997-1003.

Gupta S, Kaul S, Singh B, Vishwakarma A, Dhar MK (2016) Production of gentisyl alcohol from *Phoma herbarum* endophytic in *Curcuma longa* L. and its antagonistic activity towards leaf spot pathogen *Colletotrichum gloeosporioides*. Appl Biochem and Biotechnol 180: 1093-1109.

Hajimahdi Z, Zarghi A (2016) Progress in HIV-1 integrase inhibitor: A review of their chemical structure diversity. Iran J Pharm Sci 15:595-628.

Hamayun M, Khan SA, Khan AL, Rehman G, Sohn EY, Shah AA, Kim SK, Joo GJ, Lee IJ (2009) *Phoma herbarum* as a new gibberellin-producing and plant growth-promoting fungus. J Microbiol Biotechn 19:1244-1249.

Hazuda D, Blau CU, Felock P, Hastings J, Pramanik B, Wolfe A, Bushman F, Farnet C, Goetz M, Williams M, Silverman K, Lingham R, Singh S (1999) Isolation and characterization of novel human immunodeficiency virus integrase inhibitors from fungal metabolites. Antivir Chem Chemother 10:63–70.

Herath K, Harris G, Jayasuriya H, Zink D, Smith S, Vicente F, Bills G, Collado J, Gonzalez A, Jennifer B, Kahn N, Galuska S, Giacobbe R, Abruzzo G, Hickeys E, Liberato P, Xu D, Roemer T, Singh B (2009) Isolation, structure and biological activity of phomafungin, a cyclic lipodepsipeptide from a widespread tropical *Phoma* sp. Bioorg Med Chem 17:1361-1369.

Hoffman AM, Mayer SG, Strobel GA, Hess WM, Sovocool GW, Harper J, Arif AM, Grant DM, Swift EG (2007) Purification identification and activity of phomodione, a furandione from an endophytic *Phoma* species. Phytochemsitry 69:1049-1056.

Huang S, Xu J, Fenqi Li, Zhou D, Xu L, Chunyuan Li (2017) Identification and antifungal activity of metabolites from the mangrove fungus *Phoma* sp. Chem Nat Compd 53:1-4.

Humzah TNT, Lee SY, Hidayat A, Terhem R, Faridah-Hanum I, Mohamed R (2018) Diversity and characterization of Endophytic fungi isolated from the Tropical Mangrove species, *Rhizophora mucronata*, and identification of potential antagonists against the soil born fungus, *Fusarium solani*. Front Microbiol 9:1707.

Hussain H, Kock I, Harrasi AA, Rawahi AA, Abbas G, Green IR, Shah A, Badshah A, Saleem M, Draeger S, Schulz B, Krohn K (2014) Antimicrobial chemical constituents from endophytic fungus *Phoma* sp. Asian Pac J Trop Med 7:699-702.

Jayasuriya H, Bills HG, Zink L, Smith JL, Goetz MA, Jenkins RG, Omstead M N, Silverman KC, Bills G F, Linghum RB, Singh S B, Pelaez F, Cascales C (1995) Barceloneic acid A, a farnesyl protein transferase inhibitor from *Phoma* sp. J Nat Prod 58:986-991.

Kedar A, Rathod D, Yadav A, Agarkar G, Rai M (2014) Endophytic *Phoma* sp. Isolated from medicinal plant promote the growth of *Zea mays*. Bioscience 6:132-139.

Khan AL, Muhammad W, Hussain J, Al-Harrasi A, Al-Rawahi A, Al-Hosni K, Kim MJ, Adnan M, Lee IJ (2014) Endophytes *Aspergillus caespitosus* LK12 and *Phoma* sp. LK13 of *Moringa peregrina* produce gibberellins and improve rice plant growth. J Plant Interact 9:731–737.

Khan AL, Al-Harrasi A, Al-Rawahi A, Al-Farsi Z, Al-Mamari A, Waqas M, Asaf S, Elyassi A, Mabood F, Shin JH, Lee IJ (2016) Endophytic fungi from Frankincense tree improves host growth and produces extracellular enzymes and Indole acetic acid. PLoS ONE 11:1-19.

Kim JW, Chori HG, Song JH, Kang KS, Shim SH (2018) Bioactive secondary metabolites from an endophytes fungus *Phoma* sp PF2 derived from *Artemisia princeps pamp*. J Antibiot 72:174–177.

Kong F, Yang Y, Liu P, Dong T, Zhu W (2014) Thiodiketopiperazines from the Marine-Derived Fungus *Phoma* sp. OUCMDZ-1847. J Nat Prod 77:132–137.

Kumaran RS, Choi YK, Lee S, Jeon HJ, Jung H, Kim HJ (2012) Isolation of taxol, an anticancer drug produced by the endophytic fungus, *Phoma betae*. Afr J Biotechnol 11:950-960.

Lee MS, Wang SW, Wang GJ, Pang KL, Lee CK, Kuo YH, Cha HJ, Lin RK, Lee TH (2016) Angiogenesis Inhibitors and Anti-Inflammatory Agents from *Phoma* sp. NTOU4195. J Nat Prod 79:2983–2990.

Li X, Zhai X, Shu Z, Dong P, Ming Q, Qin L, Zheng C (2016) *Phoma glomerata* D 14: A endophytes from *Salvia miltirorrhia* that produces Salvionic acid C. Curr Microbiol 73:31-37.

Li HT, Zhou H, Duan RT, Li HY, Tang LH, Yang XQ, Yang YB, Ding ZT (2019) Inducing secondary metabolite production by co-culture of the endophytic fungus *Phoma* sp. and the symbiotic fungus *Armillaria* sp. J Nat Prod 82:1009–1013.

Liu SS, Jiang JX, Huang R, Wang YT, Jiang BG, Zheng KH, Wu SH (2019) A new antiviral 14 Nordrimane Sesquiterpanoid from an endophytic fungus *Phoma* sp. Phytochem Lett 29:75-79.

Loro M, Valero-Jiménez CA, Nozawa S, Márquez LM (2012) Diversity and composition of fungal endophytes in semiarid. Northwest Venezuela. J Arid Environ 85:46–55.

Luft L, Confortin TC, Todero I, Chaves Neto JR, Tonato D, Felimberti PZ, Zabot GL, Mazutti MA (2019) Different techniques for concentration of extracellular biopolymers with herbicidal activity produced by *Phoma* sp. Environ Technol 1-27.

Maha A, Rukachaisirkul V, Phongpaichit S, Preedanon S, Sakayaroj J (2017) tyrosine and hydantoin derivative from fungus *Phoma herbarum* PSU-H256 isolated from *Hevea brasiliensis*. Terahedrone 73:4597-4601.

Maheswari S, Rajagopal K (2013) Biodiversity of endophytic fungi in *Kigelia pinnata* during two different seasons. Curr Sci 104:515–518.

Mahish PK, Ghritlahare A (2017) Pathogenicity of *Phoma chrysanthemicola* to *Chrysanthemum* plants (Asteraceae Family) and Control of Pathogen by Chemical and Biological Approach. Biosci Biotechnol Res Asia 14:1191-1200.

Meena H, Hnamte S, Sidhhardha B (2019) Secondary metabolites from endophytic fungi: chemical diversity and application In: advances in Endophytic fungal research. Springer, Cham 1st Edition 145:169.

Mishra R, Kushveer JS, Revanthbabu P, Sarma VV (2019) Endophytic fungi and their enzymatic potential In: Advances in Endophytic Fungal Research. Springer, Cham 1st Edition 283:337.

Mohamed H, Ebrahim W, Peterson MK, Proksch P (2017) Production of Phytotoxic Polyketide Spiciferone A by *Phoma fungicola*. Mikol Fitopatol 51:1-2.

Moricca S, Ginetti B, Ragazzi A (2012) Species- and organ-specificity in endophytes colonizing healthy and declining Mediterranean oaks. Phytopathol Mediterr 51: 587-598.

Moy M, Belanger F, Duncan R, Freehoff A, Leary C, Meyer W et al. (2000) Identification of epiphyllous mycelial nets on leaves of grasses infected by *Clavicipitaceous* endophytes. Symbiosis 28:291–302.

Nalli Y, Arora P, Khan S, Malik F, Hassan SR, Gupta V, Ali A (2019) Isolation and structural modification of macophin from endophytic fungus *Phoma macrostoma* and their cytotoxic potential. Med Chem Res 28:260–266.

Oikawa H, Toshima H, Ohashi S, Konig W, Kenmoku H, Sassa T (2001) Diversity of diterpene hydrocarbon in fungus *Phoma betae*. Tetrahedron Lett 42:2329-2332.

Ostarhage C, Schwibible GM, Wright K (2000) difference between marine and terrestrial *Phoma* sp as determined by HPLC-DAD and HPLC-MS. Phytochem Anal 11:288-294.

Park YH, Chung JY, Ahn DJ, Kwon TR, Lee SK, Bae I, Yun HK, Bae H (2015) Screening and characterization of endophytic fungi of Panax ginseng Meyer for biocontrol activity against ginseng pathogens. Biological Control 91: 71–81.

Paulus B, Kanowski J, Gadek P, Hyde KD (2006) Diversity and distribution of saprobic microfungi in leaf litter of an Australian tropical rainforest. Mycol Res 110:1441–1454.

Peng X, Duan F, He Y, Gao Y, Chen J, Chang J, Ruan H (2020) Ergocytochalasin A, a polycyclic merocytochalasan from an endophytic fungus *Phoma multirotrata* XJ-2-1. Org Biomol Chem 18:4056-4062.

Pommier Y, Johnson AA, Marchand C (2005) Integrase inhibitors to treat HIV/AIDS. Nat Rev Drug Discov 4:236-249.

Qin S, Hussain H, Schulz B, Draege S, Krohn K (2010) Two New Metabolites, Epoxydine A and B, from *Phoma* sp. Helv Chim Acta 93:169-174.

Radhakrishnan R (2018) Introductory Chapter: Need of Bioherbicide for Weed Control, Biological Approaches for Controlling Weeds, Ramalingam Radhakrishnan, IntechOpen. https://doi.org/10.5772/intechopen.77958.

Radiastuti N, Bahalwan HA and Susilwati DN (2019) Phylogenetic study of endophytic fungi associated *Centella asiatica* from Bengkulu and Malaysian accessions based on the ITS rDNA sequence. Biodiversitas 20:1248-1258.

Rai MK, Rajak RC (1993) Distinguishing characteristics of some *Phoma* species. Mycotaxon 48:389–414

Rai M, Deshmukh P, Gade A, Ingle A, Kövics GJ, Irinyi L (2009) *Phoma Saccardo*: Distribution, secondary metabolite production and biotechnological applications. Crit Rev Microbiol 35:182-196.

Rai M, Gade A, Zimowska B, Ingle AP, Ingle P (2018) Marine deriverd *Phoma* - the gold mine of bioactive compound. Appl Microbiol Biotechnol 102:9053-9066.

Rai, M., Gade, A., Zimowska, B., Ingle, A.P., Ingle, P. (2020). Harnessing the potential of novel bioactive compounds produced by endophytic *Phoma* spp. – biomedical and agricultural applications. *Acta Sci. Pol. Hortorum Cultus*, 19(6), 31–45. https://doi.org/10.24326/asphc.2020.6.3

Reo NV (2002) NMR-based metabolomics. Drug Chem Toxicol 25:375-382.

Rivera-Orduña FN, Suarez-Sanchez RA, Flores-Bustamante ZR, Gracida-Rodriguez JN, Flores-Cotera N (2011) Diversity of endophytic fungi of *Taxus globosa* Mexican yew. Fungal Divers 47, 65–74.

Rodriguez RJ, White JF, Arnold AE, Redman RS (2009) Fungal Endophytes: diversity and functional roles. New Phytopathol 182:314-330.

Sanchez S, Demain AL (2011) Secondary metabolites In Comprehensive Biotechnology. Elsevier 2nd edition 155:167.

Sang XN, Chen SF, Tang MX, Wang HF, An X, Lu XJ et al., (2017) α-Pyrone derivatives with cytotoxic activities, from the endophytic fungus *Phoma* sp. YN02-P-3. Bioorg Med Chem Lett 27:3723-3725.

Schaechter M (2011). Eukaryotic Microbes. Academic Press 1st edition.

Selim KA, Elkhateeb WA, Tawila AM, El-Beih AA, Abdel-Rahman TM, El-Diwany AI, Ahmed EF (2018) Antiviral and Antioxidant potential of fungal endophytes of Egyptian medicinal plants. Fermentation 4:49.

Shi T, Qi J, Shao C, Zhao D, Hou X, Wang C (2017) Bioactive Diphenyl Ethers and Isocoumarin Derivatives from a Gorgonian-Derived Fungus *Phoma* sp. (TA07-1). Mar Drugs 15:1-7.

Shibazaki M, Taniguchi M, Yokoi T, Nagai K, Watanabe M, Suzuki K, Yamamoto T (2004) YM215343, a novel antifungal compound from *Phoma* sp. QN04621. J Antibiot 57:379-382.

Silva GBPG, Silvino KF, Bezerra J, Farias T, Araujo JM and Stamford TLM (2017) Antimicrobial activity of Phoma sp. URM 7221: An endophyte from *Schinus terebinthifolius* Raddi (Anacardiaceae). Afr J Microbiol Res 11:1-7.

Singh SB, Zink DL, Goetz MA, Dombrowski AW, Polishook JD, Hazuda DJ (1998) Equisetin and a Novel Opposite Stereochemical Homolog Phomasetin, Two Fungal Metabolites as Inhibitors of HIV-1 Integrase. Tetrahedron Lett 39:2243-2246.

Stadler M, Schulz B (2009) High energy biofuel from endophytic fungi. Trends Plant Sci 14: 353-355.

Stone JK, Polishook JD, White Jr. JF (2012) Endophytic fungi In: Biodiversity of Fungi. Elsevier Academic Press, 1st Edition 1:31.

Strobel G, Singh SK, Hassan SRU, Mitchell AM, Geary B, Sears J. (2011) An endophytic/pathogenic *Phoma* sp. from creosote bush producing biologically active volatile compounds having fuel potential. FEMS Microbiol Lett 320:87-94.

Sugano M, Sato A, Lijima Y, Oshima T, Furuya K, Kuwano H, Hata T, Hanzawa H. (1991) Phomactin A: A Novel PAF Antagonist from a Marine Fungus *Phoma* sp. J Am Chem Soc 113:5463-5464.

Sumilat D, Yamazaki H, Kanno S, Saito R, Watanabe Y, Namikosh M (2017) Biphenyl ether derivatives with protein tyrosine phosphatase 1B inhibitory activity from the freshwater fungus *Phoma* sp. J Antibiot 70:331–333.

Sun Y, Wang Q, Lu X, Kane I, Kakishima M (2012) Endophytic fungal community in stem and leaves of plant from desert area in China. Mycol Prog 11:781-790.

Szucs Z, Plaszko T, Cziaky Z, Szikszai AK, Emri T, Bertoti R et al., (2018) Endophytic fungi from roots of horseradish (*Armoracia rusticana*) and their interactions with the defensive metabolites of the glucosinolate-myrosinase-isothiocyanate system. BMC Plant Biol 18:85.

Tan XM, Zhou Y, Zhou X, Xia X, Wei Y, He L et al., (2018a) Diversity and biodiversity potential of culturable fungal endophytes of *Dysosma versipellis*; a rare medicinal plant endemic to China. Sci Rep 5:5929.

Tan XM, Li YL, Sun LY, Sun BD, Niu SB, Wang MH et al., (2018b). Spiciferones analogs from an endophytic fungus *Phoma betae* collected from desert plant in West China. J Antibiot 71:613-617.

Teimoori-Boghsani Y, Ganjeali A, Tomislav C, Muller H, Asili J and Berg (2020) Endophytic fungi of Native Salvia abrotanoides Plants reveal high taxonomic diversity and unique profiles of secondary metabolites. Front Microbiol 10:3013.

Theantana T, Hyde KD, Lumyong S (2009) Asparaginase production by endophytic fungi from Thai medicinal plants: cytotoxicity properties. Int J Integr Biol 7:1-8.

Tiwari R, Singh S, Nain PKS, Rana S, Sharma A, Pranav K, Nain L (2013) Harnessing the hydrolytic potential of phytopathogenic fungus *Phoma exigua* ITCC 2049 for saccharification of lignocellulosic biomass. Bioresour Technol 150:228-234.

Toredo I, Confortin TC, Luft L, Brun T, Ugaldo GA, Almeida TC, Arnemann JA, Zabot GL, Mazutti MA (2018) Formulation of a bioherbicide with metabolites from *Phoma* sp. Sci Hortic-Amsterdam 241:285-292.

Uzma F, Mohan CD, Siddaiah CN, Chowdappa S (2019) Endophytic fungi: Promising source of Novel Bioactive compounds In Advances in Endophytic Fungal Research. Fungal Biology. Springer Cham, 1st Edition 243:265.

Verma N, Kumar K, Kaur G, Anand S (2007) L-Asparaginase: a promising chemotherapeutic agent. Crit Rev Biotechnol 27:45-62.

Venkatasubbaiah P, Dyke CGV, Chilton WS (1992) Phytotoxic Metabolites of Phoma Sorghina, A New Foliar Pathogen of Pokeweed. Mycologia 84(5): 715–723.

Vikrant P, Verma KK, Rajak RC, Pandey AK (2006) Charcterization of Phytotoxin from *Phoma herbarum* for management of *Pathenium hysterophorus* L. J Phytopathol 154:7-8.

Wang LW, Xu BG, Wang JY, Su ZZ, Lin FC, Zhang CL, Kubicek CP (2011) Bioactive metabolites from *Phoma* species, an endophytic fungus from the Chinese medicinal plant *Arisaema erubescens*. Appl Microbiol Biotechnol 93:1231–1239.

Wang WX, Zheng MJ, Li J, Feng T, Li ZH, Huang R et al., (2019) Cytotoxic polyketides from endophytic fungus *Phoma bellidis* harbored in *Ttricyrtis maculate*. Phytochem Lett 29:41-46.

Weber RW, Stenger E, Meffert A and Hahn M (2004) Brefeldin A Production by Phoma medicanginis in dead pre-colonized plant tissue: a strategy for habitat conquest? Mycological Research, 108(6), 662-671.

Wijerante EMK, He H, Franzblau SG, Hoffman AM, Leslie AA (2013) Phomapyrrolidones A–C, Antitubercular Alkaloids from the Endophytic Fungus *Phoma* sp. NRRL 46751. J Nat Prod 76:1860-1865.

Wilson DM, Burlingame AL (1974) Deuterium and carbon-13 tracer studies of ethanol metabolism in the rat by ^{2}H, ^{1}H-decoupled ^{13}C nuclear magnetic resonance. Biochem Biophys Res Commun 56:828-835.

Wishart DS (2008) Quantitative metabolomics using NMR. Trends Anal Chem 27:228-237.

Wishart DS (2011) Advances in metabolite identification. Bioanalysis 3:1769-1782.

Wu X, Chen Z, Ding W, Liu Y, Ma Z (2018) Chemical constituent of fermentation extract of Marine fungi *Phoma* sp.CZD F11 and *Aspergillus* sp.CZD-18 from *Zhoushan Archipelago*, China. Nat Prod Res, 32:1562-1566.

Xia X, Kim S, Bang S, Lee HJ, Liu C, Park CI et al., (2014) Barceloneic acid C, a new polyketide from an endophytic fungus *Phoma* sp. JS752 and its antibacterial activities. J Antibiot 68:139–141.

Zaiyou J, Meng L, Xiqiao H (2017) An endophytic fungus efficiently producing paclitaxel isolated from *Taxus wallichiana* var. mairei. Medicine 96:e7406.

Zhang W, Krone K, Egold K, Draeger S, Schulz B (2008) Diversity of antimicrobial pyrenophorol derivatives from an endophytic fungus *Phoma* sp. Eur J Org Chem 25:4320-4328.

Zhang H, Xiong Y, Zhao H, Yi Y, Zhang C, Yu C et al., (2012) An Antimicrobial compound of from the endophytic fungi from *Phoma* sp. isolated from medicinal plant *Taraxacum mongolicum*. J Taiwan Inst Chem Eng 44:177-181.

The Genus *Phoma*: A Review of Its Potential Bioactivities, Implications, and Prospects

12

Luciana Luft, Tássia C. Confortin, Izelmar Todero, and Marcio A. Mazutti

Abbreviations

A549	Human lung cancer
CHL	Chinese hamster lung
CNS	Central nervous system
DV3	Dengue virus type 3
EPS	Exopolysaccharides
H1N1	Influenza A virus
HCT15	Human colon cancer
HeLa	Human cervical carcinoma
HIV-1	Human immunodeficiency virus type 1
HL-60	Leukemia cell line
HT-29	Colon cancer cells
L-O2 cells	Human hepatocytes
MCF-7	Breast cancer cells
MGC80-3	Gastric carcinoma cells
P-388	Leukemia cell line
RSV	Respiratory syncytial virus
SK-MEL-2	Human skin cancer
SK-OV-3	Human ovarian cancer
SMMC-772	Liver cancer cells
TeA	Tenuazonic acid
XF 498	Human CNS cancer
YCP	Yancheng polysaccharide

L. Luft · T. C. Confortin · I. Todero
M. A. Mazutti (✉)
Department of Chemical Engineering, Federal University of Santa Maria, Santa Maria, Brazil
e-mail: mazutti@ufsm.br

12.1 Introduction

The genus *Phoma* was introduced about 200 years ago, with the first description dating back to 1821 (Sutton 1980). Nevertheless, the generic name was officially introduced only in 1880 by Saccardo and later published by Boerema and Bollen in 1975 (Saccardo 1880; Boerema and Bollen 1975). In the Saccardoan system, the genus name *Phoma* applied to filamentous fungi, which were capable of forming pycnidial conidiomata with aseptate, hyaline conidia that could inhabit plant stems (Aveskamp et al. 2008). Nowadays, *Phoma* species can be identified as filamentous fungi with morphological characteristics, such as the formation of conidia (asexual spores), pycnidia (asexual fruiting bodies), and chlamydospores (enlarged, thick-walled vegetative cells within hyphae or at hyphal tips) (Rai and Rajak 1993; Boerema et al. 2004).

Phoma is a polyphyletic genus of fungi belonging to the phylum *Ascomycota*, class *Dothideomycetes*, order *Pleosporales*, and family *Didymellaceae* (Bennett et al. 2018; Valenzuela-Lopez et al. 2018). The current taxonomy is shown in Fig. 12.1. Usually, it is considered by mycologists to be taxonomically problematic due to the asexual nature of most species, the high morphological variability in vivo, and the vague generic circumscription according to the Saccardoan system. Classically, it has been grouped in the class *Coelomycetes*. However, this

M. Rai et al. (eds.), *Phoma: Diversity, Taxonomy, Bioactivities, and Nanotechnology*,
https://doi.org/10.1007/978-3-030-81218-8_12

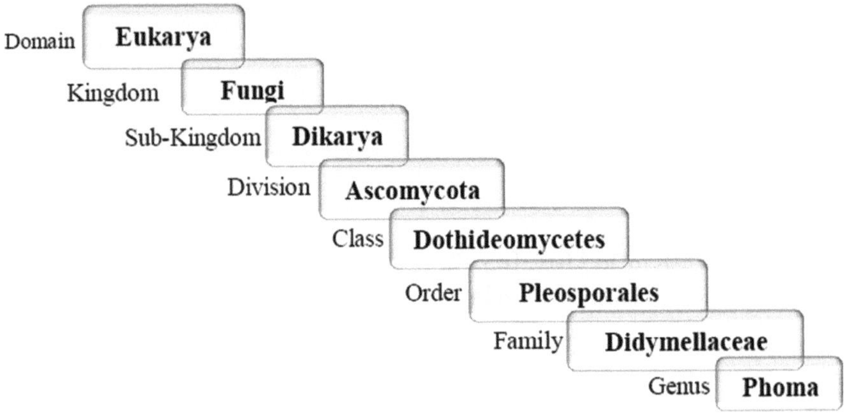

Fig. 12.1 Current taxonomy of *Phoma*

classification is considered obsolete but still used in the clinical setting (Valenzuela-Lopez et al. 2018). The controversial genus is currently linked to three different teleomorph genera. If teleomorphs are known, they reside in *Didymella*, *Leptosphaeria*, or occasionally *Pleospora*. This complexity in taxonomy is a further complicating factor for the identification and differentiation of members of the genus *Phoma* (Aveskamp et al. 2008). Although the conventional system of identification is functional, it has its limitations. For the classification and identification of these fungi, morphological studies, chemotaxonomy, secondary metabolite, and protein profiling should be assessed (Rai et al. 2014). In general, classification of fungi as a whole is under dynamic revision due to availability of modern molecular techniques for analysis of fungi at the genomic, transcriptomic, and proteomic level (Aveskamp et al. 2008; Jones et al. 2011; Bennett et al. 2018).

To date, *Phoma* is one of the largest fungal genera, with more than 3000 names recorded in MycoBank, with species occurring in more than 25 families (Valenzuela-Lopez et al. 2018), and the last published identification manual describes 223 specific and infraspecific taxa (Boerema et al. 2004). *Phoma* spp. constitute a diverse group of organisms that are ubiquitous, generally found in soil, organic matter, plants, and water sources, including numerous plant pathogens, endophytic species associated with a wide host range, saprobes, and as opportunistic humans and animals pathogens (Rai et al. 2009; Bennett et al. 2018).

As mentioned, many of the species of the genus *Phoma* can be defined as pathogens and/or endophytes (asymptomatic microbial). According to the most recent definition reformulated by Orlando Petrini, "all organisms inhabiting plant organs that at some time in their life cycle can colonize internal plant tissues without causing apparent harm to their host." This explains the endophytic organisms that have a more or less long epiphytic phase and the latent pathogens that can live without symptoms in their hosts for some time in life (Petrini 1991). In addition, it is well known that strains of some microorganisms belonging to a well-known pathogenic species of a specific plant host may even have growth-promoting effects on another plant (Hardoim et al. 2015).

An important characteristic of the symbiotic fungus-plant metabolism is the ability of the endophytic fungus to biosynthesize metabolites originally produced by the host plant. In this way, endophytic fungi can alternatively substitute plants as sources of drugs and agrochemical products, considering the advantages of biotechnological production (Canuto et al. 2012). Chemical investigations of *Phoma* have led to a great diversity of metabolites. Most of these metabolites exhibit specific bioactivity, such as cytotoxic effects (Sang et al. 2017), antivirals (Herath et al. 2009), and antimicrobials (Hussain et al. 2014), while others, such as the phytotoxins, exhibit great potential for the biological control of weeds, which can be exploited for the production of pesticides (Rai et al. 2009).

In this chapter, we provide a survey of the main bioactive compounds produced by genus *Phoma*. The major topics have been addressed concerning the use of endophytic and pathogenic *Phoma* species as biocontrol agents, highlighting in this group the features of interest for agricultural, pharmaceutical, and medical field, and reality and expectation to insert these bioproducts on the market.

12.2 *Phoma* as Biological Control Agent for Weeds

Several phytopathogenic fungi produce secondary metabolite toxins as part of a multifaceted strategy to obtain nutrients from plants. In some cases, such toxins induce plant cell death to release nutrients or subvert plant metabolism to the advantage of the fungus. Many host-selective toxins play a role in virulence, while the role of nonhost-selective toxins in virulence is more complex, and toxin production does not always correlate with virulence (Elliott et al. 2007). Usually, fungi producing toxins do not rely on growth on a host to complete their life cycle. The most likely advantage of secondary metabolites to organisms that produce such molecules is that they enhance survival (Howlett et al. 2009). According Rai et al. (2009), some species of *Phoma* secrete phytotoxin as secondary metabolites, which have great potential for the biological control of weeds and can be exploited for the production of mycopesticides, and agrophytochemicals.

A large number of phytotoxin (host and nonhost specific) have been identified from genus *Phoma*. Phomalide is an example of host-specific phytotoxin produced by causative agent of blackleg disease of brassica oilseeds, one of the highly studied *Phoma* species, the *Phoma lingam* (Pedras and Biesenthal 1998). Another host-selective phytotoxin produced by *Phoma lingam* was also reported, the phomalairdenone (Pedras et al. 1999). This species also produced important phytotoxins, and related metabolites include sirodesmin PL, phomamide, deacetylsirodesmin PL, phomapyrone A, and L-valyl-L-tryptophan anhydride (Pedras and Biesenthal 2000). Sirodesmin PL is a nonhost-selective phytotoxin and is the major phytotoxin produced by *Phoma lingam* (Pedras and Séguin-Swartz 1992), which causes chlorotic lesions on plant leaves (Rouxel et al. 1988). Other active metabolites can be isolated by changing the composition of culture medium, such as 2,4-dihydroxy-3,6-dimethylbenzaldehyde (Pedras et al. 2007) and two 2-isocyano-3-methyl-2-butenoyl substituents, which were named maculansin A and maculansin B (Pedras and Yu 2008). Maculansin A caused larger lesions on leaves of resistant brown mustard (*Brassica juncea* L.) than on susceptible canola. In addition, 2,4-dihydroxy-3,6-dimethylbenzaldehyde was isolated and shown to display significant growth inhibitory activity on roots of canola and mustard (Pedras and Yu 2008).

Zinniol, pachybasin, and epoxydon are important phytotoxins produced by *Phoma macdonaldii*, *Phoma foveata*, and *Phoma sorghina*, respectively. *P. macdonaldii* is the causal agent of the sunflower black stem disease, a widely observed disease in many countries, responsible for economic damage, which cause leaf and stem blight and withering of cut seedlings of sunflower plants (Sugawara and Strobel 1986; Roustaee et al. 2000). *Phoma foveata* is a species that causes gangrene, which is lesions on potato tubers (*Solanum tuberosum*) (Strange 1997). *Phoma sorghina* is the causative species of necrosis when spotted on pokeweed leaves and eight other weed species, indicating a nonspecific nature. Beyond epoxydon, five more phytotoxins were isolated, 6-methylsalicylic acid, desoxyepoxydon, phyllostine, 2,3′-dihydroxy-4-methoxy-5′,6-dimethyl diphenyl ether, and epoxydon 6-methylsalicylate ester. All six toxins caused necrosis, while 6-methylsalicylic acid, epoxydon, and the diphenyl ether also inhibited sorghum root growth (Venkatasubbaiah et al. 1992). This species also produces tenuazonic acid (TeA), a tetramic acid derivative, in the form of Mg and Ca chelates (Steyn and Rabie 1976).

Zhou and Qiang (2008) studied the effect of this compound on the growth and total chlorophyll contents of *Chlamydomonas reinhardtii*. The inhibition of the growth and chlorophyll contents was observed with increasing TeA concentrations (>100μg/mL).

Putaminoxin, the main phytotoxin isolated from *Phoma putaminum*, was characterized as (5*S*)5-hydroxy-9-propyl-6-nonen-9-olide, a 10-macrolide, and showed a wide range of toxicity (Evidente et al. 1995). The authors investigated putaminoxin toxicity against weed species and on nonhost-cultivated plants. It showed a range of toxicities, *Ocimum basilicum* and *Citrus aurantium* being among the less sensitive and *Mercurialis annua* the most sensitive. However, the toxicity observed on *Annual fleabane* was the most severe. Afterward, new nonenolides, isolated from *Phoma putaminum*, together with some of their natural analogs and synthetic derivatives, were used in a structure-activity relationship study. Putaminoxin proved to be the most toxic compound among putaminoxin analogs and derivatives (Evidente et al. 1998).

Due to the large number of phytotoxins produced by *Phoma* species, this genus is one of the most cited in scientific research on bioherbicides (15%), equally the genus *Fusarium* sp. (15%) and after *Colletotrichum* sp. (30%) (Klaic et al. 2015). The process of developing a bioherbicide is complex, and the crop protection research groups must consider several questions. A highly desirable property of microbial phytotoxins is the built-in species selectivity due to their isolation from host-specific plant pathogens and weed hosts (Saxena and Pandey 2001). Some attempts to evaluate these phytotoxins as bioherbicides have been reported (Table 12.1).

As can be seen in Table 12.1, *Phoma exigua*, *Phoma macrostoma*, and *Phoma herbarum* are three species extensively studied as effective herbicidal agents. *P. exigua* var. *exigua* and *P. exigua* var. *heteromorpha* produce several phytotoxic cytochalasins, viz., ascochalasin, deoxaphomin, proxiphomin, protophomin, cytochalasin A, B, C, D, E, F, H, J, T, U, V, W, Z2, Z3, Z4, Z5, Z6, and 7-O-acetylcytochalasin B (Binder and Tarnrn 1973; Scott et al. 1975; Capasso et al. 1991; Evidente et al. 1992, 1996, 2003). Cytochalasins constitute a great group of metabolites produced by a variety of fungal species, and, even though the first metabolite was discovered more than 30 years ago, these compounds are still largely studied for their attractive biological properties and potential applications as antibiotic, pharmaceutical, or herbicidal products (Evidente et al. 2003). The mycoherbicidal potential of these biocontrol agents has been studied by Cimmino et al. (2008). The phytotoxins produced were p-hydroxybenzaldehyde; cytochalasins B, F, Z2, and Z3; and deoxaphomin, and their effect against perennial weeds, viz., *Cirsium arvense* and *Sonchus arvensis*, were checked. P-hydroxybenzaldehyde was inactive when assayed on the leaves of both *C. arvense* and *S. arvensis*, whereas cytochalasin Z2 appeared to be the less toxic cytochalasin on both plants, and deoxaphomin demonstrated to be the more toxic cytochalasin on leaves of *S. arvensis*.

About two decades ago, ten isolates of *Phoma macrostoma* Montagne were discovered to have bioherbicidal effects (chlorosis and seedling death) on various broadleaf weed pests including Canada thistle, dandelion, chickweed, and scentless chamomile (Bailey and Derby 2001). Later, *P. macrostoma* 94-44B was tested against diverse species of economically important nontarget species in agriculture, horticulture, forestry, and gardening, as well as potential target weeds (Bailey et al. 2011). Again, the authors concluded that *P. macrostoma* was pathogenic to many dicotyledonous plant species but nonpathogenic to monocots. The most resistant plant families were *Poaceae*, *Pinaceae*, and *Linaceae*, whereas, the most susceptible plants belonged to *Asteraceae*, *Brassicaceae*, and *Fabaceae*. Under certain conditions, *P. macrostoma* produces phytotoxins by the mycelium, known as macrocidins, which cause growth retardation of shoots and roots along with severe photobleaching (Graupner et al. 2003, 2006).

A strain of *P. herbarum* Westend was isolated from diseased leaves of *Parthenium hysterophorus* L. (Rajak et al. 1990), which caused more than 90% inhibition in seed germination, seedling mortality, and leaf damage followed by

Table 12.1 Summary of plants in which *Phoma* species reported as biological control agents

Phoma species	Target plant	Common name	Application mode	Effects	References
P. chenopodicola	*Mercurialis annua*, *Cirsium arvense*, *Setaria viride*	Annual mercury, Canada thistle, green foxtail	Postemergent (leaf puncture)	Circular necrotic lesions	Cimmino et al. (2013)
P. commelinicola	*Commelina diffusa*	Dayflower	Inundative	Mortality (~70%) and dry weight reduction (>80%)	Boyette et al. (2015)
P. dimorpha	*Senna obtusifolia*	Bush pasture	Preemergence (germination test) and postemergence (detached leaf and aerosol applicator)	Inhibition of seed germination, leaf necrosis and death and reduction of size and fresh mass	Chaves Neto et al. (2020)
P. exigua var. *exigua*	*Sonchus arvensis*, *Cirsium arvense*	Perennial sow thistle, Canada thistle	Postemergent (leaf disk puncture)	Necrosis	Cimmino et al. (2008)
P. exigua var. *exigua*	*Taraxacum officinalis*	Dandelion	Postemergent	Great disease severity	Stewart-Wade and Boland (2005)
P. exigua var. *exigua*	*Gaultheria shallon*	Salal	Postemergent	Suppressed the growth of mature salal plant and caused 56% death of the total biomass	Zhao and Shamoun (2006)
P. exigua var. *exigua*	*Cirsium arvense*	Californian thistle	Postemergent (detached leaf)	Significant lesion development	Bithell and Stewart (2001)
P. herbarum	*A. hypochondriacus*	Prince-of-Wales	Petri dish bioassay	Inhibition of radicle growth	Rivero-Cruz et al. (2003)
P. herbarum	*Taraxacum officinalis*	Dandelion	Postemergent	100% disease severity	Neumann and Boland (2002)
P. herbarum	*Parthenium hysterophorus*	Whitetop weed	Postemergent (cut shoot, seedling, detached leaf)	Acute chlorosis and necrosis	Vikrant et al. (2006)
P. herbarum	*Parthenium hysterophorus*, *Lantana camara*, *Hyptis suaveolens*, *Sida acuta*	Whitetop weed, lantana, pignut, wireweed	Postemergent (detached leaf)	Chlorosis, curling, and complete collapse	Kalam et al. (2014)
P. macrostoma	*Cirsium arvense*	Canada thistle	Preemergent	Intense chlorosis and bleaching of emerging leaves	Graupner et al. (2003)
P. macrostoma	*Taraxacum officinale*	Dandelion	Inundative	100% chlorosis	Zhou et al. (2004)

(continued)

Table 12.1 (continued)

Phoma species	Target plant	Common name	Application mode	Effects	References
P. macrostoma	*Lactuca sativa*, *Brassica rapa*, *Taraxacum officinale*, *Gnaphalium pensylvanicum*, *Sinapis arvensis*, *Oxalis stricta*, *Polygonum convolvulus*, among others	Lettuce, canola, dandelion, wandering cudweed, wild mustard, yellow woodsorrel, wild buckwheat	Pre- and postemergent	81, 91, 92, 100, 61, 79, 95% mortality	Bailey et al. (2011)
P. macrostoma	*Senecio vulgaris*, *Lamium amplexicaule*	Groundsel, henbit	Pre- and postemergent	>80% chlorosis and >87% weed reduction	Neal et al. (2013)
P. macrostoma	*Taraxacum officinale*, *Aster subulatus*, *Malva neglecta*, *Portulaca oleracea*, *Medicago polymorpha*, *Sonchus oleraceus*	Dandelion, slender aster, common mallow, common purslane, California burclover, annual sow thistle	Postemergent (foliar granular)	Some extent of chlorosis and necrosis	Smith et al. (2015)
P. macrostoma	*Taraxacum officinale*, *Rorippa palustris*, *Cardamine flexuosa*	Dandelion, marsh yellow cress, flexuous bitter cress	Preemergent (granular treatments)	100% growth reduction	Wolfe et al. (2016)
P. macrostoma	*Taraxacum officinale*, *Senecio vulgaris*, *Cicer arietinum*	Dandelion, groundsel, chickpea	Postemergent (sprayer)	Chlorosis and mortality	Hubbard et al. (2016)
P. proboscis	*Convolvulus arvensis*	Field bindweed	Postemergent (aerosol applicator)	Germination reduction	Heiny and Templeton (1991)
P. putaminum	*Erigeron annuus*, *Mercurialis annua*, *Cynara cardunculus*, *Urtica* spp., *Sonchus oleraceus*, *Chenopodium album*, *Fragaria vesca*	Annual fleabane, annual dog's mercury, globe artichoke, nettle, annual sow thistle, fat hen, strawberry	Postemergent (leaf puncture)	Necrosis	Evidente et al. (1995)
P. putaminum	*Erigeron annus*, *Mercurialis annua*, *Cirsium arvense*, *Chenopodium album*, *Xanthium occidentale*, *Sorghum bicolor*, *Lycopersicon esculentum*, *Cucumis sativus*	Annual fleabane, annual mercury, Canada thistle, fat hen, Noogoora burr, sorghum, tomato, cucumber	Postemergent (leaf puncture)	Necrosis	Evidente et al. (1998)

reduction in height of the weed *Parthenium* (Pandey et al. 1991). Afterward, strains of *P. herbarum* with bioherbicidal potential were isolated from another weed, *Lantana camara* (Pandey and Pandey 2000). The fungus *P. herbarum* was also isolated from a special variety of *Zea mays*. Two new phytotoxic nonenolides, herbarumin I and herbarumin II, were found and caused significant inhibition of radicle growth of *A. hypochondriacus* (Rivero-Cruz et al. 2000).

Vikrant et al. (2006) reported pronounced phytotoxicity of cell-free culture filtrate of *P. herbarum* FGCC#75 against *Parthenium hysterophorus* L., which is due to a chemical 3-nitro-1,2-benzenedicarboxylic acid (3-nitrophthalic acid).

In contrast to these extensively studied species, some investigations are emerging about new species with bioherbicide potential. Chaves Neto et al. (2020) evaluated the bioherbicidal action of *Phoma dimorpha* (NRRL 43879) fermented broth in the preemergence and postemergence of *Senna obtusifolia*. The results showed in preemergence application an inhibiting of seed germination in 100%, while in detached leaves, leaf necrosis and death on the ninth day after the application. In the postemergence, the application caused moderate symptoms, such as leaf spots and reduction in the weed plant size. Prior to the identification of species, other works were developed with the same strain. First, the bioherbicide production from strain NRRL 43879 by solid-state fermentation was optimized. Results of the bioassays demonstrated that the bioherbicide presented activity toward *Cucumis sativus*, and the mode of action was based on the inhibition of carotenoid biosynthesis (Klaic et al. 2017). Later, the bioherbicide production was developed by submerged fermentation and resulted in an inhibition of 100% of *Bidens pilosa* and *A. retroflexus* seeds (Todero et al. 2019). The reduction of 100% on seed germination of *Sorghum bicolor* was also observed. The authors suggested that the bioactive compound was pyrrolo[1,2-a]pyrazine-1,4-dione, hexahydro-3-(2-methylpropyl) (Brun et al. 2016).

12.3 *Phoma* as Biological Control Agent for Microorganisms

The emergence of resistant strains to current treatments of microbial infections and the changing epidemiology require for new antimicrobial drugs with broad spectra and distinct mechanisms of action. Fungi are a rich natural source of compounds with diverse structures and bioactivities, mainly due to their highly developed and diversified secondary metabolism. The genus *Phoma* contains several species ubiquitously present in soil, water, and environment (Rai et al. 2018). Regarding to the content of secondary metabolites of this genus, a chemotaxonomic investigation showed that marine-derived species differ significantly from their terrestrial counterparts. Despite the similarity between many of these metabolites, both groups seem to be distinguished by unique metabolic features (Osterhage et al. 2000). Many investigations have been carried out on the biological activity of terrestrial- and marine-derived species and have shown that the genus of fungus *Phoma* produces many antimicrobial active compounds (Table 12.2).

A polyketide compound, 5-hydroxyramulosin, is an active compound with antimicrobial potential produced by different species of *Phoma* of origin in both marine and terrestrial environments. Osterhage et al. (2002) obtained the 5-hydroxyramulosin together with 7-methoxycoumarin from fungus *Phoma tropica* isolated from the inner tissue of the marine brown alga *Fucus spiralis*. Afterward, Santiago et al. (2012) obtained the 5-hydroxyramulosin by *Phoma* sp. isolated from the terrestrial medicinal plant *Cinnamomum mollissimum*. This compound inhibited the fungal pathogen *Aspergillus niger* and was cytotoxic against murine leukemia cells. Other polyketides with antimicrobial activity were isolated from the same strain of *Phoma* extracted from the *Cinnamomum mollissimum*. Three known compounds, namely, 4-hydroxymellein, 1-(2,6-dihydroxyphenyl)ethanone, and 4,8-dihydroxy-6-methoxy-3-methyl-3,4-dihydro-1H-isochromen-1-one (benzopyran) were isolated from the fungus and showed inhibitory activity against *Bacillus subtilis* (antibacterial), *Aspergillus niger* (antifungal), and P388 murine leukemic cells (cytotoxic) (Santiago et al. 2014).

Sirodesmin PL, a phytotoxin previously mentioned, isolated from a highly virulent toxin producing isolate of *Phoma lingam*, has also exhibited antibacterial activity against Gram-positive bacteria and particularly *Bacillus subtilis* (Boudart 1989) and antifungal activity against *Septoria nodorum* and *Botrytis cinerea* (Poiret et al. 1985). The antifungal activity of this mole-

Table 12.2 Summary of *Phoma* species reported as biological control agent for microorganisms

Phoma strain	Origin	Bioactive compound	Molecular formula	Antimicrobial activity	References
Terrestrial					
Phoma sp.	*Fagonia cretica*	Pyrenophorol (**1**, macrodiolide), 4,7-dihydroxyoctanoic acid (**2**, pyrenophorolic acid)	$C_{16}H_{24}O_6$ (**1**) $C_8H_{16}O_4$ (**2**)	Antifungal	Krohn et al. (2007)
Phoma sp. F-224,939	Soil	Phomafungin (cyclic lipodepsipeptide)	$C_{46}H_{79}N_9O_{17}$	Antifungal	Herath et al. (2009)
Phoma sp. 8874	*Lycium intricatum*	Pyrenophorol (**1**, macrodiolide), (–)-dihydropyrenophorin (**2**)	$C_{16}H_{24}O_6$ (**1**) $C_{18}H_{24}O_7$ (**2**)	Antifungal and antibacterial	Zhang et al. (2008)
Phoma sp. 25,697	*Hypoxylon* sp.	Phomadecalins A–D (**1–4**)	$C_{15}H_{20}O_3$ (**1**)	Antibacterial	Che et al. (2002)
Phoma sp.	*Cinnamomum mollissimum*	5-Hydroxyramulosin (polyketide)	$C_{10}H_{14}O_4$	Antifungal and antibacterial	Santiago et al. (2012)
Phoma sp.	*Cinnamomum mollissimum*	4-Hydroxymellein (**1**, polyketide), 4,8-dihydroxy-6-methoxy-3--methyl-3,4-dihydro-1H-isochromen-1-one (**2**, benzopyran), and 1-(2,6-dihydroxyphenyl) ethanone (**3**, polyketide)	$C_{10}H_{10}O_4$ (**1**) $C_{11}H_{12}O_5$ (**2**) $C_8H_8O_3$ (**3**)	Antifungal and antibacterial	Santiago et al. (2014)
Phoma sp. 7133	*Senecio kleinia*	Sclerodin (**1**), 8,9-dihydro-3,5,7-trihydroxy-1,8,8,9-tetramethyl-5-(2-oxopropyl)-4H-phenaleno[1,2-b]furan-4,6(5H)-dione (**2**), atrovenetinone (**3**), and sclerodione (**4**)	–	Antifungal and antibacterial	Hussain et al. (2015)
Phoma sp. NG-25	*Saurauia scaberrinae*	Usnic acid (**1**), cercosporamide (**2**), phomodione (**3**), and (–)-pseudoplacodiolic acid (**4**)	–	Antifungal	Hoffman et al. (2008)
Phoma sp. ZJWCFOO6	*Arisaema erubescens*	(3 S)-3,6,7-Trihydroxy-α-tetralone (**1**), cercosporamide (**2**), β-sitosterol (**3**), and trichodermin (**4**)	$C_{10}H_{10}O_4$ (**1**)	Antifungal	Wang et al. (2012)
Phoma sp. QN04621	Dead leaf sample (unidentified plant)	YM-215343 (apiosporamide and fischerin)	$C_{24}H_{29}NO_6$	Antifungal	Shibazaki et al. (2004)
Marine					
Phoma sp. 6282	*Fucus serratus*	Phomafuranol (**1**, dihydrofuran), phomalacton (**2**), (3R)-5-hydroxymellein (**3**), and emodin (**4**)	$C_7H_{10}O_2$ (**1**)	Antifungal and antibacterial	Hussain et al. (2014)
Phoma sp. 135	*Ectyplasia perox*	Dihydroisocoumarin derivatives (**1–2**)	$C_{11}H_{11}ClO_4$ (**1**) $C_{11}H_{11}ClO_5$ (**2**)	–	Elsebai and Ghabbour (2016)
Phoma sp. L28	*Myoporum bontioides* A. Gray (mangrove plant)	7-(γ,γ)-Dimethylallyloxymacrosporin (**1**, anthraquinone), macrosporin (**2**), 7-methoxymacrosporin (**3**), tetrahydroaltersolanol B (**4**), altersolanol L (**5**), and ampelanol (**6**)	$C_{21}H_{20}O_5$ (**1**)	Antifungal	Huang et al. (2017)
Phoma sp. Q60596	*Halichondria japonica*	YM-202204 (lactone)	$C_{37}H_{58}O_9$	Antifungal	Nagai et al. (2002)

Phoma strain	Origin	Bioactive compound	Molecular formula	Antimicrobial activity	References
Phoma tropica	*Fucus spiralis*	5-Hydroxyramulosin (**1**), 7-methoxycoumarin (**2**)	$C_{10}H_{12}O_4$ (**1**) $C_{10}H_{12}O_4$ (**2**)	Antifungal and antibacterial	Osterhage et al. (2002)
Phoma herbarum VB7	Mangrove plant	Dibutyl phthalate, mono-(2-ethylhexyl) phthalate	$C_{16}H_{22}O_4$	Antibacterial	Bhimba et al. (2012)
Phoma sp. H477	*Halichondria panicea*	Pycnidione	–	Antibacterial	Höller et al. (2000)

cule was illustrated on germination of several fungal species including *Fusarium graminearum* and *Leptosphaeria biglobosa* "brassicae," which were inhibited in the presence of a 13-day-old colony of the wild-type sirodesmin PL producing isolate when grown in vitro (Elliott et al. 2007). Other compounds with antimicrobial properties had already been isolated from the same species. These compounds are phomenolactone and phomenoic acid, which presented activity particularly against *Candida albicans* (Devys et al. 1984; Topgi et al. 1987).

Brefeldin A is a macrocyclic lactone with pronounced antifungal property, which has been isolated from *Phoma herbarum*. Brefeldin A, also named decumbin, cyanein, ascotoxin, or synergisidin, dramatically alters the Golgi apparatus and causes redistribution of Golgi proteins into the endoplasmic reticulum (Betina 1992). Cole et al. (2000) reported its effects on growth, endoplasmic reticulum, Golgi bodies, tubular vacuole system, and secretory pathway in *Pisolithus tinctorius*. The authors showed that brefeldin A reduced radial growth in *Pisolithus tinctorius* at a concentration as low as 2μM. Phomallenic acid is another important secondary metabolite with antimicrobial potential isolated from *Phoma* sp. Ondeyka et al. (2006) reported that phomallenic acid showed efficacy against wild-type strains of *Staphylococcus aureus* and methicillin-resistant *S. aureus*. However, phomallenic acid did not exhibit any activity against both Gram-negative and Gram-positive bacteria.

Some species of *Phoma* have been used as mycoparasitic agent. *Phoma glomerata* have excellent pesticidal activity against powdery mildews, which are common plant pathogens that are conspicuous by their white mycelia and powder-like conidia (Sullivan and White Jr 2000). Arie et al. (1998, 1999) reported that culture broth from an isolate of *Phoma glomerata* JCM9972 recovered from the leaves of *Viola* sp. controlled the soilborne pathogen *Plasmodiophora brassicae* which causes clubroot disease of cruciferous plants. For the first time, this effect was related to the substance epoxydon (5-hydroxy-3-(hydroxymethyl)-7-oxabicyclo[4.1.0]hept-3-en-2-one), although it was known to have antitumor activity, phytotoxicity, and antiauxin activity. The authors suggest that *P. glometara* JCM9972 strain belongs to a group called "plant growth-promoting fungi." The phytotoxicity of epoxydon isolated from *Phoma sorghina* together with desoxyepoxydon, 6-methylsalicylic acid, phyllostine, 2,3′-dihydroxy-4-methoxy-5′,6-dimethyl diphenyl ether, and epoxydon 6-methylsalicylate ester by Venkatasubbaiah et al. (1992) was reported. Epoxydon, desoxyepoxydon, and the diphenyl ether showed strong antimicrobial properties against all bacteria and fungi tested.

12.4 *Phoma* as Biological Control Agent for Viruses

An abundance of potentially active natural products have been isolated from fungi and screened for antiviral activity, even though none of them has reached the market yet (Hyde et al. 2019). Fungal compounds with antiviral activities are less extensively studied, but the number of these investigations is increasing. A group of fungi that has been a focus of interest, beyond the edible and medicinal mushrooms, are endophytic fungi, particularly those that grow in marine habitats (Linnakoski et al. 2018). Most studies are not focused solely on the antiviral activity of the compounds, and the main limitation to antiviral compound discovery is most probably related to the absence of antiviral screening systems (Rajamanikyam et al. 2017). Some antiviral compounds isolated from different species of *Phoma* sp., such as sirodesmin PL (Rouxel et al. 1988), aphidicolin (Oikawa et al. 1999, 2001), and brefeldin A (Betina 1992), have already been mentioned previously for antifungal, antibacterial, or bioherbicidal activity.

Ergocytochalasin A, phomasetin, and phomanolide are another antiviral compounds isolated from *Phoma* species. Ergocytochalasin A, the first example of a merocytochalasan which consists of one cytochalasin moiety and one ergosterol moiety, was isolated from an endophytic fungus *Phoma multirostrata* XJ2-1, and its antiviral activity against human dengue virus type 3 (DV3), influenza A virus (H1N1), and respiratory syncytial virus (RSV) was evaluated (Peng et al. 2020). Phomasetin, a tetramic acid, was isolated from the marine fungi *Phoma* sp. and exhibited in vitro inhibitory activity against the integrase enzyme of human immunodeficiency virus type 1 (HIV-1), which is an important enzyme for the HIV-1 replication cycle (Hazuda et al. 1998). Phomasetin is the opposite stereochemical homolog of equisetin and the first natural product inhibitors of integrase (Singh et al. 1998). Phomanolide, a rare 14-nordrimane-type sesquiterpenoid, was isolated together with five known compounds, (–)-6-methoxymellein, 7-hydroxy-3,5-dimethyl-isochromen-1-one, norlichexanthone, 6-methylsalicylic acid, and gentisyl alcohol, from the culture broth of an endophytic fungus *Phoma* sp. isolated from the roots of *Aconitum vilmorinianum*. Phomanolide and compound 2 showed antiviral activities against H1N1 (Liu et al. 2019).

12.5 *Phoma* as Biological Control Agent for Cancer Cells

Cancer cells are abnormal cells characterized by their uncontrolled growth, which can affect various organs of the body and lead to death. Cancer is one of the major causes of death worldwide. Endophytic fungi have been recognized as a possible source of anticancer drugs. More than 60% of the anticancer drugs and compounds are currently in clinical use and are natural products or natural product derivatives (Rajamanikyam et al. 2017). Endophytic *Phoma* is a promising source of novel metabolites and exhibits great cytotoxic potential against diverse cancer types, such as malignant melanoma, gastric cancer, colorectal cancer, liver cancer, breast cancer, and T cell leukemia, among others (Table 12.3). Wang et al. (2012) obtained four bioactive compounds from *Phoma* species (ZJWCF006) which demonstrated antimicrobial activity against four plant pathogenic fungi and two plant pathogenic bacteria, as well as in vitro antitumor activities against HT-29, SMMC-772, MCF-7, HL-60, MGC80-3, and P388 cell lines. Among the compounds are the cercosporamide and trichodermin, which have been demonstrated as valuable candidates for the discovery of agents for new drugs in cancer therapy.

Compounds previously cited as cytochalasin, tenuazonic acid, and brefeldin A also have cytotoxic potential. La Kim et al. (2012) isolated cytochalasin derivatives and cytochalasin B from *Phoma* sp. recovered from the giant jellyfish (*Nemopilema nomurai*). The compounds showed significant cytotoxicity against human lung cancer (A549), human ovarian cancer (SK-OV-3), human skin cancer (SK-MEL-2), human CNS cancer (XF 498), and human colon cancer (HCT15), and cytochalasin B showed cytotoxicity against human cervical carcinoma (HeLa) cells. Tenuazonic acid (TeA) was isolated from *Phoma sorghina* (Steyn and Rabie 1976), and its cytotoxic effect on proliferation of three mammalian normal cell lines was investigated (Zhou and Qiang 2008). The authors observed that TeA inhibited the proliferation of 3T3 mouse fibroblasts, Chinese hamster lung cells (CHL cells), and human hepatocytes (L-O2 cells) at concentrations ranging from 12.5 to 400μg/mL. Two species of *Phoma*, viz., *P. herbarum* and *Phoma medicaginis*, were obtained from medicinal parts of *A. sinensis*, and their fermented broths displayed antitumor activity against SKVO3 (Cui et al. 2011). The known compound isolated from these *Phoma* species is brefeldin A (Specian et al. 2014; Rai et al. 2018) and acts as inhibitor of protein trafficking in endomembrane system of mammalian cells (Sciaky et al. 1997).

12.6 *Phoma* as Producer of Anthraquinone Pigments

The demand for natural colors is increasing daily due to prejudicial effects of some synthetic dyes. Filamentous fungi provide an available alternative source of naturally derived pigments, which have enormous advantages including rapid growth and easy processing. Apart from dye, fungal pigments have many biological properties such as antioxidant, antimicrobial, and anticancer activity (Rao et al. 2017). An extraordinary range

Table 12.3 Summary of *Phoma* species reported as biological control agent for cancer cells

Phoma strain	Origin	Bioactive compound	Targeted cell lines	References
Phoma ZJWCF006	*Arisaema erubescens*	Cercosporamide and trichodermin	HT-29, SMMC-772, MCF-7, HL-60, MGC80-3, P388	Wang et al. (2012)
Phoma sp.	*Nemopilema nomurai*	Cytochalasin derivatives and cytochalasin B	A549, SK-OV-3, SK-MEL-2, XF 498, HCT15, HeLa	La Kim et al. (2012)
Phoma betae	*Ginkgo biloba*	Taxol	MCF-7, BT 220, HL 251	Kumaran et al. (2012)
Phoma herbarum	*Aquilaria sinensis*	Brefeldin A	SKVO3	Cui et al. (2011)
Phoma medicaginis	*Aquilaria sinensis*	Brefeldin A	SKVO3	Cui et al. (2011)
Phoma betae	*Smallanthus sonchifolius*	8-Hydroxy-6-methoxy-3--methylisocoumarin, (22E,24R)-ergosta-4,6,8(14),22-tetraen-3-one	MDA-MB435, HCT-8, SF295, HL-60	Gallo et al. (2009)
Phoma sp. QN04621	Dead leaf sample (unidentified plant)	YM-215343 (apiosporamide and fischerin)	HeLa S3	Shibazaki et al. (2004)
Phoma sp. OUCMDZ-1847	*Kandelia candel*	Phomazines A–C	K562, HL-60, HCT-116, A549, MGC-803	Kong et al. (2014)
Phoma sp. CZD-F11	Sediment (unidentified)	Cyperin-2-O-®-D-glucoside	PC-3	Wu et al. (2018a)
Phoma sp. *MF6118*	Vegetation sample (unidentified)	Fusidienol A	–	Singh et al. (1997)
Phoma sp.	Depth of Bering Sea	Ergosterol, ergosterol peroxide, and 9,11-dehydroergosterol peroxide	A549, J5, HeLa, and MCF-7	Wu et al. (2018b)
Phoma sp. PT01	*Mitragyna javanica*	–	A375, KatoIII, SW620, HepG2, BT474, Jurkat	Pharamat et al. (2013)
Phoma sp. YN02-P-3	*Sumbaviopsis* J. J. Smith	3,4-Dihydronaphthalen-1(2H)-one with spiro-butyrolactone phomol	HL-60	Sang et al. (2016)
Phoma sp. SHZK-2	Contaminated river	2-(2-Formyl-3-hydroxy-5--methylphenoxy)-5-hydroxy-3-methoxybenzoate, asterric acid, methyl asterrate, 9(Z),12(Z)-nonadecadienoic acid, orsellinic acid	HEPG2, Raji	Fang et al. (2012)
Phoma multirostrata EA-12	*Eupatorium adenophorum*	Multirostratin A, 20-oxo-deoxaphomin	HL-60, A-549, SMMC-7721, MCF-7, SW-480	Chen et al. (2015)
Phoma sp. 14,078	Soil sample	Phomacins A–C	HT-29	Alvi et al. (1997)
Phoma sp.	*Cinnamomum mollissimum*	5-Hydroxyramulosin	P388	Santiago et al. (2012)

(continued)

Table 12.3 (continued)

Phoma strain	Origin	Bioactive compound	Targeted cell lines	References
Phoma multirostrata XJ-2-1	*Parasenecio albus*	Cytochalasin	CT26, A549	Peng et al. (2020)
Phoma sp.	*Ectyplasia perox*	Epoxyphomalin A and B	A (BXF 1218 L, CNXF 498NL, GXF 251 L, LXF 629 L, MAXF 401NL, OVXF OVCAR3, PAXF 1657 L, PRXF 22RV1, PRXF PC3M, PXF 1752 L, RXF 486 L, UXF 1138 L); B (PXF 1752 L, RXF 944 L, OVXF 1619 L, MEXF 394NL, MAXF MCF7, HNXF 536 L, CNXF SF268, BXF T24)	Mohamed et al. (2009)

Table 12.4 Summary of *Phoma* species reported as producer of anthraquinone pigments

Phoma strain	Anthraquinone pigments	Color	References
P. exigua var. *foveata*	Pachybasin, chrysophanol, emodin, and phomarin	Yellow (acid condition) and red (alkaline conditions)	Bick and Rhee (1966)
P. exigua var. *exigua*	"E metabolite"	Pigment α: red-purple (pH < 10.5) and blue-green (pH > 12.5) Pigment β: yellow (pH < 3.5) and red (pH > 5.5)	Boerema and Höweler (1967)
P. exigua var. *foveata*	Pachybasin, chrysophanol, emodin, and phomarin	Pink (ammonia exposure)	Logan and Khan (1969)
Phoma sorghina	1,7-Dihydroxy-3-methyl-9,10-anthraquinone (**1**), 1,6-dihydroxy-3-methyl-9,10-anthraquinone (phomarin, **2**), 1-hydroxy-3-methyl-9,10-anthraquinone (pachybasin, **3**), and 1-7-dihydroxy-3-hydroxymethyl-9,10-anthraquinone (**4**)	Orange (**1**,**2**,**4**), yellow (**3**)	Borges and Pupo (2006)
Phoma herbarum	–	Magenta	Chiba et al. (2006)

of pigments are produced by filamentous fungi and include several chemical classes such as quinones, carotenoids, melanins, flavins, phenazines, monascins, violacein, and índigo (Dufossé et al. 2014). Fungal anthraquinones as polyketide-derived secondary metabolites occur widely in many genera of fungi (Hobson et al. 1997). Several species of *Phoma* are known to produce friendly anthraquinone pigments (Table 12.4), which have dyeing properties of dyestuffs and potential to be utilized commercially (Rai et al. 2009).

12.7 *Phoma* as Polysaccharide Producer

In general, the fungi produce various compounds that can be divided into high molecular weight compounds such as heteropolysaccharides, lipopolysaccharides, proteins, lipoproteins, or complexes of components, and a wide variety of chemical structures including glycolipids, lipopeptides, polymeric polysaccharide-protein complexes, fatty acids, and phospholipids with a low molecular weight (Luft et al. 2020).

Exopolysaccharides (EPS) are the most frequently studied fungal polysaccharides, besides cell wall polysaccharides and intracellular cytosolic polysaccharides. EPS obtained from *Ascomycota* fungal cultures are known for their antioxidant, immunostimulating, antitumor, and antimicrobial properties (Osińska-Jaroszuk et al. 2015) and for their bioemulsifying ability (Luft et al. 2019).

YCP (acronym of Yancheng polysaccharide) was isolated from the mycelium of the marine filamentous fungus *Phoma herbarum* YS4108 (Yang et al. 2005). The authors reported the YCP ability to increase phagocytic activity of mice in vitro and in vivo, indicating that it might be an immunomodulating polysaccharide, which stimulated the immune response by acting on macrophages from the systemic immune compartment. The proliferative response of IgM antibodies in B cells by YCP was studied and demonstrated the efficient immunomodulatory function of YCP in oncological immunotherapy through TLR-mediated activation of signaling pathways like MAPK and NF-kB (Zhang et al. 2013). Chen et al. (2009) described that YCP was responsible for induction of NO production in macrophages through binding of YCP to TLR 4 and CR3 pathways. Afterward, a study demonstrated that YCP acts as second signal for T cell activation, proliferation, and IFN-γ production through binding to Toll-like receptor (TLR-)-2 and TLR-4 and can be considered a good candidate for tumor-specific immunotherapy to enhance signals of antigen presentation process, antigen-specific T cell proliferation, Th1 cytokine induction, and tumor-specific killing (Chen et al. 2014). In summary, this microbial polysaccharide is a promising starting point for the development of antitumor and immunity-enhancing drugs (Sun et al. 2016).

Notwithstanding YCP is the most studied polysaccharide from the *Phoma herbarum* species; other studies have been developed. A strain of *Phoma herbarum* identified as CCFEE 5080 was isolated from continental Antarctica soil and tested for EPS production. The EPS production is up to 13.6 g/L on a variety of carbon sources, and it was maximum when the nitrogen source was $NaNO_3$ and the incubation temperature was 28 °C. The authors hypothesized a possible relationship between EPS production by the Antarctic fungus and its involvement in surviving under one of the harshest climates in the world (Selbmann et al. 2002).

Although EPS are considered potential biofertilizers and elicitors of systemic resistance of plants (Osińska-Jaroszuk et al. 2015), recent studies have shown the EPS ability to increase bioherbicidal activity of fermented broth of *Phoma dimorpha*. The EPS were concentrated through two processes (membranes and adsorption), and it was possible to observe that the higher the concentration of EPS in the fermented broth, the greater is its bioherbicidal activity. A simultaneous decrease of surface tension values with increase of EPS concentration was observed, suggesting that the increase in bioherbicidal activity may be due to the emulsifying capacity of EPS (Luft et al. 2019).

12.8 Commercial Scale of *Phoma* Products: Implications and Prospects

Microorganisms are the leading producers of useful natural products. The large number of secondary metabolites produced by fungi and their ability to produce excellent drugs, including antibiotics, anticancer agents, immunosuppressants, and products for other applications, such as antivirals, anthelmintics, enzyme inhibitors, nutraceuticals, polymers, surfactants, bioherbicides, and vaccines, make them important participants in global industry (Adrio and Demain 2003; Demain 2014). Unfortunately, fungi are relatively underexplored but are an essential, extremely interesting, and biotechnologically useful group of organisms with an incredible biotechnological potential for industrial exploitation (Hyde et al. 2019).

Another relevant feature of fungi for the global market is an increased interest in development of sustainable biotechnological processes, in order to obtain new natural products that can eventually replace traditional synthetics. As compared to plant sources, fungi have the great advantage as they are growing in large bioreactors at an industrial scale and with suitable processes for their cost-efficient fermentation (Hyde et al. 2019). However, the development of fungal products has some limitations, such as the excessive costs for clinical trials, the too short a window before the products become generics, and the difficulty in discovery of products against resistant organisms (Demain 2014).

Despite these difficulties, technology to discover new fungal products has advanced. Regarding *Phoma*, it is possible to get a sense of how great the limitations are during all stages of the development of a fungal product, ranging from screening to registration and commercialization. There are countless compounds produced by different species of *Phoma* with applications in diverse areas; however, there is only one product from it available on a commercial scale on the market. As previously seen, *Phoma macrostoma* Montagne 94–44B is an effective bioherbicide for broadleaved weed reduction. In 2016, Health Canada's Pest Management Regulatory Agency, under the authority of the Pest Control Products Act and Regulations, granted full registration for the sale and use of the bioherbicide *P. macrostoma* Montagne 94–44B to control a broad spectrum of broadleaf weeds in established turfgrass and new seeding of grasses as well as in field-grown nursery plants and trees and container-grown ornamentals (Hynes 2018).

There are a relevant number of patents deposited and registered from *Phoma*, and some of these are summarized in Table 12.5. These patents suggest a variety of applications related to the use of *Phoma* compounds, as antioxidant, antimicrobial, bioherbicidal, antitumor, immunomodulating, resistance elicitor, and contaminated water purifier. In general, patent developments based on *Phoma* show a greatest interest in antifungal compounds and bioherbicides. The main countries with contributions in descending order are China (5), USA (4), Korea (2), Russia (2), Brazil (1), Japan (1), and Italy (1).

The Brazilian patent number BR102012005283A2 refers to obtaining bioactive secondary compounds of *Phoma herbarium*, isolated from the medicinal plant *Luehea divaricata*, with broad-spectrum antimicrobial activity. The secondary metabolites produced are effective compared to a commercial fungicide against the *Alternaria alternata* phytopathogenic fungi, *Guignardia citricarpa*, and *Fusarium solani*; it has action against the bacteria *Escherichia coli*, *Micrococcus luteus*, and *Salmonella typhi*, human pathogens, and phytopathogenic bacteria *Xanthomonas* asc. *phaseoli*. The invention assists in the control of plant pathogenic fungi such as cocoa and soybean, as well as assisting in the control of human and plant pathogenic bacteria. Moreover, this invention is an alternative to the use of agrochemicals, as it is a natural product, easy to handle, and easy to produce in the environmentally friendly environment (Clemente et al. 2012).

Recently, Chinese patent number CN109997877A provided a kind of *Phoma herbarum* water-dispersible granules and its preparation and application. The product with bioherbicidal action was applied with spraying treatment in *A. retroflexus* (formulation rate is 228–236 g/acre, it is spraying primary every 7 days, it sprays three times altogether). Inventors ensure environmental security, excellent storage stability, and long shelf life of product.

Some patents have been developed for other applications in the field of agriculture, such as the patent number CN104293681B, which discloses a strain of *Phoma* endophytic fungus and application to improve the resistance of rice and wheat to salt stress. The strain can significantly promote the growth of rice seedlings on a salted MS plate, can significantly slow down or delay salt injury symptoms of wheat in saline-alkali soil, and at the same time, greatly increase the accumulation of wheat biomass (Wang et al. 2017). In summary, fungal active compound production and its numerous alternate applications have pushed the bioindustrial sector for its possible commercial-scale production, but there are still many challenges for biotechnology to reach a gold standard in this regard.

Table 12.5 Summary of some *Phoma* patents

Title	Patent number	Application	References
A kind of *Phoma herbarum* water-dispersible granules and its preparation and application	CN109997877A	Bioherbicidal	Gu et al. (2019)
The huge *Phoma cava* bacterium P2 of one plant of biological and ecological methods to prevent plant disease, pests, and erosion and its application	CN109706082A	Antifungal, bioherbicidal	Li et al. (2018)
Phoma endophytic fungus and application thereof	CN104293681B	Applied to improvement of rice or wheat salt resistance	Wang et al. (2017)
Phoma ligulicola var. *inoxydabilis* Boerema fungus strain having mycoherbicidal activity against common wormwood	RU2588470C1	Mycoherbicidal	Leonidovna et al. (2016)
Obtaining secondary bioactive compounds from the endophyte *Phoma herbarium* with broad-spectrum antimicrobial activity	BR102012005283A2	Antifungal and antibacterial	Clemente et al. (2012)
Phoma strain and application thereof	CN102154125B	Antifungal and antitumor	Wang et al. (2011)
Small spore *Phoma microspora* for controlling *Conyza sumatrensis*	CN102382776B	Bioherbicidal	Liu and Tan (2011)
Novel compounds having antioxidant activity isolated from *Phoma herbarum*	KR101107260B1	Antioxidant	Hand (2010)
Strain of fungus *Phoma glomerata* as producer of polysaccharide possessing immunomodulating activity	RU2312148C2	Immunomodulating	Pivkin et al. (2006)
Phoma macrostoma bwc05-8-3-2 having weeding efficacy against sulfonylurea-resistant *Monochoria vaginalis* and mycelia suspension prepared using this	KR20070101745A	Bioherbicidal	Park et al. (2006)
Phoma microorganism	US20040265280A1	Antifungal	Yano et al. (2004)
Compounds having antifungal activity	US20020160946A1	Antifungal	Yano et al. (2002)
Fungal isolates and biological control compositions for the control of weeds	US7772155	Bioherbicidal	Bailey and Derby (2001)
Phoma glomerata ATCC MYA-2373 for biocontrol of fungal diseases in plants	US6544512B1	Antifungal	White (2000)
Phoma hna-54f strain and cleaning of water or soil contaminated with 2-hydroxy-3-naphthoic acid derivative	JPH1028578A	Purifying water or soil contaminated	Kawanaka and Kitahatan (1996)
Process for the production and purification of cytochalasin B from *Phoma exigua* var. *heteromorpha*	IT1246345B	Bioherbicidal	Bottalico et al. (1994)

12.9 Conclusions

The wide range of possibilities for applying fungal metabolites in various fields of biotechnology has attracted the interest of many researchers. The goal of this chapter was to provide a review of the main bioactive compounds isolated from different *Phoma* species until now. The main themes explored in this chapter were the roles of *Phoma* compounds in the biocontrol of weeds, microorganisms, and viruses and the *Phoma* species potential to produce anthraquinone pigments and polysaccharides. The findings presented indicate that different pathogenic and endophytic *Phoma* species are excellent producers of bioactive secondary metabolites, but they can also stand out in the production of macromolecules. Unfortunately, the majority of the examples within these categories is still relatively at the stage of research papers and patents, and there is only one example of commercial product on the market. Despite this scenario in relation to the untapped potential of compounds produced by *Phoma* species, it is expected that in the near future, sustainable development and the focus on natural products will overcome all the challenges of biotechnology.

References

Adrio JL, Demain AL (2003) Fungal biotechnology. Int Microbiol 6: 191–199

Alvi KA, Nair B, Pu H, Ursino R, Gallo C, Mocek U (1997) Phomacins: Three novel antitumor cytochalasan constituents produced by a *Phoma* sp. J Org Chem 62: 2148–2151

Arie T, Kobayashi Y, Gen O, Kono Y, Yamaguchi I (1998) Control of soilborne clubroot disease of cruciferous plants by epoxydon from *Phoma glomerata*. Plant Pathol 47: 743–748

Arie T, Kobayashi Y, Kono Y, Gen O, Yamaguchi I (1999) Control of clubroot of crucifers by *Phoma glomerata* and its product epoxydon. Pestic Sci 55:602–604

Aveskamp MM, De Gruyter J, Crous PW (2008) Biology and recent developments in the systematics of *Phoma*, a complex genus of major quarantine significance. Fungal Divers 31: 1–18.

Bailey KL, Derby J (2001) Fungal isolates and biological control compositions for the control of weeds. Patent US7772155

Bailey KL, Pitt WM, Falk S, Derby J (2011) The effects of *Phoma macrostoma* on nontarget plant and target weed species. Biol Control 58: 379–386

Bennett A, Ponder MM, Garcia-diaz J (2018) *Phoma* Infections: Classification, Potential Food Sources, and Their Clinical Impact. Microorganisms 6:2–13

Betina V (1992) Biological effects of the antibiotic brefeldin A (Decumbin, cyanein, ascotoxin, synergisidin): A retrospective. Folia Microbiol 37, 3–11.

Bhimba V, Pushpam AC, Arumugam P, Prakash S (2012) Phthalate derivatives from the marine fungi *Phoma herbarum* VB7. Int J Biol Pharma Res 3(4):507–512

Bick IRC, Rhee C (1966) Anthraquinone pigments from *Phoma foveata* Foister. Biochenm J 98: 112–116.

Binder M, Tarnrn C (1973) Proxiphomin und Protophomin, zwei neue Cytochalasane. Helv. Chim. Acta 56(7): 2387–2396

Bithell SL, Stewart A (2001) Evaluation of the pathogenicity of *Phoma exigua* var. *exigua* on Californian thistle. N. Z. Plant Prot 54: 179–183

Boerema GH, Bollen GJ (1975) Conidiogenesis and conidial septation as differentiating criteria between *Phoma* and *Ascochyta*. Persoonia 8: 111–144

Boerema GH, Höweler LH (1967) *Phoma exigua* Desm. and its varieties. Persoonia 5: 15–28

Boerema GH, de Gruyter J, Noordeloos ME, Hamers MEC (2004) *Phoma* Identification manual differentiation of species and infra-specific taxa in culture CABI Publishing, CAB International Wallingford, Oxfordshire, UK

Borges WDS, Pupo MT (2006) Novel anthraquinone derivatives produced by *Phoma sorghina*, an endophyte found in association with the medicinal plant *Tithonia diversifolia* (Asteraceae). J Braz Chem Soc 17, 929–934

Bottalico A, Capasso R, Evidente A, Vurro M (1994) Process for the production and purification of cytocalasine B from Phoma exigua var. heteromorpha. Patent IT1246345B

Boudart G (1989) Antibacterial activity of sirodesmin PL phytotoxin: Application the selection of phytotoxin-deficient mutants. Appl Environ Microbiol 55: 1555–1559

Boyette CD, Hoagland RE, Stetina KC (2015) Biological control of spreading dayflower (*Commelina diffusa*) with the fungal pathogen *Phoma commelinicola*. Agronomy 5, 519–536

Brun T, Rabuske JE, Todero I, Almeida TC, Junior JJD, Ariotti G, Confortin T, Arnemann JA, Kuhn RC, Guedes JVC, Mazutti MA (2016) Production of bioherbicide by *Phoma* sp. in a stirred-tank bioreactor. 3 Biotech 6(2): 230

Canuto KM, Rodrigues THS, de Oliveira FSA, Gonçalves FJT (2012) Endophytic fungi: perspective of discovery and application of bioactive compounds in agriculture. Embrapa Agroindústria Tropical, ISSN 2179-8184, 154

Capasso R, Evidente A, Vurro M (1991) Cytochalasins from *Phoma exigua* var. *heteromorpha*. Phytochemistry 30: 3945–3950

Chaves Neto JR, Mazutti MA, Zabot GL, Tres MV (2020) Bioherbicidal action of *Phoma dimorpha* fermented broth on seeds and plants of *Senna obtusifolia*. Pesqui Agropecu Trop 50: 1–8

Che Y, Gloer JB, Wicklow DT (2002) Phomadecalins A-D and phomapentenone A: New bioactive metabolites from *Phoma* sp. NRRL 25697, a fungal colonist of *Hypoxylon* stromata. J Nat Prod 65:399–402

Chen S, Yin D, Yao W, Wang Y, Zhang Y, Gao X (2009) Macrophage receptors of polysaccharide isolated from a marine filamentous fungus *Phoma herbarum*. Acta Pharmacol Sin 30:1008–1014

Chen S, Ding R, Zhou Y, Zhang X, Zhu R, Gao XD (2014) Immunomodulatory effects of polysaccharide from marine fungus *Phoma herbarum* YS4108 on T cells and dendritic cells. Mediators Inflamm 2014:738631

Chen Z, Chen H, Li Y, Feng T, Liu J (2015) Cytochalasins from cultures of endophytic fungus *Phoma multirostrata* EA-12. J Antibiot 68:23–26

Chiba S, Tsuyoshi N, Fudou R, Ojika M, Murakami Y, Ogoma Y, Oguchi M, Yamanaka S (2006) Magenta pigment produced by fungus J Gen Appl Microbiol 52(4): 201–207

Cimmino A, Andolfi A, Berestetskiy A, Evidente A (2008) Production of phytotoxins by *Phoma exigua* var. *exigua*, a potential mycoherbicide against perennial thistles. J Agric Food Chem 56: 6304–6309

Cimmino A, Andolfi A, Zonno MC, Avolio F, Santini A, Tuzi A, Berestetskyi A, Vurro M, Evidente A (2013) Chenopodolin: A phytotoxic unrearranged ent-pimaradiene diterpene produced by *Phoma chenopodicola*, a fungal pathogen for chenopodium album biocontrol. J Nat Prod 76: 1291–1297

Clemente E, Pamphile JA, Wenzel JB, Sarragiotto MH, Sékua RA (2012) Obtaining secondary bioactive compounds from the endophyte Phoma herbarium with broad spectrum antimicrobial activity. Patent BR102012005283A2

Cole L, Davies D, Hyde GJ, Ashford AE (2000) Brefeldine A affects growth, endoplasmic reticulum, Golgi bodies, tubular vacuole system, and secretory pathway in *Pisolithus tinctorius* Fungal Genet Biol 29(2), 95–106

Cui J, Guo S, Xiao P (2011) Antitumor and antimicrobial activities of endophytic fungi from medicinal parts of *Aquilaria sinensis*. J Zhejiang Univ Sci B 12(5): 385–392

Demain AL (2014) Importance of microbial natural products and the need to revitalize their discovery. J Ind Microbiol Biotechnol 41, 185–201

Devys M, Fbrbzou J, Topgi RS, Barbier M, Bousquet J, Kollmann A (1984) Structure and biosynthesis of phomenoic acid, an antifungal compound isolated from *Phoma lingam* Tode. J Chem Soc Perkin Trans I 2133–2137

Dufossé L, Fouillaud M, Caro Y, Mapari SAS, Sutthiwong N (2014) Filamentous fungi are large-scale producers of pigments and colorants for the food industry. Curr Opin Biotech 26: 56–61

Elliott CE, Gardiner DM, Thomas G, Cozijnsen A, Van De Wouw A, Howlett BJ (2007) Production of the toxin sirodesmin PL by *Leptosphaeria maculans* during infection of *Brassica napus*. Mol Plant Pathol 8,: 791–802

Elsebai MF, Ghabbour HA (2016) Isocoumarin derivatives from the marine-derived fungus *Phoma* sp. 135. Tetrahedron Lett 57: 354–356

Evidente A, Lanzetta R, Capasso R, Vurro M, Bottalico A (1992) Cytochalasin U and V, two new cytochalasans, from *Phoma exigua* var. *heteromorpha*. Tetrahedron 48(30): 6317–6324

Evidente A, Lanzetta R, Capasso R, Andol A, Vurro M, Zonno MC (1995) Putaminoxin, A phytotoxic nonenolide from *Phoma putaminum*. Phytochemistry 40: 1637–1641

Evidente A, Capasso R, Vurro M, Bottalico A (1996) Cytochalasin W, a new 24-Oxa[14]cytochalasan from *Phoma exigua* var. *heteromorpha*. Nat Toxins 4: 53–57

Evidente A, Capasso R, Andolfi A, Vurro M, Zonno MC (1998) Structure-activity relationship studies of putaminoxins and pinolidoxins: Phytotoxic nonenolides produced by phytopathogenic *Phoma* and *Ascochyta* species. Nat Toxins 6: 183–188

Evidente A, Andolfi A, Vurro M, Zonno MC, Motta A (2003) Cytochalasins Z4, Z5, and Z6, Three New 24-Oxa [14] cytochalasans produced by *Phoma exigua* var. *heteromorpha*. J Nat Prod 66: 1540–1544

Fang MJ, Fang H, Li WJ, Huang DM, Wu Z, Zhao YF (2012) A new diphenyl ether from *Phoma* sp strain, SHZK-2. Nat Prod Res 26(13): 1224–1228

Gallo MBC, Chagas FO, Almeida MO, Macedo CC, Cavalcanti BC, Barros FWA, De Moraes MO, Costa-Lotufo LV, Pessoa C, Bastos JK, Pupo MT (2009) Endophytic fungi found in association with *Smallanthus sonchifolius* (Asteraceae) as resourceful producers of cytotoxic bioactive natural products. J Basic Microbiol 49: 142–151

Graupner PR, Carr A, Clancy E, Gilbert J, Bailey KL, Derby J-A, Gerwick BC (2003) The Macrocidins: novel cyclic tetramic acids with herbicidal activity produced by *Phoma macrostoma*. J Nat Prod 66: 1558–1561

Graupner PR, Gerwick BC, Siddall TL, Carr AW, Clancy E, Gilbert JR, Bailey KL, Derby J (2006) Chlorosis inducing phytotoxic metabolites: New herbicides from *Phoma macrostoma*. Natural Product for Pest Management, 3: 37–47

Gu Z, Ji M, Zou Y, Wang Y, Qi Z, He L (2019) A kind of *Phoma herbarum* water dispersible granules and its preparation and application. Patent CN109997877A

Hand (2010) Novel compounds having antioxidant activity isolated from *Phoma herbarum*. Patent KR101107260B1

Hardoim PR, van Overbeek LS, Berg G, Pirttilä AM, Compant S, Campisano A, Döring M, Sessitsch A (2015) The hidden world within plants: Ecological and evolutionary considerations for defining functioning of microbial endophytes. Microbiol Mol Biol Rev 79: 293–320

Hazuda D, Blau CU, Felock P, Hastings J, Pramanik B, Wolfe A, Bushman F, Farnet C, Goetz M, Williams

M, Silverman K, Lingham R, Singh S (1998) Isolation and characterization of novel human immunodeficiency virus integrase inhibitors from fungal metabolites. Antivir Chem Chemother 10: 63–70.

Heiny DK, Templeton GE (1991) Effects of spore concentration, temperature, and dew period on disease of field bindweed caused by *Phoma proboscis*. Phytopathology 81: 905–909

Herath K, Harris G, Jayasuriya H, Zink D, Smith S, Vicente F, Bills G, Collado J, González A, Jiang B, Kahn JN, Galuska S, Giacobbe R, Abruzzo G, Hickey E, Liberator P, Xu D, Roemer T, Singh SB (2009) Isolation, structure and biological activity of phomafungin, a cyclic lipodepsipeptide from a widespread tropical *Phoma* sp. Bioorganic Med Chem 17: 1361–1369

Hobson DK, Edwards RL, Wales DS (1997) Cynodontin: A secondary metabolite and dyestuff intermediate. J Chem Tech Biotechnol 70: 343–348

Hoffman AM, Mayer SG, Strobel GA, Hess WM, Sovocool GW, Grange AH, Harper JK, Arif AM, Grant DM, Kelley-Swift EG (2008) Purification, identification and activity of phomodione, a furandione from an endophytic *Phoma* species 69: 1049–1056

Höller U, Wright AD, Matthée GF, Konig GM, Draeger S, Aust H-J, Schulz B (2000) Fungi from marine sponges: diversity, biological activity and secondary metabolites. Mycol Res 104: 1354–1365

Howlett BJ, Fox EM, Cozijnsen AJ, Wouw AP Van De, Elliott CE (2009) The role of the toxin sirodesmin in virulence of the blackleg fungus. Online available www.gcirc.org. [Acessed: Oct. 10, 2020

Huang S, Xu J, Li F, Zhou D, Xu L, Li C (2017) Identification and antifungal activity of metabolites from the mangrove fungus *Phoma* sp. L28. Chem Nat Compd 53: 237–240

Hubbard M, Taylor WG, Bailey KL, Hynes RK (2016) The dominant modes of action of macrocidins, bioherbicidal metabolites of *Phoma macrostoma*, differ between susceptible plant species. Environ Exp Bot 132: 80–91

Hussain H, Kock I, Al-Harrasi A, Al-Rawahi A, Abbas G, Green IR, Shah A, Badshah A, Saleem M, Draeger S, Schulz B, Krohn K (2014) Antimicrobial chemical constituents from endophytic fungus *Phoma* sp. Asian Pac J Trop Med 7: 699–702

Hussain H, John M, Al-Harrasi A, Shah A, Hassan Z, Abbas G, Rana UA, Green IR, Schulz B, Krohn K (2015) Phytochemical investigation and antimicrobial activity of an endophytic fungus *Phoma* sp. J King Saud Univ-Sci 27: 92–95

Hyde KD, Xu J, Rapior S, Jeewon R, Lumyong S, Niego AGT, Abeywickrama PD, Aluthmuhandiram JVS, Brahamanage RS, Brooks S, Chaiyasen A, Chethana KWT, Chomnunti P, Chepkirui C, Chuankid B, de Silva NI, Doilom M, Faulds C, Gentekaki E, Gopalan V, Kakumyan P, Harishchandra D, Hemachandran H, Hongsanan S, Karunarathna A, Karunarathna SC, Khan S, Kumla J, Jayawardena RS, Liu JK, Liu N, Luangharn T, Macabeo APG, Marasinghe DS, Meeks D, Mortimer PE, Mueller P, Nadir S, Nataraja KN, Nontachaiyapoom S, O'Brien M, Penkhrue W, Phukhamsakda C, Ramanan US, Rathnayaka AR, Sadaba RB, Sandargo B, Samarakoon BC, Tennakoon DS, Siva R, Sriprom W, Suryanarayanan TS, Sujarit K, Suwannarach N, Suwunwong T, Thongbai B, Thongklang N, Wei D, Wijesinghe SN, Winiski J, Yan J, Yasanthika E, Stadler M (2019) The amazing potential of fungi: 50 ways we can exploit fungi industrially. Fungal Divers 97: 1–136

Hynes R, (2018) *Phoma macrostoma*: as a broad spectrum bioherbicide for turf grass and agricultural applications. CAB Rev 13(5): 1–9

Jones MDM, Forn I, Gadelha C, Egan MJ, Bass D, Massana R, Richards TA (2011) Discovery of novel intermediate forms redefines the fungal tree of life. Nature 474: 200–203

Kalam S, Khan NA, Singh J (2014) A novel phytotoxic phenolic compound from *Phoma herbarum* FGCC#54 with herbicidal potential. J Chem Nat Compd 50:644–647

Kawanaka S, Kitahatan K (1996) *Phoma* hna-54f strain and cleaning of water or soil contaminated with 2-hydroxy-3-naphthoic acid deerivative. Patent JPH1028578A

Klaic R, Kuhn RC, Foletto EL, Dal Prá V, Jacques RJS, Guedes JVC, Treichel H, Mossi AJ, Oliveira D, Oliveira JV, Jahn SL, Mazutti MA (2015) An overview regarding bioherbicide and their production methods by fermentation. In: GuptaVK, Mach RL, Sreenivasaprasad, S, eds. *Fungal Biomolecules: Sources, Applications and Recent Developments*. John Wiley & Sons, Ltda,183–199

Klaic R, Sallet D, Foletto EL, Jacques RJS, Guedes JVC, Kuhn RC, Mazutti MA (2017) Optimization of solid-state fermentation for bioherbicide production by *Phoma* sp. Brazilian J Chem Eng 34: 377–384

Kong F, Wang Y, Liu P, Dong T, Zhu W (2014) Thiodiketopiperazines from the marine-derived fungus *Phoma* sp. OUCMDZ-1847. J Nat Prod 77: 132–137

Krohn K, Farooq U, Flörke U, Schulz B, Draeger S, Pescitelli G, Salvadori P, Antus S, Kurtán T (2007) Secondary metabolites isolated from an endophytic *Phoma* sp. – Absolute configuration of tetrahydropyrenophorol using the solid-state TDDFT CD Methodology. Eur J Org Chem 3206–3211

Kumaran RS, Choi Y, Lee S, Jeon HJ, Kim HJ (2012) Isolation of taxol, an anticancer drug produced by the endophytic fungus, *Phoma betae*. Afr J Biotechnol 11: 950–960

Kim, Eun La, Li JL, Dang HT, Hong J, Lee CO, Kim DK, Yoon WD, Kim Euikyung, Liu Y, Jung JH (2012) Cytotoxic cytochalasins from the endozoic fungus *Phoma* sp. of the giant jellyfish *Nemopilema nomurai*. Bioorganic Med Chem Lett 22: 3126–3129

Leonidovna GE, Borisovna KL, Borisovich HP, Alexandrovich KI (2016) *Phoma ligulicola* var. *inoxydabilis* Boerema fungus strain having

mycoherbicidal activity against common wormwood. Patent RU2588470C1
Li G, Cheng J, Luo T, Yang L, Wu M, Zhang J (2018) The huge *Phoma cava* bacterium P2 of one plant of biological and ecological methods to prevent plant disease, pests, and erosion and its application. Patent CN109706082A
Linnakoski R, Reshamwala D, Veteli P, Cortina-Escribano M, Vanhanen H, Marjomäki V (2018) Antiviral agents from fungi: Diversity, mechanisms and potential applications. Front Microbiol 9:2325
Liu J, Tan W (2011) Small spore *Phoma microspora* for controlling *Conyza sumatrensis* Patent CN102382776B
Liu S, Jiang J, Huang R, Wang Y, Jiang B, Zheng K, Wu S (2019) A new antiviral 14-nordrimane sesquiterpenoid from an endophytic fungus. Phytochem Lett 29: 75–78
Logan C, Khan AA (1969) Comparative studies of *Phoma* sp associated with potato gangrene in Northern Ireland. Trans Br Mycol Soc 52: 9–17
Luft L, Confortin TC, Todero I, Chaves Neto JR, Tonato D, Felimberti PZ, Zabot GL, Mazutti MA (2019) Different techniques for concentration of extracellular biopolymers with herbicidal activity produced by *Phoma* sp. Environ Technol. https://doi.org/10.1080/09593330.2019.1669720
Luft L, Confortin TC, Todero I, Zabot GL, Mazutti MA (2020) An overview of fungal biopolymers: bioemulsifiers and biosurfactants compounds production. Crit Rev Biotechnol. https://doi.org/10.1080/07388551.2020.1805405
Mohamed IE, Gross H, Pontius A, Kehraus S, Krick A, Kelter G, Maier A, Fiebig H, Konig GM (2009) Epoxyphomalin A and B, prenylated polyketides with potent cytotoxicity from the marine-derived fungus *Phoma* sp. Org Lett 11: 5014–5017
Nagai K, Kamigiri K, Matsumoto H, Kawano Y, Yamaoka M, Shimoi H, Watanabe M, Suzuki K (2002) YM-202204, a new antifungal antibiotic produced by marine fungus *Phoma* sp J Antibiot (Tokyo) 55(12): 1036–1041
Neal JC, Shew B, Schiavone R (2013) Temperature and Dose influence *Phoma macrostoma* efficacy on seedling broadleaf weeds. Proc. IXth International Bioherbicide 511 Group Workshop. Beijing, China, 29-33
Neumann S, Boland GJ (2002) Influence of host and pathogen variables on the efficacy of *Phoma herbarum*, a potential biological control agent of *Taraxacum officinale*. Can J Bot 80: 425–429
Oikawa H, Ohashi S, Ichihara A, Sakamura S (1999) Biosynthesis of diterpenoid aphidicolin: Isolation of intermediates from P-450 inhibitor treated mycelia of *Phoma betae*. Tetrahedron 55, 7541–7554
Oikawa H, Toshima H, Ohashi S, Ko WA (2001) Diversity of diterpene hydrocarbons in fungus *Phoma betae*. Tetrahedron Lett 42, 2329–2332
Ondeyka JG, Zink DL, Young K, Painter R, Kodali S, Galgoci A, Collado J, Tormo JR, Basilio A, Vicente F, Wang J, Singh SB (2006) Discovery of bacterial fatty acid synthase inhibitors from a *Phoma* species as antimicrobial agents using a new antisense-based strategy. J Nat Prod 69: 377–380
Osińska-Jaroszuk M, Jarosz-Wilkołazka A, Jaroszuk-Ściseł J, Szałapata K, Nowak A, Jaszek M, Ozimek E, Majewska M (2015) Extracellular polysaccharides from Ascomycota and Basidiomycota: production conditions, biochemical characteristics, and biological properties. World J Microbiol Biotechnol 31: 1823–1844
Osterhage CM, Schwibibble GM, Konig Wright AD (2000) Differences between marine and terrestrial *Phoma* species as determined by HPLC-DAD and HPLC-MS. Phytochem Anal 11:288–294
Osterhage C, König GM, Jones PG, Wright AD (2002) 5-hydroxyramulosin, a new natural product produced by *Phoma tropica*, a marine-derived fungus isolated from the alga *Fucus spiralis*. Planta Med 68, 1052–1054
Pandey S, Pandey AK (2000) Mycoherbicidal potential of some fungi against *Lantana camara* L: A preliminary observation. J Tropical Forestry 16: 28–32
Pandey AK, Luka RS, Hasija SK, Rajak RC (1991) Pathogenicity of some fungi to *Parthenium* and *obnoxious* weed in Madhya Pardesh. J Biol Control 5: 113–115
Park S, Song Bo B, Hong S, Hong Y (2006) Phoma macrostoma bwc05-8-3-2 having weeding efficacy against sulfonylurea resistance Monochoria vaginalis and mycelia suspension prepared using this. Patent KR20070101745A
Pedras MSC, Biesenthal CJ (1998) Production of the host-selective phytotoxin phomalide by isolates of *Leptosphaeria maculans* and its correlation with sirodesmin PL production. Can J Microbiol 44: 547–553.
Pedras MSC, Biesenthal CJ (2000) HPLC analyses of cultures of *Phoma* spp: Differentiation among groups and species through secondary metabolite profiles. Can J Microbiol 46: 685–691
Pedras MSC, Séguin-Swartz G (1992) The blackleg fungus: phytotoxins and phytoalexins. Can J Plant Pathol 14:67–75
Pedras MSC, Yu Y (2008) Phytochemistry structure and biological activity of maculansin A, a phytotoxin from the phytopathogenic fungus *Leptosphaeria maculans*. Phytochemistry 69: 2966–2971
Pedras MSC, Erosa-Lopez GC, Quail JW, Taylor JL (1999) Phomalairdenone: A new host-selective phytotoxin from a virulent type of the blackleg fungus *Phoma lingam*. Bioorg Med Chem Lett 9: 3291–3294
Pedras MSC, Chumala PB, Yu Y (2007) The phytopathogenic fungi *Leptosphaeria maculans* and *Leptosphaeria biglobosa*: chemotaxonomical characterization of isolates and metabolite production in different culture media. Can J Microbiol 53: 364–371
Peng X, He Y, Gao Y, Duan F, Chen J, Ruan H (2020) Cytochalasins from an endophytic fungus *Phoma multirostrata* XJ-2-1 with cell cycle arrest and TRAIL-resistance-overcoming activities. Bioorg Chem 104: 104317

Petrini O (1991) Fungal endophyte of tree leaves. In: Andrews JH, Hirana SS (eds.) Microbial Ecology of leaves. Springer Verlag, New York, USA, pp. 179-197

Pharamat T, Palaga T, Piapukiew J, Whalley AJS, Sihanonth P (2013) Antimicrobial and anticancer activities of endophytic fungi from *Mitragyna javanica*. Afr J Microbiol Res 7: 5565–5572

Pivkin MK, Belogortsevanov NI, Lukyanov PA (2006). Strain of fungus *Phoma glomerata* as producer of polysaccharide possessing immunomodulating activity. Patent RU2312148C2

Poiret B, Kollmann A, Bousquet J-F (1985) Activités antifongiques de la sirodesmine PL et de deux analogues naturels isolés de *Phoma lingam* (Tode) Desm. Action antagoniste du zinc. Agronomie, EDP Sciences, 5(6): 533–538

Rai MK, Rajak RC (1993) Distinguishing characteristics of some *Phoma* species. Mycotaxon 48:389–414.

Rai, M.K., Tiwari, V.V., Irinyi, L. et al. (2014) Advances in taxonomy of genus *Phoma*: polyphyletic nature and role ofphenotypic traits and molecular systematics. Indian J Microbiol 54:123–128

Rai M, Deshmukh P, Gade A, Ingle A, Kövics GJ, Irinyi L (2009) *Phoma* Saccardo: Distribution, secondary metabolite production and biotechnological applications. Crit Rev Microbiol 35: 182–196

Rai M, Gade A, Zimowska B, Ingle AP, Ingle P (2018) Marine-derived *Phoma*—the gold mine of bioactive compounds. Appl Microbiol Biotechnol 102:9053–9066

Rajak RC, Farkya S, Hasija SK, Pandey AK (1990) Fungi associated with congress weed (*Parthenium hysterophorus* L). Proc Nat Acad Sci India 60, 165–168

Rajamanikyam M, Vadlapudi V, Amanchy R, Upadhyayula SM (2017) Endophytic fungi as novel resources of natural therapeutics. Braz Arch Biol Technol 60: 1–26

Rao MPN, Xiao M, Li W-J (2017) Fungal and bacterial pigments: Secondary metabolites with wide applications. Front Microbiol 8:1113

Rivero-Cruz JF, García-Aguirre G, Cerda-García-Rojas CM, Mata R (2000) Conformational behavior and absolute stereostructure of two phytotoxic nonenolides from the fungus *Phoma herbarum*. Tetrahedron 56: 5337–5344

Rivero-Cruz JF, Macías M, Cerda-García-Rojas CM, Mata R (2003) A new phytotoxic nonenolide from *Phoma herbarum*. J Nat Prod 66: 511–514

Roustaee A, Gelie B, Savy C, Dargent R, Barrault G (2000) Ultrastructural studies of the mode of penetration by *Phoma macdonaldii* in sunflower seedlings. Phytopathology 90: 915–920.

Rouxel T, Chupeau Y, Fritz R, Kollmann A, Bousque J-F (1988) Biological effects of sirodesmin PL, a phytotoxin produced by *Leptosphaeria maculans*. Plant Sci 57: 45–53

Saccardo PA (1880) Conspectus generum fungorum Italiae inferiorum nempe ad Sphaeropsideas, Melanconieas et Hyphomyceteas pertinentium systemate sporologico dispositorum. Michelia 2:1–38

Sang X, Chen S, An X, Chen G, Wang H (2016) A novel 3,4-dihydronaphthalen-1(2H)-one with spirobutyrolactone and a new isocoumarin isolated from the endophytic fungus *Phoma* sp. J Asian Nat Prod Res 6020: 1–8

Sang X, Chen S, Tang M, Wang H, An X, Lu X, Zhao D, Wang Y, Bai J, Hua H, Chen G, Pei Y (2017) α-Pyrone derivatives with cytotoxic activities, from the endophytic fungus *Phoma* sp. YN02-P-3. Bioorg Med Chem Lett 27: 3723–3725

Santiago C, Fitchett C, Munro MHG, Jalil J, Santhanam J (2012) Cytotoxic and antifungal activities of 5-hydroxyramulosin, a compound produced by an endophytic fungus isolated from *Cinnamomum mollisimum*. Evidence-based Complement Altern Med https://doi.org/10.1155/2012/689310

Santiago C, Sun L, Munro MHG, Santhanam J (2014) Polyketide and benzopyran compounds of an endophytic fungus isolated from *Cinnamomum mollissimum*: biological activity and structure. Asian Pac J Trop Biomed 4: 627–632

Saxena S, Pandey AK (2001) Microbial metabolites as eco-friendly agrochemicals for the next millennium. Appl Microbiol Biotechnol 55: 395–403

Sciaky N, Presley J, Smith C, Zaal KJM, Cole N, Moreira JE, Terasaki M, Siggia E, Lippincott-schwartz J (1997) Golgi tubule traffic and the effects of brefeldin A visualized in living cells J Cell Biol 139: 1137–1155

Scott PM, Harwig J, Chen YK, Kennedy BP (1975) Cytochalasins A and B from strains of *Phoma exigua* var. *exigua* and formation of cytochalasin B in potato gangrene. J Gen Microbiol 87:177–180

Selbmann L, Onofri S, Fenice M, Federici F, Petruccioli M (2002) Production and structural characterization of the exopolysaccharide of the Antarctic fungus *Phoma herbarum* CCFEE 5080. Res Microbiol 153: 585–592

Shibazaki M, Taniguchi M, Yokoi T, Nagai K, Watanabe M, Suzuki K, Yamamoto T (2004) YM-215343, a novel antifungal compound from *Phoma* sp. QNO4621. J Antibiot 57(6): 379–382

Singh SB, Ball RG, Zink DL, Monaghan RL, Polishook JD, Sanchez M, Pelaez F, Silverman KC, Lingham RB (1997) Fusidienol-A: A novel Ras farnesyl-protein-transferase inhibitor from *Phoma* sp. J Org Chem 62(21): 7485–7488.

Singh SB, Zink DL, Goetz MA, Dombrowski AW, Polishook JD, Hazuda DJS (1998) Equisetin and a novel opposite stereochemical homolog phomasetin, two fungal metabolites as inhibitors of HIV-1 integrase. Tetrahedron Lett 39(16): 2243–2246.

Smith J, Wherley B, Reynolds C, White R, Senseman S, Falk S (2015) Weed control spectrum and turfgrass tolerance to bioherbicide *Phoma macrostoma*. Int J Pest Manag 0874: 1–8

Specian V, Orlandelli RC, Felber AC, Azevedo JL, Pamphile JA (2014) Secondary metabolites produced by endophytic fungi of pharmaceutical interest. Cient Cienc Biol Saude 16: 345–351

Stewart-Wade SM, Boland GJ (2005) Oil emulsions increase efficacy of *Phoma herbarum* to control dandelion but are phytotoxic. Biocontrol Sci Technol 15(7): 671–681

Steyn PS, Rabie CJ (1976) Characterization of magnesium and calcium tenuazonate from *Phoma sorghina*. Phytochemistry 15: 1977–1979

Strange RN (1997) Phytotoxin associated with Ascochyta species In: Toxin in Plant Disease Development and Evolving Biotechnology Upadhyay RK & Mukherjee KG (Eds), 167–181 Oxford and IBH Publishing Co Pvt Ltd, New Delhi, India

Sugawara F, Strobel G (1986) Zinniol, a phytotoxin produced by *Phoma macdonaldii*. Plant Sci 43: 19–23

Sullivan RF, White JF Jr (2000) *Phoma glomerata* as a mycoparasite of powdery mildew. Appl Environ Microbiol 66: 425–427

Sun K, Chen Y, Niu Q, Zhu W, Wang B, Li P, Ge X (2016) An exopolysaccharide isolated from a coral-associated fungus and its sulfated derivative activates macrophages. Int J Biol Macromol 82: 387–394

Sutton BC (1980) The coelomycetes: fungi imperfecti with pycnidia, acervuli and stromata. Commonwealth Mycological Institute, Kew, Surrey, England

Todero I, Confortin TC, Soares JF, Brun T, Luft L, Rabuske JE, Kuhn RC, Tres MV, Zabot GL, Mazutti MA (2019) Concentration of metabolites from *Phoma* sp. using microfiltration membrane for increasing bioherbicidal activity, Environ Technol 40(18): 2364-2372

Topgi RS, Devys M, Bousquet JF, Kollmann A, Barbier M (1987) Phomenoic acid and phomenolactone, antifungal substances from *Phoma lingam* (Tode) Desm: Kinetics of their biosynthesis with an optimization of the isolation procedures. Appl Environ Microbiol 53(5): 966–968

Valenzuela-Lopez N, Cano-Lira JF, Guarro J, Sutton DA, Wiederhold N, Crous PW, Stchigel AM (2018) Coelomycetous *Dothideomycetes* with emphasis on the families *Cucurbitariaceae* and *Didymellaceae*. Stud Mycol 90: 1–69

Venkatasubbaiah P, Van Dyke CG, Chilton WS (1992) Phytotoxic metabolites of *Phoma sorghina*, a new foliar pathogen of pokeweed. Mycologia 84(5): 715–723

Vikrant P, Verma KK, Rajak RC, Pandey AK (2006) Characterization of a phytotoxin from *Phoma herbarum* for management of *Parthenium hysterophorus* L. J Phytopathol 154: 461–468

Wang G, Zhang C, Lin F, Wang L, Feng X (2011) *Phoma* strain and application thereof. Patent CN102154125B

Wang L, Xu B, Wang J, Su Z, Lin F, Zhang C, Kubicek CP (2012) Bioactive metabolites from *Phoma* species, an endophytic fungus from the Chinese medicinal plant *Arisaema erubescens*. Appl Microbiol Biotechnol 93: 1231–1239

Wang H, Chen Y, Lin F (2017) *Phoma* endophytic fungus and application thereof. Patent CN104293681B

White JF (2000) *Phoma glomerata* ATCC MYA-2373 for biocontrol of fungal diseases in plants. Patent US6544512B1

Wolfe JC, Neal JC, Harlow CD (2016) Selective broadleafweed control in Turfgrass with the bioherbicides *Phoma macrostoma* and Thaxtomin A. Weed Technol 30(3):688–700

Wu X, Chen Z, Ding W, Liu Y, Ma Z (2018a) Chemical constituents of the fermentative extracts of marine fungi Phoma sp. CZD-F11 and Aspergillus sp. CZD-F18 from Zhoushan, Archipelago, China. Nat Prod Res 32(13): 1562-1566

Wu H, Yang F, Li L, Rao YK, Ju T, Wong W, Hsieh C, Pivkin MV, Hua K, Wu SH (2018b) Ergosterol peroxide from marine fungus Phoma sp. induces ROS-dependent apoptosis and autophagy in human lung adenocarcinoma cells. Sci Rep 8: 17956.

Yang XB, Gao XD, Han F, Xu BS, Song YC, Tan RX (2005) Purification, characterization and enzymatic degradation of YCP, a polysaccharide from marine filamentous fungus *Phoma herbarum* YS4108. Biochimie 87: 747–754

Yano T, Inukai M, Takatsu T, Tanaka I (2002) Compounds having antifungal activity. Patent US20020160946A1

Yano T, Inukai M, Takatsu T, Tanaka I (2004) *Phoma* microorganism. Patent US20040265280A1

Zhang W, Krohn K, Egold H, Draeger S, Schulz B (2008) Diversity of antimicrobial pyrenophorol derivatives from an endophytic fungus, *Phoma* sp. Eur J Org Chem 4320–4328

Zhang X, Ding R, Zhou Y, Zhu R, Liu W, Jin L, Yao W, Gao X (2013) Toll-like receptor 2 and Toll-like receptor 4-dependent activation of B cells by a polysaccharide from marine fungus *Phoma herbarum* YS4108. PLoS One 8(3):e60781

Zhao S, Shamoun SF (2006) Effects of culture media, temperature, pH, and light on growth, sporulation, germination, and bioherbicidal efficacy of *Phoma exigua*, a potential biological control agent for salal (*Gaultheria shallon*). Biocontrol Sci Technol 16(10) 1043-1055

Zhou B, Qiang S (2008) Environmental, genetic and cellular toxicity of tenuazonic acid isolated from *Alternaria alternata*. Afr J Biotechnol 7(8): 1151–1156

Zhou L, Bailey KL, Derby J (2004) Plant colonization and environmental fate of the biocontrol fungus *Phoma macrostoma*. Biol Control 30: 634–644

Bioherbicidal Activity of *Phoma macrostoma* 13

Michelle Hubbard, Russell K. Hynes, and Karen L. Bailey

13.1 Introduction

Phoma macrostoma Montagne stands out from other *Phoma* species because of its well-researched bioherbicidal properties which result in the reduction of a range of economically important broadleaf weeds, while leaving desirable, monocotyledonous plant species unharmed. While other *Phoma* species also have bioherbicidal potential, the specific phytotoxins (compounds that are toxic to plants) and multiple modes of action (ways of harming targeted weeds) of *P. macrostoma* are unique (Hynes 2018). The potential of *P. macrostoma* for weed control was first recognized in the late 1990s by scientist Dr. Karen Bailey and research assistant Jo-Anne Derby of Agriculture and Agri-Food Canada (AAFC), Saskatoon, Saskatchewan, Canada. *P. macrostoma* strain 94-44B was originally isolated from Canada thistle (*Cirsium arvense* (L.) Scop.) plants showing mild bleaching or yellow foliar symptoms, while growing in agricultural fields in Saskatchewan. Other strains were isolated from *C. arvense* in the Canadian provinces of Nova Scotia, New Brunswick, Ontario, Quebec, and Alberta. In time, AAFC researchers found that *P. macrostoma* could induce bleaching in other broadleaf weed species such as dandelion (*Taraxacum officinale* Weber ex F.H. Wigg.), common chickweed (*Stellaria media* (L.) Vill.), and clover (*Trifolium repens* (L.) (Bailey and Derby 2010, 2012; Graupner et al. 2003). This discovery sparked excitement and marked the beginning of over two decades of work on *P. macrostoma* as a bioherbicide. This chapter will review the scientific studies that supported its registration as a bioherbicide and the efforts to complete its commercialization.

13.2 Importance of Bioherbicides and the Hurdles Encountered in Their Development

Weeds can lead to significant yield losses in agricultural crops (Oerke 2006) as well as unwanted visual features in lawns and turf (Zimdahl 2013). Numerous options exist for weed management, including synthetic herbicides; physical weeding; tillage; cultural practices, such as planting competitive, or fast-growing, non-weed species or increasing planting density of non-weeds; allelopathy: cover crops; and bioherbicides (Gage and Schwartz-Lazaro 2019; Hahn et al. 2020; Zimdahl 2013). Bioherbicides are defined as liv-

M. Hubbard (✉)
Agriculture and Agri-Food Canada (AAFC) Swift Current Research and Development Centre, Swift Current, Canada
e-mail: michelle.hubbard@agr.gc.ca

R. K. Hynes · K. L. Bailey
AAFC Saskatoon Research and Development Centre, Saskatoon, Canada

M. Rai et al. (eds.), *Phoma: Diversity, Taxonomy, Bioactivities, and Nanotechnology*,
https://doi.org/10.1007/978-3-030-81218-8_13

ing organisms, such as viruses, bacteria, or fungi, and phytotoxic metabolites produced by these organisms that can be used to control weeds (Thakore 2006). The need and demand for bioherbicides may increase as difficulties emerge with the use of synthetic herbicides such as resistance evolution in weed populations, higher synthetic herbicide development costs, and lack of new herbicide modes of action being discovered (Duke 2012; Heap 1997; Perotti et al. 2020; Umetsu and Shirai 2020; Westwood et al. 2018). The deregistration of older herbicides (Bailey 2010), potential banning of others such as glyphosate (Anonymous 2019; Peng et al. 2020), environmental concerns associated with herbicide residues, and increasing legislative and public pressure (Cordeau et al. 2016) may also contribute to bioherbicide demand. Furthermore, bioherbicides may have a role in weed control in urban areas where synthetic herbicides are not permitted (Bailey 2010) or in organic agriculture (Halde et al. 2017).

The stages in bioherbicide development include discovery, proof of concept, establishing the safety and environmental interactions of the bioherbicide, selectivity (range of weeds and crops impacted), understanding of mode(s) of action, development of production, and formulation and application processes that would facilitate integration with existing lawn, garden, horticultural, or agricultural practices, followed by registration, commercialization, and adoption/use in the marketplace (Bailey 2014; Bailey et al. 2009a; Boyetchko et al. 2009). Hurdles that impact bioherbicide success under the current business model, based on the bioherbicide development and commercialization system, can be grouped in five categories: science, industry, regulation, consumers, and markets (Bailey 2019). The fact that there are far more potential bioherbicides which have progressed through the development process to varying degrees than there are commercially available options clearly illustrates how difficult it is to overcome the barriers to bioherbicide success. Bioherbicides are underrepresented in the biopesticide market, especially considering that synthetic herbicides occupy a larger market share than synthetic insecticides or fungicides (Cordeau et al. 2016; Marrone 2019). Indeed, Bailey (2019) listed only three bioherbicides that were both registered and commercially available. Similarly, Morin (2020) identified 15 and 2 registered and commercially available bioherbicides, respectively.

Learning from obstacles faced by bioherbicides in the past and present may increase the likelihood for success for current and future products. Bailey (2019) makes recommendations for improving the probability of success, the first of which is focusing on potential bioherbicides with strong activity, relying on both biological and chemical modes of action. Indeed, Cordeau et al. (2016) noted that many bioherbicides performed poorly or inconsistently in the field and attributed lack of commercial success to these traits. Secondly, Bailey (2019) suggested focusing on workable production systems that could potentially be scaled up early in the process. Thirdly, she questioned whether all bioherbicides are well suited to a business model strictly dependent on high-profit margins or whether alternative models of distribution and economics, such as on-farm production, local distribution, and/or production systems as described by Morin (2020), might be more likely to succeed.

13.3 Modes of Action

Biopesticides tend to utilize more complex, or more numerous, modes of action than synthetic pesticides (Hubbard et al. 2014). *P. macrostoma* acts on weeds via two major mechanisms: (1) colonization of the host plant by the living fungus (Bailey et al. 2011b) and (2) the action of phytotoxic metabolites, called macrocidins (Graupner et al. 2003; Hubbard et al. 2015, 2016).

13.3.1 Plant Colonization

P. macrostoma can reproduce asexually by means of mycelial growth, fragmentation, and regrowth as well as through the production of conidia. Sexual reproduction has not been observed in *P. macrostoma*. Conidia facilitate longer-distance, rain-splash dispersal of the fungus, whereas mycelia are more stationary, remaining in the soil and plant tissues. Typically, conidia are considered to be the most infective propagules for bioherbicides (Boyette et al. 1991), and they are usually applied to plant leaves by foliar application. However, preliminary experimentation with *P. macrostoma* suggested otherwise. Bailey et al. (2011b) conducted studies to determine which of the fungal propagules (conidia or mycelium) were the best bioherbicide infective units when applied to plants through the soil or on the leaves. It was confirmed that conidia were not capable of inducing photobleaching or biomass reductions in any hosts. The most effective propagules were mycelial fragments contained within a granule that was applied to soil.

Microscopic studies of root colonization showed mycelium of *P. macrostoma* growing out of granular inoculum, through the soil, to the plant roots (Bailey et al. 2011b). Penetration occurred at sites close to root hairs without the formation of appressoria or penetration pegs. Hyphae grew between cells, as opposed to within them, until reaching the vascular system (Bailey et al. 2011b). Both susceptible and resistant plant species were colonized by *P. macrostoma*. However, the fungus penetrated the vascular tissue of susceptible hosts but was restricted to the cortical cell layer in resistant hosts (Bailey et al. 2011b). In susceptible hosts, the mycelium proliferated within the trachea releasing macrocidins, which are phloem mobile, to induce photobleaching of new leaf growth.

P. macrostoma appears to possess alternate survival strategies. It can, in some circumstances, be considered an endophyte, rather than a plant pathogen, in that it colonizes the tissues of plants in which it does not cause disease symptoms or harm. The fungus can be isolated from both susceptible and resistant or tolerant hosts (Bailey et al. 2011a, b). Similarly Evans et al. (2012) surveyed counties in the UK for the presence of white-tip disease on Canada thistle and found that *P. macrostoma* was the dominant fungus isolated with up to 100% presence in all shoots sampled, including three samples from healthy control plants. The three isolates from healthy control samples were evaluated for their disease-causing traits in greenhouse and lab studies. Two of the isolates (CAR 14B and CAR 23B) induced chlorosis and mortality in dandelion seedling assays. DNA analysis by polymerase chain reaction (PCR) showed the presence of a band amplified by primers specific to bioherbicidal *P. macrostoma* isolates described in Zhou et al. (2004) and known as Pm853L and Pm853R (Evans et al. 2012; Pitt et al. 2012). The third isolate did not induce chlorosis or cause mortality, and PCR analysis failed to produce the band amplified using primers Pm853L and Pm853R. Evans et al. (2012) suggested that *P. macrostoma* is a ubiquitous endophyte in a wide range of hosts and only becomes apparent when it sporulates on dead tissues. At some point, an endophytic strain in *C. arvense* evolved the ability to produce phytotoxins that bleached and killed the host to allow it to profusely sporulate on the dead tissues in order to propagate and disperse to new living hosts.

In recent decades, an interest in endophytes' potential to improve host plant performance or stress tolerance has emerged (Marquez et al. 2007; Papik et al. 2020). Dovana et al. (2015) isolated a fungus with 99% internal transcribed spacer (ITS) sequence similarity to *P. macrostoma* from water mint (*Mentha aquatica* L.) with the ability to promote growth in *Arabidopsis*. However, any potential benefits of *P. macrostoma* to resistant plants have yet to be demonstrated. Unpublished work has examined whether *P. macrostoma* applications to agricultural crops can result in increased biomass and/or yield in tolerant crops, such as cereals, corn (*Zea mays* L.), potatoes (*Solanum tuberosum* L.), and flax (*Linum usitatissimum* L.). However, no significant differences were found between tolerant, asymptomatic plants treated with *P. macrostoma* and untreated controls. Based on current evidence, the endophytic life phase of *P. macrostoma* does not appear to benefit the infected plants and only provides the fungus with a host venue by which to continue its survival.

13.3.2 Macrocidins

Differential colonization of hosts is only one mode of action used by *P. macrostoma*. Even after autoclaving or filtration has inactivated or removed the living fungus, bioherbicidal activity remains (Hubbard et al. 2014). This is because the phytotoxic macrocidins produced by *P. macrostoma* survive autoclaving and pass through filtration. Thus, macrocidins clearly play a critical role in the modes of action of *P. macrostoma*.

Hubbard et al. (2015) demonstrated that macrocidins partially inhibit phytoene desaturase, an enzyme involved in an early step in carotenoid biosynthesis and targeted by photobleaching herbicides (Boger 1996). Carotenoids are a class of pigments required by plants in order to carry out photosynthesis. These pigments contribute to the functioning of the light-harvesting complex and facilitate photoprotection from reactive oxygen species (reviewed by Frank and Cogdell (1996)). Thus, disruptions to their biosynthesis will negatively impact photosynthesis. However, macrocidins lead to less complete inhibition of phytoene desaturase than did the synthetic inhibitor of this enzyme, diflufenican (Boger 1996). In addition, macrocidins altered the ratio of lutein to β-carotene, two other carotenoids downstream of phytoene desaturase in the biosynthetic pathway. In order to follow up on these findings, Hubbard et al. (2016) explored the impact of macrocidins on the carotenoid profiles in dandelion, chickpea (*Cicer arietinum* L.), and groundsel (*Senecio vulgaris* L.) over time. These plant species were chosen because they developed both severe photobleaching and high mortality (in dandelion), severe photobleaching but low mortality (in chickpea), or low photobleaching and high mortality (in groundsel), when treated with macrocidins (Bailey et al. 2011a). Hubbard et al. (2016) found that macrocidins not only inhibit phytoene desaturase but also bind cations of magnesium and iron in vitro and likely impair electron transport from Q_A to Q_B and uncouple light-harvesting complex. These latter two impacts may be due to iron-binding activity and could explain the changes in the lutein to β-carotene ratio by inducing elevation of xanthophylls (including lutein) due to the photoprotective properties of xanthophylls (reviewed by Domonkos et al. (2013), Jahns and Holzwarth (2012), and Mozzo et al. (2008)). The magnesium-binding action of macrocidins may reduce the ability of the enzyme 1-deoxy-D-xylose to generate carotenoid precursors and/or phytoene synthase to catalyze the production of phytoene, leading to overall lower levels of carotenoids in macrocidin-treated plants (Cunningham and Gantt 1998; Murkin et al. 2014; Schofield and Paliyath 2005). Different modes of action are more important in different susceptible plant species with differing symptomatic responses to macrocidins implying that it would be harder for plants to develop resistance to macrocidins than to synthetic herbicides. In addition, if genetic resistance were to evolve, its transfer between plant species may be less probable.

13.4 Phylogenetics and Differentiation of Bioherbicidal and Non-bioherbicidal Isolates

Not only is *P. macrostoma* distinct functionally, in terms of its bioherbicidal properties, from other *Phoma* species, it is phylogenetically distinguishable from *P. macrostoma* isolates without biocontrol potential (Pitt et al. 2012; Zhou et al. 2005). The genetic distinctions between biocontrol and non-biocontrol isolates of *P. macrostoma* facilitated the development of primer pairs specific to *P. macrostoma* isolates with bioherbicidal properties and were used to track how long the bioherbicide persists in the environment (Zhou et al. 2004). Bioherbicidal *P. macrostoma* isolates originate only from Canada thistle (Pitt et al. 2012). Phylogenetic analysis based on DNA sequences of the ITS placed bioherbicidal and non-bioherbicidal isolates into different groups. These two groups were confirmed with amplified fragment length polymorphism (AFLP) dendogram construction (Pitt et al. 2012). Interestingly, within the monophyletic AFLP cluster of bioherbicidal isolates, there are two groups which differ at a single locus (Pitt et al. 2012).

P. macrostoma biocontrol isolates from different regions, or ecozones, within Canada did not cluster together phylogenetically (Zhou et al. 2005). Furthermore, Zhou et al. (2005) found no evidence of sexual recombination within *P. macrostoma* bioherbicidal isolates or between these isolates and *P. lingam* Tode ex Fr., the teleomorph of *Leptosphaeria maculans* (Desmaz.) Ces. and De Not., an important pathogen of canola (*B. rapa* L., *B. napus* L., or *B. juncea*(L.) Czern, a major oilseed crop in Western Canada (Zhang and Fernando 2018). The findings of Zhou et al. (2005) imply that *P. macrostoma* could be applied in Canada, the country where it was isolated, with reduced risk of unintended influences on other native populations of *P. macrostoma* lacking bioherbicide activity or other related *Phoma* species. The lack of sexual recombination between *P. macrostoma* and other *Phoma* species also bodes well for the safety of using this bioherbicide in countries outside of Canada with lower risk of the development or evolution of novel strains of phytopathogenic fungi.

Phoma species other than *P. macrostoma* have documented bioherbicidal activity. Over the past 20 years, 27 publications have reported on herbicide activity by as many *Phoma* spp. on many different weed species (reviewed in Hynes (2018)). In laboratory, greenhouse, and field studies, application of *Phoma* spp. conidia and/or mycelium to target weed species elicited herbicidal reactions including necrosis, chlorosis, and maceration on plant leaves and stems of a wide range of dicotyledonous weeds including water hyacinth (*Pontederia crassipes* Solms) (Abdel Rahim and Tawfig 1984; Hernández et al. 2007), eastern bracken fern (*Pteridium aquilinum* (L.) Kuhn var. *latiusculum* (Desf.)Underw.) (McElwee et al. 1990), field bindweed (*Convolvulus arvensis* L.) (Heiny 1994), purple loosestrife (*Lythrum salicaria* L.) (Nyvall and Hu 1997), dandelion (Neumann and Boland 2002; Stewart-Wade and Boland 2004) , thistle (Guske et al. 2004; Kluth et al. 2005), yellow star thistle (*Centaurea solstitialis* L.) (Vidal et al. 2004), salal (*Gaultheria shallon* Pursh.) (Zhao and Shamoun 2005, 2006), wild radish (*Raphanus raphanistrum* L.) (Djébali et al. 2009), sow thistle (*Sonchus arvensis* L.) (Cimmino et al. 2008), and horse purslane (*Trianthema portulacastrum* Linn. (Bishkhapra) (Ray and Vijayachandran 2013). For a complete list, see Hynes (2018).

A much smaller subset of researchers reported on phytotoxins responsible for the herbicidal activity observed on target plants. These included cucumber (*Cucumis sativus* L.) and sorghum (*Sorghum bicolor* (L.) Moench) (Brun et al. 2016), Santa Maria feverfew (*Parthenium hysterophorus* L.) (Vikrant et al. 2006), and sow thistle and lambsquarters (*Chenopodium album* L.) (Cimmino et al. 2008, 2013a, b). Herbicidal symptoms have been reported on Santa Maria feverfew from spent media of *Phoma* sp. FGCC#18 (Quereshi et al. 2010), on horse purslane from *P. herbarum* Westend. FGCC#54 (Quereshi et al. 2011), and on field bindweed from *P. crystallifera* phytotoxins (Razaghi and Zafari 2017). However, no additional information was presented on the mode(s) of action. Based on the symptoms they induce, the phytotoxins produced by other *Phoma* species likely differ in their modes of action from macrocidins. The symptoms reported in response to these phytotoxins include necrosis without photobleaching (Cimmino et al. 2008, 2013b) or inhibition of germination (Brun et al. 2016). The *Phoma* species tested by Brun et al. (2016) further differed from *P. macrostoma* in that it was active against both monocots and dicots, as opposed to dicots only.

13.5 Interactions with the Environment

13.5.1 Longevity and Residual Activity

It is important to understand the longevity of *P. macrostoma* in the environment. This information will inform application strategies and utility as a bioherbicide. Bengtsson et al. (2006) demonstrated that *P. macrostoma* can overwinter in leaf litter in the field. This, combined with the fact that *P. macrostoma* is naturally occurring in many parts of the world (Zhou et al. 2005), suggests that it could be long-lived. However, industrial *Rhizobium* inoculant products for pulse crops must be seed or soil applied each year to outcompete an indigenous, but ineffective at nodule formation, rhizobial population for root-binding sites in order to establish nitrogen fixation. Similarly, *P. macrostoma*, as a lab-grown microorganism, did not survive at high levels between growing seasons, potentially in part due to competition from already-established indigenous microbial populations. Zhou et al. (2004) demonstrated that the abundance of *P. macrostoma* DNA in inoculated soil was higher at 2 months after treatment than at either 4- or 1-month postinoculation (Zhou et al. 2004). Consistently, Wolfe et al. (2016) concluded that *P. macrostoma* had a relatively short residual life span in the soil, with biocontrol activity peaking 4 weeks after application. By 12 months after treatment, no *P. macrostoma* was detectable. In addition, more *P. macrostoma* DNA was found in soil than in plant roots (Zhou et al. 2004). The plant species influenced the level of *P. macrostoma* detected with nonmolecular microbiological techniques in rhizosphere soil or growth media.

There was no residual activity from an application of *P. macrostoma* to turf (Zhou et al. 2004). When field pea (*Pisum sativum* L.) and lentil (*Lens culinaris* Medik.) were seeded into the soil treated with *P. macrostoma* the previous year, the crops showed no signs of photobleaching or decreased yield. However, in an agricultural field setting, the impacts of *P. macrostoma* were detected the growing season following the one in which the bioherbicide was applied (Hynes et al. 2018). These impacts consisted of initial, early season, mild photobleaching symptoms on highly susceptible crops (canola, pea, and lentil) from which the crops recovered. No initial chlorosis was noted in less susceptible crops, such as flax. No losses in percent emergence, row cover, or fresh or dry biomass were recorded for canola, pea, lentil, or flax planted into barley (*Hordeum vulgare* L.), corn, canola or wheat (*Triticum aestivum* L.), or pea stubble, respectively, which had been exposed to 0, 1, or 2× dose of *P. macrostoma*. The differences between the residual activity documented by Zhou et al. (2004) and Hynes et al. (2018) point to a potential need for additional study of factors such as year-to-year variations and formulations. Bailey et al. (2009b) found that granule composition, size, and rate of disintegration influenced bioherbicide efficacy.

Moisture levels were also found to affect residual activity. In Swift Current, Saskatchewan (about 275 km south west of Saskatoon, a typically dry region of Saskatchewan), the dandelion percent coverage, but not percent coverage of all weeds, was different in July 2018 between plots that had received 0 or 1× dose of *P. macrostoma* in 2016 (Fig. 13.1). The year of application (2016) was wet, but the subsequent years (2017 and 2018) were dry. These findings suggest that, under dry conditions, prior applications of the bioherbicide may lead to weed control benefits, at least of highly susceptible weed species, for multiple years after *P. macrostoma* application. The sustained weed reductions may be due to residual impacts to the seed bank and/or harm to the perennial root stocks from the application in 2016. The performance of a susceptible crop (pea) planted in 2017 on these same plots was not negatively impacted by the 2016 dose of *P. macrostoma*, suggesting that the bioherbicide was inactive in 2017 and 2018, likely due to dry conditions.

13.5.2 Mobility, Moisture, Soil Type, and Fertility

Understanding the movement of *P. macrostoma* from the site of application has implications for bioherbicide efficacy against target plants at the application site and for off-site impacts on non-target plants. Zhou et al. (2004) found that the DNA of fungal mycelium and spores of *P. macrostoma* had low horizontal or vertical mobility in soil. However, Bailey et al. (2010) found that a major phytotoxic metabolite (macrocidin A) produced by *P. macrostoma* was leached from soils saturated above field capacity. The majority of macrocidin A, and accompanying bioherbicidal activity, was leached out in a volume of

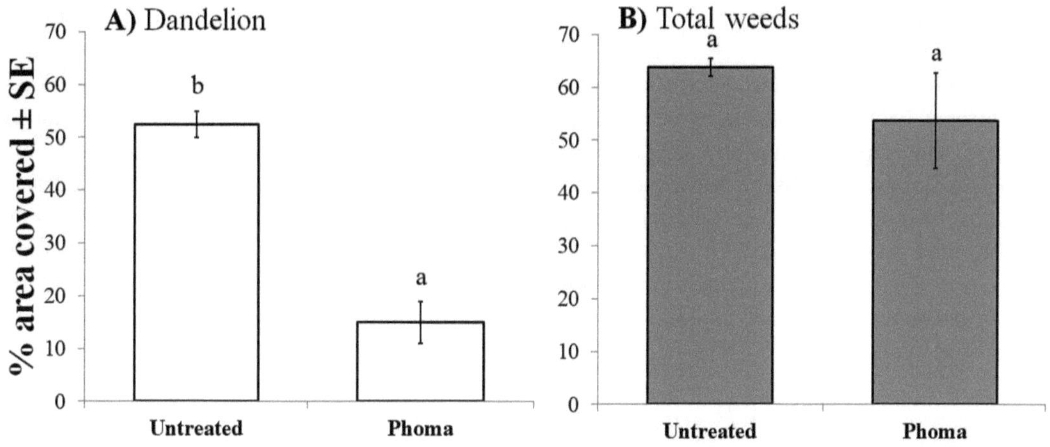

Fig. 13.1 The residual impact of *P. macrostoma* in 2018, 2 years after a bioherbicide application in 2016 in Swift Current, SK, in plots in which neither crops nor weeds were seeded, on the percent ground cover by dandelions (**a**) and all weeds (**b**). Within a panel, bars with the same letter are not significantly different (alpha = 0.05). Data analysis was done in SAS, using a Kruskal Wallis test and t-test in panels **a** and **b**, respectively

water equivalent to 128 mm of rainfall, which is likely to match what growing target weeds would take up during active growth. If this level of precipitation fell during the growing season at a reasonably gradual rate, no off-target impacts of *P. macrostoma* would be expected. However, if a larger rainfall event was to occur over a short period of time, it is possible that macrocidin A could be carried away from the target site and potentially impact locations to which the runoff moves. Highlighting this risk, extreme daily precipitation as high as 97 mm has been recorded near Saskatoon. As anthropogenic climate change progresses, the probability of short, intense rainfall events may increase, elevating the risk of runoff from *P. macrostoma.*

Well-drained (sandy) soils leached macrocidin A more readily than soils with a higher clay content (Bailey et al. 2010). In addition, soil organic matter content may bind macrocidins, reducing their mobility by some other mechanism (Bailey et al. 2010). Consistently, soil properties, such as texture, organic matter content, pH, and salinity impacted adsorption and desorption of pesticides and wastewater in soil (Aseperi et al. 2020; Ou et al. 2020; Pena et al. 2020). In addition, wetting and drying cycles of soil have been shown to impact soil aggregation and mobility of pesticides through the soil (Hochman et al. 2021). Thus, if *P. macrostoma* were applied in higher-rainfall regions, runoff is less likely on finer-textured soils, soils with higher organic matter, or increased aggregate stability. The consideration of soil properties would be particularly important when crop plants that are susceptible to *P. macrostoma* are grown in the vicinity or downslope from areas to which *P. macrostoma* is applied.

In order for *P. macrostoma* to be useful as a bioherbicide, it must retain its activity in the presence of other crop inputs, such as fertilizers. When applied to turfgrass growing in nitrogen-deficient soils, an application of nitrogen fertilizers with *P. macrostoma* reduced the number of dandelions, more than either treatment alone and more than when applied to soils not deficient in nitrogen (Bailey et al. 2013). Consistently, Neal et al. (2013) documented increased weed control under higher nitrogen fertility conditions. Other fertilizer components, such as urea, phosphate, and potassium, also led to greater reductions in the number of dandelions when applied to areas treated with *P. macrostoma* (Bailey et al. 2013). Weed control was not improved with the addition of lime, but it did not prevent bioherbicidal activity. These results confirm that fertilizer applications with *P. macrostoma* would not be detrimental to weed control efficacy.

13.5.3 Temperature

Temperature has an influence on the efficacy of *P. macrostoma*. Bioherbicidal activity peaks at temperatures that support plant and fungal growth (Dr. Karen Bailey, unpublished data). Specifically, *P. macrostoma* does not induce symptoms in susceptible host plants below about 10 °C or over 30 °C (Fig. 13.2). At 38 °C, there is no fungal growth and no expression of symptoms in dandelion seedlings. This temperature range is similar to that required by dandelion for optimal germination and growth (Stewart-Wade et al. 2002). Logically, the temperatures for plant and fungal growth must overlap in order for the bioherbicide to be effective. From the fungal side, moderate temperatures are required for mycelial growth, host colonization, and macrocidin production. Plant growth is needed for macrocidins to be taken up in the vascular system (Graupner et al.

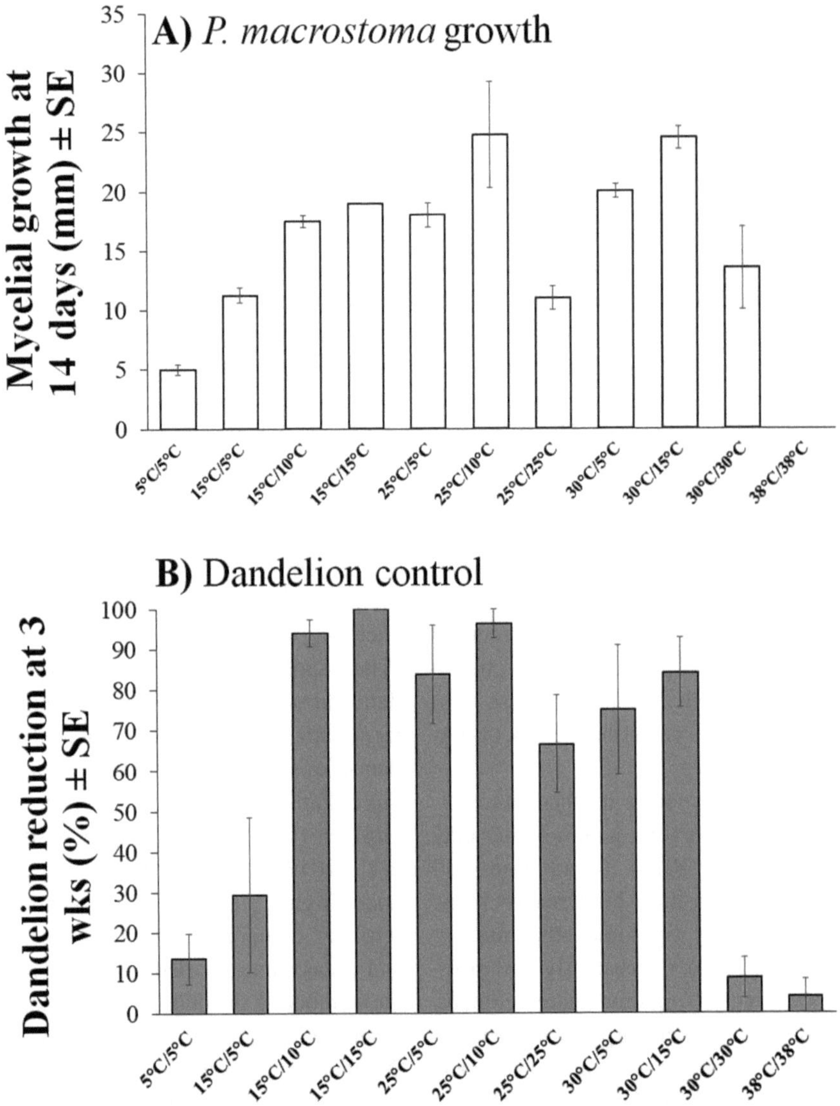

Fig. 13.2 Impact of temperature (day/night) on mycelial growth at 14 days (mm diameter) of *P. macrostoma* in vitro (**a**) and percent reduction in the number of living dandelion seedlings in a pot at 3 weeks after application of *P. macrostoma* relative to the total number of dandelion seedlings in a pot that was not treated with the fungus (**b**)

2003) and transported to leaves, where they can inhibit carotenoid biosynthesis and photosynthesis (Hubbard et al. 2015, 2016). Because *P. macrostoma* is no longer active at temperatures above 38 °C, this organism was incapable of internally infecting warm-blooded organisms such as humans as confirmed by toxicology testing.

Initial, pot-based, research conducted in North Carolina in the southern USA suggested that higher temperatures actually made *P. macrostoma* more effective against at least some weeds, such as common groundsel and henbit (*Lamium amplexicaule* L.) (Neal et al. 2013). However, subsequent field experiments in North Carolina did not support the above initial findings (Neal 2014; Wolfe 2015; Wolfe et al. 2016). Field trials found that not all broadleaf weeds, including henbit and common chickweed, were adequately controlled by *P. macrostoma* where field temperatures averaged 21.7 and 21.5 °C in the fall of 2012 and 2013, respectively, the 2 years of the study. In contrast, in September 2012 and 2013, in Saskatoon, Saskatchewan, the average mean daily temperatures were 13.0 and 15.2 °C (Environment Canada historical data). Additional research on the impacts of temperature on target and nontarget plant reactions to *P. macrostoma* are merited, both to facilitate the potential expansion of the bioherbicide to regions other than the Canadian prairies and to improve our ability to predict how climate change may alter *P. macrostoma* performance.

13.6 Host Range

It is important that *P. macrostoma* as a bioherbicide be effective on the targeted plants (weeds) but has no or minimal impact on nontarget plant species (crops, desirable or endangered plants) in the vicinity. Tolerance of plants to *P. macrostoma* may vary based on factors such as plant developmental stage, dose of bioherbicide applied, and environmental conditions. Over several years, more than 94 plant species in 34 botanical families were studied for signs of susceptibility to the bioherbicide. Six families (Graminae, Pinaceae, Linaceae, Salicaceae, Iridaceae, Cucurbitaceae) were resistant. Within most families, some species were resistant and others susceptible. The three families with the largest number of susceptible species were Asteraceae, Brassicaceae, and Fabaceae/Leguminosae. The regulators, the Canadian Pest Management Regulatory Agency (PMRA), concluded that a precautionary statement was needed on the label to avoid contact with desirable, plants, fruits, and vegetables (PMRA 2011).

13.6.1 Efficacy on Weeds

Numerous broadleaf weeds were susceptible or partially susceptible to preemergent and postemergent applications of *P. macrostoma* under controlled conditions (Bailey et al. 2011a). All monocot, grassy weeds were resistant to *P. macrostoma*, regardless of the dose or formulation applied (Bailey et al. 2011a).

While *P. macrostoma* effectively controlled many broadleaf weeds under greenhouse conditions, it was less effective against some weeds under field conditions, such as chickweed (Wolfe et al. 2016), common purslane (*Portulaca oleracea* L.), and common mallow (*Malva neglecta* Wallr.) (Smith et al. 2015). Host range testing under ideal controlled conditions provides an indication of species tolerance or susceptibility, but for commercial applications, the performance (efficacy and consistency of response over multiple years and sites) must be confirmed for each weed on the approved label when submitted for registration. For example, field studies on turfgrass confirmed greater than 80% dandelion reduction compared to untreated controls in 14 out of 21 trials at a single preemergent rate (Watson and Bailey 2013).

Generally the standard preemergent rate (1× dose) was highly effective (>80%) on dandelion, wild mustard (*Sinapis arvensis* L.), English daisy (*Bellis perennis* L.), and Dutch clover (*Trifolium repens* L.). Weeds requiring higher doses (2×–16×) were pigweed (*Amaranthus hybridus* L.), Canada thistle, cat's ear dandelion (*Hypochaeris radicata* L.), common chickweed and mouse-ear chickweed (*Cerastium vulgatum* L.), henbit, black medic (*Medicago lupulina* L.), and buckhorn and broadleaf plantain (*Plantago major* L. and *Plantago lanceolata* L.).

Because host range testing occurred over multiple years and sites, the range of doses of bioherbicide that weeds were exposed to varied. Some weeds were exposed to only one dose of *P. macrostoma* (Bailey et al. 2011a). The following species were partially resistant to a 1× dose of the bioherbicide, with low to no photobleaching but over 50% mortality: common ragweed (*Ambrosia artemisiifolia* L.), common groundsel, spotted spurge (*Chamaesyce maculata* (L.) Small), and common purslane. Wandering cudweed (*Gnaphalium pensylvanicum* Willd.) was challenged only by a 4× dose of the bioherbicide and found to be fully susceptible to this dose.

Some weeds were treated only with an inoculum mat bioassay, which applied an unquantified dose of *P. macrostoma*. In these cases, *P. macrostoma* was grown on a potato dextrose agar plate for 14 days, and the content of the plate (agar, mycelium, and spores) was spread over a pot seeded with the weed seeds. This assay showed that false cleavers (*Galium spurium* L.) and perennial sow thistle responded to *P. macrostoma* with high photobleaching but low mortality, whereas wild buckwheat (*Polygonum convolvulus* L.) and scentless chamomile (*Matricaria perforata* L.) were fully susceptible to inoculum mat treatment. Later research demonstrated that different modes of action dominated in different susceptible plant species (Hubbard et al. 2016), potentially explaining why Bailey et al. (Bailey et al. 2011a) found some plants had high photobleaching but low mortality and other species showed the reverse.

13.6.2 Tolerance in Non-weeds

The host range testing, both historically and in this chapter, concentrated on applications of *P. macrostoma* in turfgrass and agricultural plants but also included horticultural crops. Both warm and cool season grass species tolerated *P. macrostoma*, regardless of dose, application method, timing of application, moisture levels, or temperatures (Bailey et al. 2011a; Smith et al. 2015; Wolfe et al. 2016). Similarly, cereal crops, including oats (*Avena sativa* L.), barley), canary seed (*Phalaris canariensis* L.), and wheat (*Triticum aestivum*), did not develop any symptoms when treated with *P. macrostoma*. Millet (*Panicum miliaceum* L.) developed very mild photobleaching symptoms and had low levels of mortality (Bailey et al. 2011a). When tested in the field, with hand watering, but without mechanized irrigation, or with supplemental, mechanical irrigation, corn proved to be fully resistant to *P. macrostoma* (Hynes et al. 2018).

Most agricultural oilseed crops were susceptible to *P. macrostoma*. These included brassica crops such as mustard (*Brassica juncea*) and canola (*B. rapa* and *B. napus*) (Bailey et al. 2011a; Hynes 2018; Hynes et al. 2018). Sunflower (*Helianthus annuus* L.) needed a higher dose (2×) than safflower (*Carthamus tinctorius* L.) to be fully susceptible (Bailey et al. 2011a). Flax displayed photobleaching and reduced crop cover when inoculated with *P. macrostoma*, both in pot-based (Bailey et al. 2011a) and field (Bailey and Hynes 2015; Hynes et al. 2018) trials. Flax biomass was consistently reduced under dryland conditions. Under irrigation, flax biomass was diminished only in some experiments (Bailey and Hynes 2015; Hynes et al. 2018).

All pulse or oilseed legume crops tested, including lentil, field pea, faba bean (*Vicia faba* L.), and soybean (*Glycine max* L.), were susceptible to *P. macrostoma*, with the exception of chickpea, which was partially resistant (Bailey et al. 2011a; Hynes 2018; Hynes et al. 2018). Chickpea had low mortality but high rates of bleaching symptoms (Bailey et al. 2011a).

Other nitrogen-fixing forage and cover crops, such as bird's-foot trefoil (*Lotus corniculatus* L.), alfalfa (*Medicago sativum* L.), white or yellow sweet clover (*Melilotus alba* Medik. and *Melilotus officinalis* (L.) Lam.), and red clover, were susceptible. However, in field trials, alfalfa proved to be susceptible to *P. macrostoma* in the year of establishment (first year), which was consistent with Bailey et al. (2011a). However, alfalfa was largely tolerant of the bioherbicide in subsequent years (Bailey and Hynes 2015).

In the field, potato germination, biomass, or yield was not impacted by *P. macrostoma* application. However, under dryland conditions (no mechanical irrigation), potatoes did develop photobleaching symptoms at 3 weeks to roughly 2 months after bioherbicide application. As the crop developed, the symptoms became less severe. When irrigation was applied, photobleaching was milder than under dryland conditions. Irrigation and dryland production occurred at separate sites but within the same field season (Hynes et al. 2018).

Testing of edible and ornamental horticultural plants for tolerance to *P. macrostoma*, Bailey et al. (2011a) found that lettuce (*Lactuca sativa* L.) and the majority of Brassicaceae vegetables were susceptible to *P. macrostoma*, although collard greens (*Brassica oleracea* L.) and broccoli (*Brassica oleracea* L.) had mortality rates under 50% when treated with lower rates of the bioherbicide. Despite field pea being susceptible to *P. macrostoma*, garden (*Pisum sativum* L.) and sweet pea (*Lathyrus odoratus* L.) were resistant to lower doses of *P. macrostoma.* Pumpkin (*Cucurbita* L. spp.), bell pepper (*Capsicum annuum* L.), and tomato (*Solanum lycopersicum* L.) were generally resistant, although tomato did display photobleaching in response to higher doses. The edible herbs, sweet basil (*Ocimum basilicum* L.) and oregano (*Origanum vulgare* L.), proved to be resistant and partially susceptible, respectively (Bailey et al. 2011a). Berries and fruits tested for tolerance to *P. macrostoma* were watermelon (*Citrullus lanatus* (Thunb.) Matsum. and Nakai) and strawberry (*Fragaria x ananassa Duchesne*). The former was resistance and latter was susceptible but only to higher doses (Bailey et al. 2011a). Testing of other woody berry and fruit crops may uncover additional crops in which *P. macrostoma* could be safely used for weed control. Numerous ornamental and flowering plants—including pink periwinkle (*Catharanthus roseus* (L.) G. Don), garden petunia (*Petunia* x *hybrida* (Hook.) Vilm.), crimson sage (*Salvia coccinea* Buchoz ex Etl.), and rhododendron (*Rhododendron* L. P.J.M)—were largely resistant to *P. macrostoma*, although a few were susceptible to higher doses. See Bailey et al. (2011a) for a complete list. Additional testing of horticultural plants would be needed to determine the lowest effective rate of application under field conditions before determining the specific situations where the bioherbicide could be applied.

The few tree species that were tested, black spruce (*Picea mariana* (Mill.) Britt., Ster. and Poggen.) and pine (*Pinus* L. spp.), were resistant to an 8× dose of *P. macrostoma* (Bailey et al. 2011a). Thus, it seems likely that the bioherbicide could be safely used for weed control in forestry/silviculture.

13.7 *Phoma macrostoma* Bioherbicide Application

In order to obtain regulatory approval as a bioherbicide in Canada, evidence is required to establish the lowest effective rate of application using a standardized method of application for defined patterns of use such as the specific target weeds, safety of the nontarget crops in the area of application, timing of application, and number of applications (PMRA 2001, 2011). The market potential of *P. macrostoma* as a bioherbicide was very strong as it could provide broad-spectrum control of many broadleaf weeds in turfgrass and agriculture. Agriculture and Agri-Food Canada partnered with The Scotts Company to submit a registration that focused on the use of *P. macrostoma* in turfgrass situations. Using a granular formulation, there were 129 efficacy trials and 69 phytotoxicity trials conducted across Canada and the USA over 7 years (Bailey and Falk 2011). Efficacy data were generated on 14 weed species and 10 grass species using either a single preemergent application or up to three postemergent applications spaced 1 month apart during the growing season. A minimum of four rates of application (0×, 1/2×, 1×, and 2×) were evaluated. The lowest effective rate listed on the label was 1.6 kg per 100 m^2 for a preemergent application and 3.2 kg per 100 m^2 for a postemergent application. Others also confirmed that lower doses, on a per area basis, of *P. macrostoma* were needed when the bioherbicide was applied before weed emergence (Neal et al. 2013; Wolfe 2015).

It was observed that *P. macrostoma* did not reduce germination but resulted in seedlings emerging with photobleaching symptoms and died very quickly. This finding corroborates Bailey et al. (2010) and makes sense since macrocidins, produced by *P. macrostoma*, are water-soluble, phloem-mobile phytotoxins and act on actively growing plants. The smaller the plant, the faster the action and the lower the plant's reserves in the root to support continued growth. Postemergent applications were intended to target perennial weeds with established root systems and thus required larger doses and multiple doses to eventually cause lethal damage.

13.8 Commercialization of *Phoma macrostoma* as a Bioherbicide: Status, Barriers, and Potential

P. macrostoma has many features that make it an attractive candidate for development as a bioherbicide. It has a broad market potential, broad-spectrum weed control, high efficacy in pot and field trials, both biological and chemical modes of action, long shelf life, and low risks to the environment and health (Bailey and Falk 2011). It received full registration in 2016 from PMRA for broadleaf weed control on turfgrass. Despite partnerships with industry, the bioherbicide has yet to reach its commercial potential.

The major barrier to its commercialization has not been *P. macrostoma*'s performance but rather the economics of production. Typically, industry uses submerged liquid fermentation to produce microbial biomass, usually spores, at a low cost (Bailey et al. 2009a). Unfortunately, this method has not worked for *P. macrostoma* because mycelial biomass is required to elicit macrocidin production and fermenters with rotary impellers that result in shearing are not conducive to mycelial growth. Solid-state fermentation, using a grain-based substrate to produce granules containing living mycelial fragments and macrocidins, was used for all research required for bioherbicide registration (Bailey and Falk 2011; PMRA 2011). Although the preemergent rate of application was economical for broadleaf weed control in turf, the postemergent rate for turf and agricultural applications was not economical (Bailey 2019).

To remove this barrier to commercialization, a more efficient, reliable, and economical production system needs to be developed. There are ways to achieve this. One is to investigate other fermentation systems such as those using airlift technology instead of rotary impellers or using a two-stage production process that involves submerged fermentation of mycelium followed by drying the biomass on a solid matrix (Bailey et al. 2009a). Research on how to modulate macrocidin production using genetics or epigenetics could take advantage of the triggers *P. macrostoma* requires to initiate and continue macrocidin production, leading to higher concentrations of the bioherbicidal ingredients. Exploring alternative business and partnership models that seek more moderate returns on investment and provide alternatives or supplementation to synthetic herbicides may also contribute to the eventual release of *P. macrostoma* as a bioherbicide (Bailey 2019).

13.9 Conclusions

Over two decades of work on *P. macrostoma* has resulted in extensive knowledge of this organism's biology; phylogenetics; impacts on weeds, crops, turf, and other plants; how it is impacted by the environment; bioherbicidal modes of action; and application. Nonetheless, many questions remain unanswered. Despite its strong potential as a bioherbicide, including its high level of activity; multiple, complex modes of action; favorable interactions with biotic and abiotic components of the environment; backing from the Canadian federal government; and interest from multiple for-profit companies, *P. macrostoma* is still not available commercially. However, as weed pressures and resistance to synthetic herbicides continue to mount and ever-increasing knowledge of bioherbicides in general and *P. macrostoma* in particular, the possibility that *P. macrostoma* will eventually be successfully launched remains.

References

Abdel Rahim AM, Tawfig S (1984) Pathogenicity of fungi and bacteria from the Sudan to water hyacinth. Weed Res 24:233-238.

Anonymous (2019) Austria takes the lead to ban glyphosate, bringing human health and the environment to the forefront. Brussels, Belgium. Pesticide Action Network, Europe. December 3, 2019. https://www.pan-europe.info/press-releases/2019/12/austria-takes-lead-ban-glyphosate-bringing-human-health-and-environment.

Aseperi AK, Busquets R, Hooda PS, Cheung PCW, Barker J (2020) Behaviour of neonicotinoids in contrasting soils. J Environ Manage 276:111329.

Bailey KL (2010) Canadian innovations in microbial biopesticides. Can J Plant Pathol 32:113-121.

Bailey KL (2014) The bioherbicide approach to weed control using plant pathogens. In: Integrated Pest Management: Current Concepts and Ecological Perspective. Ed: Abrol DP. Chapter 13 245-266.

Bailey KL (2019) Navigating the bioherbicide success pathway. In: XV International Symposium on Biological Control of Weeds. Engelberg, Switzerland. Ed: Hintz HI, Bon M-C, Bourdot G, Cristofaro M, Desurmont G, Kurose D, Müller-Schärer H, Rafter M, Schaffner U, Seier M, Storza RFH, Smith L, Stutz S, Thomas S, Weyl P and Winston R. 87-90.

Bailey KL, Derby J, inventors (2010) Her Majesty the Queen in right of Canada, as represented by the Minister of Agriculture and Agri-Food Canada (Saskatoon, CA). Continuation in Part., assignee. Fungal isolates and biological control compositions for the control of weeds. United States Patent # 7,772,155. August 10, 2010.

Bailey KL, Derby J, inventors (2012) Her Majesty the Queen in right of Canada, as represented by the Minister of Agriculture and Agri-Food Canada (Saskatoon, CA), assignee. Fungal isolates and biological control compositions for the control of weeds. United States Patent # 8,211,830. July 3, 2012.

Bailey KL, Falk S (2011) Turning research on microbial bioherbicides into commercial products - a *Phoma* story. Pest Technology 5:73-79.

Bailey KL, Hynes RK (2015) Final report: Broadleaved weed control in alfalfa using the bioherbicide *Phoma macrostoma*. Agriculture & Agri-Food Canada. Western Grains Research Foundation. Agriculture & Agri-Food Canada, 107 Science Place, Saskatoon, SK, Canada, S7N 0X2.

Bailey KL, Boyetchko S, Peng G, Hynes R, Taylor W, Pitt WM (2009a) Developing weed control technologies with fungi. In: Advances in fungal biotechnology. Ed: Rai M. New Dehli - 110016 India. 8189866532 1 1 1-44.

Bailey KL, Boyetchko S, Peng G, Hynes R, Wolf TM (2009b) Delivering bioherbicides with improved efficacy. In: Microbial products: exploiting microbial diversity for sustainable plant production. Ed: Zydenbos SM and Jackson TA. New Zealand Plant Protection Society. 39-49.

Bailey KL, Pitt WM, Derby J, Walter S, Taylor W, Falk S (2010) Efficacy of *Phoma macrostoma,* a bioherbicide, for control of dandelion (*Taraxacum officinale*) following simulated rainfall conditions. Am J Plant Sci Biotech 4:35-42.

Bailey KL, Pitt W, Falk S, Derby J (2011a) The effects of *Phoma macrostoma* on nontarget plant and target weed species. Biol Control 58:379-386.

Bailey KL, Pitt W, Leggett F, Sheedy C, Derby J (2011b) Determining the infection process of *Phoma macrostoma* that leads to bioherbicidal activity on broadleaved weeds. Biol Control 59:268-276.

Bailey KL, Falk S, Derby J-A, Melzer M, Boland GJ (2013) The effect of fertilizers on the efficacy of the bioherbicide, *Phoma macrostoma*, to control dandelions in turfgrass. Biol Control 65:147-151.

Bengtsson M, Green H, Lerouli N, Pedersen HL, Hockenhull J (2006) Effect of autumn application of urea on saprotrophic fungi in off-season leaf litter of sour cherry and evaluation of fungal isolates to reduce primary inoculum of *Blumeriella jaapii*. J Plant Dis Prot 113:107-112.

Boger P (1996) Mode of action of herbicides affecting carotenogenesis. J Pesticide Sci Commentaries. Agrochemistry and Biology Series. 21:473-478.

Boyetchko SM, Bailey KL, De Clerck-Floate RA (2009) Current biological weed control agents—their adoption and future prospects. Prairie Soils Crops 2:38-45.

Boyette CD, Qiumby Jr. PC, Connick Jr. WJ, Daigle DJ, Fulgham FE (1991) Progress in the production, formulation, and application of mycoherbicides. In: Microbial Control of Weeds. Ed: D.O. T. Boston, MA, USA. 209-222.

Brun T, Rabuske JE, Todero I, Almeida TC, Junior JJ, Ariotti G, Confortin T, Arnemann JA, Kuhn RC, Guedes JV, Mazutti MA (2016) Production of bioherbicide by Phoma sp. in a stirred-tank bioreactor. 3 Biotech 6:230.

Cimmino A, Andolfi A, Berestetskiy A, Evident A (2008) Production of phytotoxins by *Phoma exigua* var. *exigua*, a potential mycoherbicide against perennial thistles. J Agric Food Chem 56:6304-6309.

Cimmino A, Andolfi A, Zonno MC, Avolio F, Berestetskiy A, Vurro M, Evidente A (2013a) Chenopodolans A-C: phytotoxic furopyrans produced by *Phoma chenopodiicola*, a fungal pathogen of *Chenopodium album*. Phytochemistry 96:208-213.

Cimmino A, Andolfi A, Zonno MC, Avolio F, Santini A, Tuzi A, Berestetskyi A, Vurro M, Evidente A (2013b) Chenopodolin: a phytotoxic unrearranged *ent*-pimaradiene diterpene produced by *Phoma chenopodicola*, a fungal pathogen for *Chenopodium album* biocontrol. J Nat Prod 76:1291-1297.

Cordeau S, Triolet M, Wayman S, Steinberg C, Guillemin J-P (2016) Bioherbicides: Dead in the water? A review of the existing products for integrated weed management. Crop Protection 87:44-49.

Cunningham FX, Gantt E (1998) Genes and enzymes of carotenoid biosynthesis in plants. Annu Rev Plant Physiol Plant Mol Biol 49:557-583.

Djébali N, Scott J, Jourdan M, SouissiI T (2009) Fungi pathogenic on wild radish (*Raphanus raphanistrum* L.) in northern Tunisia as potential biocontrol agents. Phytopathol Mediterr 48:205-213.

Domonkos I, Kis M, Gombos Z, Ughy B (2013) Carotenoids, versatile components of oxygenic photosynthesis. Prog Lip Res 52:539-61.

Dovana F, Mucciarelli M, Mascarello M, Fusconi A (2015) *In vitro* morphogenesis of Arabidopsis to search for novel endophytic fungi modulating plant growth. PLoS One 10:e0143353.

Duke SO (2012) Why have no new herbicide modes of action appeared in recent years? Pest Manag Sci 68:505-12.

Evans H, Seier M, Derby J, Falk S, Bailey K (2012) Tracing the origins of white tip disease of *Cirsium arvense* and its causal agent, *Phoma macrostoma.* Weed Res 53:42-52.

Frank HA, Cogdell RJ (1996) Carotenoids in photosynthesis. Photochem Photobiol 63:257-264.

Gage KL, Schwartz-Lazaro LM (2019) Shifting the paradigm: An ecological systems approach to weed management. Agriculture 9:179-195.

Graupner PR, Carr A, Clancy E, Gilbert J, Bailey KL, Derby J-A, Gerwick BC (2003) The macrocidins: Novel cyclic tetramic acids with herbicidal activity produced by *Phoma macrostoma.* J Nat Prod 66:1558-1561.

Guske S, Schulz B, Boyle C (2004) Biocontrol options for *Cirsium arvense* with indigenous fungal pathogens. Weed Res 44:107-116.

Hahn D, Sallenave R, Pornaro C, Leinauer B (2020) Managing cool-season turfgrass without herbicides: Optimizing maintenance practices to control weeds. Crop Sci 60:2204-2220.

Halde C, Gagné S, Charles A, Lawley Y (2017) Organic no-till systems in eastern Canada: A review. Agriculture 7:36-42.

Heap IM (1997) The occurrence of herbicide-resistant weeds worldwide. Pestic Sci 51:235-243.

Heiny DK (1994) Field survival of *Phoma proboscis* and synergism with herbicides for control of field bindweed. Plant Disease 78:1156-1164.

Hernández MC, Pildain MB, Novas MV, Sacco J, Lopez SE (2007) Mycobiota associated with larval mines of *Thrypticus truncatus* and *T. sagittatus* (Diptera, Dolichopodidae) on water hyacinth, *Eichhornia crassipes*, in Argentina. Biol Control 41:321-326.

Hochman D, Dor M, Mishael Y (2021) Diverse effects of wetting and drying cycles on soil aggregation: Implications on pesticide leaching. Chemosphere 263:127910.

Hubbard M, Hynes RK, Erlandson M, Bailey KL (2014) The biochemistry behind biopesticide efficacy. Sustain Chem Proc 2:18.

Hubbard M, Hynes R, Bailey K (2015) Impact of macrocidins, produced by *Phoma macrostoma*, on carotenoid profiles of plants. Biol Control 89:11-22.

Hubbard M, Taylor WG, Bailey KL, Hynes RK (2016) The dominant modes of action of macrocidins, bioherbicidal metabolites of *Phoma macrostoma*, differ between susceptible plant species. Environ Exp Bot 132:80-91.

Hynes RK (2018) *Phoma macrostoma*: as a broad spectrum bioherbicide for turfgrass and agricultural applications. CAB Reviews 13:9.

Hynes RK, Sheedy C, Wahab J, Bailey KL (2018) *Phoma macrostoma*, bioherbicide for broadleaf weed management in agricultural field and horticultural crops. Agriculture & Agri-Food Canada. Western Grains Research Foundation.

Jahns P, Holzwarth AR (2012) The role of the xanthophyll cycle and of lutein in photoprotection of photosystem II. Biochim Biophys Acta 1817:182-193.

Kluth S, Kruess A, Tscharntke T (2005) Effects of two pathogens on the performance of *Cirsium arvense* in a successional fallow. Weed Res 45:261–269.

Marquez LM, Redman RS, Rodriguez RJ, Roossinck MJ (2007) A virus in a fungus in a plant: three-way symbiosis required for thermal tolerance. Science 5811:513-515.

Marrone PG (2019) Pesticidal natural products - status and future potential. Pest Manage Sci 75:2325-2340.

McElwee M, Irvine JJM, Burge MN (1990) A mycoherbicidal approach to bracken control. FRI Bulletin – New Zealand Ministry of Forestry, Forest research Institute 155:74-79.

Morin L (2020) Progress in biological control of weeds with plant pathogens. Annu Rev Phytopathol 58:201-223.

Mozzo M, Dall'Osto L, Hienerwadel R, Bassi R, Croce R (2008) Photoprotection in the antenna complexes of photosystem II: role of individual xanthophylls in chlorophyll triplet quenching. J Biol Chem 283:6184-6192.

Murkin AS, Manning KA, Kholodar SA (2014) Mechanism and inhibition of 1-deoxy-D-xylulose-5-phosphate reductoisomerase. Bioorg Chem 57:171-185.

Neal JC (2014) Evaluations of biologically based alternative for broadleaf weed control in turf and ornamentals http://ir4.rutgers.edu/Biopesticides/bioFinalReport/2014-18A.pdf. 2014-18A.

Neal JC, Shew B, Schiavone R (2013) Temperature and dose influence *Phoma macrostoma* efficacy on seedling broadleaf weeds. In: The XIth International Bioherbicide Group Workshop. Nanjing, China. Ed: Qiang S and Kuang J. 19-23.

Neumann S, Boland GJ (2002) Influence of host and pathogen variables on the efficacy of *Phoma herbarum*, a potential biological control agent of *Taraxacum officinale*. Can J Bot 80:425-429.

Nyvall RF, Hu A (1997) Laboratory evaluation of indigenous North American fungi for biological control of Purple Loosestrife. Biol Control 8:37-42.

Oerke EC (2006) Crop losses to pests. J Agric Sci 144:31-43.
Ou J, Li H, Ou X, Yang Z, Chen M, Liu K, Teng Y, Xing B (2020) Degradation, adsorption and leaching of phenazine-1-carboxamide in agricultural soils. Ecotoxicol Environ Saf 205:111374.
Papik J, Folkmanova M, Polivkova-Majorova M, Suman J, Uhlik O (2020) The invisible life inside plants: Deciphering the riddles of endophytic bacterial diversity. Biotechnol Adv 44:107614.
Pena A, Delgado-Moreno L, Rodriguez-Liebana JA (2020) A review of the impact of wastewater on the fate of pesticides in soils: Effect of some soil and solution properties. Sci Total Environ 718:134468.
Peng W, Lam SS, Sonne C (2020) Support Austria's glyphosate ban. Science Letters 367:257-258.
Perotti VE, Larran AS, Palmieri VE, Martinatto AK, Permingeat HR (2020) Herbicide resistant weeds: A call to integrate conventional agricultural practices, molecular biology knowledge and new technologies. Plant Sci 290:110255.
Pitt WM, Bailey KL, Fu Y-B, Peterson GW (2012) Biological and genetic characterisation of *Phoma macrostoma* isolates with bioherbicidal activity. Biocontrol Sci Technol 22:813-835.
PMRA (2001) Regulatory Directive: Guidelines for the Registration of Microbial Pest Control Agents and Products. Health Canada, Pest Management Regulatory Agency. DIR2001-02.
PMRA (2011) Evaluation report ERC2011-09, *Phoma macrostoma* strain 94-44B. Health Canada, Pest Management Regulatory Agency. Ottawa, Canada. Catalogue number H113-26/2011-9E. 59 pages.
Quereshi S, Panjwani R, Pandey AK, Singh AK (2010) Formulation of phytotoxins from *Phoma* sp. FGCC#18 for the management of *Parthenium hysterophorus* L. Arch Phytopathol Plant Protect 43:1260-1267.
Quereshi S, Khan NS, Pandey AK (2011) Anthraquinone pigment with herbicidal potential from *Phom herbarum* FGCC#54. Chem Nat Compd 47:521-523.
Ray P, Vijayachandran LS (2013) Evaluation of indigenous fungal pathogens from horse purslane (*Trianthema portulacastrum*) for their relative virulence and host range assessments to select a potential mycoherbicidal agent. Weed Sci 61:580-585.
Razaghi P, Zafari D (2017) *Phoma crystallifera* with phytotoxic effects and pathogenic potential against field bindweed (*Convolvulus arvensis* L.) in Iran. J Appl Microbiol 122:1275-1285.
Schofield A, Paliyath G (2005) Modulation of carotenoid biosynthesis during tomato fruit ripening through phytochrome regulation of phytoene synthase activity. Plant Physiol Biochem 43:1052-1060.
Smith J, Wherley B, Reynolds C, White R, Senseman S, Falk S (2015) Weed control spectrum and turfgrass tolerance to bioherbicide *Phoma macrostoma*. Int J Pest Manage 61:91-98.
Stewart-Wade SM, Boland GJ (2004) Selected cultural and environmental parameters influence disease severity of dandelion caused by the potential bioherbicidal fungi, *Phoma herbarum* and *Phoma exigua*. Biocontrol Sci Technol 14:561-569.
Stewart-Wade SM, Neumann S, Collins LL, Boland GJ (2002) The biology of Canadian weeds. 117. *Taraxacum officinale* G. H.Weber ex Wiggers. Can J Plant Sci 82:825-853.
Thakore Y (2006) The biopesticide market for global agricultural use. BCC Research. The New Biopesticide Market, Industrial Biotechnology. 194-208.
Umetsu N, Shirai Y (2020) Development of novel pesticides in the 21st century. J Pestic Sci 45:54-74.
Vidal K, Guermache F, Widmer TL (2004) *In vitro* culturing of yellow starthistle (*Centaurea solstitialis*) for screening biological control agents. Biol Control 30:330-335.
Vikrant P, Verma KK, Rajak RC, Pandey AK (2006) Characterization of a phytotoxin from *Phoma herbarum* for management of *Parthenium hysterophorus* L. J Phytopathol 154:461-468.
Watson A, Bailey KL (2013) *Taraxacum officinale* FH Wigg, Dandelion (Asteraceae). In: Biological Control Programmes in Canada 2001–2012. Ed: Mason PG and Gillespie DR. Chapter 58. 383-391.
Westwood JH, Charudattan R, Duke SO, Fennimore SA, Marrone P, Slaughter DC, Swanton C, Zollinger R (2018) Weed management in 2050: Perspectives on the future of weed science. Weed Sci 66:275-285.
Wolfe JC (2015) Selective weed control in turfgrass using the bioherbicides *Phoma macrostoma*, Thaxtomin A and FeHEDTA. M.Sc. North Carolina State University Horticultural Science Raleigh, NC, USA.
Wolfe JC, Neal JC, Harlow CD (2016) Selective broadleaf weed control in turfgrass with the bioherbicides *Phoma macrostoma* and Thaxtomin A. Weed Technology 30:688-700.
Zhang X, Fernando WGD (2018) Insights into fighting against blackleg disease of *Brassica napus* in Canada. Crop Pasture Sci 69:40-47.
Zhao S, Shamoun SF (2005) Effects of potato dextrose broth and gelatin on germination and efficacy of *Phoma exigua*, a potential biocontrol agent for salal (*Gaultheria shallon*). Can J Plant Pathol 27:234-244.
Zhao S, Shamoun SF (2006) Effects of culture media, temperature, pH, and light on growth, sporulation, germination, and bioherbicidal efficacy of *Phoma exigua*, a potential biological control agent for salal (*Gaultheria shallon*). Biocontrol Sci Technol 16:1043-1055.
Zhou L, Bailey K, Derby J (2004) Plant colonization and environmental fate of the biocontrol fungus *Phoma macrostoma*. Biol Control 30:634-644.
Zhou L, Bailey K, Chen C, Keri M (2005) Molecular and genetic analyses of geographic variation in isolates of *Phoma macrostoma* used for biological weed control. Mycologia 97:612-620.
Zimdahl R (2013) Fundamentals of weed science. Academic Press, Inc. Elsevier. San Diego, CA, USA. Third Edition. ISBN 978-0-12-372518-9. 689 pages.

New Technologies for the Formulation of Secondary Metabolites Produced by *Phoma* sp. for Biological Control of Weeds

14

José R. C. Neto, Marcio A. Mazutti, Giovani L. Zabot, Maicon S. N. dos Santos, and Marcus V. Tres

14.1 Introduction

Unwanted organisms cause serious damages to agricultural production. Over the years there is an increase in the occurrence of diseases, pests, and weeds in agricultural fields. In this context, the main method of control is the use of chemical products. However, the excessive use of agrochemicals and the lack of suitable technical assistance do not solve the problem. Otherwise, it can cause productivity loss and the contamination of food, soil, water, animals, and human beings. The excessive use of agrochemicals can also cause other problems, such as the emergence of pathogens, pests, and invasive species.

However, the current model of agricultural production has explored the development of sustainable products, which can be characterized as "alternative agriculture" or "sustainable agriculture." For this purpose, it is essential to take some precautions and measures, which aim at reducing the impact of conventional agriculture on the natural environment (Schwan-Estrada et al. 2000; Confortin et al. 2019). Also, the pathogens can become resistant to chemical molecules, which can cause a biological imbalance because of the elimination of beneficial organisms (Kleinman et al. 2016; Lamichhane et al. 2017; Tahmasebi et al. 2018; Neve et al. 2018; Hawkins et al. 2019).

In this context, biological control is a promising method to reduce the use of agrochemicals (Chagas et al. 2014; Mannazzu et al. 2019). This method consists of manipulating the production environment to modify the survival rate of pest-natural enemies. The bioactive compounds produced by biological control agents can be used since they have antifungal, herbicide, and insecticide activity to reduce insect populations, pathogens, and weeds (Parsaa et al. 2018). Among the biological control agents, microorganisms and natural substances are the most studied (Confortin et al. 2018; Aita et al. 2019). Regarding microorganisms, fungi are the most investigated, since several endophytic fungi can produce a wide variety of toxic substances (Cordeau et al. 2016; Zhang et al. 2019; Reis et al. 2019).

The use of secondary metabolites is considered a potential form of biological control. Currently, the chemical composition of compounds produced by microorganisms for their use in biological control is seldom known. To obtain secondary metabolites from endophytic fungi, several processes have been considered,

J. R. C. Neto · G. L. Zabot · M. S. N. dos Santos
M. V. Tres (✉)
Laboratory of Agroindustrial Processes Engineering (LAPE), Federal University of Santa Maria (UFSM), Cachoeira do Sul, Brazil
e-mail: marcus.tres@ufsm.br

M. A. Mazutti
Department of Chemical Engineering, Federal University of Santa Maria (UFSM), Santa Maria, Brazil

M. Rai et al. (eds.), *Phoma: Diversity, Taxonomy, Bioactivities, and Nanotechnology*,
https://doi.org/10.1007/978-3-030-81218-8_14

such as plant tissue culture, emulsification, and fermentation. However, in some cases, the secondary metabolites present in the fermented product have hydrophobic molecules of low solubility in water and sensitivity to light, which makes it impossible for their use in the composition of bioproducts (Brun et al. 2016).

Furthermore, considering many fungi related to the biological control, *Phoma* sp. is an important agent against unwanted vegetable species and has demonstrated satisfactory effects regarding weed control. *Phoma* is a genus belonging to phylum *Ascomycota* that comprises many species (approximately 220 species), severely pathogenic to plants, humans, and animals (Rai 1989; Kövics et al. 2013; Rai et al. 2015). Some species are plant pathogens, while others are saprophytes, located in vegetable tissues and soil. In plant species, these organisms cause incident diseases in different parts and result in drastic injuries, such as necrosis and leaves fall. This genus includes organisms found in different environments, such as soil and soil residues, plants, and humid locations (Bennet et al. 2018).

Among the endophytic fungi with the potential to produce secondary metabolites with herbicidal activity, *Phoma* sp. has shown promising results (Bailey and Falk 2011; Cimmino et al. 2013; Todero et al. 2018a, b). In these studies, the submerged fermentation process was used to produce bioactive molecules, where the crude extracts (cell-free) are being applied for biological control.

The secondary metabolites have a great influence on fungal development and interactions with other organisms. The secondary metabolites are produced by specific fungi species, mostly by filamentous fungi belonging to different fungi classes, such as *Ascomycota*. These active molecules originated in specific central metabolic pathways and primary metabolite synthesis sites, where acyl-CoAs are the element that originated the synthesis of secondary metabolites and some amino acids (e.g., penicillin), polyketide (e.g., aflatoxin), and terpene (e.g., carotene). Various species of *Phoma* produce bioactive secondary metabolites that are phytotoxic, antimicrobials, and mycoherbicides (Keller 2019). The biosynthetic pathways are related to the synthesis of these molecules, which comprehends an overly complex process. These structures are synthesized by a specific number of building blocks. In general, the nomenclature of the pathways is held after enzymes and is commonly applied to classify the secondary metabolites (Pickens et al. 2011). In general, these molecules are biosynthesized and excreted by different pathways. These metabolites are present in the site where microorganisms grow but are not essential for fungal growth and survival.

After the production, characterization, and concentration of metabolites, the use of suitable adjuvants in combination with broth is necessary to allow the proper use of such secondary metabolites. The addition of adjuvants is intended to mitigate the adverse effects of the environment on the fermented medium during storage and after application. Among the adjuvants that have been used in the composition of bioproducts, the most prominent ones are surfactants, vegetable oil, and nitrogen salts (Oliveira et al. 2015).

Emulsification is another technique to be adopted in the process of bioproduct formulations. It consists of two immiscible and dispersed liquid systems, in which the first phase is distributed as a drop-in another phase. This technique enables the distribution of lipid molecules for obtaining bioproducts with suitable texture and chemical properties (Charcosset 2009). A relatively novel application to produce emulsions is the emulsification with membranes. This process is simple and with potential applications in several areas, allowing the production of droplets in a controlled manner. A valuable advantage of membrane emulsification is that droplet size distributions can be carefully and easily controlled because the droplet size is causally related to the membrane pore size distribution (Cheng et al. 2008).

The emulsification with membranes allows producing microemulsions and nanoemulsions, which are differentiated as a function of the membrane used in the process and of the pore size. The emulsification is causally related to the size of the dispersed phase droplets formed and distributed in the continuous phase. Micro- and nanoemulsions are thermodynamically stable, optically transparent, isotropic, and low viscosity systems formed by manometric droplets. The advantage of these emulsions is the high stability of droplets in suspension (Fanun 2010).

In this context, studies are needed to associate technologies to provide a suitable formulation of bioproducts. It can allow maximum expression of their potential in biological control. Therefore, this review aims to present a critical and constructive discussion about the current findings that describe the efficiency of the use of secondary metabolites from fermentative processes of *Phoma* sp. Future perspectives for the formulation of bioproducts using membrane emulsification that justifies the use of metabolites produced by fungi for different purposes are also provided.

14.2 Biological Control of Weeds

Biological control or biocontrol is currently used in modern agriculture, which is also called sustainable or alternative agriculture. Such an option of control enables the reduction of using agrochemicals, to minimize the damage already caused and to avoid future harm to the natural environment (Schwan-Estrada et al. 2000). The use of biocontrol in pests, weeds, and phytopathogenic fungi consists in manipulating the production environment to increase the survival and performance of the natural enemies as well as the use of bioactive compounds from microorganisms or medicinal plants that reduce the number of species but not their eradication (Villaverde et al. 2014).

Many endophytic microorganisms used in this type of control are fungi and specific bacteria present inside plant organs and tissues. These organisms do not cause any damage to their hosts, which is different from phytopathogenic microorganisms, since they are harmful to plants, causing countless diseases (Hussain et al. 2014). Studies show that most vascular plants are hosts for several microorganisms, bacteria, and endophytic fungi. These microorganisms spent their whole life span or part of it colonizing the host plants' healthy tissues and most of the time do not show any apparent symptoms (Hussain et al. 2012).

Among these microorganisms, the group of endophytic fungi is the most applied in biological control. These microorganisms are an important source of a selection of biological-control agents to suppress pests from plants (insects and pathogens) and to overcome abiotic stress (drought, salt, and heat). Studies reaffirm that endophytic fungi are a rich source of bioactive compounds with increased medicinal and agricultural potential (Wijeratne et al. 2013; Daniel et al. 2018; Rai et al. 2018, 2020).

Considering the diversity of fungi and bacteria present in the environment, there are several biological activities developed by these organisms, which can be explored by different research fields. The bioactive compounds produced by endophytic fungi present significant biological activity and a high degree of biodiversity. Therefore, these compounds could be used in the development of new pharmaceutical products or the formulation of natural products within the agricultural sphere (Kennedy 2016; Da Rosa et al. 2019).

Bioherbicides are presented as integrated tools for weed management, because they offer several benefits, such as higher target specificity and rapid degradation. Nevertheless, only a few bioherbicides are currently available on the market. Further studies are necessary to identify bioactive compounds and manufacture efficient bioherbicides. Besides, it is necessary to demonstrate that the organisms applied to biological control are sufficiently specified for the target species and a large number of these organisms do not represent a risk to the species existing in the places where this control is applied. This evaluation is one of the preliminary steps for the development and later registration of possible bioproducts (Cordeau et al. 2016).

14.2.1 *Phoma* spp.: Its Role in Biological Control

The *Phoma* spp. are regarded as one of the largest in the fungal kingdom. These fungi are isolated from different plants and are considered to be a promising source of biologically active secondary metabolites, which can be used for human medicine and also in agriculture for biological control of weeds, phytopathogenic fungi, and insects (Singh et al. 2017). Generally, these microorganisms are subjected to different procedures, such as periodic transfers, as means of preservation and mycelial growth (Neto et al. 2019). One of the main species of *Phoma* is *Phoma dimorpha* (Fig. 14.1), which has been potentially used to control weeds, especially due to the high bioherbicidal action of its metabolites. Some species of *Phoma* secrete secondary metabolites such as phytotoxin pigments and anthraquinone, which hold the high potential of biological control and can be explored in the production of bioproducts used in agriculture (Kim et al. 2012; Luft et al. 2019).

Considering the high production of phytotoxins from the *Phoma* sp., this genus has been widely explored recently. Several researchers are investigating the biotechnological potential of *Phoma* spp. to produce biologically active metabolites and the formulation of natural products designed for biological control of weeds (Table 14.1). Different toxins produced by *Phoma* fungi affect the growth of different species of weeds since their seed germination has typically a rate of inhibition of approximately 90%. It corroborates the bioherbicidal potential of these compounds (Wang et al. 2007; Bailey et al. 2011a). These phytotoxins present great potential to penetrate the plant cells and influence the biochemical activities and cellular structure, which results in the occurrence of devastating injuries to the plants.

The potential for biological control of weeds was also examined by Cimmino et al. (2008). The referenced authors assessed the effects of phytotoxins produced by strains of *Phoma exigua* var. *exigua* on *C. arvense* and *Sonchus*

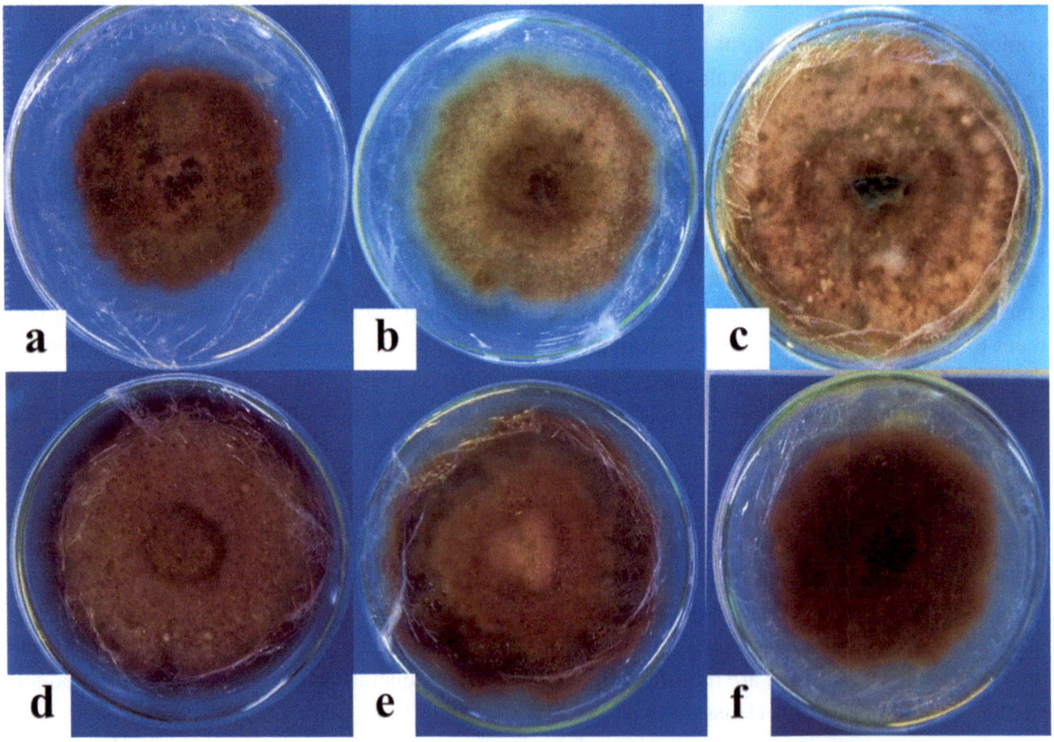

Fig. 14.1 Isolates of *P. dimorpha* preserved by periodic transfer at 7 (**a, b**), 15 (**c, d**), and 30 (**e, f**) days after transfer

Table 14.1 Identification of bioactive compounds of metabolites produced by submerged

Fungus	Bioactive compounds	Reference
Phoma macrostoma	Macrocidins A and B	Graupner et al. (2003)
Phoma macrostoma	Macrocidins A and Z	Zhou et al. (2004)
Phoma herbarum	3-Nitro-1,2-benzenedicarboxylic acid (3-nitrophthalic acid)	Vikrant et al. (2006)
Phoma exigua	Oxaphomin	Cimmino et al. (2008)
Phoma macrostoma	Macrocidins A and Z	Bailey et al. (2011b)
Phoma chenopodiicola	Chenopodolans A–C	Cimmino et al. (2013)
Phoma chenopodiicola	Chenopodolan D, chenisocoumarin, chenopodolin B	Graupner et al. (2003)
Phoma sp. NRRL 43879	Pyrrolo[1,2-a]pyrazine-1,4-dione, hexahydro-3-(2-methylpropyl)	Brun et al. (2016)
Phoma chenopodiicola	Chenopodolans E and F	Evidente et al. (2016)

Fermentation from different species of *Phoma* for bioherbicide activity [source, authors]

arvensis weeds. The compound oxaphomin presented the strongest toxic effect on *S. arvensis*. *Phoma macrostoma* Montagne has a known bioherbicidal action, and it is efficient for the biological control of weeds (Hubbard et al. 2014). Mycelial extracts and lyophilized fermented broth extracts from six strains of *P. macrostoma* isolated in Canada showed similar effects. When they were applied in preemergence in the soil, they caused intense chlorosis and bleaching of emergent leaves of the Canadian thistle (*C. arvense* L (scop)). When applied in postemergence on leafy plants, they caused intense bleaching, which resulted in a slight growth inhibition after 4–6 days but was lethal after 14 days for sunflower plants (Graupner et al. 2003).

Efficacies of the secondary metabolites produced by *P. macrostoma* for biological control of other weeds are reported, such as Canadian thistle (*C. arvense* (L.) Scop.), dandelion (*Taraxacum officinale* Weber ex FH Wigg.), clown (*Stellaria media* (L.) Vill.), and unscented chamomile (*Matricaria perforata* Merat). Otherwise, no effect was observed on monocotyledonous plants such as wheat (*Triticum* spp.), oats (*A. sativa*), and millet (*Pennisetum glaucum*), which is interesting because such crops should not be controlled (Zhou et al. 2004) . Furthermore, the biomolecules produced by *Phoma* sp. are promising metabolites for the development of natural products (Table 14.2).

As reported by Bailey and Falk (2011), only the phytotoxins by *P. macrostoma* are registered as a bioherbicide for its use only in vegetable gardens but not for agriculture. *P. macrostoma* generates two secondary metabolites that show phytotoxicity known as macrocidins A and Z. These compounds inhibit carotenoid biosynthesis in susceptible plants, severe chlorosis whitening, and occasional death of broadleaf weeds (Hubbard et al. 2014). Secondary metabolites with high phytotoxicity rates produced by *P. macrostoma* are known as macrocidins A and Z. The macrocidins inhibit the biosynthesis of carotenoids in sensitive plants, inducing chlorosis, whitening of pants, and occasional necrosis (Bailey et al. 2011b).

With the chemical characterization and evaluation of the potential bioherbicides of metabolites produced in submerged culture of the fungus *Phoma chenopodiicola*, it identified three new metabolites, which form chenopodolan D, chenisocoumarin, and chenopodolin B (Evidente et al. 2015). When applied on detached leaves of nonhost weeds, effective phytotoxicity was verified, causing necrosis in leaf tissues of *Stellaria media*, *Urtica dioica*, *Sonchus arvensis*, *Parietaria officinalis*, *Lactuca serriola*, and *Helianthus annuus*.

Fermented broth without cells from *Phoma* sp. was applied in preemergence, postemergence, and in isolated leaves of *C. sativus* and *Sorghum bicolor* (Brun et al. 2016). The metabolites produced by this fungus are highly bioherbicidal. They have completely inhibited the growth of these monocotyledon (*S. bicolor*) and dicotyledon (*C. sativus*) plant seeds.

Table 14.2 Reality and future perspectives of studies related to bioherbicides from secondary metabolites produced by fungi (data from Cordeau et al. 2016)

Reality				
Commercial product	Registration/ authorization	Fungus	Weed(s) controlled	
Devine™	1981	*Phytophthora palmivora* Butler	*Morrenia odorata* (Hook. & Arn.) Lindl	
Collego™	1982	*Colletotrichum gloeosporioides*, strain 20,358	*A. virginica* (L.) B.S.P.	
BioMal®	1992	*Colletotrichum Gloeosporioides*, strain ATCC 20767	*Malva pusilla* Sm.	
LockDown®	1997	*Colletotrichum gloeosporioides*, strain *ATCC 20358*	*A. virginica* (L.) B.S.P.	
Warrior®	2002	*Puccinia thlaspeos* C. Shub.	*Isatis tinctoria* L.	
Mycotech™	2002/2005	*Chondrostereum purpureum*, strain HQ1	*Prunus serotina* Ehrh; *Populus euramericana*	
Chontrol®	2004/2005	*Chondrostereum purpureum*, strain PFC 2139	*Prunus serotina* Ehrh; *Populus euramericana*	
Smoulder®	2005	*Alternaria destruens* L. Simmons, strain 059	Several weed species	
Sarritor®	2007/2010	*Sclerotinia minor* Jagger, strain IMI 344141	Dicotyledonous weeds	
Phoma	2012	*Phoma macrostoma*, strain 94-44B	Dicotyledonous weeds	
Future perspectives				
Product/ metabolite	**Technology**	**Applications**	**Reality**	**New technologies**
Secondary metabolites	Submerged fermentation	Fungicidal, bactericidal, and herbicidal action	Formulation with adjuvants and surfactants	Use of membrane concentration processes and membrane emulsification

Indeed, more studies are necessary for the investigation of secondary metabolites produced by endophytic fungi isolated from medicinal plants, making it possible to find other natural bioactive compounds and new secondary metabolites. Likewise, it is important to research fermentation processes to scale up production. There is a necessity for tests focusing on the efficacy of macrobiotic action of secondary metabolites produced by endophytic fungi from *Phoma* sp. (Fig. 14.2).

14.2.2 Fermentation Processes

In addition to the development of tools to identify and select microorganisms that produce secondary metabolites with potential for biological control, it is necessary to establish the most suitable bioprocesses and determine the best growing conditions for the increase of production of these compounds. Fermentation processes may be an alternative to the production of secondary metabolites (Brun et al. 2016; Daniel et al. 2019). Fermentation processes are labeled as a sum of steps or phases where the mediums of culture provide all the necessary nutrients. In a fermentative process, the conditions of the fungal growth, such as temperature, pH, agitation, and concentration of nutrients, are controlled to offer favorable conditions to the development of microorganisms and, consequently, an increase in the yield of secondary metabolites. Furthermore, fermentation is characterized as a process of easy manipulation. The optimal conditions for the growth of microorgan-

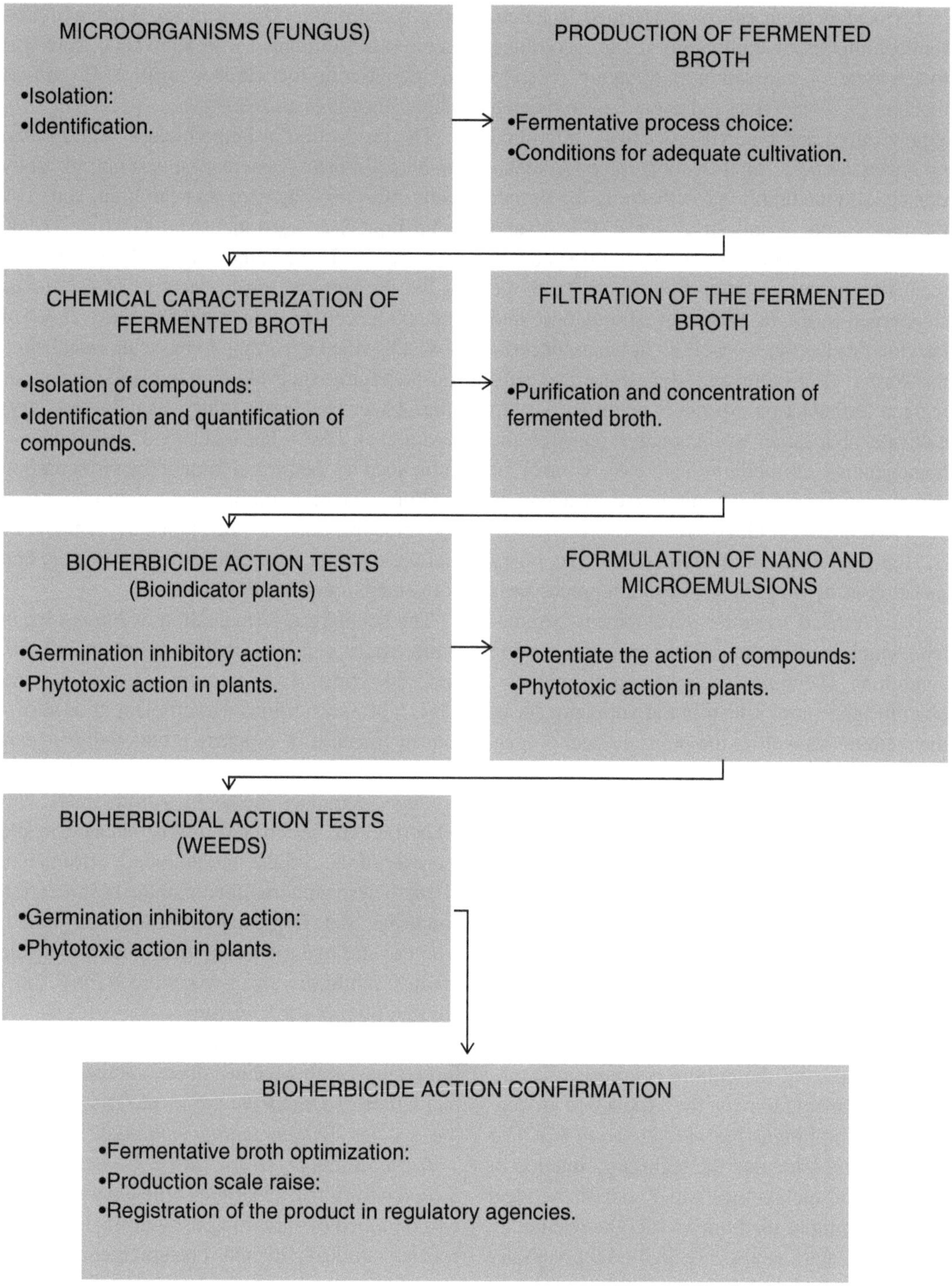

Fig. 14.2 Steps and processes used in the development of a bioherbicide from *Phoma* sp.

isms during the fermentative process, such as the substrate, the amount of biomass, nutritional variety, and the incubation period, can be easily modified, benefiting fungal growth (Arumugam et al. 2014; Brun et al. 2016). After the process, the fermented broth is submitted to the final treatment to separate products and subproducts (Dernain 2006; Karahalil et al. 2019).

Fermented broth can be performed in batch, semicontinuous or continuous mode, according to the type of cultivation (cultivation in solid medium or submerged) and according to oxygen supply (aired and unaired processes). Microbial processes are performed on submerged fermentation or solid medium of growth, being the former the most used at industrial scale (Castro and Ferreira 2010; Qian et al. 2019). Solid-state fermentation processes differ significantly from the submerged ones in terms of sporulation and enzyme production, as well as in what concerns the forms of mixing and diffusion. The submerged fermentation process is relatively easy to cultivate at a larger scale, once it ensures the homogeneity of medium and easy control of parameters, especially if monitored by adequate sensors (Daryaei et al. 2016).

Higher production of biomass and secondary metabolites by fungi in fermentation processes is causally related to some development parameters, which involve the medium and the growing conditions. The main parameters are the pH, carbon/nitrogen ratio, and mineral supplements of the medium, as well as the physiological properties of microorganisms (such as inoculum age). These parameters must be carefully selected to optimize the growth and production process of the secondary metabolites of endophytic fungi, maintaining viability and efficacy (Klaic et al. 2017).

Several studies have been conducted in recent years aiming at determining the fermentation process and the conditions of cultivation suitable to produce metabolites by endophytic fungi. When optimizing the culture conditions in submerged fermentation for the production of pigments by the fungus *Phoma herbarum* FGCC # 54, it was reported that the highest production of biomass and metabolites was found at the ideal pH of the culture medium as 4.0 (Quereshi et al. 2010). The authors also highlighted the viability of the commercial production of this pigment as a potential bioherbicide. Evaluating the influence of the formulation of metabolites produced by the fungus *Diaporthe* sp. in solid-state fermentation with adjuvants, it was found that solid-state fermentation allowed the production of metabolites by the fungus *Diaporthe* sp. When a bioherbicide was formulated with adjuvants, there was an increase in the efficacy against *C. sativus* plants (Bastos et al. 2017).

The production of bioherbicides from *Phoma* sp. by solid-state fermentation was optimized by using residues (bagasse, soybean bran, and corn steep liquor) as agro-industrial substrate (Klaic et al. 2017). These authors attested high efficacy of the bioherbicide on the control of *C. sativus* and evidenced an injury phytotoxicity level of 40%, classified according to the scale established for the phytotoxic evaluation in plants, as a moderate effect (moderate injury, usually recovery) (Frans et al. 1986). The intensity of the effect was influenced by the formulation of the fermentation medium. The best optimization condition for bioherbicide production was moisture content of 70.0%, soybean bran content of 30.0 wt.%, and corn steep liquor of 20.0 wt.%.

The submerged fermentation of *Phoma* sp. in a bioreactor under controlled agitation (40, 50, and 60 rpm) and aeration (1.0, 2.0, and 3.0 L min^{-1}) conditions was efficient, thus allowing an increase in biomass productivity. It evidences that process optimization methods to produce secondary metabolites from fungi with bioherbicidal action are fundamental for the increase of the content of compounds (Brun et al. 2016). These authors found that the best aeration condition for fermentation was 3 L min^{-1}, whereas the agitation condition that caused the greatest inhibitory effect was 40 rpm for *C. sativus* and 60 rpm for *S. bicolor*.

When evaluating the herbicidal action of the fermented broth obtained from submerged fermentation of *Diaporthe* sp. in preemergence of the species *Lolium multiflorum*, *Glycine max*, *Triticum aestivum*, *Oryza sativa*, *C. sativus*, and *S. halepense*, the maximum efficiency of 100% was obtained by inhibiting the germination of all species studied. In the postemergence, plant growth was suppressed for the species *Conyza* sp. and *Echinochloa* sp. (Pes et al. 2016).

A recent study was developed to concentrate the fermented broth of *Phoma* sp. in a hollow fiber membrane, to increase the herbicidal activity of this broth, and to evaluate the phytotoxic effect of the different concentrations (using separately both retentate and permeate fractions); absorption bioassays were developed in leaves of *C. sativus* and two preemergence tests (chamber germination and substrate germination) of weeds of *B. pilosa* and *A. retroflexus*. The highest phytotoxicity (80.7%) was obtained to a concentration of 30% in the fraction of retentate, presenting severe symptoms of the leaves of *C. sativus* when compared with the raw fermented broth. The preemergence had 100% control for both species in the germination chamber, while for the germination in the substrate, the control rate was influenced by the applied concentration of bioherbicide (Todero et al. 2018b).

14.2.3 Emulsions: Concepts, Types, Application, and Formation

Emulsions are defined as heterogeneous systems, thermodynamically unstable, composed of two or more immiscible liquids, usually oil and water. The emulsions are composed of three steps: oily step, aqueous step, and surfactants. In the first step or phase (disperse step), the formation of small and spherical drops takes place. These drops are suspended in another step (continuous). The size of the dimensions depends on the emulsification process, which can vary between 0.1μm and 0.1 mm (De Luca et al. 2008; Lam and Nickerson 2013; Zabot et al. 2019). Emulsions are classified into different types, depending on the definition of which is the continuous phase and which is the dispersed phase and according to its droplet size. For the phases, they are classified into single emulsions and multiple emulsions. For droplet size, they are classified in macroemulsions, microemulsions, and nanoemulsions (Van Der Graaf et al. 2005).

The simple emulsions are divided into two classes, according to the relative distribution of the different phases that make up the system. The simple oil-in-water (O/W) emulsions are typically systems formed by oil droplets dispersed in an aqueous phase. However, water-in-oil emulsions (W/O), less frequent, are systems formed by water droplets dispersed in an oil phase (Pereira and Garcia-Rojas 2015; Zanatta et al. 2017).

The multiple or double emulsions present more complex structures and are defined as multicompartmentalized systems composed of three steps, usually with different compositions, in which emulsions of oil in water (O/W) and water in oil (W/O) coexist. The aqueous droplets are dispersed inside oily drops, and these oily drops are dispersed in an external aqueous step, which allows uniform distribution of the lipid molecules. The most common multiple emulsions are water/oil/water (W/O/W), and depending on the application, oil/water/oil (O/W/O) can be prepared (Bouyer et al. 2012). Multiple emulsions have promising applications in the food, cosmetics, and pharmaceutical industries. Other fields of industry are also using emulsions, as for agricultural product preparation and the production of multicompartmental microspheres (Van Der Graaf et al. 2005).

The simple O/W emulsions are the most used and of larger interest for the food industry, to produce yogurt, creams, and mayonnaise. The simple W/O emulsions are used in the margarine and butter food industry (Zanatta et al. 2017). Multiple emulsions have a high potential for application in the food industry (low-calorie products, improved sensory characteristics, taste masking), cosmetic industry (easily diffusible creams with ingredients encapsulated in the water and oil phases), and in the pharmaceutical industry (Bouyer et al. 2012). Recent studies show these emulsions are also being used in the agricultural product industry, being applied in the formulation of products by the production of microspheres, which leads to the innovation of products destined for the control of fungi, bacteria, and weeds (Bastos et al. 2017; Hazrati et al. 2017).

Among the methods for forming an emulsion, the most used are i) physical emulsification by drop rupture, which depends on the mechanical nature of the process and rheologic and chemical properties of the compounds, ii) emulsification by phase inversion, and iii) "spontaneous" emulsification, which depends primarily on the nature of the chemical additives, the relation between the two steps, and the temperature (Schmidts et al. 2010; Anton and Vandamme 2011).

Some aspects to take into consideration when preparing an emulsion are the selection of the continuous and the dispersed phases and the study of factors that determine the rheology and stability of emulsified systems (Liu et al. 2011). The emulsification can be produced by conventional methods or using the technique known as membrane emulsification. It is a process that is energetically efficient since it has a low demand for energy (the content of energy by the cubic meter of produced emulsion) when compared to other conventional emulsification techniques (Van Der Graaf et al. 2005; Zanatta et al. 2017).

The emulsions may also be sized for droplet size formed in the dispersed phase and distributed in the continuous phase, which may be called micro- or nanoemulsions. Microemulsions are thermodynamically stable and isotropically translucent systems of two immiscible liquids, usually, W/O stabilized by an interfacial surfactant film and co-surfactant with droplets in the range from 10μm to 100 nm. For the formation of microemulsions, it is necessary to combine three to five components: oil, water, surfactant, co-surfactant, and electrolyte (Damasceno et al. 2011). Nanoemulsions are kinetically stable, transparent, or translucent systems of two immiscible liquids stabilized by an interfacial surfactant film with droplets mainly covering <100 nm (Wang et al. 2007). This small size of the droplets is responsible for the stability of nanoemulsions, avoiding sedimentation (Anton and Vandamme 2011).

Unlike the conventional emulsions, in micro- and nanoemulsions, which are systems consisting of an aqueous phase, a lipophilic phase, and a surfactant, it is necessary the addition of a co-surfactant for the preparation of micro- and nanoemulsions. The addition of this co-surfactant turns the system to be homogeneously dispersed, which can diffuse easily, being thermodynamically stable unlike conventional emulsions (Van Der Graaf et al. 2005; Anton and Vandamme 2011).

Antimicrobial activity of nanoemulsions formulated from basil oil with low droplet diameter associated with nonionic surfactant Tween 80 was confirmed to preserve food against microbial deterioration (Ghosh et al. 2013). The results highlight the potential of formulating nanoemulsions to allow higher biological expression of the compound metabolites produced by plants or even by microorganisms. Emulsions were formulated to produce a bioherbicide from the mixture of secondary metabolites from the solid-state fermentation of *Diaporthe* sp. with adjuvants to improve herbicidal activity toward the target organism (Bastos et al. 2017). In such a study, the authors verified that the fungus *Diaporthe* sp. presented the potential to produce molecules with herbicidal activity and the emulsions produced with the addition of adjuvants improved three times their efficiency against *C. sativus* plants. In addition to providing the final product with characteristics such as stability, they also increase wettability, scatter ability, and penetration, which these aspects are highly attractive for agricultural product industries (Solans and Solé 2012).

The main advantages of nanoemulsions over conventional emulsions include high kinetic stability, improved disposability, allow the carrying of active compounds, high permeability of low-solubility substances, and low turbidity (McClements 2011; Ghosh et al. 2013). With the optimization of process parameters for the formulation of basil oil nanoemulsion with a small droplet diameter and high kinetic stability, it was observed that the nanoemulsion was stable and had a droplet diameter of 29.6 nm. Also, bactericidal activity was demonstrated by the emulsions formulated against *Escherichia coli* (Ghosh et al. 2013).

The properties of nanoemulsions including antimicrobial activity are highly dependent on the type of essential oil used in the formulation and the droplet size obtained at the end of the process. When evaluating the formulation of nanoemulsions containing essential oils from different plant species, it was reported that the nanoemulsions showed a higher bactericidal action in vitro against *E. coli*, especially for the oil obtained from the clove (Salvia-Trujillo et al. 2015). In another work, antimicrobial activity and stability of free basil oil nanoemulsions were seen, especially against *C. albicans*, *E. coli*, *Streptococcus* sp., and *Staphylococcus aureus* (Bolzan et al. 2015).

Considering that geranium oil showed majoritarian constituents such as citronellol and geraniol, a nanoemulsion containing geranium oil was developed to assert its accelerated stability and antifungal activity for the species *Candida* spp. (Cannas et al. 2016). The study reported that nanoemulsions presented stability for a period of 60 days, also showing significant antifungal activity for the species *C. albicans*, *Candida glabrata*, and *C. tropicalis*. In a recent study, an O/W nanoemulsion was developed composed of essential oil extracted from *Satureja hortensis* L. and evaluated the bioherbicidal activity in pre- and postemergence of two weeds of global importance (*A. retroflexus* L. and *C. album* L.) (Hazrati et al. 2017). The tests in preemergence showed that the nanoemulsion reduced the levels of germination and presented a higher inhibitory effect when applied a dose of 800μL L^{-1}.

Membrane Emulsification

Membrane emulsification is a recent application of the technology of membranes, which has been receiving higher attention in recent years with potential applications in several areas. The main advantage of using membranes compared to conventional emulsification techniques is the formation of micro- and nanodroplets with little dispersion and uniform size. Furthermore, it allows controlling the droplet average diameter by the average size of the membrane pores (Table 14.3). The size of the droplets and the distribution of size are among the most important properties that must be considered when preparing a certain type of emulsion. It results in higher interest in the use of the membrane emulsification process (Mello et al. 2010).

Apart from these factors, the process of membrane emulsification is a promising technique, especially for offering mildness to the process, easy operation, and simple and changeable equipment for large-scale production. Membrane emulsification is a technique that has been widely used in different areas but especially important in fine applications in the medical industry, since it allows the control of the size and distribution of particles, producing multiple emulsions of uniform size, which adds more value to the medicine obtained (Liu et al. 2011).

There are two distinct methods for performing emulsion formation by the membrane emulsification process. In the direct emulsification method, the dispersed phase permeates through the membrane to the continuous phase in which the droplets are formed. The premix emulsification method is generally performed in cycles, where the dispersed phase passes successively through the membrane, to reduce the size of the droplets more and more. Initially, a primary emulsion is prepared by the conventional method (simple mechanical agitation), which is predetermined to permeate through a microporous membrane, promoting the reduction of droplet size of the dis-

Table 14.3 Advantages and drawbacks of membrane emulsification compared to conventional methods. Reproduced from Charcosset et al. (2009) [Reprinted with permission]

Advantages	Drawbacks
Membrane contactors (general)	
High and constant specific interfacial area	Additional resistance to mass transfer created by the membrane
Independence of the fluid phases in contact	Relatively low membrane lifetime; high replacement costs
No moving parts, use of plastic equipment (absence of corrosion)	
High modularity and compatibility, straight-forward scale-up, easy control	
Membrane emulsifiers	
Lower energy input with respect to conventional emulsifiers	Low permeation rates associated to narrow droplet size distribution and reduced productivity
No foaming-reduced coalescence phenomena; narrow droplet size distributions; various configurations: premix emulsification, rotating devices to reduce fouling and to increase productivity	Fouling phenomena on the membrane surface and/or in the pores

persed phase. This technique is usually performed in cycles, which are successive membrane passages, to reduce droplet size more and more (Piacentini et al. 2014).

The direct membrane emulsification method has the advantages of low shear rates, low energy consumption, uniform droplet size, and ease of scale up (Van Der Graaf et al. 2005). However, the premix emulsification method requires less energy consumption and allows permeate fluxes higher than the other method. Overall, the experimental apparatus is simple and easy to control (Nazir et al. 2010).

The methods of membrane emulsification have some limitations. The direct emulsification presents low permeate flux of the dispersed phase through the membrane and requires large areas of the membrane (Nazir et al. 2010). In the case of premix emulsification, in most cases, several cycles are required. However, the major drawbacks are the fouling occurring on the surface and in the pores of the membrane, considerably reducing permeate flux and may become a serious problem depending on the emulsion formulation (Charcosset et al. 2004; Van Der Graaf et al. 2008).

Two methods of operation are used in the process of membrane emulsification: crossflow membrane emulsification and membrane emulsion pre-blend. In membrane emulsification, the speed-flow in the continuous phase and kinetics of surfactant adsorption or emulsifiers used has a significant effect on the distribution of size and on the droplet average size of the emulsion. The size of the droplet is determined mainly by the membrane type, pore size, continuous and dispersed phase flow, the pressure applied to the membrane, the relative viscosity of the phases, applied shear rate, crossflow velocity, pressure transmembrane, and interfacial tension (Hancocks et al. 2013). Furthermore, the type of surfactant and the concentration to be used are important for the membrane emulsification process. In most of the applications in different areas of industry, blended surfactants are employed in emulsion formulations (Fanun 2010).

To optimize the membrane emulsification process, it is necessary to use a suitable model that allows maintaining the time of production of emulsion in the minimum, without harming its quality. During the drop formulation process, the molecules adsorb the surfactant into the interface oil/water just formed and reduce the interfacial tension, facilitating droplet formation and avoiding coalescence immediately after the formation (Schmidts et al. 2010).

14.3 Concluding Remarks and Future Perspectives

The biomolecules produced by *Phoma* spp. are favorable components for the efficiency of weed control. Biological control is becoming a reality in the agricultural area. In the last years, researches have been carried out to select biological agents and to develop novel technologies to produce biologically active metabolites to evaluate laboratory-scale efficiency, including the purification, concentration, and formulation of bioproducts. Therefore, these compounds are an alternative against weeds, even at a gradual substitution of synthetic herbicides. They contribute to the development of control strategies for sustainable agriculture.

There are challenges regarding secondary metabolites produced by microorganisms, mostly those produced by fungi. One of the challenges is the low productivity, which is influenced by conditions of cultivation, type of fermentation, and factors inherent to the processes (e.g., increases in the temperature and aeration). Therefore, it is necessary to carry out more research focusing not only on the enhancement of the production scale of these compounds but also on the development of new technologies or even on the adaptation of the existing ones that make it possible to demonstrate the high potential of these compounds. One of these cases is membrane emulsification, which makes it possible to produce emulsions with higher kinetic stability, malleability, spreadability, penetration, and bioavailability. Additionally, membrane emulsification can extend the storage period of bioproducts.

Biological control of weeds is one of the main challenges for organic and conventional agriculture. Accordingly, endophytic *Phoma* spp. are microorganisms that produce a large variety of secondary metabolites, which are still poorly investigated. The outlook for the near future is that more biocontrol products with high potential can be launched and marketed worldwide. Correspondingly, technologies can be confirmed to concentrate and to formulate secondary metabolites produced by endophytic fungi, and processes can be optimized for developing the productions at a larger scale.

Acknowledgments We appreciate the support from the Coordination for the Improvement of Higher Education Personnel (CAPES), National Council Scientific and Technological Development (CNPq: 428180/2018-3), and the Research Support Foundation of the State of Rio Grande do Sul (FAPERGS: 16/2551-0000522-2; 17/2551-0000893-6) for the financial support. M. V. Tres, M. A. Mazutti, and G. L Zabot thank CNPq (306241/2020-0, 308936/2017-5; 304882/2018-6) for the productivity grant.

References

Aita BC, Spannemberg SS, Schmaltz S, Zabot GL, Tres MV, Kuhn RC (2019) Production of Cell-wall degrading enzymes by solid-state fermentation using agroindustrial residues as substrates. J Environ 7.

Anton N, Vandamme TF (2011). Nano-emulsions and micro-emulsions: clarifications of the critical differences. Pharm Res 28: 978-985.

Arumugam GK, Srinivasan SK, Joshi G, Gopal D, Ramalingam K (2014) Production and characterization of bioactive metabolites from piezotolerant deep sea fungus Nigrospora sp. in submerged fermentation. J Appl Microbiol 118: 99-111.

Bailey KL, Falk S (2011). Turning research on microbial bioherbicides into commercial products - a *Phoma* story. Pest Technol 5: 73-79.

Bailey KL, Pitt WM, Leggett F, Sheedy C, Derby J (2011a) Determining the infection process of *Phoma macrostoma* that leads to bioherbicidal activity on broadleaved weeds. Biol Control 59: 268-276.

Bailey KL, Pitt WM, Falk S, Derby J (2011b) The effects of *Phoma macrostoma* on nontarget plant and target weed species. Biol Control 58: 379-386.

Bastos BO, Deobald GA, Brun T, Dal Prá V, Junges E, Kuhn RC (2017) Solid-state fermentation for production of a bioherbicide from *Diaporthe* sp. and its formulation to enhance the efficacy. 3 Biotech 7: 135-143.

Bennet A, Ponder MM, Garcia-Diaz J (2018) *Phoma* infections: classification, potential food sources, and their clinical impact. Microorganisms 6.

Bolzan AA, Lopes LQ, Marin LS, Santos RCV, Raffin R, Fagan SB (2015) Avaliação da atividade antimicrobiana do óleo de orégano livre e em nanoemulsões. RDS 16: 325-332.

Bouyer E, Mekhloufi G, Rosilio V, Grossiord JL, Agnely F (2012) Proteins, polysaccharides, and their complexes used as stabilizers for emulsions: alternatives to synthetic surfactants in the pharmaceutical field? Int J Pharm 436: 359-378.

Brun T, Rabuske JE, Todero I, Almeida TC, Junior JJD, Ariotti G (2016) Production of bioherbicide by *Phoma* sp. in a stirred-tank bioreactor. 3 Biotech 6: 230-239.

Cannas S, Usai D, Tardugno R, Benvenuti S, Pellati F, Zanetti S (2016) Chemical composition, cytotoxicity, antimicrobial and antifungal activity of several essential oils. Nat Prod Res 30: 332-339.

Castro AM, Ferreira JR (2010) Produção, propriedades e aplicação de celulases na hidrolise de resíduos agroindustriais. Quim Nova 33: 181-188.

Chagas HA, Basseto MA, Rosa DD, Toppa EVB, Furtado EL, Zanotto MD (2014) Evaluation of fungicides, essential oils and biological agents on *Amphobotrys ricini* control in castor bean (*Ricinus communis* L.). Summa Phytopathol 40: 42-48.

Charcosset C (2009) Preparation of emulsions and particles by membrane emulsification for the food processing industry. J Food Eng 92: 241-249.

Charcosset C, Limayem I, Fessi H (2004) The membrane emulsification process – a review. J Chem 79: 209-218.

Cheng CJ, Chu LY, Zhang J, Zhou MY (2008) Preparation of monodisperse poly (N-isopropylacrylamide) microspheres and microcapsules via Shirasu-porous-glass membrane emulsification. Desalination 234: 184-194.

Cimmino A, Andolfi A, Berestetskiy A, Evidente A (2008) Production of phytotoxins by *Phoma exigua* var exigua, a potential mycoherbicide against perennial thistles. J Agric Food Chem 56: 630-634.

Cimmino A, Andolfi A, Zonno MC, Avolio F, Berestetskiy A, Vurro M (2013) Chenopodolans A–C: phytotoxic furopyrans produced by *Phoma chenopodiicola*, a fungal pathogen of *Chenopodium album*. Phytochem 96: 208-213.

Charcosset C (2009) Preparation of emulsions and particles by membrane emulsification for the food processing industry. J Food Eng 92: 241–249.

Confortin TC, Todero I, Luft L, Soares JF, Mazutti MA, Zabot GL (2018) Importance of *Lupinus albescens* in agricultural and food-related areas: a review. 3 Biotech 8: 448-458.

Confortin TC, Todero I, Luft L, Teixeira AL, Mazutti MA, Zabot GL (2019) Valorization of Solanum viarum dunal by extracting bioactive compounds from roots and fruits using ultrasound and supercritical CO_2. Braz J Chem Eng 36.

Cordeau S, Triolet M, Wayman S, Steinberg C, Guillemin J-P (2016) Bioherbicides: dead in the water? A review of the existing products for integrated weed management. J Crop Prot 87: 44-49.

Da Rosa BV, Kuhn KR, Ugalde GA, Zabot GL, Kuhn RC (2019) Antioxidant compounds extracted from

Diaporthe schini using supercritical CO_2 plus cosolvent. Bioproc Biosyst Eng 43: 133-141.

Damasceno BPGL, Silva JA, Oliveira EE, Silveira WLL, Araújo IB, Oliveira AG (2011) Microemulsão: um promissor carreador para moléculas insolúveis. Rev Cien Farm Básica Apl 32: 9-18.

Daniel J, Zabot GL, Tres MV, Harakava R, Kuhn RC, Mazutti MA (2018) *Fusarium fujikuroi*: a novel source of metabolites with herbicidal activity. Biocatal Agric Biotechnol 14: 314-320.

Daniel J, Brun T, Todero I, Almeida TC, Confortin TC, Kuhn, KR (2019) Association of adjuvants with culture filtrate from *Fusarium fujikuroi* for increasing the control of *Conyza* sp. Biointerface Res Appl Chem 10: 6481-6487.

Daryaei A, Jones E, Glare TR, Falloon RE (2016) Nutrient amendments affect *Trichoderma atroviride* conidium production, germination and bioactivity. Biol Control 93: 8-14.

De Luca G, Di Maio FP, Di Renzo A, Drioli E (2008) Droplet detachment in cross-flow membrane emulsification: Comparison among torque- and force-based models. Chem Eng Process 47: 1150-1158.

Dernain AL (2006) From natural products discovery to commercialization: a success story. J Ind Microbiol Biotechnol 33: 486-495.

Evidente M, Cimmino A, Zonno MC, Mais M, Berestetskyi A, Santoro E (2015) Phytotoxins produced by *Phoma chenopodiicola*, a fungal pathogen of *Chenopodium album*. Phytochem 117: 482-488.

Evidente M, Cimmino A, Zonno MC, Mais M, Santoro E, Vergura S (2016) Chenopodolans E and F, two new furopyrans produced by *Phoma chenopodiicola* and absolute configuration determination of chenopodolan B. Tetrahedron 72: 8502-8507.

Fanun M (2010) Microemulsions with mixed nonionic surfactants and flavor oil. J Surfactants Deterg 13: 321-328.

Frans R, Talbert R, Marx D, Crowley H (1986) Experimental design and techniques for measuring and analysing plant responses to weed control practices. In: Camper ND, editor. Research Methods in Weed Science. Champaign: Southern Weed Science Society. 29-46.

Ghosh V, Mukherjee A, Chandrasekaran N (2013) Ultrasonic emulsification of food-grade nanoemulsion formulation and evaluation of its bactericidal activity. Ultrason Sonochem 20: 338-344.

Graupner PR, Carr A, Clancy E, Gilbert J, Bailey KL, Derby JA (2003) The macrocidins: novel cyclic tetramic acids with herbicidal activity produced by *Phoma macrostoma*. J Nat Prod 66: 1558-1561.

Hancocks RD, Spyropoulos F, Norton IT (2013) Comparisons between membranes for use in cross flow membrane emulsification. J Food Eng 116: 382-389.

Hawkins NJ, Bass C, Dixon A, Neve P (2019) The evolutionary origins of pesticide resistance. Biol Rev 94: 135-155.

Hazrati H, Saharkhiz MJ, Niakousari M, Moein M (2017) Natural herbicide activity of *Satureja hortensis* L. essential oil nanoemulsion on the seed germination and morphophysiological features of two important weed species. Ecotox Environ Safe 142.

Hubbard M, Hynes RK, Erlandson M, Bailey KL (2014) The biochemistry behind biopesticide efficacy. Sustain Chem Process 2: 1-8.

Hussain H, Krohn K, Pescitelli G, Ahmed, I, Draeger S (2012) Phomopsinones A–D: four new pyrenocines from endophytic fungus *Phomopsis* sp. Eur J Org Chem 5: 1783-1789.

Hussain H, Kock I, Al-Harrasi A, Al-Rawahi A, Abbas G, Green IR (2014) Antimicrobial chemical constituents from endophytic fungus *Phoma* sp. Asian Pac J Trop Med 7: 669-702.

Karahalil E, Coban HB, Turhan I (2019) A current approach to the control of filamentous fungal growth in media: microparticle enhanced cultivation technique. Crit Rev Biotechnol 39: 192-201.

Keller NP (2019) Fungal secondary metabolism: regulation, function and drug discovery. Nat Rev Microbiol 17: 167-180.

Kennedy AC (2016) Pseudomonas fluorescens strains selectively suppress annual bluegrass (*Poa annua* L.). Biol Control 103: 210-217.

Kim EL, Li JL, Dang HT, Hong J, Lee CO, Kim DK (2012) Cytotoxic cytochalasins from the endozoic fungus *Phoma* sp. of the giant jellyfish *Nemopilema nomurai*. Bioorg Med Chem Lett 22: 3126-3129.

Klaic R, Sallet D, Foletto EL, Jacques RJS, Guedes JVC, Kuhn RC (2017) Optimization of solid-state fermentation for bioherbicide production by *Phoma* sp. Braz J Chem. Eng 34: 377-384.

Kleinman Z, Ben-Ami G, Rubin B (2016) From sensitivity to resistance – factors affecting the response of *Conyza* spp. to glyphosate. Pest Manag Sci 72: 1681-1688.

Kövics, G.J., Sándor, E., Rai, M. and Irinyi, L. (2013). *Phoma*-like fungi on soybeans. *Critical Reviews in Microbiology* 40(1):49-62. https://doi.org/10.3109/1040841X.2012.755948

Lam RSH, Nickerson MT (2013) Food proteins: a review on their emulsifying properties using a structure-function approach. Food Chem 141: 975-984.

Lamichhane JR, Devos Y, Beckie HJ, Owen MD, Tillie P, Messéan A (2017) Integrated weed management systems with herbicide-tolerant crops in the European Union: lessons learnt from home and abroad. Crit Rev Biotechnol 37: 459-475.

Liu W, Yang XL, Ho WSW (2011) Preparation of uniform-sized multiple emulsions and micro/nano particulates for drug delivery by membrane emulsification. J Pharm Sci 100: 75-96.

Luft L, Confortin TC, Todero I, Chaves Neto JR, Tonato D, Felimberti PZ (2019) Different techniques for concentration of extracellular biopolymers with herbicidal activity produced by *Phoma* sp. Environ Technol

Mannazzu I, Domizio P, Carboni G, Zara S, Zara G, Comitini F (2019) Yeast killer toxins: from ecological significance to application. Crit Rev Biotechnol 39: 603-617.

McClements DJ (2011) Edible nanoemulsions: fabrication, properties, and functional performance. Soft Matter 7: 2297-2316.

Mello BCBS, Petrus JCC, Hubinger MD (2010) Desempenho do processo de concentração de extratos de própolis por nanofiltração. Ciencia Tecnol Aliment 30: 166-172.

Nazir A, Schroën K, Boom R (2010) Premix emulsification: a review. J Membr Sci 362: 1-11.

Neto JRC, Mazutti MA, Zabot GL, Tres MV (2019) Avaliação de diferentes métodos de preservação do fungo *Phoma dimorpha*. Colloq Agrariae 15: 1-10.

Neve P, Barney JN, Buckley Y, Cousens RD, Graham S, Jordan NR (2018) Reviewing research priorities in weed ecology, evolution and management: a horizon scan. Weed Res 58: 250-258.

Oliveira RB, Antuniassi UR, Gandolfo MA (2015) Spray adjuvant characteristics affecting agricultural spraying drift. Eng Agric 35: 109-116.

Parsaa S, Ortiza V, Gómez-Jiménez MI, Kramer M, Veja FE (2018) Root environment is a key determinant of fungal entomopathogen endophytism following seed treatment in the common bean, *Phaseolus vulgaris*. Biol Control 116: 74-81.

Pereira LJB, Garcia-Rojas EE (2015) Multiple emulsions: formation and application in microencapsulation of bioactive componentes. Cienc Rural 45: 155-162.

Pes MP, Mazutti MA, Almeida TC, Curioletti LE, Arrué AM, Guedes JVC (2016) Bioherbicide based on *Diaporthe* sp. secondary metabolites in the control of three tough weeds. Afr J Agric Res 11: 4242-4249.

Piacentini E, Drioli E, Giorno L (2014) Membrane emulsification technology: twenty-five years of inventions and research through patent survey. J Membr Sci 468: 410–422.

Pickens LB, Tang Y, Chooi Y (2011) Metabolic Engineering for the production of natural products. Annu Rev Chem Biomol Eng 2: 211-236.

Qian X, Yan W, Zhang W, Dong W, Ma J, Ochsenreither K (2019) Current status and perspectives of 2-phenylethanol production through biological processes. Crit Rev. Biotechnol 39: 235-248.

Quereshi S, Pandey AK, Singh J (2010) Optimization of fermentation conditions for red pigment production from *Phoma herbarum* (FGCC#54) under submerged cultivation. J Phytol 2: 1-8.

Rai, M. (1989). *Phoma sorghina* infection in human being. Mycopathologia 105:167-170.

Rai M, Tiwari V and Balis E (2015). *Phoma* as Opportunistic Fungal Pathogens in humans, In: Molecular Biology of Food and water borne mycotoxigenic and mycotic fungi (Eds: R. Russell M. Patterson and Nelson Lima), CRC Press-Taylor and Francis group, USA, pp451-462

Rai M, Gade A, Zimowska B, Ingle AP, Ingle P (2018). Marine-derived *Phoma*-the gold mine of bioactive compounds, *Appl Microbiol Biotechnol*, 102:9053–9066. https://doi.org/10.1007/s00253-018-9329-2

Rai, M., Gade, A., Zimowska, B., Ingle, A.P., Ingle, P. (2020). Harnessing the potential of novel bioactive compounds produced by endophytic *Phoma* spp. – biomedical and agricultural applications. *Acta Sci. Pol. Hortorum Cultus*, 19(6), 31–45. https://doi.org/10.24326/asphc.2020.6.3

Reis CER, Bento HBS, Carvalho AKF, Rajendran A, Hu B, Castro HFD (2019) Critical applications of *Mucor circinelloides* within a biorefinery context. Crit Rev Biotechnol 39: 555-570.

Salvia-Trujillo L, Rojas-Graü A, Soliva-Fortuny R, Martín-Belloso O (2015) Physicochemical characterization and antimicrobial activity of foodgrade emulsions and nanoemulsions incorporating essential oils. Food Hydrocoll 43: 547-556.

Schmidts T, Dobler D, Guldan AC, Paulus N, Runkel F (2010) Multiple W/O/W emulsions - using the required HLB for emulsifier evaluation. Colloids Surf A Physicochem Eng Asp 372: 48-54.

Schwan-Estrada KRF, Stangarlin JR, Cruz MES (2000) Uso de extratos vegetais no controle de fungos fitopatogênicos. Floresta 30: 129-137.

Singh R, Kuma M, Mitta A, Mehta PK (2017) Microbial metabolites in nutrition, healthcare and agriculture. 3 Biotech 7: 1-14.

Solans C, Solé I (2012) Nano-emulsions: formation by low-energy methods. Curr Opin Colloid Interface Sci 17: 246-254.

Tahmasebi BK, Alebrahim MT, Roldán-Gómez RA, Silveira HMD, Carvalho LBD, Cruz RA-D (2018) Effectiveness of alternative herbicides on three *Conyza* species from Europe with and without glyphosate resistance. J Crop Prot 112: 350-355.

Todero I, Confortin TC, Soares JF, Brun T, Luft L, Rabuske JE (2018a) Concentration of metabolites from *Phoma* sp. using microfiltration membrane for increasing bioherbicidal activity. Environ Technol 22: 1-9.

Todero I, Confortin TC, Luft L, Brun T, Ugalde GA, Almeida TCD (2018b) Formulation of a bioherbicide with metabolites from *Phoma* sp. Sci Hortic 241: 285-292.

Van Der Graaf S, Schroen CGPH, Boom RM (2005) Preparation of double emulsions by membrane emulsification - a review. J Membr Sci 251: 7-15.

Van Der Graaf S, Schroen CGPH, Boom RM (2008) Premix membrane emulsification by using a packed layer of glass beads. AICHE J 54: 2190-2197.

Vikrant P, Verma KK, Rajak RC, Pandey AK (2006) Characterization of phytotoxin from Phoma herbarum for management of *Parthenium hysterophorus* L. J Phytopathol 154: 461-468.

Villaverde JJ, Sevilla-Morán B, Sandín-España P, López-Goti C, Alonso-Prados JL (2014) Biopesticides in the framework of the European pesticide regulation (EC) no. 1107/2009. Pest Manag Sci 70: 2-5.

Wang L, Li X, Zhang G, Dong J, Eastoe J (2007) Oil-in water nanoemulsions for pesticide formulations. J Colloid Interf Sci 314: 230-235.

Wijeratne EMK, He H, Franzblau SG, Hoffman AM, Gunatilaka AAL (2013) Phomapyrrolidones A-C, antitubercular alkaloids from the endophytic fungus *Phoma* sp. NRRL 46751. J Nat Prod 76: 1860-1865.

Zabot GL, Silva EK, Emerick LB, Felisberto MHF, Clerici MTPS, Meireles MAA (2019) Physicochemical, morphological, thermal and pasting properties of a novel native starch obtained from annatto seeds. Food Hydrocoll 89: 321-329.

Zanatta V, Rezzadori K, Penha FM, Zin G, Lemos-Senna E, Petrus JCC (2017) Stability of oil-in-water emulsions produced by membrane emulsification with microporous ceramic membranes. J Food Eng 195: 73-84.

Zhang B-B, Guan Y-Y, Hu P-F, Chen L, Xu G-R, Liu L (2019) Production of bioactive metabolites by submerged fermentation of the medicinal mushroom *Antrodia cinnamomea*: recent advances and future development. Crit Rev Biotechnol 39: 541-554.

Zhou L, Bailey KL, Derby J (2004) Plant colonization and environmental fate of the biocontrol fungus *Phoma macrostoma*. Biol Control 30: 634-644.

Biological Management of Plant Diseases by Non-pathogenic *Phoma* spp.

15

Md. Motaher Hossain

Abbreviations

AMF	Arbuscular mycorrhizal fungi
BCAs	Biological control agents
BTH	Benzothiadiazole
CMV	*Cucumber mosaic virus*
DAB	Diaminobenzidine tetrahydrochloride
ET	Ethylene
FHB	*Fusarium* head blight
JA	Jasmonic acid
PGPF	Plant growth-promoting fungi
SA	Salicylic acid

15.1 Introduction

Plants constitute the majority of the earth's living surroundings. Humans and animals directly or indirectly rely on the plant as a source of food and other needs. Since the birth of agriculture, the unwavering cultivation of plants remains a long-cherished goal for human beings. Unfortunately, diseases are the greatest threats to uninterrupted cropping. The majority of plant diseases occurs wherever the host plants are grown. Plant diseases hinder plant growth and cultivation by their presence, destroy parts or all of the plants, and reduce their productivity. Diseases are responsible for the significant decline of crop yields and quality in the field and increase damages of produce in storage (Hossain et al. 2014a, b). Food shortage, higher food prices, and mass starvation are historically associated with plant diseases. Plant diseases are annually accountable for intense crop losses throughout the world. It has been estimated that 14.1% of the crops are lost annually due to plant diseases alone. The total amount of worldwide crop loss per year from plant diseases is about $220 billion (Agrios 2005). Protection against such enormous losses caused by plant diseases can play a significant role in meeting the growing demand for food quality and quantity and improving food security worldwide. Various approaches have been developed to protect crop plants from becoming diseased or curing them after becoming diseased. The common approaches to managing plant diseases are regulatory, cultural, biological, physical, or chemical methods. These methods have different efficacy levels, advantages, and disadvantages.

Synthetic fungicides have been the mainstay for plant disease control since the beginning of the green revolution. Fungicides of diverse arrays have successfully controlled most diseases that cause rapid and extensive losses of crops. Fungicide use is highly prevalent worldwide. They are cheaper and easy to apply with a wide

Md. M. Hossain (✉)
Department of Plant Pathology, Bangabandhu Sheikh Mujibur Rahman Agricultural University, Gazipur, Bangladesh
e-mail: hossainmm@bsmrau.edu.bd

M. Rai et al. (eds.), *Phoma: Diversity, Taxonomy, Bioactivities, and Nanotechnology*,
https://doi.org/10.1007/978-3-030-81218-8_15

spectrum of activity and high efficacy. Undoubtedly, their use is responsible for much of the unparalleled increase in crop productivity. However, many of the fungicides and their major metabolites are stable and persistent in the environment for a long period after application and pose a substantial hazard. Moreover, their continued use may result in the emergence of resistance to pesticides. Several countries, including the United States, Canada, France, and Australia, have reported fungal isolates resistant to fungicides (Szkolnik and Gilpatrick 1969; Sutton 1978; Thind et al. 1986; Carisse and Pelletier 1994). The banning of methyl bromide, pentachloronitrobenzene, and several other persistent fungicides has also contributed to the current revival of interest in alternative disease control systems. Idyllically, the alternate pest control practice is expected to reduce the disease damages at efficiency as compared to that achieved with chemical fungicides but not to pose a threat to either the environment or the non-target organisms (Pandey et al. 2001; Hossain et al. 2014a, b; Akhter et al. 2015). With these considerations, the biological management of plant diseases by microbial inoculants appears to be a highly viable and promising alternative to chemical control. In recent decades, many microorganisms have been identified as biological control agents (BCAs) for plant diseases. Microbial BCAs can offer several advantages over synthetic fungicides: they are non-toxic for mammals and the environment, do not leave toxic residues, have no adverse effect on soil microflora and microfauna, and do not negatively affect the natural biocontrol population. Thus, the use of BCAs may lower the number of risks to the environment.

Among various microbial inoculants, non-pathogenic fungi belonging to different species of *Phoma* have long been demonstrated to be generally effective as BCAs against plant pathogens. A few previous reports have identified *Phoma* spp. directly inhibitory to both airborne and soilborne plant pathogens. Other reports have characterized them as a modulator of plant defences against a wide spectrum of pathogens (Hossain et al. 2008a; Sultana et al. 2008, 2009; Elsharkawy and El-Khateeb 2019). Many of the *Phoma* spp. are known to produce bioactive metabolites, which have shown great potential for antifungal and antibacterial activities (Logan and O'Neill 1970; Huang et al. 2017). *Phoma* spp. have also been well-known for their plant growth-promoting activities (Hossain et al. 2017a; Hossain and Sultana 2020). Amendment of soil with *Phoma* inoculum exerts a significant positive effect on plant growth and yield, making them well fitted to sustainable agriculture. The dynamic ability of non-pathogenic *Phoma* spp. to promote plant growth, antagonize pathogens, and induce defences against diseases gives rise to the hope that they can be made available to develop potential BCAs of many plant diseases (Fig. 15.1).

This chapter is attempted to focus on the recent findings related to the use of non-pathogenic *Phoma* for the biological management of plant diseases caused by various groups of pathogens. Special attention has been paid to the underlying biocontrol mechanisms, the interaction of *Phoma* with arbuscular mycorrhizal fungi (AMF) on disease suppression, and factors affecting the *Phoma* biocontrol activity. A review has also been included on the genes, metabolites, and signalling pathways involved in this process. The information produced from this review could be very valuable to those who are interested in environmental protection and agricultural sustainability.

15.2 The Nature and Diversity of Non-pathogenic *Phoma* spp.

The anamorphic genus *Phoma* Sacc. emend. Boerema et G.J. Bollen is an omnipresent taxon and found worldwide. The genus represents one of the largest fungi groups belonging to the phylum *Ascomycota*, class *Dothideomycetes*, order *Pleosporales*, and family *Didymellaceae*. Although the genus initially included more than 3000 species (Monte et al. 1991), the descriptions of 223 specific and infraspecific taxa of *Phoma* are formally recognized in the *Phoma*

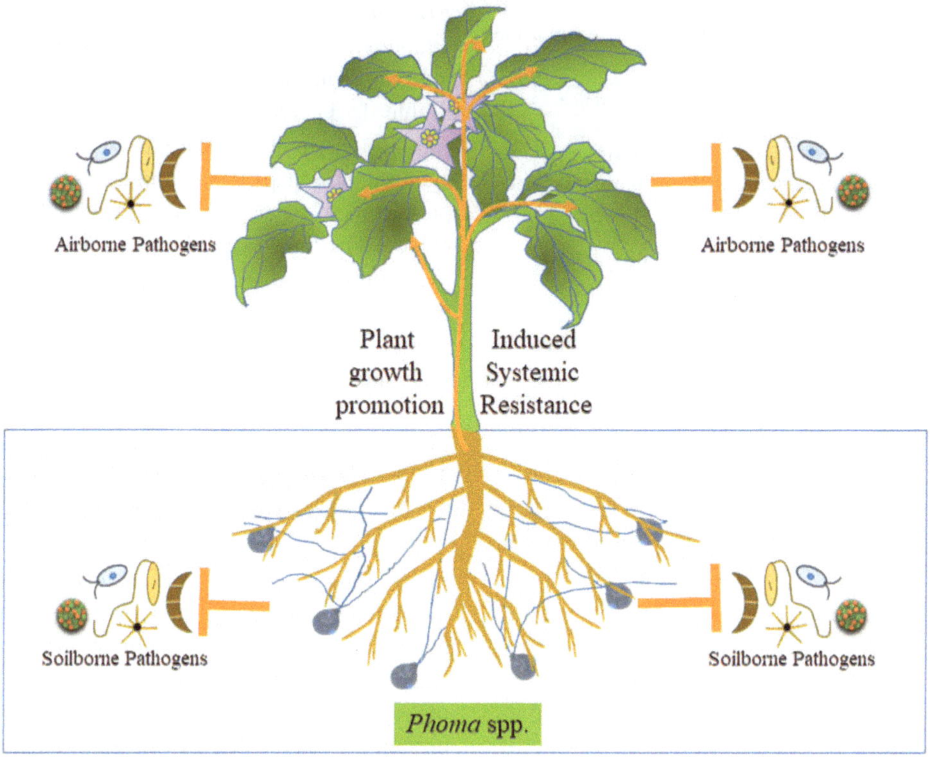

Fig. 15.1 Effect of non-pathogenic *Phoma* spp. on disease suppression and growth promotion of plants. *Phoma* spp. antagonize the deleterious microorganisms in the rhizosphere and control the soilborne diseases. *Phoma* spp. play a role in the protection of plants against a wide spectrum of pathogens by inducing systemic resistance. *Phoma* spp. also stimulate shoot growth, root growth, flowering, and yield of plants

Identification Manual (Boerema et al. 2004). It is perceived that the actual number of taxa within this genus may be much higher. The terrestrial *Phoma* are notorious phytopathogens, including quarantine species that cause significant loss in economically important crops (Zimowska 2007; Badillo-Vargas et al. 2008). They are also implicated as human and animal opportunistic pathogens (Monte et al. 1991). However, numerous species of *Phoma* are harmless and non-pathogenic to plants. Rather they are intimately linked to plant growth and health due to their roles in suppressing pathogens, decomposing organic residues, and providing nutrients (Hossain et al. 2017a, b; Hossain and Sultana 2020). Among these species, *Phoma* sp., *P. glomerata*, *P. etheridgei*, *P. tropica*, and *P. betae* are, by far, the most extensively reported as plant growth promoters and agents for biological control of plant diseases (Table 15.1). It is apparent that *Phoma* spp. have a widespread distribution and occur in a wide range of ecological niches. There is a large body of descriptive work on the occurrence of *Phoma* species on natural substrata. Some *Phoma* is common rhizosphere-associated, soilborne, and saprobic fungi (Meera 1994; Shivanna et al. 1996; Aveskamp et al. 2008; Liermann et al. 2009). The non-pathogenic *Phoma* also contains endophytic species associated with a wide range of hosts such as aspen (*Populus tremuloides* Michx.) (Hutchinson et al. 1994), *Viola* sp. (Arie et al. 1999), wheat (*Triticum aestivum*) (Diamond and Cooke 2003; Comby et al. 2017), mangrove plant (*Myoporum bontioides*) (Huang et al. 2017), Quebec orchard (Ouimet et al. 1997), Guinea plant (*Saurauia scaberrinae*) (Hoffman et al. 2008), *Salsola oppositifolia* (Qin et al. 2010), Malaysian medicinal plant *Cinnamomum mollissimum* (Santiago et al. 2012), and Chinese medicinal plant

Table 15.1 Non-pathogenic species of *Phoma* reported as biological control agent of plant diseases with their original source of isolation

Strain	Source of isolation	Reference
Phoma sp. GS8-1, GS8-2, GS8-3, GS6-1, GS6-2, GS7-3, GS7-4, GS10-2, GS12-2	Zoysiagrass (*Zoysia tenuifolia*) rhizosphere	Meera et al. (1995a, b); Shivanna et al. (1996); Sultana et al. (2008); Elsharkawy et al. (2014)
Phoma glomerata	Sycamore tree (*Platanus occidentalis* L.) leaf	Sullivan and White (2000)
Phoma glomerata strain JCM9972	*Viola* sp. leaf	Arie et al. (1999)
Phoma etheridgei	Aspen (*Populus tremuloides* Michx.) stem	Hutchinson et al. (1994)
Phoma etheridgei	Black galls of Aspen (*Populus tremuloides* Michx.)	Ayer and Jimenez (1994)
	Lower crown of Guinea plant (*Saurauia scaberrinae*)	Hoffman et al. (2008)
Phoma sp. L28	Mangrove plant *Myoporum bontioides* roots	Ding et al. (2014); Huang et al. (2017)
Phoma sp. (P59A), *Phoma* sp. (P66A)	Quebec orchard	Bernier (1995); Ouimet et al. (1997)
Phoma betae	Wheat (*Triticum aestivum*)	Diamond and Cooke (2003)
Phoma spp. WF4	Finger millet (*Eleusine coracana* L. Gaertn.) roots	Mousa et al. (2015)
Phoma glomerata	Wheat (*Triticum aestivum* cv. Apache and Caphorn) inner tissues	Comby et al. (2017)
Phoma sp. strain 8889	*Salsola oppositifolia*	Qin et al. (2010)
Phoma sp.	French alpine soil	Liermann et al. (2009)
Phoma sp.	*Cinnamomum mollissimum* leaf and stem	Santiago et al. (2012)
Phoma sp.	*Arisaema erubescens* tubers	Wang et al. (2012)

Arisaema erubescens (Wang et al. 2012). These endophytes are known to occur on the leaf, stem, lower crown, roots, tuber, and plant inner tissues (Table 15.1). Other non-pathogenic species of *Phoma* are being reported as mycoparasitic agents and perhaps feed on fungal saprophytes or pathogens of diseased tissue (Sutton 1980). It is evident that the non-pathogenic *Phoma* spp. due to their ubiquitous occurrence, high diversity, and miscellaneous modes of action can be assigned to provide multiple ranges of benefits to plants, including disease suppression.

15.3 Non-pathogenic *Phoma* spp. as a Plant Growth Promoter

Many fungi isolated from diverse ecological niches can stimulate plant growth and yield. In general, the promotion of plant growth is mostly caused by a heterogeneous group of non-pathogenic fungi. They are called plant growth-promoting fungi (PGPF) (Hossain and Sultana 2020). Over the past few decades, a wide range of PGPF has been studied, including those that belong to the genus *Phoma*. *Phoma* spp. mediate both short- and long-term effects on the plant. The most common of which is their effect on germination, seedling vigour, plant growth, and yield. Treatment of bacterial wilt susceptible tomato cultivar with *P. multirostrata* resulted in early seedling emergence and enhanced vigour compared to untreated controls (Jogaiah et al. 2013). Similarly, isolates of *Phoma* spp. improved seed germination and seedling vigour in sunflower (Nagaraju et al. 2012). Stimulated shoot and root growth by different isolates of *Phoma* spp. were reported in cucumber, soybean, tobacco, and *Suaeda* (Chandanie et al. 2005; Shivanna et al. 2005; Hamayun et al. 2009; Zhou et al. 2014). Soil treatment with isolates *Phoma* spp. significantly improved earhead length, ear weight, and seed number in wheat (Shivanna et al. 1996). Again, *Phoma* spp. significantly increased fruit number of cucumbers until 10-week post-planting in greenhouse trials (Shivanna et al. 2005). Hence, there is strong evidence for a proven role of *Phoma* spp. as plant growth promoters in different crops. The process of plant growth promotion by *Phoma* spp. is multifaceted and often cannot be credited to a single mechanism. Various mechanisms such as phosphate solubilization, mineralization of organic substrate, phytohormone and enzyme production, suppression of deleterious microorganisms, and amelioration of abiotic stresses are known to modulate plant growth and development by *Phoma* spp. (Hossain and Sultana 2020). With one or more of these mechanisms, a particular *Phoma* isolate may influence plant growth and development.

15.4 Non-pathogenic *Phoma* spp. as Biological Control Agents of Plant Diseases

Considerable emphasis has been given to using non-pathogenic *Phoma* spp. as a potential means of protecting plants from various diseases. The evidence for the role played by *Phoma* spp. as BCAs for plant disease suppression has come from several pot and field experiments. The results of these experiments often indicated that *Phoma*-associated plants had less disease than their non-treated counterparts did. The *Phoma* BCAs successfully protected a wide range of plants from both root-infecting and shoot-infecting diseases caused by fungi, bacteria, viruses, protists, and peronosporomycetes (Table 15.2). In many cases, the outcome of using *Phoma* spp. for biological control is as dramatic as the use of pesticides and other control measures.

15.4.1 Biological Control of Fungi by *Phoma* spp.

Fungi form the largest group of plant pathogens and are responsible for the majority of serious plant diseases. Fungicides are widely applied to control these diseases. However, the increased knowledge of farmers and the strict thresholds in pesticide residues set by many food vendors are also inspiring the shift from chemical to biopesticides. Many microorganisms have been tried, and a few hundred strains have shown efficacy against fungal diseases. Some of these microorganisms are the bioactive strains of non-pathogenic *Phoma* spp. These strains were successfully applied against a wide spectrum of soilborne and airborne fungal pathogens. Trunk decay caused by an obligate fungal agent *Phellinus tremulae* is the most serious aspen disease (*Populus tremuloides* Michx.) in Western Canada, where *Phellinus tremulae* accounts for approximately 39% of trunk decay volume in aspen (Thomas et al. 1960). Presently, no other means of control is effective against this disease. A strain of *P. etheridgei* isolated from aspen stem has shown potential for biological control of the pathogen. Pre-inoculation of *P. etheridgei* onto aspen wood chips inhibited colonization and decay by *Phellinus tremulae* (Hutchinson et al. 1994), indicating that the inoculation of *P. etheridgei* into stems of young aspen stands could be used as an approach of controlling possible infection and decay by *Phellinus tremulae.*

Certain *Phoma* spp. can provide a practical alternative of chemicals for the control of diseases in wheat. Take-all caused by *Gaeumannomyces graminis* var. *tritici* and common root rot and seedling blight caused by *Cochliobolus sativus* are economically important wheat diseases (*Triticum aestivum* L.). Due to the absence of genetic resistance and effective chemical controls, importance has been given to amending soil with BCAs for protecting the roots of wheat. Isolates of *Phoma* spp. from the zoysiagrass rhizosphere were investigated for their capacity to suppress take-all and common root rot of wheat. Amending soil with barley grain inoculum of *Phoma* sp. GS6-1 and GS7-4 suppressed take-all and common root rot incidence in wheat compared to disease caused by pathogen alone. In this study, *Phoma* isolates were more effective as BCAs than the sterile fungus (Shivanna et al. 1996). *Fusarium* head blight (FHB) is another most serious disease on wheat crops worldwide. A complex of several *Fusarium* species causes FHB. *Fusarium graminearum* Schwabe and *F. culmorum* (Smith) Saccardo are

Table 15.2 Use of non-pathogenic *Phoma* spp. for biological control of various diseases in different host plants

Pathogen category	Pathogen name	Disease name	Host plant	*Phoma* strain	Effect	Reference
Fungi	*Phellinus tremulae*	Wood decay	Aspen	*Phoma etheridgei*	Inhibition of wood decay	Hutchinson et al. (1994)
	Ophiostoma crassivaginata	Blue strain	Aspen	*Phoma etheridgei*	Suppression of blue stain	Ayer and Jimenez (1994)
	Colletotrichum orbiculare	Anthracnose	Cucumber	*Phoma* sp. GS8-1, GS8-2, GS8-3	Reduction in lesion number and lesion size	Meera et al. (1994); Meera et al. (1995a, b); Chandanie et al. (2006)
	Gaeumannomyces graminis	Take-all	Wheat	*Phoma* sp. GS6-1, GS7-4	Reduction in disease severity	Shivanna et al. (1996)
	Cochliobolus sativus	Common root rot and seedling blight	Wheat	*Phoma* sp. GS6-1, GS7-4	Reduction in disease severity	Shivanna et al. (1996)
	Sclerotium cepivorum	White rot	Onion	*Phoma* sp. GS8-1, GS8-3	Reduction in disease severity and incidence	Elsharkawy and El-Khateeb (2019)
	Fusarium moniliforme	Bakanae disease	Rice	*Phoma* sp. GS12-2	Reduction in disease severity	Elsharkawy et al. (2014)
	F. culmorum	Fusarium head blight	Wheat	*Phoma betae*	Reduction in Fusarium head blight	Diamond and Cooke (2003)
	F. culmorum	Fusarium head blight	Wheat	*Phoma glomerata*	Reduction in Fusarium head blight	Comby et al. (2017)
	F. oxysporum f.sp. *cucumerinum*	Fusarium wilt	Cucumber	*Phoma* sp. GS8-2	Reduction in disease severity	Koike et al. (2001)
	F. semitectum (F-48)	Fusarium blight	Rice	*Phoma* sp. GS8-2, GS6-4, GS10-2	Reduction in disease severity	Elsharkawy et al. (2014)
	Rhizopus sp. (R16)	Rhizopus blight	Rice	*Phoma* spp. *GS8-2, GS6-4, GS12-2*	Reduction in disease severity	Elsharkawy et al. (2014)
Bacteria	*Pseudomonas syringae* pv. tomato	Leaf speck	*Arabidopsis*	*Phoma* sp. GS8-1	Restricted disease development and inhibited pathogen proliferation	Hossain et al. (2008a)
	Ps. syringae pv. tomato	Leaf speck	*Arabidopsis*	*Phoma* sp. GS8-3	Restricted disease development and inhibited pathogen proliferation	Sultana et al. (2009)

	Ps. syringae pv. *tomato*	Leaf speck	*Arabidopsis*	*Phoma* sp. GS6-2; GS7-3	Restricted disease development and inhibited pathogen proliferation	Sultana et al. (2008)
	Ps. lachrymans	Angular leaf spot	Cucumber	*Phoma* sp. GS8-2	Reduction in lesion number and size	Koike et al. (2001)
Viruses	*Cucumber Mosaic Virus*	*Cucumber mosaic virus* disease	*Arabidopsis*, cucumber	*Phoma* sp. GS8-2	Reduction in severity and virus accumulation	Elsharkawy (2018).
	Cucumber Mosaic Virus	*Cucumber mosaic virus* disease	Cucumber, tobacco	*Phoma* sp. GS8-3	Reduction in severity and virus accumulation	Elsharkawy et al. (2020).
Protista	*Plasmodiophora brassicae*	Clubroot	Chinese cabbage, turnip, cabbage, broccoli	*Phoma glomerata* strain JCM9972	Suppression of root galls	Arie et al. (1999)
Peronosporomycetes	*Phyllactinia guttata*	Powdery mildew	Oak	*Phoma glomerata*	Suppression of powdery mildew	Sullivan and White (2000)

the most pathogenic and most dominant causes of FHB. The pathogens can cause severe diseases and produce toxins that are phytotoxic and dangerous to animals and humans. A strain of *P. betae* isolated from wheat plants was found promising as a BCA against FHB. When wheat-ears were pre-inoculated with *P. betae* and challenged with *F. culmorum* during the flowering time under glasshouse conditions, FHB symptoms decreased by 60% relative to the control treatment (Diamond and Cooke 2003). Recently a study conducted by Comby et al. (2017) reported that *P. glomerata* isolated from inner tissues of wheat also effectively inhibited the growth of *F. graminearum* and *F. culmorum* in vitro and reduced the level of FHB infection in a detached wheat spikelet showing the potential of *Phoma* species in controlling FHB in wheat.

Some rhizosphere fungal isolates belonging to different PGPF were evaluated to control *Fusarium* blight, *Rhizopus* blight, and bakanae disease (Elsharkawy et al. 2014). Among these isolates, *Phoma* sp. GS8-2, GS6-4, and GS10-2 caused a 74–89% reduction of *Fusarium* blight severity in rice seedlings. The PGPF *Phoma* sp. GS8-2, GS6-4, and GS12-2 reduced *Rhizopus* blight disease severity to 81–93%, while *Phoma* sp. GS12-2 appeared the most effective isolate in controlling bakanae disease, showing a 90% reduction in disease severity (Elsharkawy et al. 2014). Treatment with *Phoma* isolates can also be used successfully to reduce the *Fusarium* wilt caused by *F. oxysporum* f.sp. *cucumerinum* and white rot caused by *Sclerotium cepivorum*. Cucumber plants grown in potting medium amended with barley grain inoculum of *Phoma* sp. GS8-2 showed significantly lower disease severity (1.33) of *Fusarium* wilt than that of untreated control plants (2.80) (Koike et al. 2001). Similarly, onion seedlings treated with barley grain inoculum of *Phoma* sp. isolate GS8-1 and GS8-3 before pathogen inoculation significantly reduced the severity and incidence of white rot compared to control in the greenhouse and field (Elsharkawy and El-Khateeb 2019). Therefore, inoculation of soil with *Phoma* inoculum appears to be an effective way of controlling soilborne fungal diseases in plants.

The PGPF *Phoma* spp. have widely been investigated for their ability to confer resistance against fungal leaf pathogens. *Colletotrichum orbiculare* is an economically important pathogen causing anthracnose disease on cucumber in Japan, particularly in greenhouses. When cucumber roots were pretreated with 16 PGPF isolates as barley grain inoculum, mycelial inoculum, and culture filtrate, only *Phoma* sp. isolate GS8-1, GS8-2, and GS8-3 protected plants against *Colletotrichum orbiculare* (Meera et al. 1994; Koike et al. 2001). They significantly decreased the lesion number and lesion diameter on leaves caused by *C. orbiculare*. Moreover, the average rate of expansion of lesion diameter on leaves due to *C. orbiculare* was also slower in treated plants than the untreated control plants. The protection afforded by these isolates with a single prior inoculation lasted until 9 weeks when tested in the greenhouse (Meera et al. 1995a) and for 6 weeks in the field (Meera 1994). This result proves that *Phoma* spp. can protect plants over a prolonged duration.

15.4.2 Biological Control of Bacteria by *Phoma* spp.

Plant pathogenic bacteria are associated with many important diseases of plants all over the world. Various reports on applications of *Phoma* spp. as BCAs of plant bacterial diseases have been documented. The bacterium *Pseudomonas syringae* made the strongest appearance on scientific and economic grounds in a recent survey on 'Top 10 plant pathogenic bacteria in molecular plant pathology' (Mansfield et al. 2012). The economic impact of *Ps. syringae* has been growing, with a reappearance of old diseases, including bacterial speck of tomato pathogen *Ps. syringae* pv. *tomato* (Shenge et al. 2007). Biocontrol potential of *Phoma* against bacterial speck disease was tested in a classical *Arabidopsis* and *Ps. syringae* pv. *tomato* pathosystem. *Phoma* sp. GS8-1 protects *A. thaliana* against bacterial leaf speck pathogen *Ps. syringae* pv. *tomato*. The one-hour root treatment with *Phoma* sp. GS8-1 culture filtrate is significantly sufficient to reduce the leaf surface area invaded by the pathogen and the extent of pathogen proliferation in inoculated *Arabidopsis leaves* (Hossain et al. 2008a). The GS8-1-induced protection resulted in an average 58% reduction in disease severity and a sixfold less pathogen population than in the control plants. The observed suppression of disease symptoms and bacterial proliferation in *Phoma*-treated plants was phenotypically similar to those achieved by treatment with commercial plant defence activator benzothiadiazole (BTH) (Hossain et al. 2008a). Similarly, the root treatment of *Arabidopsis* with a culture filtrate from *Phoma* sp. GS8-3 conferred protection against the bacterial speck pathogen (Sultana et al. 2009). The reduction in disease severity and pathogen population in GS8-3-treated plants over control plants was 59% and threefold growth, respectively. In another independent study, *Arabidopsis* plants were treated with *Phoma* GS6-2 and GS7-3 culture filtrates, which reduced the number of bacterial speck symptomatic leaves, restricted disease severity, and suppressed *Ps. syringae* pv. *tomato* proliferation compared to control plants (Sultana et al. 2008). *Phoma* sp. can also protect cucumber plants significantly against bacterial angular leaf spots caused by *Ps. lachrymans* (Koike et al. 2001). Cucumber roots treated with *Phoma* sp. GS8-2 barley grain inoculum and culture filtrate decreased the total lesion number and diameter of the angular leaf spot significantly compared to the untreated control plants. The level of disease protection induced by barley grain inoculum and culture filtrate of this isolate was 31.6% and 75.5%, respectively (Koike et al. 2001). Treatment with *Phoma* spp. is thus often successful for biological control of bacterial diseases in plants.

15.4.3 Biological Control of Viruses by *Phoma* spp.

The infectious pathogen viruses cause many serious diseases in plants and are responsible for partial or total crop losses. They are one of the problematic plant pathogens and cannot directly be controlled by chemical applications. An applicable and effective management strategy is needed for controlling virus infection. Biological control involving the use of microbial agents is thought to be a practical option for viral diseases. In several studies, the *Phoma* spp. were successfully used to enhance resistance in plants against viruses. *Cucumber mosaic virus* (CMV) (genus *Cucumovirus*) is considered the most serious pathogen, causing enormous damage to various plants worldwide. The PGPF *Phoma* sp. GS8-2 and GS8-3 were evaluated for their ability to control CMV. *Arabidopsis* and cucumber plant treated with GS8-2 barley grain inoculum and culture filtrate showed decreased disease severity relative to the control plant that displayed serious symptoms (Elsharkawy 2018). The treatment with GS8-2 also reduced virus accumulation in the treated plants compared to the control (Elsharkawy 2018). Another study evaluated the resistance stimulated by *Phoma* sp. GS8-3 against CMV under pot and field conditions (Elsharkawy et al. 2020). Treatments with GS8-3 barley grain inoculum and culture filtrate showed a significant decrease in CMV severity and titre in tobacco and cucumber under pot and field conditions, respectively. *Phoma* sp. defends plants from viruses by activating varied defence responses (Elsharkawy et al. 2020). Their ability to protect plants from virus infection makes them a significant resource for developing BCAs against plant viruses.

15.4.4 Biological Control of Protists by *Phoma* spp.

The Protista is a big composite group of unicellular organisms from the eukaryotic lineage. Some protists are pathogenic to plants and cause massive damage to crops. Clubroot caused by the biotrophic protist *Plasmodiophora brassicae* Woronin is an important disease of cruciferous crops. The disease has a worldwide significance and causes substantial damage to these crops. The infected root cells are subjected to a process of abnormal cell proliferation and enlargement, leading to the formation of gall-like, spherical, knobby, or club-shaped swellings (Kageyama and Asano 2009). Away from liming, disease management strategies are limited to this disease. Very few fungicides are effective against clubroot (Hwang et al. 2017). Some antagonistic microorganisms, including *Phoma* spp., have been employed to control clubroot disease. A *P. glomerata* strain isolated from *Viola* sp. leaf sampled in Saitama, Japan, was found to control clubroot disease in brassica crops. Mixing mycelial inoculum of *P. glomerata* cultured on soil-bran medium or pouring culture broth of the fungus into the infested soil inhibited clubroot disease on the root of Chinese cabbage, turnip, cabbage, and broccoli (Arie et al. 1998, 1999). Both treatments suppressed the development of galls on the root. Tian et al. (2019) indicated significant differences between the endophyte communities in the healthy roots and clubroots. *Phoma* was found to be one of the common fungal genera in the healthy roots of tumorous stem mustard compared to those infected with clubroot disease (Tian et al. 2019). *Phoma* in the healthy roots of tumorous stem mustard may accomplish an inhibitory function to *Plasmodiophora brassicae*.

15.4.5 Biological Control of Peronosporomycetes by *Phoma* spp.

The peronosporomycetes, often referred to as oomycetes, are fungus-like heterotrophs placed in the kingdom Straminipila (Islam and Hossain 2013). The main economic impact of the peronosporomycetes results from the phytopathogenic genera. As the causative agents of powdery mildew and various other diseases, they are a significant impairment to agriculture. Powdery mildews are widespread diseases that occur on a wide range of host plants. Despite the effectiveness, there remains a robust interest in evaluating alternative methods to chemical control of powdery mildews. However, only a few microbial inoculants have been demonstrated to be effective as a biocontrol agent of powdery mildews (Sundheim and Tronsmo 1988; Falk et al. 1995). Several pycnidial fungi, other than *Ampelomyces*, show the potential to parasitize powdery mildew fungi. For instance, pycnidial fungi isolated from powdery mildew colonies collected from the sycamore leaves (*Platanus occidentalis* L.) were characterized and confused with *P. glomerata* isolates (Sullivan and White 2000). Such isolates were detected as 'fast-growing *Ampelomyces* isolates' in previous works (Rudakov 1979; Kiss and Nakasone 1998), usually derived from sessile pycnidia from mildew-infected leaves, while the so-called slow-growing *Ampelomyces* isolates came from intracellular pycnidia typical of the 'true' *Ampelomyces*. The molecular systematics using ITS1 sequence analysis grouped the 'fast-growing isolates' in a clade specified by *Phoma* species and the 'slow-growing isolates' in different species (Sullivan and White 2000). The fast-growing pycnidial fungi confused with *Ampelomyces* in earlier works are, in fact, *P. glomerata*. They are natural mycoparasites of powdery mildew mycelia (Sullivan and White 2000). When a conidial suspension of *P. glomerata* was used to inoculate epiphyllous mycelia of *Phyllactinia guttata* on intact oak leaves, treated leaves developed abundant pycnidia on and around the inoculation sites. However, they did not produce powdery mildew cleistothecia (Sullivan and White 2000). None of the control leaves showed the development of *P. glomera*ta pycnidia, and cleistothecia developed normally. This result indicated that *P. glomerata* could colonize and suppress powdery mildew development on oak (Sullivan and White 2000). Hence, *P. glomerata* is a potential new mycoparasitic agent of powdery mildews. The mycoparasite *P. glomerata* appears to have a widespread distribution. In addition to isolation in the United States (Sullivan and White 2000), several mycoparasitic strains of *P. glomerata* were isolated from species of the powdery mildew genus *Microsphaera* in Russia (as *Ampelomyces quercinus*) (Rudakov 1979). The mycoparasitic strains of *P. glomerata* have also been found in the downy mildew of grapes (*Plasmopara viticola*) and the species of the genus *Sphaerotheca* (as *Ascochyta humuli*) in Russia (as *Ascochyta heraclei*) (Rudakov 1979).

15.5 Interaction of *Phoma* spp. and Arbuscular Mycorrhizal Fungi on Disease Protection

The plant rhizosphere harbours a diverse array of microorganisms. The complex interaction among them is common in nature. This interaction appears to be a key determinant for plant growth and health. The ongoing plant-microbe interaction in nature can be mimicked by artificially introducing two or more microbial inoculants in the plant rhizosphere. However, the interaction between two particular microorganisms in the rhizosphere is very specific. It can be either additive (inhibit the beneficial effect of individual species), neutral (no influence on the beneficial effect of individual species), or antagonistic (inhibit the beneficial effect of individual species). Before exploiting the interaction effects of two microorganisms for the benefit of sustainable agriculture, it is necessary to characterize the nature of the interaction between them. Arbuscular mycorrhizal fungi (AMF) are obligate endophytes that establish symbiotic relationships with plant roots. They are known to improve the plant nutritional status (Smith and Read 1997) and protect plants against plant pathogens (Azcon-Aguilar and Barea 1996). Likewise, PGPF *Phoma* spp. are root-associated saprophytes that provide a wide range of benefits to plants, including plant growth promotion and disease suppression (Hossain et al. 2017a, b). Since both AMF and *Phoma* are beneficial to a plant, it is important to know whether their interaction effect could benefit the plant. Interaction between AM fungus *Glomus mosseae* and two isolates of PGPF *Phoma* sp. (GS8-2 and GS8-3) was studied in the cucumber rhizosphere (Chandanie et al. 2005). The interaction effect of *G. mosseae* and *Phoma* sp. on the extent of their respective root colonization ability was determined. It was observed that the root length colonized by *G. mosseae* was not adversely affected by the presence of *Phoma* isolates. However, the root colonization ability of *Phoma* sp. was significantly reduced (80–84%) by the inoculation of *G. mosseae* in cucumber plants. The interaction effect of *G. mosseae* and two isolates of *Phoma* sp. (GS8-2 and GS8-3) on plant growth and suppression of anthracnose disease caused by *C. orbiculare* has also been investigated. Evidence suggests a significant effect of interaction between *G. mosseae* and *Phoma* sp. on plant growth and protection against anthracnose disease (Chandanie et al. 2005, 2006). Dual inoculation of *Phoma* with *G. mosseae* was found inhibitory to cucumber growth. In contrast, stimulated plant growth was observed when applied alone (Chandanie et al. 2005). A similar interaction effect between them was also detected in the suppression of cucumber anthracnose disease. Inoculation of *Phoma* spp. GS8-2 or GS8-3 alone into the cucumber root system provided considerable protection against *C. orbiculare* (Chandanie et al. 2006). However, individual inoculation of *G. mosseae* into cucumber roots had no significant effect on the disease suppression. Accordingly, combined inoculation of *Phoma* sp. and *G. mosseae* decreased the level of disease protection induced by a single inoculation of each *Phoma* isolate. These reduced disease protection levels in the dual-inoculated plants might be related to the mycorrhizosphere effect observed on *Phoma* colonization. Root colonization ability is an essential characteristic of rhizosphere fungi for rendering beneficial effects on the plant. The reduced root colonization ability of *Phoma* due to two species' interaction effects may ultimately affect plant growth and performance under stress. AMF are known for their ability to incite alterations in the host's physiology and root exudates (Linderman 1988). AM fungi can produce substances capable of interfering with microorganisms and support a highly specified microbial community around mycorrhizal root. The root colonization by *Phoma* sp. could have been affected by the substances released by the AM fungi and/or root exudates modified by AMF symbiosis.

15.6 Factors Affecting the Biocontrol Potential of *Phoma* spp.

The reliable performance of the introduced BCAs has been the greatest challenge to developing successful commercial products. In many cases, biological control of plant diseases by BCAs is very difficult to achieve under natural field conditions, despite showing promising activity in laboratory, pots, or field microplots. Variation in the abiotic environment in the field is the primary reason for the variable performance of BCAs. Understanding the influence of different abiotic factors on the disease suppressive activity of introduced BCAs will facilitate adapting them for use in fields.

Soil properties may influence the biocontrol activity of *Phoma* spp. by affecting the biological availability of essential elements for microbial growth. The biological control activities of *P. glomerata* were tested against the clubroot pathogen *Plasmodiophora brassicae* in three types of soils, i.e. horticultural, alluvium, and diluvium soil (Arie et al. 1999). Both soil-bran mycelial inoculum and culture broth of *P. glomerata* suppressed the pathogen in brassica crops in all three soil types tested, demonstrating that the biocontrol ability of *P. glomerata* against *Plasmodiophora brassicae* was not affected by different soil conditions.

The pH of a substrate may influence the antagonistic properties of *Phoma* spp. Culture pH showed a strong effect on *Phoma* sp. (P66A) inhibitory activity against an apple scab pathogen *Venturia inaequalis* (Cke.) Wint. (Ouimet et al. 1997). *Phoma* sp. (P66A) reduced the mycelial growth of *V. inaequalis* by more than 50–80% in a dual culture test at pH 5.0–6.5. The mycelial growth inhibition of *V. inaequalis* by *Phoma* sp. (P66A) increased up to 100% when the dual culture test was done at pH 7.0. The observed enhanced antagonistic properties of *Phoma* sp. (P66A) against *V. inaequalis* at pH 7.0 suggest the involvement of potent antifungal substance(s) produced by *Phoma* in a neutral environment.

The light regime may also have a significant effect on the antagonistic activity of biocontrol agents. The inhibition of *V. inaequalis* by *Phoma* sp. (P66A) was reduced by 20% under continuous light (Ouimet et al. 1997). Continuous light may likely be detrimental to the production and activity of antifungal compounds by *Phoma* spp. as it was observed with an antibiotic, chaetomin, produced by *Chaetomium globosum*. Chaetomin inhibited *V. inaequalis* in the dark, but the antibiotic was rapidly degraded by light (Boudreau and Andrews 1987).

Temperature is an important factor that significantly affects the inhibition of pathogen growth by *Phoma* spp. The inhibition of mycelial growth of *V. inaequalis* by *Phoma* spp. was 100% at 10–25 °C, but it decreased to 55% at 5 °C (Ouimet et al. 1997). Although *Phoma* spp. can act as a BCA at a wide range of temperatures, the production of antifungal metabolites might be adversely affected by cold temperature.

Variations in the carbon sources often result in an alteration in the antagonistic activity. *Phoma* sp. (P66A) exhibited a high degree of inhibition of *V. inaequalis* in the presence of starch, fructose, glucose, inositol, maltose, mannose, pectin, peptone, and sucrose tested. However, the inhibitory effect fluctuated with the different carbon sources, indicating a wider *Phoma* ability to adapt to diverse nutritional conditions. *Phoma* sp. (P66A) inhibited the growth of *V. inaequalis* in carbon source media by 46.8–100%, where the highest and lowest inhibition was found in the presence of inositol and peptone, respectively (Ouimet et al. 1997). Inositol is an essential nutrient used for building phosphatidylinositol and its derivatives in eukaryotes (Reynolds 2009). The BCA *Phoma* sp. (P66A) may acquire inositol to proliferate and inhibit the pathogen.

The cultivars' disease resistance level may mask fungal biocontrol potential (Hossain and Sultana 2015). However, a study conducted with *Phoma* isolates showed that the host level of basal resistance might not influence the ability of *Phoma* spp. to protect plants. Inoculation of soil with *Phoma* sp. isolate GS8-1 and GS8-2 significantly decreased the disease caused by *C. orbiculare* in all five tested cucumber cultivars, including one susceptible (cv. Suyo), one moderately susceptible (cv. Gibai), two resistant (cv.

Shogoin fushinari and Sagami hanjiro), and one moderately resistant (cv. Ochiai fushinari) cultivars in the greenhouse (Meera et al. 1995b). However, the resistant cultivar responded more effectively than the susceptible cultivar to induction with *Phoma* against *C. orbiculare* (Hassan et al. 2014). The observed protection in both susceptible and resistant cultivars by *Phoma* sp. against the disease excludes the possibility that the biocontrol potential of *Phoma* sp. is masked by the host disease resistance level.

15.7 Mechanisms of Disease Suppression

Understanding the mechanisms through which biocontrol of plant disease occurs is crucial to further improvement and wider use of BCAs. The course of plant disease suppression by BCAs, such as non-pathogenic *Phoma* spp., often cannot be attributed to a single mechanism. A highly diverse and complex network of mechanisms is involved in the biocontrol activity by *Phoma* spp. It has been reported that *Phoma* species may either provide direct or indirect protective activity against the pathogens (Fig. 15.2). The direct protection works predominantly through direct inhibition of pathogen activity by *Phoma* isolates using one or more antagonistic mechanisms. However, the indirect pathogen inhibition by *Phoma* BCAs is largely succeeded by the plant defence activities without direct antagonistic interaction with the pathogen.

15.7.1 Direct Protective Activity Against Plant Diseases by *Phoma* spp.

Direct pathogen suppression by *Phoma* spp. has frequently been reported in plant roots and shoots. Direct suppression generally results from the physical contact between the biocontrol agent and the pathogen. The primary direct protective modes include antibiosis, mycoparasitism, and competition for resources like nutrients or space that ultimately retard pathogen growth in plants (Fig. 15.2). Hence, the *Phoma* isolates selected as BCAs for direct disease suppression might be antagonists, parasites, or competitors of the targeted pathogens.

Antibiosis

Antibiosis is a class of biological relationships between two or more microorganisms that is inhibitory to at least one of them. Antibiosis develops by specific or non-specific metabolites, lytic enzymes, volatile compounds, or other toxic microbial substances. Using *Phoma* antagonists against the plant pathogen is an encouraging alternative since it is potentially less harmful to the environment. A few reports have identified a correlation between the in vitro antagonism of *Phoma* isolates against the pathogen and the in vivo disease control activities (Table 15.3). Two isolates, *Phoma* sp. (P66A) and (P59A), significantly reduced radial growth and conidial germination of apple scab (*V. inaequalis*) (Ouimet et al. 1997). The inhibition was stronger closer to the antagonist, which is expected since the metabolites excreted are probably more concentrated in its immediate vicinity. In inhibiting the mycelial growth and conidial germination of *V. inaequalis*, *Phoma* sp. has the desired effect (Ouimet et al. 1997). Another species, *P. etheridgei* sp. nov. Hutchison and Hiratsuka, was isolated from black galls of Canadian aspen trees (Ayer and Jimenez 1994). This fungus showed antagonistic activity in agar culture and on aspen chips against the aspen decay fungus *Phellinus tremulae* and the blue strain fungus *Ophiostoma crassivaginata* (Hutchinson et al. 1994; Ayer and Jimenez 1994). Liquid cultures of *P. etheridgei* also showed antifungal activity against *Phellinus tremulae*. The antagonism observed was chemical as no surface parasitism of *Phellinus tremulae* by *P. etheridgei* was observed microscopically (Ayer and Jimenez 1994). The extract from endophyte *Phoma* sp. WF4 had the widest antifungal target spectrum (7/26 fungal pathogens tested). It inhibited the growth of four tested *Fusarium* pathogens, such as *F. graminearum*, *F. lateritium*, *F. sporotrichioides*, and *F. avenaceum* (Mousa et al. 2015). *Phoma* sp. GS8-1 and GS 8-3 inhibited the mycelial growth and sclerotia formation

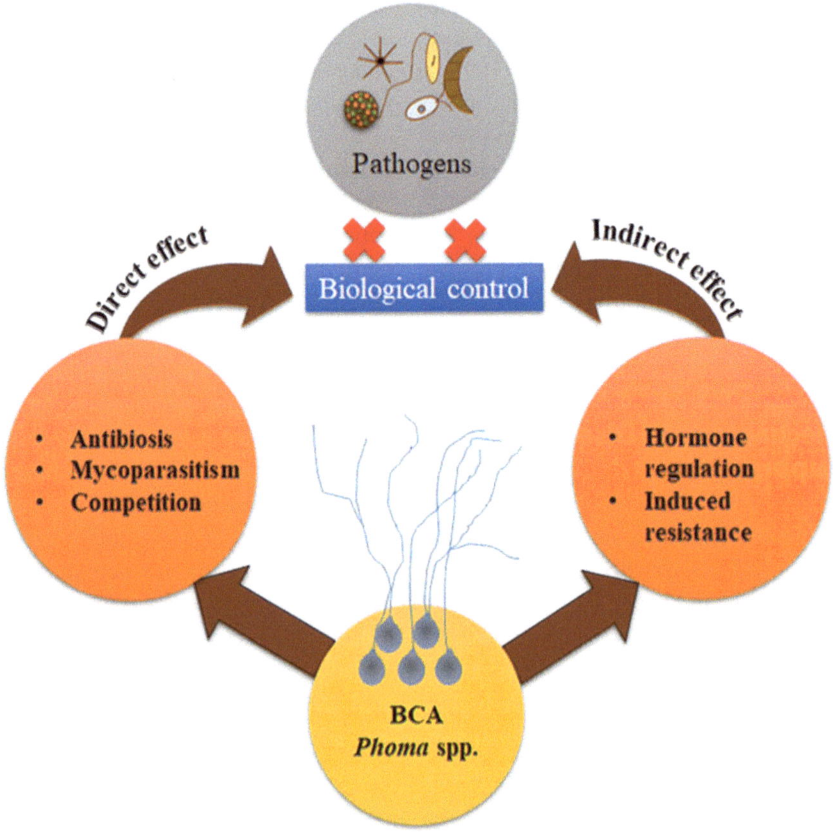

Fig. 15.2 The direct or indirect effects of BCA *Phoma* spp. on plant disease control

by white rot pathogen *Sclerotium cepivorum* (Elsharkawy and El-Khateeb 2019). Structural alternation and lysis in the mycelia of the pathogen were observed in the presence of *Phoma* sp. GS8-1 and GS8-3. The germination of sclerotia was also reduced after soaking in culture filtrates of *Phoma* sp. isolates (Elsharkawy and El-Khateeb 2019). The results show that *Phoma* isolates acting through antibiosis are directly interfering with the pathogen growth, leading to clear biocontrol effects on the disease. Such interference is a highly regulated cascade of metabolic events involving one or more active compounds. Evidence suggests that the antimicrobial compounds produced by *Phoma* spp. often relate to their in situ antagonism against pathogens.

Hyperparasitisms/Mycoparasitism

Biological control with *Phoma* antagonists relies on different mechanisms, including mycoparasitism. Mycoparasitism is the most promising nature of antagonism involving direct physical contact between the antagonist fungal agent and the target pathogen. In mycoparasitic interaction, one living fungus acts as a nutrient source for another. The mycelium of mycoparasite fungus grows towards the target pathogen, extensively coils around it, and secrets various hydrolytic enzymes leading to the dissolution of the pathogen's cell wall or membrane (Tiwari 1996). Indeed, the whole process occurs in four steps: target location, recognition, contact and penetration, and nutrient acquisition. At first, the mycoparasite mycelium shows chemotropic growth towards the host fungus (Chet et al. 1981). This tropic reaction is instigated by a chemical stimulus secreted by the host fungus. The direction of the concentration gradient dictates the mycoparasite growth direction (Barak et al. 1985). When a mycoparasite gets direct attachment with a fungal

Table 15.3 Different mechanisms of biological control by various *Phoma* spp.

Mechanism	Strain	Pathogen	Specific activities against pathogen	Reference
Antibiosis	*Phoma etheridgei*	*Phellinus tremulae*	Inhibition of mycelial growth	Hutchinson et al. (1994)
	Phoma etheridgei	*Ophiostoma crassivaginata*	Inhibition of mycelial growth	Ayer and Jimenez (1994)
	Phoma sp. GS8-1, GS8-3	*Sclerotium cepivorum*	Inhibition of mycelial growth, formation and germination of sclerotia, lysis and abnormal alternation in the mycelium	Elsharkawy and El-Khateeb (2019)
	Phoma sp. (P66A), (P59A)	*Venturia inaequalis*	Inhibition of radial growth and conidial germination	Ouimet et al. (1997)
	Phoma etheridgei	*Phellinus tremulae*, *Ophiostoma crassivaginata*	Showed antagonistic activity in agar culture and on aspen chips	Hutchinson et al. (1994); Ayer and Jimenez (1994)
	Phoma sp. WF4	*F. graminearum*, *F. lateritium*, *F. sporotrichioides*, *F. avenaceum*	Inhibition of the growth and breakage of hyphae	Mousa et al. (2015).
Hyperparasitism	*Phoma glomerata*	*Phyllactinia guttata*	Inhibition of production of cleistothecia	Sullivan and White (2000)
	Phoma betae	*Fusarium culmorum*	Less pathogen population in *Phoma*-colonized roots	Diamond and Cooke (2003)
Competitive root colonization	*Phoma* sp. GS6-1, GS7-4	*Gaeumannomyces graminis* *Cochliobolus sativus*	Inhibit root colonization by pathogen	Shivanna et al. (1996)
Hormonal regulation	*Phoma glomerata* strain JCM9972	*Plasmodiophora brassicae*	Production of anti-auxin compound	Arie et al. (1999)
Induced systemic resistance	*Phoma* sp. GS8-1, GS8-2, GS8-3	*Colletotrichum orbiculare*	Protected distantly parts of plants	Meera et al. (1995a, b); Chandanie et al. 2006
	Phoma sp. GS8-1	*Pseudomonas syringae* pv. *tomato*	Protected distantly parts of plants. Higher transcript levels of *PR-1*, *PR-2*, *PR-5*, *ChitB*, and *Hel* genes	Hossain et al. (2008a)
	Phoma sp. GS8-3	*Ps. syringae* pv. *tomato*	Protected distantly parts of plants. Higher transcript levels of *PR-1* and *PDF1.2* genes	Sultana et al. (2009)
	Phoma sp. GS6-2, GS7-3	*Ps. syringae* pv. *tomato*	Protected distantly parts of plants. β-Glucuronidase reporter gene expression of *PR-1a* and *PDF 1.2* constructs	Sultana et al. (2008)
	Phoma sp. GS8-2	*Cucumber mosaic virus*	Protected distantly parts of plants. Stimulated the transcription levels of β-1,3-glucanase, chitinase, *PR1*, *PAL1*, and *LOX1*	Elsharkawy (2018)
	Phoma sp. GS8-3	*Cucumber mosaic virus*	Protected distantly parts of plants. Upregulation of defence-related genes, especially heat-shock protein	Elsharkawy et al. (2020)

host, both recognize each other. The mutual recognition is achieved by binding the cell wall carbohydrate of mycoparasite with the host fungus surface lectins (Inbar et al. 1996). The next step involves the rapid growth of mycoparasite hypha on the host fungus and coiling around the host fungus hypha, followed by the host fungal cell penetration. Ultimately, the mycoparasitic fungus utilizes the mycelium of host fungi as a nutrient source. Whether the host fungus remains alive or dies depends on the nature of parasitism. In biotrophic mycoparasitism, the host fungus remains alive. The mycoparasitic fungus acquires nutrients from inside the host fungus mycelium in a well-adjusted relationship. On the contrary, necrotrophic mycoparasitism results in host death, from which the mycoparasite obtains nutrients. The significance of mycoparasitism in the biological control of plant diseases is undoubtedly remarkable. The BCA *Phoma* spp. employ mycoparasitism to kill phytopathogenic fungi. *Phoma glomerata* exhibits tremendous mycoparasitic activity against oak powdery mildew fungus *Phyllactinia guttata* (Sullivan and White 2000). *Phoma glomerata* colonizes and naturally infects powdery mildews, thereby suppressing the disease (Table 15.3). In another study, *P. glomerata* was found to inhibit *Fusarium* growth and control FHB in wheat but did not form an inhibition zone between *P. glomerata* and *Fusarium* spp. (Comby et al. 2017). The absence of an inhibition zone corresponded to the inability of the endophyte strain to produce antimicrobial compounds might likely attack as mycoparasite. *Phoma glomerata* seems to be a key member of a mycoparasitic guild of fungi that might parasitize *Fusarium* spp.

Competitive Root Colonization

One of the disease suppression mechanisms by *Phoma* spp. is competition with pathogens for nutrients and infection sites on the root surface. Competition for essential substrates and the excretion sites on roots is the key to controlling pathogens in the root zone. Rhizosphere fungi with fast growth rates with large population numbers and extensive distribution throughout the soil show a unique advantage of niche colonization and substrate utilization. The distinctive ability of root colonization often helps the BCAs to outcompete pathogens in the root zone. The *Phoma* BCAs may aggressively occupy infection sites and exclude pathogens because of their physical covering morphology (Table 15.3). The suppression of take-all and common root rot in wheat by *Phoma* sp. GS6-4 and GS7-4 appeared to be due to their competitive root colonization ability (Shivanna et al. 1996). These isolates do not produce a zone of inhibition against *Co. sativus* or *Ga. graminis* var. *tritici* in a dual culture, excluding the possibility that these *Phoma* isolates might produce antibiotics or toxins. Both isolates aggressively colonized wheat roots and re-isolated at higher frequencies from roots. As a result, when the *Phoma* inoculum was added to soil at a high concentration (2.0% w/w), both fungal pathogens were isolated less frequently from roots than the non-treated roots. Conversely, when *Co. sativus* and *Ga. graminis* var. *tritici* inoculum were added to soil at high concentration, the *Phoma* isolates were isolated in high frequencies from roots (Shivanna et al. 1996). These show that *Phoma* can competitively colonize plant roots even when the pathogen inoculum level is very high in the soil. Hence, prior colonization of wheat roots by *Phoma* isolates reduces the availability of infection sites for pathogens. Owing to this reason, when wheat was grown in soil amended with GS7-4 inoculum for 10 days, later transferred to soil containing *Ga. graminis* var. *tritici* inoculum, and grown for 4 weeks in the greenhouse, the seedlings showed less disease (Shivanna et al. 1996.). The colonization of the rhizosphere and successful establishment in the root zone by *Phoma* at the early developmental stage of host plants may provide a critical competitive advantage to these plants, potentially improving their ability to cope with a pathogen and/or utilize limited resources.

15.7.2 Indirect Protective Activity Against Plant Diseases by *Phoma* spp.

Indirect protection results from activities or actions that do not involve sensing or targeting a pathogen by the BCAs *Phoma* spp. Although many non-pathogenic *Phoma* spp. directly interact with pathogens, others indirectly control plant diseases. Phytohormone-mediated regulation of disease development and stimulation of plant defence systems by *Phoma* spp. are the most common indirect disease suppression forms (Fig. 15.2).

Hormonal Interplay

Phytohormone regulates various systems of plants. Interaction among various phytohormones also plays an important role in plant defence against pathogens. Phytohormone, such as auxin, has a complicated behaviour in plant-pathogen interactions. Auxin promotes disease in many plants. Treatment with the auxin analogues 2,4-dichlorophenoxyacetic acid (2,4-D) or 1-naphthalacetic acid (NAA) increases disease symptoms by *Ps. syringae* pv. *tomato* DC3000 in *Arabidopsis* (Navarro et al. 2006; Chen et al. 2007). *Arabidopsis* plants inoculated with *Pseudomonas syringae* pv. *maculicola*, along with NAA treatment, show its augmented growth and higher susceptibility to this pathogen (Wang et al. 2007). Exogenous treatment of indole-3-acetic acid (IAA) (most important auxin) or 2,4-D results in increased lesion areas of bacterial blight disease caused by *Xanthomonas oryzae* pv. *oryzae* in resistant rice lines (Ding et al. 2008). External IAA application exacerbates the disease progress caused by *X. oryzae* pv. *oryzicola* and *Magnaporthe oryzae* in rice (Fu et al. 2011). These results show that auxin is a key character in pathogenesis and plant defence. Additional evidence shows that biological control of plant disease by *Phoma* is conferred by suppressing auxin in the plant. The suppression of clubroot disease in brassica crops by *P. glomerata* JCM9972 is associated with an anti-auxin compound epoxydon produced by the fungus (Arie et al. 1999). Epoxydon constitutes a new group of plant protection agents that showed very poor antimicrobial activity. Moreover, the compound did not elicit the induction of acquired resistance in the plant (Arie et al. 1999). However, epoxydon has been reported to have an anti-auxin activity (Sakai et al. 1970). *Plasmodiophora brassicae*, as a gall-inducing species, produces excessive amounts of auxin in the infected galls compared to healthy roots (Raa 1971; Ludwig-Mueller et al. 1993). Exogenous application of epoxydon or treatment with an anti-auxin compound such as 2,3,5-triiodobenzoic acid showed significant control of clubroot disease (Arie et al. 1999), suggesting that the clubroot suppression by *P. glomerata* is due to the anti-auxin activity (Table 15.3). Auxin promotes cell elongation of plant tissues by increasing the cell walls (Cosgrove 1993). The long-term extension of the cell wall is caused by breaking the hydrogen bonds between cellulose microfibrils and polysaccharides in the wall (McQueen-Mason and Cosgrove 1995). This process can make the plant cell wall, a natural barrier against pathogens, vulnerable and create an opportunity for pathogen attack.

Induced Resistance

The host plant's enhanced ability to mobilize its innate immunity or defence system against perceived plant pathogen is referred to as priming or induced systemic resistance (ISR) (Hossain et al. 2007, 2008b, 2014a, b; Kojima et al. 2013). ISR is considered one of the most efficient mechanisms of plant defence against invading plant pathogens. *Phoma* application to elicit ISR in plants against various leaf pathogens has been widely studied (Table 15.3). Cucumber plants that were grown in soil containing *Phoma* sp. GS8-1, GS8-2, and GS8-3 showed significant suppression of anthracnose disease caused by *C. orbiculare* at a distance above the treatment site. The *Phoma* isolates only colonized the cucumber roots but not the plant's aerial portions (Meera et al. 1995a, b). These isolates were not antagonistic and did not produce antibiotic substances in vitro against *C. orbiculare* (Meera et al. 1994). Pieces of evidence support the view that *Phoma* isolates elicited ISR in cucumber against *C.*

orbiculare (Meera et al. 1995a, b). Similar suppression of anthracnose disease was observed in the third leaf of the cucumber plant when the second leaf of the plant was treated with the cell-free filtrate of GS8-1, reducing the total lesion number and total lesion diameter per infected leaf (Fig. 15.3a). Suppression of the disease appeared to be systemic, as second leaves were treated with the cell-free filtrate. At the same time, the pathogen was challenged on third leaves. Diaminobenzidine tetrahydrochloride (DAB) staining of treated leaves showed that the H_2O_2 generation was detected as a response to pathogen infection of GS8-1-treated plants (Fig. 15.3b). In the chemiluminescence assay, a massive accumulation of H_2O_2 was also triggered by GS8-1 treatment (Fig. 15.3c). The rapid release of H_2O_2 is assumed to be an important event in the orchestration of cellular defence responses against pathogens in plants by molecules with eliciting activities (Alvarez et al. 1998). Known that H_2O_2 is a key element in plant defence, the ability of GS8-1 culture filtrate to stimulate host-active oxygen species suggests ISR as the possible mode of action against *C. orbiculare*. Lignins were produced in the epidermal tissues of cell-free filtrate-treated cucumber plants' surrounding spores or appressoria of challenge fungi, thus inhibiting the fungal penetration in the systemically protected plant parts (Fig. 15.3d). Lignification explains, at least in part, the GS8-1-mediated resistance of cucumber to *C. orbiculare*. A possible relationship of induced lignification to an induced systemic protection mechanism against cucumber anthracnose has already been detected (Shimizu et al. 2013).

The ISR mechanisms in the suppression of CMV infection by *Phoma* sp. (GS8-2 and GS8-3)

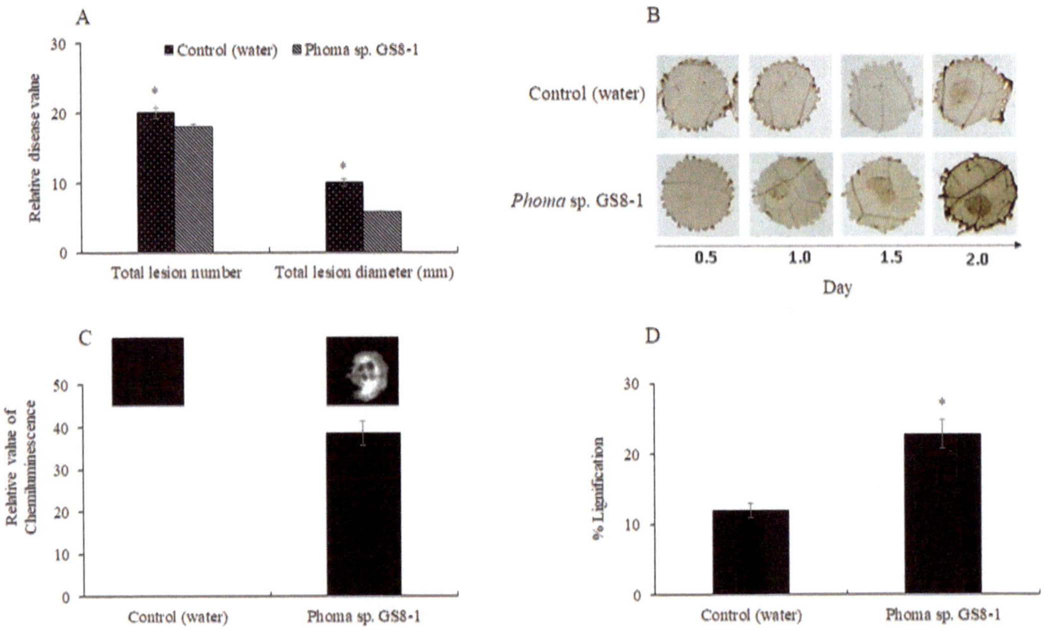

Fig. 15.3 Effect of plant growth-promoting fungi (PGPF) *Phoma* sp. GS8-1 on anthracnose disease caused by *Colletotrichum orbiculare* and generation of cellular responses in cucumber plants. (**a**) Cucumber plants receiving leaf treatment with culture filtrate from GS8-1 exhibited clear systemic resistance against subsequent infection with *C. orbiculare*, reducing disease lesions' number and size. (**b**) Time course changes for production of H_2O_2 in leaves of cucumber plants in response to culture filtrate of GS8-1 treatment followed by *C. orbiculare* inoculation as indicated by the intensity of diaminobenzidine tetrahydrochloride (DAB) staining. (**c**) Emission of chemiluminescence from cucumber fruit disks induced with culture filtrate of GS8-1 relative to control treatment. (**d**) Lignification of hypocotyls of cucumber seedlings induced by culture filtrate of GS8-1 following challenge inoculation with *C. orbiculare*. The degree of lignin deposition was estimated by microscopic determination of the percentage of germinated spores together with aspersoria around which lignin depositions were induced

have been reported in *Arabidopsis*, tobacco, and cucumber (Elsharkawy 2018; Elsharkawy et al. 2020). Roots of these plants were treated with *Phoma* inoculum, and leaves were inoculated with CMV, indicating that reduction of diseases took place without direct contact between the two microorganisms. This observation is further reinforced by the fact that *Arabidopsis* treatment with the cell-free filtrate of *Phoma* sp. GS6-2, GS7-3, GS8-1, and GS8-3 also protected leaves against bacterial leaf speck pathogen *Ps. syringae* pv. *tomato* DC3000 (Hossain et al. 2008b; Sultana et al. 2008, 2009). Since the *Phoma* culture filtrates did not directly affect the growth and proliferation of *Ps. syringae* pv. *tomato*, the observed disease suppression seemed to be neither a direct parasitic nor an antibiotic activity of *Phoma* sp. Seemingly, signals originating in the *Phoma* sp.-treated roots reached the upper part of the plant and triggered the plant to activate defence responses (Hossain et al. 2007, 2008a). The *Phoma* BCAs, with the help of the ISR mechanism, can act against a wide range of fungal, bacterial, and viral plant pathogens in a variety of host plants.

15.8 Activated Signalling and Defence Genes in *Phoma* spp.-Induced Resistance

Plants utilize a complex but sophisticated defence mechanism during the induction of resistance to plant diseases. A diverse array of external and internal signals is involved in recognizing and responding to plant pathogens. The interaction of elicitor molecules with membrane receptors triggers a series of cellular responses. It transduces a comprehensive network of signalling pathways. These responses may ultimately lead to increased gene transcription and an altered physiological state of the host cell against the pathogen. It is known that signalling pathways involved in these processes rely on endogenous regulators, such as salicylic acid (SA), ethylene (ET), and jasmonic acid (JA) (Hossain et al. 2007). Each of these signalling pathways has a corresponding set of inducible defence genes. In most cases, the signalling networks and the activated defence compounds are specific to a particular elicitor of induced resistance. Plants can activate two different types of induced resistance depending on the type of elicitor. The state of pathogen-induced classical systemic-acquired resistance (SAR) is characterized by an early increase in endogenously produced SA and activation of pathogenesis-related (*PR*) genes (van Loon et al. 1998). On the contrary, integrated JA and ET signalling pathways controlled ISR against *Ps. syringae* pv. *tomato* in *Arabidopsis* stimulated by plant growth-promoting rhizobacteria (PGPR) such as *Ps. fluorescence* (Pieterse et al. 1998). Although ISR and SAR act through different signalling pathways, a discrepancy in activated signalling pathways was reported depending on the pathogen and the elicitor. The culture filtrate of *Phoma* sp. GS8-1 was fully protective against the bacterial speck pathogen *P. syringae* pv. *tomato* in jasmonate-resistant mutant *Arabidopsis jar1* and ethylene insensitive mutant *ein2* similar to wild-type plants (Hossain et al. 2008a). However, its efficacy was reduced in salicylic acid-deficient *Arabidopsis* plants containing transgene NahG, indicating a partial SA pathway requirement in GS8-1-mediated ISR against the bacterium. Examination of systemic gene expression exhibited that GS8-1 treatment modulates both SA-inducible expression (*PR-1*, *PR-2*, and *PR-5* genes) and JA/ET-inducible (*ChitB* and *Hel* genes) genes. Besides, the induction of these genes was potentiated after an attack by the pathogen (Hossain et al. 2008a). These suggested that in addition to a partial requirement for SA, the JA and ET signals might also play a role in defence responses during *Phoma* sp. GS8-1-mediated ISR in *Arabidopsis* (Hossain et al. 2008a). Similarly, *Arabidopsis* mutant plants *jar1* and *ein2* and transgenic NahG plants were protected by a culture filtrate from *Phoma* sp. GS8-3 against *Ps. syringae* pv. *tomato* DC3000. Gene expression studies indicated that GS8-3 treatment modulated the systemic expression of the SA-inducible *PR-1* and JA/ET-inducible *PDF1.2* genes. The pathogenic challenge to GS8-3-treated plants resulted in potentiated expression of the *PR-1* gene and downregulated expression of the *PDF1.2* gene. These data confirmed that *Phoma* sp. GS8-3 induces resistance in *Arabidopsis* in a manner where priming and both SA and JA/ET signals may play a role in defence responses (Sultana et al. 2009). In a different study, treatment with *Phoma* sp. GS7-3, which induced hypersensitive reaction-like (HR) cell death and β-glucuronidase reporter gene expression of *PR-1a* and *PDF 1.2* constructs, showed the ability to induce resistance against *Ps. syringae* pv. *tomato* DC3000 in *Arabidopsis* plants *jar1*, *ein2*, and NahG (Sultana et al. 2008). Accordingly, the GS7-3-mediated ISR does not require SA, JA, or ET signalling. In contrast, GS6-2 culture filtrate induced HR-like cell death and β-glucuronidase reporter gene expression of *PR-1a* construct and failed to induce resistance in NahG plants, suggesting that GS6-2-mediated ISR requires SA accumulation in plants (Sultana et al. 2008). It is likely that different *Phoma* spp. vary in their reliance on SA, JA, and ET signalling pathways during ISR against plant pathogens.

The suppression of white rot disease in onion by *Phoma* sp. isolates GS8-1 and GS8-3 is partly associated with the fungal induction of defence compounds. Hence, peroxidase and polyphenol oxidase levels increased in the *Phoma*-treated onion plants than in the untreated onion plants 7 days after *Phoma* treatment. Likewise, early and strong expressions of defence genes *PR-1* and *PR-2* were reported in plants treated with the isolates of *Phoma* sp. (Elsharkawy and El-Khateeb 2019). Reduction in anthracnose disease in susceptible and resistant cultivars of cucumber (cv. Suyo and cv. Shogoin fushinari, respectively) by *Phoma* sp. GS8-1 was accompanied by increased endochitinase activities, β-1,3-endoglucanase, peroxidase, polyphenol oxidase, and phenylalanine ammonia lyase (Hassan et al. 2014). Microarray results showed that treatment with *Phoma* sp. GS8-3 caused an upregulation of 19,417 genes, especially those related to heat-shock proteins in tobacco plants (Elsharkawy et al. 2020). The gene expression changes induced by GS8-3 are different from those altered by CMV, indicating that ISR gene expression signatures are quite distinctive. It is apparent that *Phoma* spp. differentially modulate diverse signalling pathways and defensive gene expressions in plants during the elicitation of ISR against pathogens.

15.9 Bioactive Compounds Produced by Biocontrol Agent *Phoma* spp.

The purification and identification of active compounds produced by BCAs remain an important issue in understanding their mode of action. The disease reduction activity observed between the antagonist *Phoma* spp. and the plant pathogen is generally chemical in nature. Various studies have detected abundant amounts of biogenic metabolites in *Phoma* culture (Table 15.4). Some species are known to produce toxic metabolites that are strongly inhibitory to plant pathogens. Other species produce active metabolites that have other disease suppression mechanisms. These signature compounds give a clear idea about the sources of biocontrol activity of *Phoma* against different plant pathogens. An antibiotic (antibiotic E) was isolated from *P. exigua* Desm. var. *exigua* Desm. and *P. exigua* var. *foveata* (Foister) Boerema, which was antagonistic to several gram-positive and gram-negative bacteria and two fungal pathogens, *F. coeruleum* (Lib.) Sacc. and *Phytophthora infestans* (Mont.) de Bary (Logan and O'Neill 1970). An antibacterial antibiotic was isolated from *P. pigmentivora* Massee *(Phoma herbarum* Westd.) (McGahren et al. 1972; Chandler et al. 1992). A previously undescribed antifungal compound phomalone, purified from *P. etheridgei*, conferred strong antifungal activity against the aspen decay fungus *Phellinus tremulae* (Ayer and Jimenez 1994). Phomalone is a natural product with the same carbon skeleton as the *P. pigmentivora* metabolite LL-D253α (Ayer and Jimenez 1994). An anti-auxin compound epoxydon (5-hydroxy-3-(hydroxymethyl)-7-oxabicyclo[4.1.0]hept-3-en-2-one) was found in the culture broth of *Phoma glomerata* no. 324, which was attributed to the clubroot disease suppression in crucifers (Arie et al. 1998). Two new

Table 15.4 Microbial products and their biological activity against plant pathogens

Bioactive compounds	Source	Biological activity	Target organism	Reference
Antibiotic E	*Phoma exigua* var. *exigua*, *Phoma exigua* var. *foveata*	Antibacterial and antifungal	Gram-positive, gram-negative bacteria, *Fusarium coeruleum*, *Phytophthora infestans*	Logan and O'Neill (1970)
Phomodione, usnic acid, and cercosporamide	*Phoma pinodella*	Antimicrobial	*Pythium ultimum*, *Sclerotinia sclerotiorum*, and *Rhizoctonia solani*	Hoffman et al. (2008)
7-(γ, γ)-Dimethylallyloxymacrosporin, macrosporin, 7-methoxymacrosporin, tetrahydroaltersolanol B, altersolanol L, and ampelanol	*Phoma* sp. L28	Antifungal activity	*Colletotrichum musae*, *Co. gloeosporioides*, *F. graminearum*, *Penicillium italicum*, *F. oxysporum* f. sp. *lycopersici*, *Rhizoctonia solani*	Huang et al. (2017)
Epoxydon	*Phoma glomerata* no. 324	Anti-auxin activity (poor antimicrobial activity)	*Plasmodiophora brassicae*	Arie et al. (1999)
Phomalone	*P. etheridgei*	Antifungal activity	*Phellinus tremulae*	Ayer and Jimenez (1994)
Viridicatol, tenuazonic acid, alternariol, and alternariol monomethyl ether	*Phoma* sp. WF4	Antifungal activity	*F. graminearum*	Mousa et al. (2015)
Epoxydines A and B	*Phoma* sp.	Antifungal	*Microbotryum violaceum*	Qin et al. (2010)
Xanthepinone	*Phoma* sp.	Antifungal	*Pyricularia oryzae*, *Botrytis cinerea*, *Phytophthora infestans*	Liermann et al. (2009)
(3S)-3,6,7-trihydroxy-α-tetralone	*Phoma* sp.	Antifungal	*F. oxysporum*, *R. solani*	Wang et al. (2012)
5-Hydroxyramulosin	*Phoma* sp.	Antifungal	*Aspergillus niger*	Santiago et al. (2012)

metabolites, epoxydine A ((4R,4′S,6S,6′S)-4,4′-oxybis(6-methylcyclohex-2-en-1-one)) and epoxydon B ((1S,5R,6R)-5,6-Dihydroxy-3-(hydroxymethyl)-2-oxocyclohex-3-en-1-yl furan-2-carboxylate) along with four known and related metabolites, were purified from a fungal endophyte *Phoma* sp. isolated from the halotolerant plant *Salsola oppositifolia* (Amaranth family) (Qin et al. 2010). These compounds were shown to have antifungal activities against the fungus *Microbotryum violaceum* (Qin et al. 2010). A polyketide metabolite xanthepinone was isolated from a fungus closely related to *P. medicaginis*, collected from French alpine soil (Liermann et al. 2009). Xanthepinone inhibited the fungal conidia germination and suppressed *Pyricularia oryzae*, *Botrytis cinerea*, and *Phytophthora infestans* (Liermann et al. 2009). Another polyketide, 5-hydroxyramulosin, was isolated from a fungal endophyte morphologically related to *Phoma*, inhabiting the tree *Cinnamomum mollissimum* from Malaysia (Santiago et al. 2012). The purified compound exhibited antifungal activity against *Aspergillus niger* (Santiago et al. 2012). In another study, an α-tetralone derivative, (3S)-3,6,7-trihydroxy-α-tetralone, was purified from a *Phoma* endophyte, isolated from the Chinese medicinal plant *Arisaema erubescens*, and shown to have antifungal activities against *Fusarium oxysporum* and *Rhizoctonia solani* (Wang et al. 2012). An endophyte *Phoma* sp. L28, isolated from mangrove plant *Myoporum bontioides* roots, was known for its antifungal activity against plant pathogens (Ding et al. 2014). A new anthraquinone derivative, 7-(γ,γ)-dimethylallyloxymacrosporin, and with five known analogues, macrosporin, 7-methoxymacrosporin, tetrahydroaltersolanol B, altersolanol L, and ampelanol, were identified from this fungus and shown antifungal activity against *C. musae* (Berk. & M. A. Curtis) Arx., *F. graminearum* Schw., *C. gloeosporioides* (Penz) Sacc., *Penicillium italicum* Wehme, *F. oxysporum* Schlecht. f. sp. *lycopersici* (Sacc.) W.C. Snyder and H. N. Hansen, and *Rhizoctonia solani* Kuhn (Huang et al. 2017). A new phenolic compound phomodione, isolated from the culture of an endophyte *P. pinodella*, showed fungicidal activity against oomycete, ascomycete, and basidiomycete such as *Pythium ultimum*, *Rhizoctonia solani*, and *Sclerotinia sclerotiorum* (Hoffman et al. 2008). Along with phomodione, two other compounds, usnic acid and cercosporamide with known antibiotic activity, were also isolated from *P. pinodella* and showed antifungal activity against the tested fungi (Hoffman et al. 2008). Four antifungal compounds viridicatol, tenuazonic acid, alternariol, and alternariol monomethyl ether were purified from *Phoma* sp. WF4 isolated from *Eleusine coracana* (L.) Gaertn. All the purified compounds caused intense breakage of *F. graminearum* hyphae in vitro (Mousa et al. 2015). Except for tenuazonic acid, these compounds have previously neither been reported to be produced by *Phoma* nor have previously been reported to exhibit anti-*Fusarium* activity. Thus, the *Phoma* BCAs are novel sources of potent antimicrobial compounds that may help develop a new agrochemical group for efficient plant disease management.

15.10 Conclusions

Biological management of plant diseases by using non-pathogenic *Phoma* spp. has the potential to be a part of an overall integrated pest management program to reduce the legal, environmental, and public safety hazards of chemicals. When other classical practices are insufficient or cannot be applied, the microbial biocontrol agent *Phoma* can effectively be employed to control a broad range of plant diseases. Despite wider applicability, great environmental incentives, and robust research efforts, the ongoing commercialization of *Phoma* agents as disease management tools has been slow to advance. Cost and non-stability of the beneficial traits have hindered *Phoma* spp. from apprehending their full commercialization and end-user acceptance. Technological difficulties and implementation challenges that make them less consistent or inferior to chemical pesticides must be addressed. Adequate solutions to key field application problems require comprehensive research on biology and techniques for mass production, shelf life, formulation, and cost-effective delivery systems of the candidate agent. The genetic engineering tools can be applied to selectively address the non-stability of beneficial characters and discover markers of beneficial genes. Subsidies and logistic support should be expanded to make *Phoma* BCAs more attractive to commercial companies and growers. Strong collaboration and linkage between academia and industry should be attempted to swiftly transfer the developed BCAs to the end users. Different aspects of *Phoma* research and applications need to be enriched to optimize *Phoma* for improved plant disease control efficacy.

Acknowledgements The support from the Ministry of Education is greatly appreciated (LS2019878).

References

Agrios GN (2005) Plant Pathology. Fifth Edition. Academic Press.

Akhter W, Bhuiyan MKA, Sultana F, Hossain MM (2015) Integrated effect of microbial antagonist, organic amendment and fungicide in controlling seedling mortality (*Rhizoctonia solani*) and improving yield in pea (*Pisum sativum* L.). Comptes Rendus Biologies 338:21-28.

Alvarez ME, Pennell RI, Meijer PJ, Ishikawa A, Dixon RA, Lamb CJ (1998) Reactive oxygen intermediates mediate a systemic signal network in the establishment of plant immunity. Cell 92:773–784.

Arie T, Kobayashi Y, Okada G, Kono Y, Yamaguchi I (1998) Control of soilborne clubroot disease of cruciferous plants by epoxydon from *Phoma glomerata*. Plant Pathol 47:743–748.

Arie T, Kobayashi Y, Kono Y et al. (1999) Control of clubroot of crucifers by *Phoma glomerata* and its product epoxydon. Pesticide Science 55:602–604.

Aveskamp MM, de Gruyter J, Crous PW (2008) Biology and recent developments in the systematics of *Phoma*, a complex genus of major quarantine significance. Fungal Divers 31:1–18

Ayer WA, Jimenez LD (1994) Phomalone, an antifungal metabolite of *Phoma etheridgei*. Can. J. Chem. 72:2326–2332. https://doi.org/10.1139/v94-296

Azcon-Aguilar C, Barea JM (1996) Arbuscular-mycorrhizas and biological control of soilborne plant pathogens: an overview of the mechanisms involved. Mycorrhiza 6:457–464.

Badillo-Vargas IS, Rivera-Vargas LI, Calle-Bellido J (2008) Morphological, pathogenic and molecular characterization of *Phoma* spp. isolated from onion field soils in Puerto Rico. Journal of Agriculture-University of Puerto Rico 92:73–86.

Barak R, Elad Y, Mirelman D, Chet I (1985) Lectins: a possible basis for specific recognition in the interaction of *Trichoderma* and *Sclerotium rolfsii*. Phytopathology 75 (4): 458–462.

Bernier J (1995) A study of fungal leaf decomposition in relation to biological control of the apple scab pathogen *Venturia inaequalis*. M.Sc. thesis, Department of Plant Science, University of McGill, Montreal, Que.

Boerema GH, de Gruyter J, Noordeloos ME, Hamers MEC (2004) *Phoma* Identification Manual. Differentiation of Specific and Infra-specific Taxa in Culture. CABI Publishing, CAB International, Wallingford, UK. 448 pp.

Boudreau MA, Andrews JH (1987) Factors influencing antagonism of *Chaetomium globosum* to *Venturia inaequalis*: a case study in failed biological control. Phytopathology 77: 1470–1475.

Carisse, O., and Pelletier, J.R. 1994. Sensitivity distribution of *Venturia inaequalis* to fenarimol in Quebec apple orchards. Phytoprotection 75: 35-43.

Chandanie WA, Kubota M, Hyakumachi M (2005) Interaction between arbuscular mycorrhizal fungus *Glomus mosseae* and plant growth promoting fungus *Phoma* sp. on their root colonization and growth promotion of cucumber. Mycosci. 46: 201–204.

Chandanie WA, Kubota M, Hyakumachi M (2006) Interactions between plant growth promoting fungi and arbuscular mycorrhizal fungus *Glomus mosseae* and induction of systemic resistance to anthracnose disease in cucumber. Plant Soil 286: 209–217.

Chandler IM, McIntyre CR, Simpson TJ (1992) Structural revision and synthesis of LL-D253a and related chromanone fungal metabolites. J Chem Soc Perkin Trans I 18: 2271-2284.

Chen Z, Agnew JL, Cohen JD, He P, Shan L, Sheen J, Kunkel BN (2007). *Pseudomonas syringae* type III effector AvrRpt2 alters *Arabidopsis thaliana* auxin physiology. Proc. Natl. Acad. Sci. U.S.A. 104:20131–20136.

Chet I, Harman GE, Baker R (1981) *Trichoderma hamatum*: Its hyphal interactions with *Rhizoctonia solani* and *Pythium* spp. Microbial Ecology 7 (1): 29-38.

Comby M, Gacoin M, Robineau M, Rabenoelina F, Ptas S, Dupont J, Profizi C, Baillieul F (2017) Screening of wheat endophytes as biological control agents against *Fusarium* head blight using two different in vitro tests. Microbiological Research 202: 11-20.

Cosgrove DJ (1993) Wall extensibility: its nature, measurement, and relationship to plant cell growth. New Phytol 124:1–23https://doi.org/10.1111/j.1469-8137.1993.tb03795.x

Diamond H, Cooke BM (2003) Preliminary studies on biological control of the *Fusarium* ear blight complex of wheat. Crop Protection 22:99-107.

Ding X, Cao Y, Huang L, Zhao J, Xu C, Li X, Wang S (2008) Activation of the indole-3-acetic acid-amido synthetase GH3-8 suppresses expansin expression and promotes salicylate- and jasmonate-independent basal immunity in rice. Plant Cell 20:228-240https://doi.org/10.1105/tpc.107.055657

Ding WJ, Zhang SQ, Gong B, Li CY, Wang XF (2014) Isolation and inhibitory activity of endophytic fungi from the Semi-mangrove plant *Myoporum bontioides* A.Gray. Guangdong Agric Sci 3: 74-78.

Elsharkawy MM (2018) Induced systemic resistance against Cucumber mosaic virus by *Phoma* sp. GS8-2 stimulates transcription of pathogenesis-related genes in Arabidopsis. Pest Management Science 75: 859-866.

Elsharkawy MM, El-Khateeb NMM (2019) Antifungal activity and resistance induction against *Sclerotium cepivorum* by plant growth-promoting fungi in onion plants. Egyptian Journal of Biological Pest Control 29:68.

Elsharkawy MM, Hassan N, Ali M, Mondal SN, Hyakumachi M (2014) Effect of zoysiagrass rhizosphere fungal isolates on disease suppression and growth promotion of rice seedlings. Acta Agriculturae Scandinavica, Section B-Soil & Plant Science 64:135-140.

Elsharkawy MM, Suga H, Shimizu M (2020) Systemic resistance induced by Phoma sp. GS8-3 and nanosilica against Cucumber mosaic virus. Environ Sci Pollut Res 27:19029–19037. https://doi.org/10.1007/s11356-018-3321-3.

Falk SP, Gadoury DM, Cortesi P, Pearson RC, Seem RC (1995) Partial control of grape powdery mildew by the mycoparasite *Ampelomyces quisqualis*. Plant Dis 79:483–490.

Fu J, Liu H, Li Y, Yu H, Li X, Xiao J, Wang S (2011) Manipulating broad-spectrum disease resistance by suppressing pathogen-induced auxin accumulation in rice. Plant Physiol 155:589–602https://doi.org/10.1104/pp.110.163774

Hamayun M, Khan SA, Khan AL, Rehman G, Sohn EY, Shah AA, et al. (2009) *Phoma herbarum* as a new gibberellin producing and plant growth-promoting fungus. Journal of Microbiology and Biotechnology 19:1244-1249. https://doi.org/10.4014/jmb.0901.030

Hassan N, Elsharkawy MM, Shivanna MB, Meera MS, Hyakumachi M (2014) Elevated expression of hydrolases, oxidase, and lyase in susceptible and resistant cucumber cultivars systemically induced with plant growth-promoting fungi against anthracnose. Acta Agric Scand B Soil Plant Sci 64:155–164.

Hoffman AM, Mayer SG, Strobel GA, Hess WM, Sovocool GW, Grange AH, et al., Purification, identification and activity of phomodione, a furandione from an endophytic *Phoma* species, Phytochemistry 69 (2008) 1049-1056.

Hossain MM, F. Sultana 2015. Genetic variation for induced and basal resistance against leaf pathogen *Pseudomonas syringae* pv. *tomato* DC3000 among Arabidopsis thaliana accessions. Springer Plus 4:296.

Hossain MM, Sultana F (2020) Application and mechanisms of plant growth promoting fungi (PGPF) for phytostimulation. In: Das SK, Editor. Organic Agriculture, ISBN 978-1-78984-669-0. IntechOpen, London, UK, https://doi.org/10.5772/intechopen.92338

Hossain MM, Sultana F, Kubota M, Koyama H, Hyakumachi M (2007) The plant growth-promoting fungus *Penicillium simplicissimum* GP17-2 induces resistance in *Arabidopsis thaliana* by activation of multiple defense signals. Plant and Cell Physiology 48 (12), pp. 1724-1736.

Hossain MM, Sultana F, Kubota M, Koyama H, Hyakumachi M (2008a) Systemic resistance to bacterial leaf speck pathogen in Arabidopsis thaliana induced by the culture filtrate of a plant growth-promoting fungus (PGPF) *Phoma* sp. GS8-1. Journal of General Plant Pathology 74 (3), pp. 213-221.

Hossain MM, Sultana F, Kubota M, Hyakumachi M (2008b) Differential inducible defense mechanisms against bacterial speck pathogen in *Arabidopsis thaliana* by plant-growth-promoting-fungus *Penicillium* sp. GP16-2 and its cell free filtrate. Plant and Soil 304 (1-2), pp. 227-239.

Hossain MM, Sultana F and Rahman AHM (2014a) A comparative screening of hybrid, modern varieties and local rice cultivar for brown leaf spot disease susceptibility and yield performance. Archives of Phytopathology and Plant Protection, 47: 795-802.

Hossain MM, Sultana F, Miyazawa M, Hyakumachi M (2014b) The plant growth-promoting fungus *Penicillium* spp. GP15-1 enhances growth and confers protection against damping-off and anthracnose in the cucumber. Journal of Oleo Science 63(4): 391-400.

Hossain MM, Sultana F, Islam S (2017a) Plant Growth-Promoting Fungi (PGPF): Phytostimulation and Induced Systemic Resistance. In: D.P. Singh et al. (eds.), Plant-Microbe Interactions in Agro-Ecological. Springer Nature Singapore Pte Ltd.

Hossain MM, Sultana F, Islam S (2017b) Plant growth-promoting fungi (PGPF): Phytostimulation and induced systemic resistance. In: Singh D, Singh H, Prabha R, editors. Plant-Microbe Interactions in Agro-Ecological Perspectives, Volume 2: Microbial Interactions and Agro-Ecological Impacts. Singapore: Springer. pp. 135-191. https://doi.org/10.1007/978-981-10-6593-4.

Huang S, Xu J, Li F, Zhou D, Xu L, Li C (2017) Identification and antifungal activity of metabolites from the mangrove fungus *Phoma* sp. L28. Chem Nat Compd 53(2):237-240.

Hutchinson LJ, Chakravarty P, Kawchuck LM, Hirasuka Y (1994) *Phoma etheridgei* sp. nov. from black galls and cankers of trembling aspen (*Populus tremuloides*) and its potential role as a bioprotectant against the aspen decay pathogen *Phellinus tremulae*. Can J Bot 72:1424–1431.

Hwang SF, Ahmed HU, Strelkov SE, Zhou Q, Gossen BD, McDonald MR, Peng G, Turnbull GD (2017) Suppression of Clubroot by Dazomet Fumigant. Canadian Journal of Plant Science 98:172-182.

Inbar J, Menendez A, Chet I (1996) Hyphal interaction between *Trichoderma harzianum* and *Sclerotinia sclerotiorum* and its role in biological control. Soil Biology and Biochemistry 28 (6): 757–763.

Islam MT, Hossain MM (2013) Biological control of peronosporomycete phytopathogen by bacterial antagonist. In: Maheshwari DK (ed.) Bacteria in Agrobiology: Disease Management. Springer-Verlag Berlin Heidelberg. pp. 167-218.

Jogaiah S, Abdelrahman M, Tran LSP, Shin-ichi I (2013) Characterization of rhizosphere fungi that mediate resistance in tomato against bacterial wilt disease. J Exp Bot 64:3829-3842.

Kageyama K, Asano T (2009) Life cycle of *Plasmodiophora brassicae*. Journal of Plant Growth Regulation 28(3):203–211.

Kiss L, Nakasone KK (1998) Ribosomal DNA internal transcribed spacer sequences do not support the species status of *Ampelomyces quisqualis*, a hyperparasite of powdery mildew fungi. Current Genetics 33:362–367.

Koike N, Hyakumachi M, Kageyama K. et al. (2001) Induction of Systemic Resistance in Cucumber against Several Diseases by Plant Growth-promoting Fungi: lignification and Superoxide Generation. European Journal of Plant Pathology 107:523–533.

Kojima H, Hossain MM, Kubota M, Hyakumachi M (2013) Involvement of the salicylic acid signaling pathway in the systemic resistance induced in Arabidopsis by plant growth-promoting fungus Fusarium equiseti GF19-1. Journal of Oleo Science; 62(6):415-26.

Liermann JC, Kolshorn H, Opatz T, Thines E, Anke H (2009) Xanthepinone, an antimicrobial polyketide from a soil fungus closely related to *Phoma medicaginis*. J Nat Prod 72:1905-1907. https://doi.org/10.1021/np9005234

Linderman RG (1988) Mycorrhizal interactions with the rhizosphere microflora: the mycorhizosphere effect. Phytopathology 78:366–370

Logan C, O'Neill R (1970) Production of an antibiotic by *Phoma exigua*. Trans Br Mycol Soc 55: 67-75.

Ludwig-Mueller J, Bendel U, Thermann P, Ruppel M, Epstein E, Hilgenberg W (1993) Concentrations of indole-3-acetic acid in plants of tolerant and susceptible varieties of Chinese cabbage infected with *Plasmodiophora brassicae* Woron. New Phytologist 125:763-9.

Mansfield J, Genin S, Magori S, Citovsky V, Sriariyanum M, Ronald P. et al. (2012) Top 10 plant pathogenic bacteria in molecular plant pathology. Molecular Plant Pathology 13:614-629.

McGahren WJ, Ellestad GA, Morton GO, Kunstmann MP (1972) LL-D253.alpha., -.beta., and -.gamma., novel chromanones from the fungus *Phoma pigmentivora*. The Journal of Organic Chemistry 37(10):1636-1639.

McQueen-Mason SJ, Cosgrove DJ (1995) Expansin mode of action on cell walls. Analysis of wall hydrolysis, stress relaxation, and binding. Plant Physiol 107:87–100.

Meera MS (1994) Induction of systemic resistance in cucumber against anthracnose using plant growth promoting fungi. Ph.D. Thesis, Gifu University, Japan pp. 154.

Meera MS, Shivanna MB, Kageyama K and Hyakumachi M (1994) Plant growth promoting fungi from zoysiagrass rhizosphere as potential inducers of systemic resistance in cucumbers. Phytopathology 84:1399-1406.

Meera MS, Shivanna MB, Kageyama K and Hyakumachi M (1995a) Persistence of induced systemic resistance in cucumber in relation to root colonization by plant growth promoting fungal isolates. Crop Protection 14:123-130

Meera MS, Shivanna MB, Kageyama K and Hyakumachi M. (1995b) Responses of cucumber cultivars to induction of systemic resistance against anthracnose by plant growth promoting fungi. European Journal of Plant Pathology 101: 421-430.

Monte E, Bridge PD, Sutton BC (1991) An integrated approach to *Phoma* systematics. Mycopathologia 115:89–103.

Mousa WK., Schwan A, Davidson J, Strange P, Liu H, Zhou T, Auzanneau F, Raizada MN (2015) An endophytic fungus isolated from finger millet (*Eleusine coracana*) produces antifungal natural products. Frontiers in Microbiology 6:1157.

Nagaraju A, Murali M, Sudisha J, Amruthesh KN, Murthy SM (2012) Beneficial microbes promote plant growth and induce systemic resistance in sunflower against downy mildew disease caused by *Plasmopara halstedii*. Current Botany 3(5):12-18.

Navarro L., Dunoyer P., Jay F., Arnold B., Dharmasiri N., Estelle M., Voinnet O., Jones J. D. (2006). A plant miRNA contributes to antibacterial resistance by repressing auxin signaling. Science 312:436–439https://doi.org/10.1126/science.1126088

Ouimet A, Carisse O, Neumann P (1997) Environmental and nutritional factors affecting the in vitro inhibition of the vegetative growth of *Venturia inaequalis* by five antagonistic fungi. Can J Bot 75:632–639.

Pandey A, Palni LMS, Hebbar KP (2001) Suppression of damping-off in maize seedlings by *Pseudomonas corrugate*. Microbiol Res 156:191-194.

Pieterse CMJ, van Wees SCM, van Pelt JA, Knoester M, Laan R, Gerrits H, Weisbeek PJ, van Loon LC (1998) A novel signaling pathway controlling induced systemic resistance in Arabidopsis. Plant Cell 10:1571–1580.

Qin S, Hussain H, Schulz B, Draeger S, Krohn K (2010) Two new metabolites, epoxydine A and B, from *Phoma* sp. Helv. Chim. Acta 93:169–174. https://doi.org/10.1002/hlca.200900199

Raa J (1971) Indole-3-acetic acid levels and the role of indole-3-acetic oxidase in the normal root and club root of cabbage. Physiologia plantarum 25:130-134.

Reynolds TB (2009) Strategies for acquiring the phospholipid metabolite inositol in pathogenic bacteria, fungi and protozoa: making it and taking it. Microbiology 155:1386-1396. https://doi.org/10.1099/mic.0.025718-0

Rudakov OL (1979) Fungi of the genus *Ampelomyces* Ces. Ex. Schlect. Mikologiya i Fitopatologiya 13:104–110.

Sakai R, Sato R, Niki H, Sakamura S (1970) Biological activity of phyllosinol, a phytotoxic compound isolated from a culture filtrate of *Phyllosticta* sp. Plant and Cell Physiology 11:907-20.

Santiago C, Fitchett C, Munro MH, Jalil J, Santhanam J (2012) Cytotoxic and antifungal activities of 5-hydroxyramulosin, a compound produced by an endophytic fungus isolated from *Cinnamomum mollissimum*. Evid. Based Complement. Alternat Med 689310. https://doi.org/10.1155/2012/689310

Shenge KC, Mabagala RB, Mortensen CN, Stephan D, Wydra K (2007) First report of bacterial speck of tomato caused by *Pseudomonas syringae* pv. *tomato* in Tanzania. Plant Dis 91:462.

Shimizu K, Hossain MM, Kato K, Kubota M, Hyakumachi M (2013) Induction of defense responses in cucumber plants by using the cell-free filtrate of the plant growth-promoting fungus *Penicillium simplicissimum* GP17-2. Journal of Oleo Science, 62(8):613-21.

Shivanna MB, Meera MS, Hyakumachi M (1996) Role of root colonization ability of plant growth promoting fungi in the suppression of take-all and common root rot of wheat. Crop Protection 15(6): 497-504.

Shivanna MB, Meera MS, Kubota M, Hyakumachi M (2005) Promotion of growth and yield in cucumber by zoysiagrass rhizosphere fungi. Microbes Environ 20(1):34-40

Smith SE, Read DJ (1997) Mycorrizal symbiosis, 2nd edn. Academic Press, Cambridge

Sullivan RF, White JR JF (2000) *Phoma glomerata* as a mycoparasite of powdery mildew. Applied and Environmental Microbiology 66:425–427.

Sultana F, Hossain MM, Kubota M, Hyakumachi M (2008) Elicitation of systemic resistance against the bacterial speck pathogen in *Arabidopsis thaliana* by culture filtrates of plant growth-promoting fungi. Canadian Journal of Plant Pathology 30 (2):196-205.

Sultana F, Hossain MM, Kubota M, Hyakumachi M (2009) Induction of systemic resistance in *Arabidopsis thaliana* in response to culture filtrate from a plant growth-promoting fungus, *Phoma* sp. GS8-3. Plant Biology 11 (1):97-104.

Sundheim L, Tronsmo A (1988) Hyperparasites in biological control, p. 53–70. In: Mukerji KG, Garg KL (ed.). Biocontrol of plant diseases. CRC Press, Boca Raton, Fla.

Sutton TB (1978) Failure of combination of benomyl with reduced rates of non-benzimidazoles fungicides to control *Verlrurin irlaequalis* resistant to benomyl and the spread of resistant strains in North Carolina. Plant Dis Rep 62: 830-834.

Sutton BC (1980) The coelomycetes. Commonwealth Mycological Institute, Kew, United Kingdom.

Szkolnik M, Gilpatrick JD (1969) Apparent resistance of *Venturia inaequalis* to dodine in New York apple orchards, Plant Dis Rep 53:86.

Thind T, Olivier JM, Clerjeau M (1986) Tavelure du pommier: mise en evidence d'une rtsistance aux fongicides inhibiteurs de la biosynthese de I'ergosttrol. Phytoma 381:13- 16.

Thomas GP, Etheridge DE, Paul G (1960) Fungi and decay in aspen and balsam poplar in the boreal forest region, Alberta. Can. J. Bot. 38: 459-466.

Tian X, Wang D, Mao Z, Pan L, Liao J, Cai Z (2019) Infection of *Plasmodiophora brassicae* changes the fungal endophyte community of tumourous stem mustard roots as revealed by high-throughput sequencing and culture-dependent methods. PLoS ONE 14(6): e0214975.

Tiwari AK (1996) Biological controls of chickpea wilt complex using different formulations of *Gliocladium virens* through seed treatment. Ph.D. Thesis submitted to G. B. Pant University of Agriculture and Technology, Pantnagar India, pp. 167.

Van Loon LC, Bakker PAHM, Pieterse CMJ (1998) Systemic resistance induced by rhizosphere bacteria. Annu Rev Phytopathol 36:453-483.

Wang D, Pajerowska-Mukhtar K, Culler AH, Dong X (2007) Salicylic acid inhibits pathogen growth in plants through repression of the auxin signaling pathway. Curr Biol 17:1784. https://doi.org/10.1016/j.cub.2007.06.067

Wang LW, Xu BG, Wang JY, Su ZZ, Lin FC, Zhang CL, et al. (2012) Bioactive metabolites from *Phoma* species, an endophytic fungus from the Chinese medicinal plant *Arisaema erubescens*. Appl Microbiol Biotechnol 93:1231–1239. https://doi.org/10.1007/s00253-011-3472-3

Zhou Z, Zhang C, Zhou W, Li W, Chu L, Yan J, Li H (2014) Diversity and plant growth-promoting ability of endophytic fungi from the five flower plant species collected from Yunnan, Southwest China. Journal of Plant Interactions 9(1):585-591`.

Zimowska B (2007) Fungi colonizing and damaging different parts of peppermint (*Mentha piperita* L.) cultivated in south-eastern Poland. Herba Polonica 53:97–105.

Plant Growth-Promoting *Phoma* spp.

16

Mohsen Mohamed Elsharkawy

16.1 Introduction

There is overwhelming evidence for the notion that crop production should be increased to feed the increasing populations all over the world. One of the solutions could be the intensification of agriculture. This will increase the global need for fertilizers to meet the agricultural intensification. The utilization of chemical fertilizers increases the harmful effects on human health and the environment. Organic farming is considered one of the alternative strategies to support agroecosystem health. Beneficial microorganisms are the key factors in developing organic farming. These microorganisms are attracted by root exudates produced by plants. Solubilizing soil nutrients is one of the most important effects of these microorganisms.

Nonpathogenic fungi that colonize plant roots intercellularly and/or intracellularly are called endophytes (Wilson 1995). Endophytic populations vary greatly based on the host and climatic conditions. Several fungal endophytes were isolated from the rhizosphere such as *Trichoderma*, *Fusarium*, *Penicillium*, and *Phoma* (Hyakumachi 1994; Shivanna et al. 2005; Hossain et al. 2014; Elsharkawy et al. 2014). Different species of *Phoma* spp. were isolated from the rhizosphere of many hosts (Table 16.1). Growth-promoting bioactive compounds were produced by endophytes with the ability to improve both plant growth and health (Lee et al. 2004; Das and Varma 2009). *Phoma tropica* was found to be an endophytic fungus (Bezerra et al. 2012). Plants colonized with *Phoma* spp. exhibited faster and greater shoot and root growth resulting in a hasty increase in plant biomass. Stimulated shoot growth by different isolates of *Phoma* spp. was reported in numerous hosts (Shivanna et al. 2005; Hamayun et al. 2010; Zhou et al. 2014). *Phoma* sp. GS6-1 and GS7-4 enhanced plant height and biomass, earhead length and weight, and seed number at harvest time (Shivanna et al. 1996). Moreover, *Phoma* sp. treatment resulted in a significant increase in dry biomass and the number of cucumber (cv. Jibai) fruits at 3 months after planting in the greenhouse experiments (Shivanna et al. 2005). Also, the isolates increased the yield of cucumber plants under field conditions. *Phoma* sp. (GS8-2 and GS8-3) enhanced shoot dry weight of cucumber plants (Chandanie et al. 2005).

The mechanisms of growth promotion are complicated and include direct and indirect effects of these microorganisms such as production of phytohormones, enzymes, and volatile organic compounds and control plant diseases (Elsharkawy et al. 2012, 2013; Hossain et al. 2017). The production of growth substances and

M. M. Elsharkawy (✉)
Agricultural Botany Department, Faculty of Agriculture, Kafrelsheikh University, Kafr El Sheikh, Egypt
e-mail: mohsen.abdelrahman@agr.kfs.edu.eg

M. Rai et al. (eds.), *Phoma: Diversity, Taxonomy, Bioactivities, and Nanotechnology*,
https://doi.org/10.1007/978-3-030-81218-8_16

Table 16.1 Different growth-promoting *Phoma* isolates with their original source of isolation

Isolate of *Phoma* sp.	Original source of isolation	References
Phoma sp.	Zoysiagrass	Hyakumachi (1994)
Phoma herbarum	Soybean	Hamayun et al. (2009)
Phoma sp. GAH7	Cucumber	Hamayun et al. (2010)
Phoma glomerata LWL2	Cucumber	Waqas et al. (2012)
Phoma multirostrata	Sunflower, tomato	Jogaiah et al. (2013)
Phoma sp.	*Rosa rugosa*, *Camellia japonica*, *Delonix regia*, *Dianthus caryophyllus*	Zhou et al. (2014)

the availability of nutrients due to the presence of fungi in the soil are considered direct mechanisms, while suppression of plant pathogens is considered an indirect mechanism. Production of enzymes such as pectinases, xylanases, and proteases was reported using different isolates of *Phoma* spp. such as *Phoma tropica* (Bezerra et al. 2012). The production of phytases and phosphatases enzymes by fungi converts insoluble phosphorus in the soil to soluble form for the plants (Nahas 1996; Rajankar et al. 2007; Gaind and Nain 2015). Improved production of marketable fruits and increased accumulation of minerals such as Ca, Mg, and K in cucumber shoots treated with *Phoma* spp. were observed (Shivanna et al. 2005). Phytohormones such as auxins (IAA), zeatin, and gibberellins were produced by *Phoma* spp. resulting in increased shoot elongation and root biomass and development (Hamayun et al. 2009, 2010; Saxena 2015). Isolates of *Phoma* empower plants to resist different plant pathogens leading to better growth of these plants (Hossain et al. 2008; Sultana et al. 2008; Elsharkawy et al. 2020). Utilization of appropriate formulation from *Phoma* spp. could certainly improve plant growth and yield under field conditions for sustainable agricultural intensification. It can be used in biofertilizer and mycofungicide formulations.

16.2 Existence of Rhizospheric Fungi in the Soil

Rhizosphere is the zone of soil around plant roots. It is the natural environment for many microbes that affect nutrient cycling and plant growth (Hossain et al. 2007; Mendes et al. 2013). Rhizospheric microbiota includes all microbes associated with plant roots such as fungi, bacteria, viruses, oomycetes, algae, arthropods, and protozoa. Rhizospheric fungi have several roles for environmental sustainability and ecological adaptability. It can be distinguished to endophytic fungi that live inside plant roots and epiphytic fungi that live outside plant roots (Hossain et al. 2017). Recently, the composition, biological activities, and unlimited potential of rhizospheric fungi were explored. Among their biological activities, nutrient recycling and conservation are the most prevalent (Zhang et al. 2017). Nutrient cycling and biodegradation are necessary to increase the availability of nutrients in the ecosystem. Plants need almost 16 elements for growth and development which are presented in the surrounding environment. Endophytes increase nutrient uptake such as phosphorus and nitrogen (Malinowski et al. 2000). Colonization of plant roots with rhizospheric fungi is useful for both the plant and the fungus. Rhizodeposits produced by the plants are used as a main source of energy for fungal growth (Denef et al. 2007; Gahan and Schmalenberger 2014). The growth of plants colonized with endophytes was faster than non-colonized plants (Hamayun et al. 2010). However, rhizospheric fungi enhanced nutrient uptake and acquisition and plant development and yield (Shaikh and Mokat 2018). Several fungal species were isolated from the rhizosphere such as *Penicillium*, *Aspergillus*, *Trichoderma*, *Fusarium*, *Rhizoctonia*, *Piriformospora*, and *Phoma* (Shoresh et al. 2010; Elsharkawy et al. 2012; Hossain et al. 2014). Decomposition of organic components in the soil was reported using different rhizospheric fungi such as *Chaetomium* sp., *Cladosporium* sp., and *Guignardia* sp. (He et al. 2012). Growth traits such as germination; seedling vigor; biomass; the number of leaves; number, size, and timing of

Table 16.2 Effect of *Phoma* isolates on seed germination, plant growth, and yield in different plants

Host	Isolates of *Phoma* sp.	Effect on plant growth	References
Wheat	*Phoma* sp. GS6-1 and GS7-4	Plant length and root length	Shivanna et al. (1996)
Cucumber	*Phoma* sp. GS8-2 and GS8-3	Shoot dry weight	Chandanie et al. (2005)
Cucumber	*Phoma* sp. GS8-2 and GS8-3	Root length	Chandanie et al. (2006)
Soybean	*Phoma herbarum*	Plant length and biomass	Hamayun et al. (2009)
Cucumber	*Phoma* sp. GAH7	Plant length, shoot length, plant fresh weight, and plant dry weight	Hamayun et al. (2010)
Cucumber	*Phoma glomerata* LWL2	Shoot height, shoot fresh weight, dry weight, chlorophyll content, and leaf area	Waqas et al. (2012)
Tomato	*Phoma multirostrata*	Seed germination, seedling vigor, plant height, stem girth, shoot and root lengths, and fruit weight	Jogaiah et al. (2013)
Onion	*Phoma* sp. GS8-1 and GS8-3	Plant height, root length, bulb perimeter, dry weight	Elsharkawy and El-Khateeb (2019)
Tobacco, cucumber	*Phoma* sp. GS8-2	Plant height, fresh weight, dry weight, yield	Elsharkawy (2019)
Cucumber	*Phoma* sp. GS8-3		Elsharkawy et al. (2020)

flowers; and yield were all increased due to fungal treatments. *T. harzianum* T-22 treatment increased the yield of several hybrids and inbred lines (Harman et al. 2004). The yields of sugarcane, tomato, strawberry, lettuce, chickpea, and pea and mustard were improved due to treatment with *Trichoderma* spp. (Bal and Altintas 2006; Elad et al. 2006; Srivastava et al. 2006; Haque et al. 2012; Hossain et al. 2013; Akhter et al. 2015). Colonization with *Phoma* sp. significantly improved growth attributes in *Arabidopsis*, tobacco, and cucumber (Elsharkawy 2019; Elsharkawy et al. 2020). The inoculum of *Phoma* sp. was applied to the roots, but shoot biomass and number of leaves were increased explaining that the effects of inoculum are systemic (Table 16.2). Similarly, *P. herbarum* isolated from soybean rhizosphere promoted plant length and biomass. The rhizospheric isolate (*Phoma* sp. GS8-2) was greatly effective in improving the growth rate and yield of cucumber (Elsharkawy 2019). Treatment with *Phoma* sp. GS8-2 significantly raised the dry biomass of cucumber plants (~50%). Similarly, cucumber plants colonized with *Phoma* sp. GS8-3 grew better and yielded ~2.0 times more than the control group (Elsharkawy et al. 2020). Germination of seeds grown in soil inoculated with the rhizospheric fungi was increased. Suppression of seed-borne pathogens could be one of the involved mechanisms of increased germination by rhizospheric fungi (Szopińska et al. 2010). Moreover, rhizospheric fungi may help to overcome and break seed dormancy (Delgado-Sánchez et al. 2011). A preferential interaction between *Phoma* spp. and a particular host was observed. For example, *Phoma* spp. promoted the growth of cucumber cultivars (Aodai kyuri and Shogoin fushinari) by increasing plant length, while the plant length in cultivars Jibai and Ochiai fushinari was not affected by inoculation with GS6-4, GS6-1, GS7-4, GS8-6, and GS10-2 (Shivanna et al. 2005). Further research is needed to elucidate the effect of *Phoma* on plant genotypes for growth promotion. These results revealed that *Phoma* spp. hold great potential in the development of crop yields.

16.3 Production of Volatile Organic Compounds (VOCs)

Several factors (biotic or abiotic) affect plant growth. The photosynthetic rate is affected by the gaseous atmosphere (Haisel et al. 1999). VOCs are mixtures of gas-phase and carbon-based compounds. VOCs produced by living organisms are important gaseous components in the atmosphere (Shiojiri et al. 2006). VOCs have a big contribution to the interaction between plants and microorganisms compared with nonvolatile compounds. Plants have the ability to release and detect volatiles in the surrounding environment. VOCs are produced by several rhizospheric fungi including *Phoma* sp. with stimulatory effects on numerous plants (Table 16.3). Plant growth was improved due to volatile compounds emitted by beneficial microorganisms (Ryu et al. 2003; Yamagiwa et al. 2011). Volatile organic compounds such as 2-methyl-propanol and 3-methyl-butanol produced by *Phoma* sp. GS8-3 were separated using gas chromatography (Naznin et al. 2013). These VOCs were identified as C4-C8 hydrocarbons with different numbers and quantities using mass spectrometry. The volatiles obtained from *Phoma* sp. GS8-3 cultures after 3 days of incubation resulted in significant stimulation of tobacco growth in vitro. Lower concentrations achieved better results than higher concentrations. Shoot growth was improved in plants grown in the presence of VOCs emitted by *Phoma* sp. (Naznin et al. 2013). Similar results were obtained by Lee et al. (2016) in *Arabidopsis* treated with *Trichoderma*. The effect of VOCs on plant growth could be due to CO2 enrichment during cocultivation (Kai et al. 2009). Additionally, VOCs increased lateral root formation (Felten et al. 2009). The effect varies greatly based on the duration of exposure, growth condition, plant stage, and fungal species (Hung et al. 2013; Lee et al. 2015). Photosynthetic efficiency

Table 16.3 Mechanisms of plant growth promotion utilized by *Phoma* spp.

Mechanism	Activity	Species of *Phoma*	References
Availability of nutrients	Availability of ammonium nitrogen	*Phoma* sp. (isolates GS6-1, GS6-2, GS7-3, GS7-4, GS8-6, GS10-1, and GS10-2	Shivanna et al. (1994)
		Phoma sp. (isolates GS8-1, GS8-2, and GS8-3)	Shivanna et al. (2005)
	Phosphate solubilization	*Phoma multirostrata* and *Phoma putaminum*	Jogaiah et al. (2013)
Phytohormones	GAs production	*Phoma herbarum* TK-2-4	Hamayun et al. (2009)
	Production of GA3, GA4, GA9, and GA19	*Phoma* sp. GAH7	Hamayun et al. (2010)
	Production of GA1, GA3, GA4, GA7, and indole acetic acid (IAA)	*Phoma glomerata* LWL2	Waqas et al. (2012)
Volatile organic compounds (VOCs)	Production of 2-methyl-propanol and 3-methyl-butanol	*Phoma* sp. GS8-3	Naznin et al. (2014)
Suppression of plant pathogens	Take-all and common root rot diseases (*Gaeumannomyces graminis* var. *tritici*)	*Phoma* sp. (isolates GS6-1 and GS7-4)	Shivanna et al. (1996)
	Bacterial leaf speck (*Pseudomonas syringae* pv. *tomato* DC3000)	*Phoma* sp. (isolates GS6-2 and GS7-3)	Sultana et al. (2008)
	Damping off (*Rhizoctonia solani* AG-4) and root rot (*Fusarium oxysporum* f. sp. *radicis-lycopersici*)	*Phoma* sp. GS8-2	Saldajeno et al. (2012)
	Bacterial wilt (*Ralstonia solanacearum*)	*Phoma multirostrata*	Jogaiah et al. (2013)
	White rot (*Sclerotium cepivorum*)	*Phoma* sp. GS8-1 and GS8-3	Elsharkawy and El-Khateeb (2019)
	Cucumber mosaic virus	*Phoma* sp. GS8-2	Elsharkawy (2019)
		Phoma sp. GS8-3	Elsharkawy et al. (2020)

was improved in *Phoma*-treated plants. This could be due to increased number of leaves in the inoculated plants relative to the control. VOCs emitted by *A. alternata* resulted in improved photosynthesis efficiency (Sánchez-López et al. 2016).

Phenotypic plasticity of flowering was affected by treatment with rhizospheric fungi (Forrest and Miller-Rushing 2010). The number and size of flowers were improved in fungal-treated plants. Colonization of plant roots with endophytes also affected flowering time and the numbers of flower buds (Ousley et al. 1994; Das et al. 2012). The early flowering phenotype was observed in plants exposed to volatile compounds produced by rhizospheric fungi (Jogaiah et al. 2013; Hung et al. 2013; Sánchez-López et al. 2016). It is well documented that the endophytes increase the availability of nutrients in the soil especially phosphorus. The importance of phosphorus and cytokines in plant growth and yield was reported (Poulton et al. 2002; Riefler et al. 2006; D'Aloia et al. 2011).

16.4 Production of Phytohormones

Fungal endophytes have the capacity to promote plant growth and yield. Growth hormones are believed to be the most common factor affecting plant growth. Production of phytohormones such as gibberellin and indole-3-acetic acid (IAA) by fungal endophytes was reported. Recently, more than 100 gibberellins have been recognized, and most of them were produced by plant-associated microorganisms (Kawaide 2006; Vandenbussche et al. 2007). Gibberellin secretion by different fungal endophytes was revealed (Ogas 2000; MacMillan 2002; Kawaide 2006; Khan et al. 2008). Different gibberellins were found in the culture filtrates of the isolate TK-2-4 of *P. herbarum* including the gibberellic acids GA4 and GA7 resulting in increased growth of soybean plants (Table 16.3). The potential of *Phoma* sp. GAH7 to promote plant growth and production of different gibberellins including GA1, GA3, and GA4 was reported (Hamayun et al. 2010). The levels of gibberellins in the culture filtrate were higher than the wild type of *Fusarium fujikuroi* that is able to produce industrial GAs (Takahashi et al. 1991; O'Donnell et al. 1998). Moreover, bioactive gibberellins were secreted by *Phoma glomerata* isolated from the spotted bellflower and viola plants (Rim et al. 2007). Similarly, Kang et al. (2015) reported high levels of GA4, GA9, GA34, phosphate, and IAA in fungal cultures which were suggested to be the reason for increased shoot growth. Phytohormones emitted by the fungi enhanced seed germination and seedling vigor (Gupta and Chakrabarty 2013). Moreover, *Pi. indica* promoted *Arabidopsis* growth through activation of cytokine production (Vadassery et al. 2008). Lateral root emergence was also affected by auxin production (Dubrovsky et al. 2008). Increasing the root surface area increases nutrient uptake and plant growth. A recent study has pointed to the role of auxin in tobacco response to *Phoma* sp. GS8-3 explaining growth promotion in the treated plants. Similarly, the expression of auxin-regulated genes was stimulated in *Arabidopsis* plants treated with *T. virens* and in barley and Chinese cabbage treated with *Piriformospora indica* (Contreras-Cornejo et al. 2009; Schäfer et al. 2009; Lee et al. 2011). Ethylene was closely associated with plant growth and development. Growth promotion effect due to the rhizospheric fungi treatments was not noticed in tobacco plants transformed to silence ethylene production (Barazani et al. 2005).

16.5 Suppression of Plant Pathogens

The isolates of endophytic *Phoma* sp. was utilized as effective biocontrol agents. Induced systemic resistance is the most widely reported mechanism of disease suppression by *Phoma* sp. (Sultana et al. 2008; Elsharkawy 2019; Elsharkawy et al. 2020). *Phoma* sp. could successfully protect cucumber plants from *cucumber mosaic virus* infection by reducing disease severity and virus titer (Table 16.3). The transcription levels of defense-related genes were improved

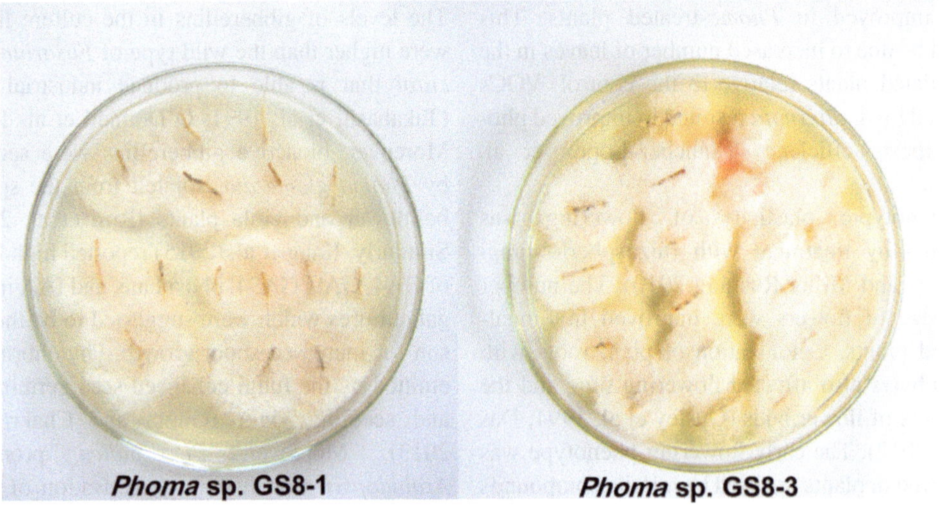

Fig. 16.1 Re-isolation of *Phoma* sp. GS8-1 and GS8-3 at high frequencies from the roots of *Nicotiana tabacum* after 6 weeks from germination

(Elsharkawy 2019). *Phoma* sp. contends with the pathogen for colonization position on plant roots (Shivanna et al. 1996). Antibiosis and competition for nutrients and space were reported by *Phoma* sp. GS8-1 and GS8-3 against white rot disease in onion plants (Elsharkawy and El-Khateeb 2019). *Phoma* sp. GS8-2 enhanced growth of cucumber by counteracting *Fusarium oxysporum* f. sp. *radicis-lycopersici* and *Rhizoctonia solani* AG-4 pathogens (Saldajeno et al. 2012). *Phoma* sp. could compete with *Fusarium oxysporum* and *Rhizoctonia solani* for the infection sites on cucumber roots. Seed priming with *Phoma multirostrata* PhoM_JSB17 promoted seedling vigor and resistance to *Ralstonia solanacearum*, the causal pathogen of bacterial wilt disease in tomato (Jogaiah et al. 2013). The mechanisms governing disease resistance by *Phoma* sp. remain elusive. The mechanisms of growth promotion could be similar to the mechanisms of disease resistance by *Phoma* sp.

16.6 Colonization of Plant Roots with *Phoma* sp.

Colonization of plant roots with rhizospheric fungi is necessary to stimulate plant growth. Colonization of plant roots exhibits fungal potential to proliferate along growing roots. The required period for root colonization depends on the growth rate of both *Phoma* sp. and plant roots. Colonization of cucumber roots with *Phoma* sp. GS8-2 was reported at 10 weeks after planting (Shivanna et al. 2005). *Phoma* sp. was selective in root colonization of plant roots. *Phoma* sp. could not efficiently colonize soybean roots, and then plant growth was not stimulated (Shivanna et al. 1996). Re-isolation frequency of root-associated fungi displays its colonization capability. In general, colonization of the upper roots is higher than the colonization in the lower ones because of the fast growth of the roots. *Phoma* sp. GS8-2 and GS8-3 were re-isolated in high frequencies from plant roots (Fig. 16.1).

16.7 Conclusion

Due to the current concerns over the harmful impact of pesticides on human health and environment, it is important to improve the usage of beneficial microbes in agriculture. The rhizospheric fungi display great ability in sustainable agriculture through increasing plant growth and health. Research has provided ample support for the assertion that *Phoma* isolates could be used as an eco-friendly and sustainable tool to increase the crop yield. The potential fungal isolates from the genus *Phoma* for plant growth promotion could reduce the negative impact of chemical fertilizers. Biotechnological tools could help in engineering of *Phoma* sp. to widen the application under field conditions. Research may be required on the field of genetic transformation and stability to improve the performance of *Phoma* sp. on plant growth and yield.

References

Akhter W, Bhuiyan MKA, Sultana F, Hossain MM (2015) Integrated effect of microbial antagonist, organic amendment and fungicide in controlling seedling mortality (*Rhizoctonia solani*) and improving yield in pea (*Pisum sativum* L.) C R Biol 338:21–28.

Bal U, Altintas S (2006) Effects of *Trichoderma harzianum* on yield and fruit quality characteristics of tomato (*Lycopersicon esculentum* Mill) grown in an unheated greenhouse. Aust J Exp Agric 46:131–136.

Barazani O, Benderoth M, Groten K, Giovannini O, Gessler G et al (2005) *Piriformospora indica* and *Sebacina vermifera* increase growth performance at the expense of herbivore resistance in *Nicotiana attenuata*. Oecologia 146:234–243.

Bezerra JDP, Santos MGS, Svedese VM et al (2012) Richness of endophytic fungi isolated from *Opuntia ficus-indica* Mill. (Cactaceae) and preliminary screening for enzyme production. World Journal of Microbiology and Biotechnology 28 (5): 1989–1995,

Chandanie WA, Kubota M, Hyakumachi M (2005) Interaction between arbuscular mycorrhizal fungus *Glomus mosseae* and plant growth promoting fungus *Phoma* sp. on their root colonization and growth promotion of cucumber (*Cucumis sativus* L.). Mycoscience 46:201-204.

Chandanie WA, Kubota M, Hyakumachi M (2006) Interactions between plant growth promoting fungi and arbuscular mycorrhizal fungus *Glomus mosseae* and induction of systemic resistance to anthracnose disease in cucumber. Plant Soil 286: 209–217.

Contreras-Cornejo HA, Macías-Rodríguez LI, Cortés-Penagos C, López-Bucio J (2009) *Trichoderma virens*, a plant beneficial fungus, enhances biomass production and promotes lateral root growth through an auxin-dependent mechanism in *Arabidopsis*. Plant Physiol 149:1579–1592.

D'Aloia M, Bonhomme D, Bouché F (2011) Cytokinin promotes flowering of *Arabidopsis* via transcriptional activation of the FT paralogue TSF. Plant J 65:972–979.

Das K, Das S, Tiwari RKS (2012) Bioprospecting of Endophytic Fungi for Bioactive Natural Products: Recent Trends and Future Perspectives. Microbiology Application (eds) pp. 346–364.

Das A, Varma A (2009). Symbiosis: the art of living, in Symbiotic Fungi Principles and Practice, A. Varma and A. C. Kharkwal, Eds., pp. 1–28, Springer, Berlin, Germany.

Delgado-Sánchez P, Ortega-Amaro MA, Jiménez-Bremont JF, Flores J (2011) Are fungi important for breaking seed dormancy in desert species? Experimental evidence in *Opuntia streptacantha* (Cactaceae). Plant Biol 13:154–159.

Denef K, Bubenheim H, Lenhart K et al (2007) Community shifts and carbon translocation within metabolically-active rhizosphere microorganisms in grasslands under elevated CO2. Biogeosciences 4:769–779.

Dubrovsky JG, Sauer M, Napsucialy-Mendivil S (2008) Auxin acts as a local morphogenetic trigger to specify lateral root founder cells. Proc Natl Acad Sci U S A 105:8790–8794.

Elad Y, Chet I, Henis Y (2006) Biological control of *Rhizoctonia solani* in strawberry fields by *Trichoderma harzianum*. Plant Soil 60:245–254.

Elsharkawy MM (2019) Induced systemic resistance against *Cucumber mosaic virus* by *Phoma* sp. GS8-2 stimulates transcription of pathogenesis-related genes in *Arabidopsis*. Pest Manag Sci 75: 859–866.

Elsharkawy MM, El-Khateeb NM (2019) Antifungal activity and resistance induction against *Sclerotium cepivorum* by plant growth-promoting fungi in onion plants. Egyptian Journal of Biological Pest Control 29, 68. https://doi.org/10.1186/s41938-019-0178-9.

Elsharkawy MM, Shimizu M, Takahashi H, Hyakumachi M (2012) Induction of systemic resistance against *Cucumber mosaic virus* by *Penicillium simplicissimum* GP17–2 in *Arabidopsis* and tobacco. Plant Pathol 61:964–976.

Elsharkawy MM, Shimizu M, Takahashi H et al. (2013) Induction of systemic resistance against *Cucumber mosaic virus* in *Arabidopsis thaliana* by *Trichoderma asperellum* SKT-1. Plant Pathol J 29:193–200.

Elsharkawy MM, Hassan N, Ali M et al (2014) Effect of zoysiagrass rhizosphere fungal isolates on disease suppression and growth promotion of rice seedlings. Acta Agric Scand Sect B Soil Plant Sci 64:135–140.

Elsharkawy MM, Suga H, Shimizu M (2020) Systemic resistance induced by *Phoma* sp. GS8-3 and nanosilica against Cucumber mosaic virus. Environ Sci Pollut Res 27: 19029–19037. https://doi.org/10.1007/s11356-018-3321-3.

Felten J, Kohler A, Morin E et al (2009) The ectomycorrhizal fungus *Laccaria bicolor* stimulates lateral root formation in poplar and *Arabidopsis* through auxin transport and signaling. Plant Physiol 151:1991–2005.

Forrest J, Miller-Rushing JA (2010) Toward a synthetic understanding of the role of phenology in ecology and evolution. Philos Trans R Soc B 365:3101–3112.

Gahan J, Schmalenberger A (2014) The role of bacteria and mycorrhiza in plant sulfur supply. Front Plant Sci 5:723.

Gaind S, Nain L. (2015). Soil-phosphorus mobilization potential of phytate mineralizing fungi. Journal of Plant Nutrition 38(14):2159-2175. https://doi.org/10.1080/01904167.2015.1014561.

Gupta R, Chakrabarty SK (2013) Gibberellic acid in plant: still a mystery unresolved. Plant Signal Behav. 8(9): e25504. https://doi.org/10.4161/psb.25504.

Haisel D, Pospíšilová J, Synková H, Čatský J, Wilhelmová N, Plzáková Š (1999) Photosynthetic pigments and gas exchange of *in vitro* grown tobacco plants as affected by CO2 supply. Biologia Plantarum 42:463–468.

Hamayun M, Khan SA, Khan AL, Rehman G, Sohn EY, Shah AA, et al. (2009) *Phoma herbarum* as a new gibberellin producing and plant growth-promoting fungus. Journal of Microbiology and Biotechnology 19: 1244-1249. https://doi.org/10.4014/jmb.0901.030.

Hamayun M, Khan SA, Khan AL et al. (2010) Growth promotion of cucumber by pure cultures of gibberellin-producing *Phoma* sp. GAH7. World J Microbiol Biotechnol 26:889–894.

Haque M, Ilias GNM, Molla AH (2012) Impact of Trichoderma-enriched bio-fertilizer on the growth and yield of mustard (*Brassica rapa* L.) and Tomato (*Solanum lycopersicon* Mill.) Agriculturists 10:109–119.

Harman GE, Howell CR, Viterbo A et al (2004) Trichoderma species-opportunistic avirulent plant symbionts. Nat Rev Microbiol 2:43–56.

He X, Han G, Lin Y et al (2012) Diversity and decomposition potential of endophytes in leaves of a *Cinnamomum camphora* plantation in China, Ecological Research 27(2): 273–284.

Hossain MM, Sultana F, Kubota M et al (2007) The plant growth-promoting fungus *Penicillium simplicissimum* GP17-2 induces resistance in *Arabidopsis thaliana* by activation of multiple defense signals. Plant Cell Physiol 48:1724–1736.

Hossain MM, Sultana F, Kubota M, Koyama H, Hyakumachi M (2008) Systemic resistance to bacterial leaf speck pathogen in *Arabidopsis thaliana* induced by the culture filtrate of a plant growth-promoting fungus (PGPF) *Phoma* sp GS8-1. Journal of General Plant Pathology 74: 213–221.

Hossain MM, Hossain N, Sultana F et al (2013) Integrated management of Fusarium wilt of chickpea (*Cicer arietinum* L.) caused by *Fusarium oxysporum* f.sp. *ciceris* with microbial antagonist, botanical extract and fungicide. Afr J Biotechnol 12:4699–4706

Hossain MM, Sultana F, Miyazawa M, Hyakumachi M (2014) The plant growth promoting fungi *Penicillium* spp. GP15-1 enhances growth and confers protection against damping-off and anthracnose in the cucumber. J Oleo Sci 63(4):391–400

Hossain MM, Sultana F, Islam S (2017). Plant growth-promoting fungi (PGPF): Phytostimulation and induced systemic resistance. In: Singh D, Singh H, Prabha R, editors. Plant Microbe Interactions in AgroEcological Perspectives, Volume 2: Microbial Interactions and Agro-Ecological Impacts. Singapore: Springer; pp. 135-191. https://doi.org/10.1007/978-981-10-6593-4.

Hung R, Lee S, Bennett JW (2013) *Arabidopsis thaliana* as a model system for testing the effect of Trichoderma volatile organic compounds. Fungal Ecol 6:19–26.

Hyakumachi M (1994) Plant growth promoting fungi from turfgrass rhizosphere with potentials for disease suppression. Soil Microorganisms 44: 53-68.

Jogaiah S, Abdelrahman M, Tran LSP, Shin-ichi I (2013) Characterization of rhizosphere fungi that mediate resistance in tomato against bacterial wilt disease. Journal of Experimental Botany 64:3829-3842.

Kai M, Haustei M, Molina F, Petri A, Scholz B, et al. (2009) Bacterial volatiles and their action potential. Appl Microbiol Biotechnol 81: 1001–1012.

Kang JW, Lee NY, Cho KC et al (2015) Analysis of nitrated proteins in *Saccharomyces cerevisiae* involved in mating signal transduction. Proteomics 15(2–3):580–590.

Kawaide H (2006) Biochemical and molecular analysis of gibberellin biosynthesis in fungi. Biosci Biotechnol Biochem 70: 583-590.

Khan SA, Hamayun M, Yoon HJ, Kim HY, Suh SJ, Hwang SK, et al (2008) Plant growth promotion and *Penicillium citrinum*. BMC Microbiol 8: 231.

Lee S, Flores-Encarnacion M, Contreras-Zentella M, Garcia-Flores L, Escamilla JE, Kennedy C (2004) Indole-3-acetic acid biosynthesis is deficient in *Gluconacetobacter diazotrophicus* strains with mutations in Cytochrome c biogenesis genes. Journal of Bacteriology. 186 (16): 5384–5391.

Lee YC, Johnson JM, Chien CT et al (2011) Growth promotion of Chinese cabbage and *Arabidopsis* by *Piriformospora indica* is not stimulated by mycelium-synthesized auxin. Mol Plant-Microbe Interact 24:421–431.

Lee SM, Chung J, Ryu C-M (2015) Augmenting plant immune responses and biological control by microbial determinants. Res Plant Dis 21:161–179.

Lee S, Yap M, Behringer G et al (2016) Volatile organic compounds emitted by Trichoderma species mediate plant growth. Fungal Biol Biotechnol 3:7.

MacMillan J (2002) Occurrence of gibberellins in vascular plants, fungi and bacteria. J. Plant Growth Reg. 20: 387-442.

Malinowski DP, Alloush GA, Belesky DP (2000) Leaf endophyte *Neotyphodium coenophialum* modifies mineral uptake in tall fescue. Plant and Soil. 227 (1-2): 115–126.

Mendes R, Garbeva P, Raaijmakers JM (2013) The rhizosphere microbiome: significance of plant beneficial, plant pathogenic, and human pathogenic microorganisms. FEMS Microbiol Rev 37:634–663.

Nahas E (1996) Factors determining rock phosphate solubilization by microorganisms isolated from soil. World Journal of Microbiology and Biotechnology 12:567-572.

Naznin HA, Kimura M, Miyazawa M, Hyakumachi M (2013) Analysis of volatile organic compounds emitted by plant growth-promoting fungus *Phoma* sp. GS8-3 for growth promotion effects on tobacco. Microbes Environ 28:42–49.

Naznin HA, Kiyohara D, Kimura M et al (2014) Systemic resistance induced by volatile organic compounds emitted by plant growth promoting fungi in

Arabidopsis thaliana. PLoS One 9: e86882. https://doi.org/10.1371/journal.pone.0086882.

O'Donnell K, Cigelnik E, Nirenberg HL (1998) Molecular systematics and phylogeography of the *Gibberella fujikuroi* species complex. Mycol 90:465–493.

Ogas J (2000) Gibberellins. Curr. Biol. 10: R48.

Ousley MA, Lynch JM, Whipps JM (1994) The effects of addition of Trichoderma inocula on flowering and shoot growth of bedding plants. Scientia Horticulturae. 59:147–155.

Poulton JL, Bryla D, Koide RT, Stephenson AG (2002) Mycorrhizal infection and high soil phosphorus improve vegetative growth and the female and male functions in tomato. New Phytol 154:255–264.

Rajankar PN, Tambekar DH, Wate SR. (2007) Study of phosphate solubilization efficiencies of fungi and bacteria isolated from saline belt of Purna river basin. Research Journal of Agriculture and Biological Sciences 3 (6):701-703

Riefler M, Novak O, Strnad M, Schmülling T (2006) Arabidopsis cytokinin receptor mutants reveal functions in shoot growth, leaf senescence, seed size, germination, root development, and cytokinin metabolism. Plant Cell 18:40–54.

Rim SO, Lee JH, Khan SA, Lee IJ, Rhee IK, Lee KS, Kim JG (2007) Isolation and identification of fungal strains producing gibberellins from the root of plants. Kor J Microbiol Biotechnol 35: 357-363.

Ryu CM, Farag MA, Hu CH, Reddy MS, Wei HX, Pare PW, Kloepper JW (2003) Bacterial volatiles promote growth in *Arabidopsis*. Proc. Natl. Acad. Sci. U.S.A. 100:4927–4932.

Saldajeno MGB, Ito M, Hyakumachi M (2012) Interaction between the plant growth-promoting fungus *Phoma* sp. GS8-2 and the arbuscular mycorrhizal fungus *Glomus mosseae*: impact on biocontrol of soil-borne diseases, microbial population, and plant growth. Australasian Plant Pathol. 41: 271–281. https://doi.org/10.1007/s13313-011-0101-7

Sánchez-López AM, Bahaji A, De Diego N et al (2016) *Arabidopsis* responds to *Alternaria alternata* volatiles by triggering plastid phosphoglucose isomerase-independent mechanisms. Plant Physiol 172:989–2001

Saxena S (2015) Applied Microbiology. India: Springer Pvt. Ltd; p. 190.

Schäfer P, Pfiffi S, Voll LM et al (2009) Manipulation of plant innate immunity and gibberellin as factor of compatibility in the mutualistic association of barley roots with *Piriformospora indica*. Plant J 59:461–474

Shaikh MN, Mokat DN (2018) Role of rhizosphere fungi associated with commercially explored medicinal and aromatic plants: A Review. Curr Agr Res J 6(1):72–77 1:36–43

Shiojiri K, Kishimoto K, Ozawa R, et al. (2006) Changing green leaf volatile biosynthesis in plants: An approach for improving plan resistance against both herbivores and pathogens. Proc. Natl. Acad. Sci. U.S.A. 103:16672–16676.

Shivanna MB, Meera MS, Hyakumachi M (1994) Sterile fungi from zoysiagrass rhizosphere as plant growth promoters in spring wheat. Can J Microbial 40:637–644.

Shivanna MB, Meera MS, Hyakumachi M (1996) Role of root colonization ability of plant growth promoting fungi in the suppression of take-all and common root rot of wheat. Crop Protection 15(6):497-504.

Shivanna MB, Meera MS, Kubota M, Hyakumachi M (2005) Promotion of growth and yield in cucumber by zoysiagrass rhizosphere fungi. Microbes Environ 20:34–40.

Shoresh M, Harman GE, Mastouri F (2010) Induced systemic resistance and plant responses to fungal biocontrol agents. Annu Rev Phytopathol 48:21–43.

Srivastava S, Singh V, Gupta PS et al (2006) Nested PCR assay for detection sugarcane grassy shoot phytoplasma in the leafhopper vector *Deltocephalus vulgaris*: a first report. Plant Pathol 55:25–28.

Sultana F, Hossain MM, Kubota M, Hyakumachi M (2008) Elicitation of systemic resistance against the bacterial speck pathogen in *Arabidopsis thaliana* by culture filtrates of plant growth promoting fungi. Canadian Journal of Plant Pathology 30(2):196-205

Szopińska D, Jensen B, Knudsen IMB et al (2010) Non-chemical methods for controlling seedborne fungi in carrot with special reference to *Alternaria radicina*. J Plant Protec Res 50(2):184–192.

Takahashi N, Phinney BO, MacMillan J (1991) Gibberellins. Springer, New York.

Vadassery J, Ritter C, Venus Y et al (2008) The role of auxins and cytokinins in the mutualistic interaction between *Arabidopsis* and *Piriformospora indica*. Mol Plant-Microbe Interact 21:1371–1383

Vandenbussche F, Fierro AC, Wiedemann G, Reski R, Van Der Straeten D (2007). Evolutionary conservation of plant gibberellin signalling pathway components. BMC Plant Biol. 7: 65.

Waqas M, Khan AL, Kamran M, Hamayun M, Kang S-M, Kim Y-H, Lee I-J (2012) Endophytic fungi produce gibberellins and indoleacetic acid and promotes host–plant growth during stress. Molecules 17:10754–10773.

Wilson D (1995) Endophyte: the evolution of a term, and clarification of its use and definition. Oikos 73 (2): 274–276.

Yamagiwa Y, Toyoda K, Inagaki Y, Ichinose Y, Hyakumachi M, Shiraishi T (2011) *Talaromyces wortmannii* FS2 emits β-caryophyllene, which promotes plant growth and induces resistance. J Gen Plant Pathol 77:336–341.

Zhang X, Ferris H, Mitchell J, Liang W (2017) Ecosystem services of the soil food web after long term application of agricultural management practices. Soil Biol Biochem 111:36-43

Zhou Z, Zhang C, Zhou W, Li W, Chu L, Yan J, et al. (2014) Diversity and plant growth-promoting ability of endophytic fungi from the five flower plant species collected from Yunnan, Southwest China. Journal of Plant Interactions.;9(1):585-591.

Part IV

Nanotechnology and Fossil Records of *Phoma*

17 Potential Role of *Phoma* spp. for Mycogenic Synthesis of Silver Nanoparticles

Aniket Gade, Sudhir Shende, and Mahendra Rai

17.1 Introduction

The term "nanotechnology" was derived from the Greek word "nano," meaning "dwarf," and is defined as the fabrication, observation, measurement, and maneuver at a scale of 1–100 nm. Nanotechnology (NT) has been playing a crucial role in the twenty-first century in solving various problems particularly in the field of agriculture, medicine, and electronics. Nanotechnology is a broad and interdisciplinary area of research and development activity that has been growing rapidly worldwide. Due to small size, surface tailor ability, improved solubility, and multifunctionality of nanoparticles, it may open up new research avenues. Moreover, it offers an exciting possibility to study matter, which is smaller bulk and bigger than atoms or molecules, as well as the effect of spatial confinement on electron behavior. It also provides an opportunity to explore the problems related to matter with higher surface area to volume ratio or interface because of their interfacial nature (Zhang et al. 2007).

Nanoparticles are being viewed as fundamental building blocks of nanotechnology. The utilization of microorganisms such as bacteria and yeast (Mehra and Winge 1991) in the synthesis of nanoparticles is a relatively recent phenomenon. The microorganisms minimize the toxicity by reduction of the metal ions or by formation of insoluble comples with metal ions (e.g., metal sulfides) in the form of colloidal particles. An amalgamation of curiosity, environmental compulsions, and understanding that nature has evolved the processes for synthesis of inorganic materials on nano- and micro-length scale have created great interest among material scientists for the utilization of microorganisms in the synthesis of nanomaterials.

Moreover, there has been a growing need to develop scalable, low-cost, nontoxic, and environmentally benign procedures for synthesis of metallic nanoparticles. Therefore, the biological approach for synthesis of metal nanoparticle becomes important. Moreover the biologically synthesized nanoparticles are more biocompatible over the other processes and occur close to ambient temperature, pressure, and also at neutral pH. Variety of biological resources available in nature including plants (plant products), algae, fungi, bacteria, and viruses could be employed for the fabrication of nanoparticles.

Fabrication of silver nanoparticles (AgNPs) till date has been reported by several biological

A. Gade (✉) · S. Shende
Nanobiotechnology Laboratory, Department of Biotechnology, Sant Gadge Baba Amravati University Amravati, Amravati, Maharashtra, India

M. Rai (✉)
Nanobiotechnology Laboratory, Department of Biotechnology, Sant Gadge Baba Amravati University Amravati, Amravati, Maharashtra, India

Department of Microbiology, Nicolaus Copernicus University, Torun, Poland
e-mail: mahendra.rai@v.umk.pl

M. Rai et al. (eds.), *Phoma: Diversity, Taxonomy, Bioactivities, and Nanotechnology*,
https://doi.org/10.1007/978-3-030-81218-8_17

agents. Among all the biological systems used for the fabrication of AgNPs, fungal system has several advantages as compared to other biological system. When focusing on the synthesis of nanoparticles using fungi also known as mycogenic synthesis, it was observed that nanoparticles of good monodispersity and well dimensions could be obtained by mycogenic synthesis. As fungi are found to secrete high amount of biomolecules in general and proteins in particular, they might result in the significant mass productivity of nanoparticles. The fungal proteins are capable of hydrolyzing metal ions and possess high affinity for metal ions (Rai et al. 2010). In addition to this, fungi are easy to isolate and culture. Moreover, the downstream processing and the handling of fungal biomass are less complex than the other synthetic methods. Fungi can accumulate metal ions by physicochemical and biological mechanisms including extracellular binding by metabolites and polymers, binding to specific polypeptides, and metabolism-dependent accumulation (Volesky and Holan 1995). The possible use of fungi has gained much importance Also, the extracellular secretion of proteins has an added advantage in the downstream processing and handling of biomass (Gade et al. 2008) when compared to the bacterial fermentation process which involves use of sophisticated instruments to obtain clear filtrate from the colloidal broth. Moreover, fungi are excellent secretors of protein compared to bacteria and actinomycetes, resulting into higher yield of nanoparticles (Sastry et al. 2003). Thus, using these dissimilatory properties of fungi, it could be extensively used for the rapid and eco-friendly mycogenic synthesis of metal nanoparticles.

The mycogenic synthesis of nanoparticles and their subsequent application, particularly in medicine, are studied under myconanotechnology. It is the interface between "mycology" and "nanotechnology" and has considerable potential, due to the wide range and diversity of the fungi (Rai et al. 2009a). The use of fungi for catalyzing specific reactions leading to inorganic nanoparticles is a new and rational mycogenic strategy. Extracellular synthesis of nanoparticles by filamentous fungi offers an advantage of obtaining large quantities in a relatively pure state and at a rapid rate. Furthermore, fungal biomass can be easily separated by filtration, which can be easily achieved by filter press of similar simple equipment (Gade et al. 2010).

Phoma is a cosmopolitan fungus; species of this occurs in wide variety of substrates ranging from various parts of plants, soil, water, air, and even human and animals (Rai 2004; Kövics et al. 2013). Several species of the genus live parasitically or saprophytically on plants and are often associated with distinct disease symptoms, viz., leaf spots; lesions on stem, fruits, or even on roots (tubers); damping off dieback; seed and fruit rots; seedling blight; etc. Despite of severe phytopathogenic potential shown by various species and varieties of this genus, *Phoma* species can be selected for the biosynthesis of nanoparticles because these are easy to isolate and can be easily grown on simple medium like potato dextrose agar and most important is that it has stable biochemical characteristics.

Silver has been in use since time immemorial for the treatment of burns, wounds, and several bacterial infections. But after the emergence of several antibiotics, the use of these silver compounds has been declined remarkably. AgNPs have made a remarkable comeback as a potential antimicrobial agent. The use of AgNPs is also important, as several pathogenic bacteria have developed resistance against various antibiotics. Hence, AgNPs have emerged up with diverse medical applications ranging from silver nanoparticle-containing dressings and silver nanoparticle-coated medicinal formulations, such as nanogels, nanolotions, etc. AgNPs have been studied and applied in many areas including the biomedical, agricultural, electronic fields, etc. Several products of colloidal silver are already on the market. Research on new, eco-friendly, and cheaper methods has been initiated. Biological production of metal nanoparticles has been studied by many researchers due to the convenience of the method that produces small particles stabilized by protein (Gaikwad et al. 2013).

17.2 *Phoma* as a Ubiquitous Genus

The genus *Phoma* belongs to the order *Pleosporales* of *Ascomycota*. It occurs ubiquitously and has been reported from a wide variety of hosts and substrates, particularly from plants and soil. It has also been recovered from aquatic and aerial environment (Rai 1985) and marine environment (Sugano et al. 1991) and is entomopathogenic (Narendra and Rao 1974) and has been found to cause diseases in human beings (Shukla et al. 1984; Zaitz et al. 1997; Rai 1989, 2004). This fungus can be easily differentiated from its related genus *Phyllosticta* by the formation of relatively smaller pycnidiospores. In addition, the latter produces a hyaline sheath around pycnidiospores. A special appendage is also produced by the pycnidiospores, which can be visible in aqueous condition. The fungus is characterized by the formation of single-celled, hyaline pycnidiospores borne inside a fruiting body referred to as pycnidium which varies from globose, subglobose, to coalesce forms. The typical pycnidium contains ostiole and pycnidiospores inside it (Fig. 17.1).

The species of *Phoma* are common plant pathogens causing disease in crucifers, celery, beets, tomatoes, and peppers. Indoor *Phoma* species are secondary colonizers of building materials. They can range from slow growing and have a powdery to velvety texture. The colony color of *Phoma* varies with species but is commonly gray to olive green.

Species of the genus *Phoma* are commonly found as soil inhabitants and may often be isolated from dead or decaying plants, as primary or secondary parasites. Although over 2000 *Phoma* species have been reported throughout the world, many of these have been characterized by host substrate alone. *Phoma* is recognized by its cellular and more or less flash-shaped fruiting structures (pycnidia) containing masses of one-celled colorless to yellow or pink spores (pycnidiospores). The pycnidiospores are borne from inconspicuous peg like phialides lining the inner wall of the pycnidium. Species of *Phoma* having spines or setae on their pycnidia are sometimes confused with those of *Pyrenochaeta*. However, the two genera can be distinguished on the basis of their conidium-bearing structures: pycnidiospores of *Phoma* species are produced from simple phialides, while those of *Pyrenochaeta* arise from phialides occurring along the sides of elongated conidiophores.

Phoma species are reported from plants as pathogens (Rai and Misra 1981; Agrawal and Mishra 1981; Rao and Thirumalachar 1981; Davari and Hajieghrari 2008; Aveskamp et al. 2008; Lopez and Gomez-Gomez 2009; Pellegrino et al. 2010; Garibaldi et al. 2010; Quiroz et al. 2014; Gilardi et al. 2017; Colmán et al. 2018; Basavand et al. 2020; Yarmalovich and Siaredzich

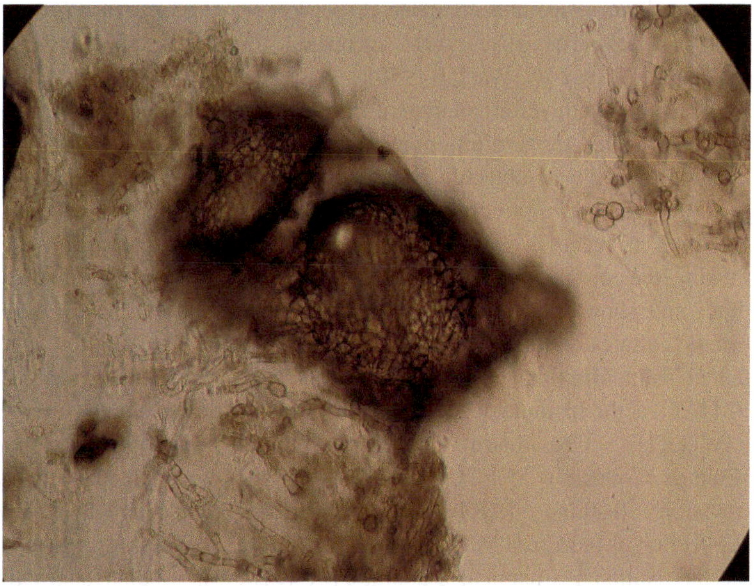

Fig. 17.1 Typical ostiolate pycnidia of *Phoma* sp. with chlamydospores

2019), endophytes (Rai et al. 2018, 2020), and soil (Mathur and Thirumalachar 1959; Dorenbosch 1970). There are few cases of *Phoma* infection in human beings (Shukla et al. 1984; Rai 1989, 2004; Suh 2005; Balis et al. 2006; Tullio et al. 2010; Kukhar et al. 2020). Recently, an elaborated account on *Phoma* diseases their epidemiology and control was reported by Deb et al. (2020).

17.3 Isolation, Procurement, and Maintenance of *Phoma* spp.

The isolations of *Phoma* were made from the infected portions of the hosts present at the junctions of healthy and diseased region, followed by surface sterilization with 70% ethanol and placing the cuttings onto the petri plate containing sterilized potato dextrose agar (PDA). Isolations from soil were performed by serial dilution and Warcup methods. In the latter, the soil samples were taken with the help of micro-spatula and were transferred aseptically in the bottom of a sterilized petri dish. The soil aggregates were crushed and spread with the help of the micro-spatula after adding two to three drops of sterile water, followed by addition of 10–15 mL of sterilized and cooled medium into petri dish and manipulated before setting of agar medium to assume complete dispersion of soil particles. The plates were incubated for 5–7 days at 26 ± 2 °C (Warcup 1950).

The other method for isolation of *Phoma* from the soil is the serial dilution method. In this method the soil was subjected to serial dilution so that the number of microorganisms gradually declines in successive dilutions. Therefore, on plating, the colonies appear well isolated and distinct and show their characteristic forms. The axenic cultures of other *Phoma* species were procured from Microbial Type Culture Collection (MTCC) at the Institute of Microbial Technology (IMTECH), Chandigarh; National Fungal Culture Collection (NFCC) at the Agharkar research Institute (ARI), Pune; National Collection of Industrial Microorganisms (NCIM) at the National Chemical Laboratory (NCL), Pune; and Indian Type Culture Collection at the Indian Agriculture Research Institute (IARI), New Delhi.

The isolated and procured cultures of *Phoma* were maintained on potato dextrose agar (PDA) and malt agar slants at 25 °C. Stock cultures were maintained by subculturing at monthly intervals. They were grown at pH 6.0 and 25 °C for 5–6 days; the slants were preserved at 4 °C. From an actively growing stock culture, subcultures were made on fresh slants and, after 5–6 days incubation at pH 6.0 and 25 °C, were used as the starting material for the synthesis of nanoparticles.

17.4 Mycosynthesis of Silver Nanoparticles

The research on biotechnological exploitation of the fungus *Phoma* for the production of pharmaceutically active secondary metabolites, their exploitation in antibiotic production, and biological control of the weeds is gaining attention (Rai et al. 2009b). The application of microbial secondary metabolites of Phoma in various fields of biotechnology has attracted the interests of a broad array of researchers. The *Phoma* species, which are a rich but underrecognized source of many metabolites, and the screening of different species of *Phoma* for the synthesis of AgNPs may help in understanding the unsolved puzzle of synthesis of AgNPs, thus representing a potentially fruitful idea, which may open up new vistas in the field of nanotechnology in general and bionanotechnology in particular. Recently, an elaborative account on a large number of structurally unique metabolites with potential biological and pharmacological activities isolated from the marine *Phoma* species was reviewed in Rai et al. (2018).

Mycosynthesis was the other term used for mycogenic synthesis. The term "mycosynthesis" was for the first time used by Ingle et al. in 2008 to describe synthesis of nanoparticles by *Fusarium acuminatum.* After the extensive literature survey, it was clear that mycogenic synthesis is a better alternative as compared to other biological system for the synthesis of metal nanoparticles (Suryawanshi et al. 2017; Ramos et al. 2020; Win et al. 2020; Shende et al. 2021; Rai et al. 2021). Many fungal species in general and *Phoma* spp. in particular have been explored for the mycogenic synthesis of different metal nanoparticles of different shapes and size.

Shivshankar et al. (2003) reported use of geranium leaves (*Pelargonium graveolens*) and its endophytic fungus in mycogenic synthesis of gold nanoparticles. Endophytic fungus (*Colletotrichum* sp.) growing in the leaves was treated with aqueous chloroaurate ions. Rapid reduction of the metal ions was observed which resulted in the formation of stable gold nanoparticles of variable size. In case of gold nanoparticles synthesized using geranium leaves, the reducing and capping agents appear to be polypeptides/enzymes in case of *Colletotrichum* sp. The mycosynthesized gold nanoparticles were spherical in shape. Chen et al. (2003) studied the extracellular formation of AgNPs using *Phoma* species. Some soilborne fungi like *Aspergillus fumigatus* reported to produce the AgNPs extracellularly, when the cell extract was challenged with aqueous silver ions (Bhainsa and D'souza 2006). Gade et al. (2008) reported mycogenic synthesis of AgNPs from the *Aspergillus niger* isolated from soil and also proposed the mechanism for the action of AgNPs on the *E. coli*. Out of different fungal genera used for the synthesis of nanoparticles, the genus *Fusarium* was found to be used by many investigators, whereas genus *Phoma* was found to be most versatile. Ahmad et al. (2003) used the *Fusarium oxysporum* for the mycogenic synthesis of AgNPs and gave a new vision to the mycogenic synthesis. They have reported that silver ions, precursor salt solution when exposed to the fungus *Fusarium oxysporum*, are reduced in solution, thereby leading to the fabrication of an extremely stable silver hydrosol. The AgNPs were in the range of 5–15 nm in dimensions and stabilized in solution by proteins secreted by the fungus. It is believed that the reduction of the metal ions occurs by an enzymatic process.

Later, Bansal et al. (2004) found that exposure of *Fusarium oxysporum* to an aqueous solution of K2ZrF6 results in the protein-mediated extracellular hydrolysis of the zirconium hexafluoride ions at room temperature. It leads to the formation of crystalline zirconia nanoparticles. Duran et al. (2005) studied the extracellular production of metal nanoparticles by several strains of the fungus *Fusarium oxysporum*. They found that aqueous silver ions when exposed to several *Fusarium oxysporum* strains are reduced in the solution, thereby leading to the formation of silver hydrosol. The AgNPs were in the range of 20–50 nm in size.

Similarly, other *Fusarium* species like *F. oxysporum* strain 5115 (Mohammadian et al. 2007), *F. semitectum* (Basavaraja et al. 2008), *F. acuminatum* (Ingle et al. 2008), *F. solani* (Ingle et al. 2009), and *F. culmorum* (Bawaskar et al. 2010) were reported for the synthesis of AgNPs. Other fungal species used for the production of AgNPs include *Pleurotus* sp. (Gade et al. 2007), *A. niger* (Gade et al. 2008; Jaidev and Narasimha 2010), *Phoma glomerata* (Birla et al. 2009), *Trichoderma viride* (Fayaz et al. 2010), *Trichoderma reesei* (Vahabi et al. 2011), *Penicillium fellutanum* (Kathiresan et al. 2009), *Neurospora crassa* (Castro-Longoria et al. 2011), and *Pythium* sp. (Gade et al. 2016).

Mycogenic synthesis of AgNPs was observed extracellularly in silver-tolerant yeast strains MKY3 when challenged with 1 mM silver precursor salt in the log phase of growth (Kowshik et al. 2003). A low-cost *Saccharomyces cerevisiae*-mediated biosynthesis of CdS nanoparticles was reported. Individual nanoparticles as well as a few aggregate having the size of 2.5–5.5 nm were observed (Prasad and Jha 2010). Rai and Gade (2009) and Gade et al. (2010) enlisted several advantages of filamentous fungi over bacteria for the synthesis of metal nanoparticles.

17.5 Silver Nanoparticle Synthesis Using *Phoma* spp.

First report on mycogenic synthesis of AgNPs by Indian *Phoma* sp. was made by Birla et al. (2009). They reported extracellular synthesis of silver nanoparticles using *Phoma glomerata* (MTCC-2010) procured from the Microbial Type Culture

Collection (MTCC) center, IMTECH, Chandigarh. They also evaluated AgNPs' antibacterial efficacy against *Escherichia coli*, *Staphylococcus aureus*, and *Pseudomonas aeruginosa*. The AgNPs showed remarkable sensitivity against these drug-resistant pathogens, when used in combination of antibiotics. Silver nanorods were synthesized extracellularly by *Phoma sorghina* (MTCC-2906), whereas other *Phoma* sp. could synthesize spherical AgNPs. The synthesis of silver nanorods was possible due to the secretion of anthraquinone derivatives by *Phoma sorghina* (Gade et al. 2011). AgNPs synthesised using *Phoma* sp. not only showed antibacterial and antifungal potential but also demonstrated antiviral activity. *Phoma* sp. isolated from the infected plants and identified on the basis of morphological and molecular characteristics was used for the synthesis of AgNPs. This showed a consistent decrease in replication efficiency for HSV-1 and HPIV-3 and a minor effect on the replication of HSV-2 at a concentration of 10 mg/mL (Gaikwad et al. 2013). They also observed that AgNPs' ability of controlling viral infectivity was most likely dependent on the size and zeta potential of the AgNPs, which acts by blocking interaction of the virus with the cell.

Gade et al. (2013) have screened 18 different *Phoma* spp. for the synthesis of AgNPs. They have collected six different isolates from infected plant parts and soil samples, which were identified on the basis of comparative cultural and morphological characters on potato dextrose agar and malt agar, whereas 12 cultures of *Phoma* were procured from the Microbial Type Culture Collection (MTCC), Chandigarh. All the 18 cultures of *Phoma* spp. were screened for the synthesis of AgNPs. All the 18 *Phoma* spp. showed the color change of the reaction mixture from yellowish to dark-brown and were further confirmed using UV-Vis spectrometer (Perkin Elemer, Lambda-950) by scanning the absorbance spectra from 200 to 800 nm of colloidal sample. The UV-Vis spectrum of 17 *Phoma* species showed characteristic absorbance peak around 420 nm, and the position of the absorbance peak depends on the nanoparticle size and shape. Only one species, *Phoma tropica* (MTCC-7129), did not show the clear absorption peak as shown by other samples; this might be due to insufficient secretion of the biomolecules, which are responsible for the reduction of silver ions to form AgNPs. All the absorbance peaks shown were symmetric, except for *P. sorghina* (MTCC-2096), which showed asymmetric absorption peak with tailing. Further characterization revealed the synthesis of silver nanorods by *P. sorghina*. UV-Vis absorbance maxima, shift, and peak nature of mycofabricated AgNPs have been shown in Table 17.1.

They also obtained additional information from the analysis of nanoparticles using different characterization techniques: Fourier transform infrared (FTIR) analysis was used to obtain information about the biomolecules involved in the mycogenic synthesis and capping of the AgNPs; X-ray diffraction (XRD) pattern revealed information about the crystalline nature; electron microscopy (scanning electron microscopy (SEM) and transmission electron microscopy (TEM)) analysis provided information about the shape, size, and arrangement of the nanoparticles; energy dispersive X-ray spectroscopy (EDX) provided with elemental composition of the sample; nanoparticle tracking analysis (NTA) estimated size, size distribution, and concentration of nanosize particles in colloids and Zetasizer by measuring zeta potential providing insights into the stability of nanoparticles.

Rapid and green approach of AgNPs synthesis by *Phoma glomerata* (MTCC-2210) procured from IMTECH, Chandigarh, was also reported by Gade et al. (2014). They reported rapid synthesis in the presence of sunlight, apart from the physical conditions like temperature, pH, and light intensity. The authors also optimized the concentration of silver nitrate and amount of fungal filtrate required for the optimum synthesis of AgNPs. The synthesis of AgNPs by *Phoma gardeniae* (ITCC-4554) was performed, and their efficacy was evaluated against human disease-causing bacteria and fungi like *Candida albicans*, *Salmonella choleraesuis*, *Pseudomonas aeruginosa*, *Staphylococcus aureus*, and *Escherichia coli* (Rai et al. 2015a). In another study, *Phoma capsulatum* (ITCC-2086), *Phoma putaminum* (ITCC-2164), and *Phoma citri* (ITCC-

Table 17.1 UV-Vis absorbance maxima, shift, and peak nature of synthesis of AgNPs by *Phoma* spp.

S. no.	*Phoma* spp.	Absorbance maxima (nm)	Absorbance shift	Peak nature
1	*P. arachidicola* (MTCC-401)	437	Red	Broad and symmetric
2	*P. pomorum* (MTCC-467)	434	Red	Broad and symmetric
3	*P. herbarum* (MTCC-2319)	433	Red	Broad and symmetric
4	*P. glomerata* (MTCC-2210)	435	Red	Broad and symmetric
5	*P. betae* (MTCC-2100)	438	Red	Broad and asymmetric
6	*P. sorghina* (MTCC-2096)	416	Blue	Sharp and asymmetric with tailing
7	*P. medicaginis* (MTCC-2313)	430	Red	Broad and symmetric
8	*P. exigua* (MTCC-2314)	416	Blue	Broad and asymmetric
9	*P. macrostoma*(MTCC-2320)	434	Red	Broad and symmetric
10	*P. exigua* var. *exigua* (MTCC-2321)	427	Red	Broad and asymmetric
11	*P. fimeti* (MTCC-2323)	420	–	Broad and asymmetric
12	*P. tropica* (MTCC-7129)	NA[a]	–	–
13	*P. fimeti* (DBT-64)	440	Red	Broad and symmetric
14	*P. multirostrata* (DBT-65)	429	Red	Broad and symmetric
15	*P. destructiva* (DBT-66)	434	Red	Broad and symmetric
16	*P. herbarum* (DBT-67)	439	Red	Broad and symmetric
17	*P. exigua* var. *exigua.* (DBT-8)	444	Red	Broad and symmetric
18	*Phoma* sp. (DBT-69)	442	Red	Broad and symmetric

[a]NA No clear absorbance peak was observed

4913) procured from the Indian Type Culture Collection, Indian Agriculture Research Institute, New Delhi, India, were used for the synthesis of AgNPs. These AgNPs were used for minimal inhibitory concentration (MIC) determination as per the method suggested by the National Committee for Clinical Laboratory Standards (NCCLS 2003) using a tissue culture test plate (96 wells). *Aspergillus niger*, *Candida albicans*, *S. choleraesuis*, *P. aeruginosa*, *S. aureus*, and *E. coli* were used for the determination of their sensitivity to AgNPs. For bacteria, chloramphenicol, ciprofloxacin, and gentamicin sulfate were maintained as control, and for fungi, fluconazole, itraconazole, and nystatin (Merck, USA) were used as control (Rai et al. 2015b).

The AgNPs are most promising as they show significant antimicrobial potential due to their large surface area-to-volume ratio. This has generated huge interest among the researchers about AgNPs and its use in several consumer products. Due to a large number of consumer products containing engineered AgNPs in the market, databases have been constructed to keep track of these products, such as Project on Emerging Nanotechnologies (PEN), Nanotechnology Consumer Products Inventory (CPI), and Nanodatabase (Vance et al. 2015). This has led to the huge demand of AgNPs, which in turn has encouraged large-scale production methods. AgNP synthesis by fungi in general and *Phoma* sp. in particular has limitation in large-scale production. Since quantity of fungal biomass is directly correlated with synthesis of AgNPs, to obtain the sufficient fungal biomass for the synthesis requires considerable time. Practical solution to the problem was provided by reuse of fungal biomass, which was possible by encapsulation of fungal biomass in calcium alginate beads. AgNP synthesis using encapsulated biomass beads of *Phoma exigua* var. *exigua* (MTCC-2321) was procured from the Institute of Microbial Technology (IMTECH), Chandigarh, India, which was reported by Shende et al. (2017). They found that the immobilized *Phoma exigua* biomass could secrete appreciable amount of biomolecules, which can be sufficient for the synthesis for up to ten batches of AgNPs, without much difference among the particles size and shape. The immobilization of fungal biomass made it easy to be removed from the fungal secretion and reuse the immobilized biomass for the next set of synthesis. Moreover, the immobilization of biomass could lead to the development of continuous process for large-scale synthesis of AgNPs. The immobilized biomass in the calcium alginate beads can be packed into the column for the continuous synthesis as well.

17.6 Mechanistic Approach

Since the study of mycogenic synthesis of AgNPs offers valuable contribution into material synthesis advancement of nanoscale devices using mycogenic AgNPs, their use in a wide range of applications has recently fascinated researchers toward the understanding of mechanistic aspects of synthesis. Several hypothetical mechanisms reveal the involvement of nitrate reductase in the mycogenic synthesis process. Toward the understanding of mechanistic aspects, the effect of several factors like silver nitrate, fungal filtrate, pH, and temperature was studied on the mycogenic synthesis of AgNPs (Gade et al. 2014).

Three-step mechanism for the mycogenic synthesis of silver nanorods has been proposed by Gade et al. (2014). The first step is nucleation, which involves the role of proteins acting as capping agent to initiate the silver nanorod formation. Second step is elongation in which photosensitized anthraquinone derivative acts as electron shuttle, which takes up the electron donated by inorganic nitrate and transfers it to silver ions and thereby reduces them to form silver particles (Ag0). Metal nitrates can be decomposed thermally to the respective oxides. Anthraquinone derivatives play a key role in elongation of silver nanorod synthesis. The third and the final step is termination of silver nanorod synthesis process. The synthesis process will be terminated once the anthraquinone molecule involved in synthesis is either recruited by other nucleation center for the elongation or is unable to act as an electron shuttle (Gade et al. 2011), Fig. 17.2.

The effect of substrate (silver nitrate) concentration and enzyme (fungal filtrate) does not follow Michaelis-Menton kinetics, which is a model of enzyme catalyzed reaction. The mycosynthesis of AgNPs was evident even in the boiled fungal filtrate, which made us to doubt the involvement of enzyme in the mycogenic process, which was suggested by many researchers. The two-step mechanism of mycogenic synthesis of spherical AgNPs involves the photosensitization of the aromatic compounds secreted by the *Phoma* sp. in the presence of bright sunlight and secondly the role of excited aromatic compounds in the reduction of Ag^+ to Ag^0 (Gade et al. 2014).

Three-step mechanism for the mycofabrication of spherical AgNPs by *P. glomerata* was put forth by Gade et al. (2014) (Fig. 17.3). The three steps for the mycofabrication of spherical AgNPs are as follows: the first step is activation by photosensitization of aromatic compounds in the fungal filtrate on exposure to bright sunlight. Second step is nucleation, which involves the role of photosensitized aromatic compounds or proteins acting as capping agent to initiate the mycogenic synthesis of spherical AgNPs, and third step includes the actual mycofabrication by reduction of silver ions to form AgNPs. This step includes the donation of electron either by inorganic nitrate or by photosensitized aromatics compounds from fungal filtrate, by donating an electron to silver ions and leading to the reduction of them to form spherical AgNPs. Metal nitrates can be decomposed due to the heat generated on exposure to bright sunlight, i.e., thermally to the respective oxides (Gade et al. 2014).

Phoma sp. can be exploited for the mycosynthesis of spherical AgNPs and silver nanorods (*P. sorghina*) extracellularly. The mycosynthesized AgNPs exhibit significant antimicrobial and synergistic activity (with antibiotics) against the pathogens. The process of mycosynthesis of extracellular AgNPs is rapid, eco-friendly, simple, safe, and scalable.

17.7 Future Perspectives

There are a number of issues from the nanotechnology and synthesis point of view which require to be addressed to understand and compete with the existing biological systems like bacteria,

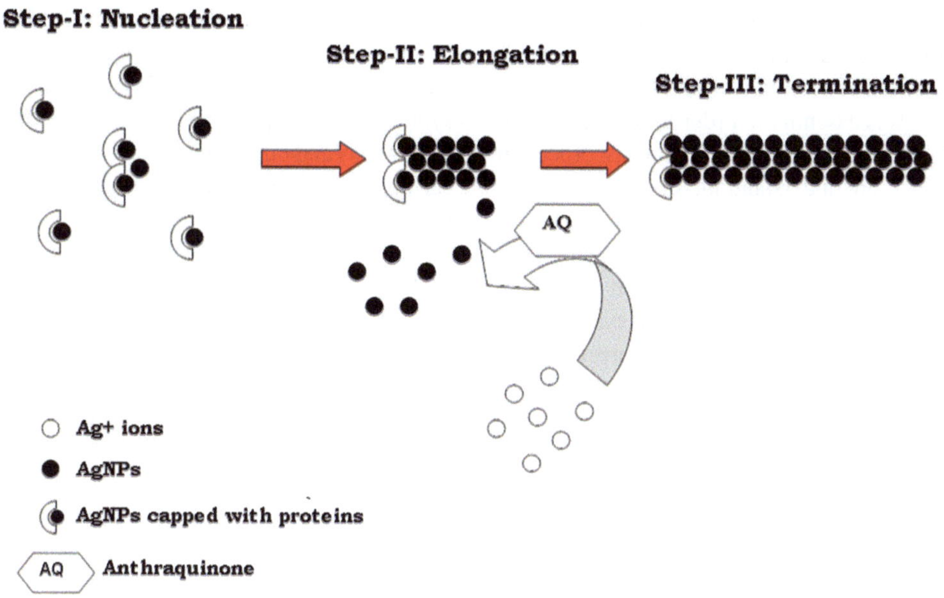

Fig. 17.2 Three-step mechanism of silver nanorod mycofabrication by *P. sorghina* (Gade et al. 2011)

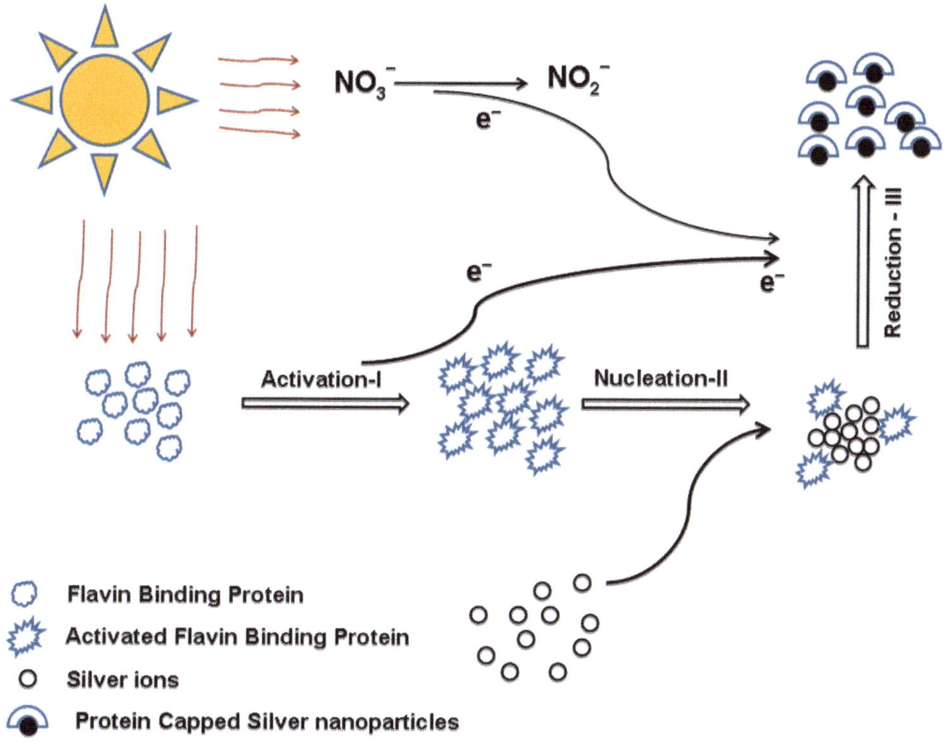

Fig. 17.3 Three-step mechanism of spherical AgNP mycofabrication by *P. glomerata*. Reproduced from Gade et al. (2014) with permission from Elsevier

plant, and algae for the biosynthesis protocols. The revelation of biochemical pathways leading to metal ion reduction and mycogenic synthesis of AgNPs is essential in order to develop a rational bio-based nanoparticle synthesis procedure. Similarly, an understanding of the stability of mycogenic AgNPs due to capping agent (i.e., the nature of capping surfactants/peptides/proteins) would be equally important, which could then lead to the possibility of fungal system to be genetically engineered to overexpress specific reducing molecules and capping agents, thereby controlling not only the size of the nanoparticles but also their shape. The rational use of constrained environments outside the fungal cells such as the pH, temperature, light intensity, concentration of salt, and quantity of secreted proteins to modulate nanoparticle size and shape is an exciting possibility. The range of chemical composition of nanoparticles currently synthesized by mycogenic approach is extremely limited and confined to metals, metal sulfides, and iron oxide. Extension of the synthesis protocols to enable the reliable mycogenic synthesis of nanocrystals of other oxides and nitrides, carbides, etc. could make this synthesis protocol a commercially viable proposition.

The mycogenic synthesis of AgNPs has many advantages such as ease with which the process can be scaled up, economic viability, possibility of easily covering large surface areas by suitable growth of the mycelia, etc.; fungal broth can be easily filtered by simple filtration process with the simple equipment, thus making the process economical. This shift from bacteria to fungi as a means of developing natural "nanofactories" was due to the advantages like easy downstream processing and handling of the biomass, reusability of fungal biomass for a number of synthesis batches, and ease of converting colloidal nanoparticles into powder form and vice versa. Further, compared to bacteria, fungi being eukaryotic system are known to secrete much higher amount of biomolecules, thereby significantly increasing the productivity of nanoparticle synthesis.

The use of specific proteins secreted by fungi in the extracellular mycogenic synthesis of AgNPs is exciting for the following reasons. The synthesis of nanoparticles would be of important in catalysis and other applications such as nonlinear optics. The exact mechanism of synthesis of mycogenic nanoparticles needs to be worked out. Better understanding of the mechanisms of AgNPs synthesis will enable us to achieve better control over size, shape, and monodispersity, which will lead to the development of high precision in the production level and the application of nanoparticles for commercial-scale applications. Thus, solving the different issues and reducing agents related to the synthesis of AgNPs would help in developing the mycogenic route as the most efficient method for the synthesis of nanoparticles.

17.8 Conclusion

It can be concluded that the ability of different *Phoma* spp. to reduce the inorganic metals is due to the extracellular secretions of reducing metabolites such as anthraquinones *Phoma* spp. whether terrestrial or marine are known to secret diverse metabolites which have already demonstrated applications in the field of agriculture, medicine, and pharmacy. Therefore, these metabolites can be harnessed and explored for the synthesis of nanoparticles of different sizes and shapes. *Phoma* spp. have undoubtedly opened up a new exciting, rapid, cost-effective, scalable, and eco-friendly approach toward the synthesis of AgNPs and development of *Phoma* spp. as natural "nanofactories" that are able to synthesize the nanoparticles at ambient temperature and pressure.

Acknowledgements MR is thankful to the Polish National Agency for Academic Exchange (NAWA) for financial support (Project No. PPN/ULM/2019/1/00117/A/DRAFT/00001) to visit the Department of Microbiology, Nicolaus Copernicus University, Toruń, Poland.

References

Agrawal, S. C.and Mishra, J. K. (1981). *Phoma lukhnownsis* sp. Nov. From Indian alkaline soils. *Current Science*, 50 (13): 600.

Ahmad, A., Senapati, S., Khan, M. I., Kumar, R., Ramani, R., Srinivas, V. and Sastry, M. (2003). Intracellular synthesis of gold nanoparticles by a novel alkalotolerant actinomycete, *Rhodococcus* species. *Nanotechnology*, 14: 824-828

Aveskamp, M.M., Gruyter, J. and Crous, P.W. (2008). Biology and recent developments in the systematics of *Phoma*, a complex genus of major quarantine significance. *Fungal Diversity,* 31:1-18.

Balis, E., Velegraki, A., Fragou, A., Pefanis, A., Kalabokas, T. and Mountokalakis, T. (2006). Lung mass caused by *Phoma exigua. Scandinavian Journal of Infectious Diseases*, 38(6): 552–555.

Bansal, V., Ahmad, A., Sastry, M. and Rautray, D. (2004) Biosynthesis of zirconia nanoparticles using the fungus *Fusarium oxysporum. Journal of Material Chemistry*, 14: 3303–3305.

Basavand, E., Babaeizad, V., Mirhosseini, H.A. and Niri, M.D., (2020). Occurrence of leaf spot disease caused by *Phoma herbarum* on oregano in Iran. *Journal of Plant Pathology*, 102: 575-576.

Basavaraja, S., Balaji, S. D., Legashetty, A., Rasab, A. H. and Venkatraman, A. (2008). Extracellular biosynthesis of silver nanoparticles using the fungus *Fusarium semitectum. Materials Research Bulletin*, 43(5): 1164-1170.

Bawaskar, M., Gaikwad, S., Ingle, A., Rathod, D., Gade, A., Duran, N., Marcato, P. and Rai, M. (2010). A new report on mycosynthesis of silver nanoparticles by *Fusarium culmorum. Current Nanoscience*, 6: 376-380.

Bhainsa, K. C. and D'souza, S. K. (2006). Extracellular biosynthesis of silver nanoparticles using the fungus *Aspergillus fumigatus. Colloids and Surfaces B: Biointerfaces*, 47: 160-164.

Birla, S. S., Tiwari, V. V., Gade, A. K., Ingle, A. P., Yadav, A. P and Rai, M. K. (2009). Fabrication of silver nanoparticles by *Phoma glomerata* and its combined effect against *Escherichia coli, Pseudomonas aeruginosa* and *Staphylococcus aureus. Letters in Applied Microbiology,* 48: 173-179.

Castro-Longoria, E., Vilchis-Nestor, A.R. and Avalos-Borja, M. (2011). Biosynthesis of silver, gold and bimetallic nanoparticles using the filamentous fungus *Neurospora crassa. Colloids and Surfaces B: Biointerfaces*, 83(1):42-48.

Chen, J. C, Lin, Z. H. and Ma, X. X. (2003). Evidence of the production of silver nanoparticles via pretreatment of *Phoma* sp. 3.2883 with silver nitrate. *Letters in Applied Microbiology*, 37:105-08.

Colmán, A.A., Alves, J.L., da Silva, M. and Barreto, R.W., (2018). *Phoma destructiva* causing blight of tomato plants: a new fungal threat for tomato plantations in Brazil?. *Tropical Plant Pathology*, 43(3): 257-262.

Davari, M. and Hajieghrari, B. (2008). *Phoma negriana,* a new invasive pathogen for Moghan's vineyards in Iran. *African Journal of Biotechnology*, 7(6):788-791.

Deb, D., Khan, A. and Dey, N., (2020). *Phoma* diseases: Epidemiology and control. *Plant Pathology,* 69(7):1203-1217.

Dorenbosch, M. M. J. (1970). Key to nine ubiquitous soil-borne *Phoma* like fungi. *Persoonia,* 6:1-14.

Duran, N., Marcato, P. D., Alves, O. L., D'Souza, G. and Esposito, E. (2005). Mechanistic 'aspects of biosynthesis of silver nanoparticles by several *Fusarium oxysporum* strains. *Journal of Nanobiotechnology*, 3: 8-14.

Fayaz, M., Tiwary, C.S., Kalaichelvan, P.T. and Venkatesan, R. (2010). Blue orange light emission from biogenic synthesized silver nanoparticles using *Trichoderma viride. Colloids and Surfaces B: Biointerfaces,* 75: 175–178.

Gade, A., Rai, M., Karwa, A., Bonde, P and Ingle, A. (2007). Extracellular biosynthesis of silver nanoparticles by *Pleurotus* species. In International Medical Mushroom Conference – 4 (IMMC-4). Ljubljana (Slovenia) from 23 to 37th Sept. 2007. *International Journal of Medicinal Mushroom.* 9(3-4): 298-299.

Gade, A. K, Bonde, P. P., Ingle, A. P., Marcato, P. D., Duran, N. and Rai, M. K. (2008). Exploitation of *Aspergillus niger* for fabrication of silver nanoparticles. *Journal of Biobased Material and Bioenergy,* 2 (3): 243-247.

Gade, A. K., Ingle, A. P., Whiteley, C. and Rai, M. (2010). Mycogenic metal nanoparticles: progress and applications. *Biotechnology Letters*, 32 (5): 593-600.

Gade, A. K. Rai, M. K. and Kulkarni, S. K. (2011). *Phoma sorghina*, a phytopathogen mediated synthesis of unique silver rods. *International Journal of Green Nanotechnology,* 3 (3): 153-159.

Gade, A. K., Gaikwad, S. C., Duran, N. and Rai, M. K. (2013) Screening of different species of *Phoma* for the synthesis of silver nanoparticles, *Biotechnology and applied Biochemistry*, 60 (5): 482-493

Gade, A. K., Gaikwad, S. C., Duran, N. and Rai, M. K. (2014) Green synthesis of silver nanoparticles by *Phoma glomerata, Micron* 59: 52-59.

Gade A, Adams J, Britt DW, Shen FA, McLean JE, Jacobson A, Kim Y-C, Anderson A J (2016) Ag nanoparticles generated using bio-reduction and-coating cause microbial killing without cell lysis. *BioMetals*, 29 (2): 211-213.

Gaikwad, S., Ingle, A., Gade, A., Rai, M., Falanaga, A., Incoronato, N., Galdiero, S. and Galdiero, M. (2013). Antiviral activity of mycosynthesized silver nanoparticles against herpes simplex virus and human parainfluenza virus type 3. *International Journal of Nanomedicine,* 8: 4303-4314.

Garibaldi, A., Gilardi, G., Gullino, M.L. (2010). First Report of Leaf Spot Caused by *Phoma multirostrata* on Fuchsia × hybrida in Italy. *Plant Diseases*, 94(3):382.1

Gilardi, G., Matic, S., Gullino, M.L. and Garibaldi, A., (2017). First report of *Phoma herbarum* causing leaf spot of woodland sage (*Salvia nemorosa*) in northern Italy. *Plant Disease*, 101(10):1824.

Ingle, A., Gade, A., Pierrat, S., Sonnichsen, C. and Rai, M. (2008). Mycosynthesis of silver nanoparticles using the fungus *Fusarium acuminatum* and its activity against some human pathogenic bacteria. *Current. Nanoscience*, 4: 141–144.

Ingle, A., Gade, A., Bawaskar, M. and Rai, M. (2009). *Fusarium solani*: A novel biological agent for the extracellular synthesis of silver nanoparticles. *Journal of Nanoparticle Research*, 11 (8): 2079-2085.

Jaidev, L. R. and Narasimha, G. (2010). Fungal mediated biosynthesis of silver nanoparticles, characterization and antimicrobial activity. *Colloids and Surfaces B: Biointerfaces*, 81(2):430-433.

Kathiresan, K., Manivannan, S., Nabeel, M. A. and Dhivya, B. (2009). Studies on silver nanoparticles synthesized by a marine fungus, *Penicillium fellutanum* isolated from coastal mangrove sediment. *Colloids and Surfaces B: Biointerfaces*, 71: 133–137.

Kövics, G.J., Sándor, E., Rai, M. and Irinyi, L. (2013). *Phoma*-like fungi on soybeans. *Critical Reviews in Microbiology*, 40(1):49-62.

Kowshik, M., Ashtaputre, S., Kharrazi, S., Vogel, W., Urban, J., Kulkarni, S. K. and Paknikar, K. M. (2003). Extracellular synthesis of silver nanoparticles by a silver-tolerant yeast strain MKY3. *Nanotechnology*, 14: 95-100.

Kukhar, E., Smagulova, A. and Kiyan, V., (2020). Biological properties of *Phoma macrostoma* related to non-dermatophyte onychomycosis. *Medical Mycology Case Reports*, 27:55-58.

Lopez, R.C. and Gomez-Gomez, L. (2009). Isolation of a new fungi and wound-induced chitinase class in corms of *Crocus sativus*. *Plant Physiology and Biochemistry*, 47(5):426-434.

Mathur, P. N. and Thirumalachar, M. J. (1959). Studies on some Indian soil fungi I some new or noteworthy Sphaeropsidales. *Sydowia*, 13: 143–147.

Mehra, R. K. and Winge, D. R. (*1991*). Metal ion resistance in fungi: Molecular mechanisms and their regulated expression. *Journal of Cellular Biochemistry*, 45(1): 30-40.

Mohammadian, A., Shojaosadati, S. A. and Habibi-Rezaee, M. (2007). *Fusarium oxysporum* mediates photogeneration of silver nanoparticles. *Scientia Iranica*, 14: 323-326.

Narendra, D.V. and Rao, V.G. (1974). A new entomogenous species of *Phoma*. *Mycopathologia et mycologia applicata*, 54:135-140.

Pellegrino, C, Gilardi, G, Gullino, M. L. and Garibaldi, A. (2010). Detection of *Phoma valerianellae* in lamb's lettuce seeds. *Phytoparasitica*, 38:159-165.

Prasad, K. and Jha, A. K. (2010). Biosynthesis of CdS nanoparticles: An improved green and rapid procedure. *Journal of Colloid and Interface Science*, 342: 68–72.

Quiroz, F.J., Molina, J.E. and Dosio, G.A.A., (2014). Black stem by *Phoma macdonaldii* affected ecophysiological components that determine grain yield in sunflower (*Helianthus annuus* L.). *Field Crops Research*, 160: 31-40.

Rai MK. (1985). Taxonomic studies of species of Phoma isolated from air. *J Econ Taxon Bot*. 7:645–647.

Rai, M. K. (1989). *Phoma sorghina* infection in human being. *Mycopathologia*, 105(3):167–170.

Rai, M. K. (2004). New and emerging *Phoma* species as human and animal pathogens, In; Fungi in human and animal health, Scientific Publisher, Jodhpur, Rajasthan, pp 69-80.

Rai M. K. and Gade A. K. (2009). Strategic Synthesis of Nanoparticles by Mycetes Biotechnology of Industrial Microorganisms – An Industrial approach (Ed. Maheshwari), I K International, New Delhi, pp 322-332.

Rai, J. N. and Misra, J. K. (1981). A new species of *Phoma* from Indian alkaline soil. *Current Science*, 50: 377.

Rai, M., Yadav, A., Bridge, P. and Gade, A. (2009a). Myconanotechnology: a new and emerging science. In *Applied Mycology*, CAB international. 258-267.

Rai, M. K., Deshmukh, P., Gade, A., Ingle, A., Kövics, G. J. and Irinyi, L. (2009b). *Phoma* Saccardo: Distribution, secondary metabolite production and biotechnological applications. *Critical Reviews in Microbiology*, 35 (3): 182-196.

Rai, M., Yadav, A. and Gade, A. (2010). Mycofabrication, mechanistic aspects and multifunctionality of metal nanoparticles-where are we? And where should we go? In: Current Research, Technology and Education Topics in Applied Microbiology and Microbial Biotechnology (Ed. Mendez-Vilas, A.) pp 1343-1354.

Rai, M., Ingle, A., Gade, A. and Duran, N. (2015a). Synthesis of silver nanoparticles by *Phoma gardeniae* and in vitro evaluation of their efficacy against human disease-causing bacteria and fungi. *IET Nanobiotechnology*, 9(2), 71-75.

Rai M, Ingle A. P., Gade A. K., Duarte M.C.T., and Duran N (2015b) Three *Phoma* spp. synthesised novel silver nanoparticles that possess excellent antimicrobial efficacy. *IET Nanobiotechnology*, 9 (5), 280-287.

Rai, M., Gade, A., Zimowska, B., Ingle, A.P. and Ingle, P., (2018). Marine-derived *Phoma*—the gold mine of bioactive compounds. *Applied Microbiology and Biotechnology*, 102(21): 9053-9066.

Rai, M., Gade, A., Zimowska, B., Ingle, A.P., Ingle, P. (2020). Harnessing the potential of novel bioactive compounds produced by endophytic *Phoma* spp. – biomedical and agricultural applications. *Acta Sci. Pol. Hortorum Cultus*, 19(6), 31–45. https://doi.org/10.24326/asphc.2020.6.3

Rai, M., Bonde, S., Golinska, P., Trzcińska-Wencel, J., Gade, A., Abd-Elsalam, K., Shende, S., Gaikwad, S. and Ingle, A., (2021). *Fusarium* as a Novel Fungus for the Synthesis of Nanoparticles: Mechanism and Applications. *Journal of Fungi*, 7(2):139.

Ramos, M.M., Morais, E.D.S., Sena, I.D.S., Lima, A.L., de Oliveira, F.R., de Freitas, C.M., Fernandes, C.P., de Carvalho, J.C.T. and Ferreira, I.M., (2020). Silver nanoparticle from whole cells of the fungi

Trichoderma spp. isolated from Brazilian Amazon. *Biotechnology letters*, 1-11.

Rao, S. and Thirumalachar, U. (1981). *Phoma exigua* infecting brinjal leaves. *Indian Phytopathology*, 34: 37.

Sastry, M., Ahmad, A., Khan, M. I. and Kumar, R. (2003). Biosynthesis of metal nanoparticles using fungi and actinomycetes. *Current Science*, 85 (2): 162-170.

Shende, S., Gade, A., and Rai, M. (2017). Large-scale synthesis and antibacterial activity of fungal-derived silver nanoparticles. *Environmental Chemistry Letters,* 15(3): 427-434.

Shende, S., Bhagat, R., Raut, R., Rai, M. and Gade, A., (2021). Myco-fabrication of Copper nanoparticles and its effect on crop pathogenic fungi. *IEEE Transactions on Nanobioscience.* https://doi.org/10.1109/tnb.2021.3056100.

Shivshankar, S., Ahmad, A., Pasricha, R. and Sastry, M. (2003). Bioreduction of chloroaurate ions by geranium leaves and its endophytic fungus yields gold nanoparticles of different shapes. *Journal of Material Chemistry*, 13: 1822-1826.

Shukla, N. P., Raja, R. K., Agrawal, G. P. and Gupta, D. K. (1984). *Phoma minutispora* as a human pathogen. *Mykosen*, 27:255-258.

Sugano, M., Sato, A. L., Oshima, T., Furuya, K., Kuwano, H., Hata, T. and Hanzawa, H. (1991). Phomactin A: A novel PAF antagonist from a marine fungus *Phoma* sp. *Journal of American Chemical Society*, 113 (14):5463-5464.

Suryavanshi, P., Pandit, R., Gade, A., Derita, M., Zachino, S., and Rai, M. (2017). *Colletotrichum* sp.-mediated synthesis of sulphur and aluminium oxide nanoparticles and its *in vitro* activity against selected food-borne pathogens. *LWT-Food Science and Technology*, 81, 188-194.

Suh, M. K. (2005) Phaeohyphomycosis in Korea. *Japanese Journal of Medical Mycology*, 46:67-70.

Tullio, V., Banche, G., Allizond, V., Roana, J., Mandras, N., Scalas, D., Panzone, M., Cervetti, O., Valle, S., Carlone, N. and Cuffini, A. M. (2010). Non-dermatophyte moulds as skin and nail foot mycosis agents: *Phoma herbarum, Chaetomium globosum* and *Microascus cinereus*. *Fungal Biology*, 114(4):345-349.

Vahabi, K., Mansoori, G. A., and Karimi, S. (2011). Biosynthesis of Silver Nanoparticles by Fungus *Trichoderma reesei* (A Route for Large-Scale Production of AgNPs). *Insciences Journal,* 1(1): 65-79.

Vance ME, Kuiken T, Vejerano EP, McGinnis SP, Hochella MF Jr, Rejeski D, Hull MS (2015) Nanotechnology in the realworld: redeveloping the nanomaterial consumer products inventory. *Beilstein J Nanotechnol* 6:1769–1780.

Volesky, B. and Holan, Z. R. (1995). Biosorption of heavy metals. *Biotechnology Progress*, 11: 235–250.

Warcup, J. H. (1950). The soil-plate method for isolation of fungi from soil [L]. *Nature,* 166: 117-118.

Win, T.T., Khan, S. and Fu, P., (2020). Fungus-(Alternaria sp.) Mediated Silver Nanoparticles Synthesis, Characterization, and Screening of Antifungal Activity against Some Phytopathogens. *Journal of Nanotechnology*, 2020.

Yarmalovich, V.A. and Siaredzich, M.O., (2019). *Phoma* Blight of Planting Stock of *Pinus sylvestris* L. and *Picea aAbies* L. in Forest Nurseries of Belarus. Bulletin of the Transilvania University of Brasov. Forestry, Wood Industry, Agricultural Food Engineering. Series II, 12(2), pp. 27-36.

Zaitz, C., Heins-Vaccari, E. M., De Freitas, R. S., Arriagada, G. L. H., Ruiz, L., Totoli, S. A. S., Marques, A. C., Rezze, G. G., Múller, H., Valente, N. S. and Lacaz, C. (1997). Subcutaneous Pheohypomycosis caused by *Phoma cava*. Report of a case and review of the literature. *Revista do Instituto de Medicina Tropical de São Paulo*, 39: 43-48.

Zhang, W., Qiao, X. and Chen, J. (2007). Formation of silver nanoparticles in SDS Microemulsion. *Materials Chemistry and Physics*, 109: 411-416.

In Situ Occurrence of *Phomites* Fritel in the Phyllosphere of Ancient Siwalik Forests of Eastern Himalaya During the Mio-Pleistocene

18

Meghma Bera, Mahasin Ali Khan,
Krishnendu Acharya,
Arkamitra Vishnu (née Mandal), and Subir Bera

18.1 Introduction

Phoma Sacc., first reported by Saccardo in 1884, is a notorious plant pathogen having cosmopolitan distribution and is represented by a large number of species (Aveskamp et al. 2008). At present, *Phoma* is described as a common anamorphic fungus belonging to class *Ascomycetes*, order *Dothideal*es and family *Pleosporaceae* (Gruyter et al. 2009). They are ubiquitous in the environment and occupy numerous ecological niches (Fitt et al. 2006; Aveskamp et al. 2008). Due to its ubiquitous nature, they have been reported in multiple natural habitats including aquatic environments, water distribution systems, soil and air (Zhang et al. 2016; Babič et al. 2017). Besides other fungi (Sullivan and White Jr. 2000) and lichens (Kocourkova 2000; Hawksworth and Cole 2004), they have been also reported from a large number of diverse angiosperms (Boerema 1976; Singh et al. 2004; Fitt et al. 2006; Thaung 2008; Pharamat 2009; Zhang et al. 2015). This fungus shows pycnidial fruit body with definite ostiolar opening and scattered conidiospores.

Abundant leaf megafossils having diverse angiosperm affinities are known from the eastern Himalayan Siwalik foreland basin (Joshi et al. 2003; Khan et al. 2011, 2014a, b, 2015, 2016, 2017a, b, 2019; Khan and Bera 2014), but little attention has been paid to the micro-ecological study of the fossil phylloplane. The in situ occurrences of other epiphyllous fungi such as *Asterina, Meliola* and *Vizella* have been recorded from the Neogene Siwalik sediments of eastern Himalaya (Mitra and Banerjee 2000; Mitra et al. 2002; Das et al. 2007; Mandal et al. 2011; Khan et al. 2015; Vishnu (née Mandal) et al. 2017; Khan et al. 2018a, b; Bera et al. 2019), but there are only three known records of fossil fungi resembling modern-day *Phoma* from this area (Mandal et al. 2009;

M. Bera
Centre of Advanced Study, Department of Botany, University of Calcutta, Kolkata, India

Department of Botany, Vidyanagar College, Vidyanagar, West Bengal, India

M. A. Khan
Department of Botany, Sidho-Kanho-Birsha University, Purulia, India

K. Acharya · S. Bera (✉)
Centre of Advanced Study, Department of Botany, University of Calcutta, Kolkata, India

A. Vishnu (née Mandal)
Centre of Advanced Study, Department of Botany, University of Calcutta, Kolkata, India

Centre for Ecological Sciences, Indian Institute of Science, Bangalore, India

M. Rai et al. (eds.), *Phoma: Diversity, Taxonomy, Bioactivities, and Nanotechnology*,
https://doi.org/10.1007/978-3-030-81218-8_18

Bera et al. 2018; Vishnu et al. 2019). So, the present study reporting in situ occurrence of *Phomites* from different compressed angiosperm leaf remains recovered from the Siwalik sediments (middle Miocene, ~13 Ma to early Pleistocene, ~1 Ma) of eastern Himalaya (Fig. 18.1) is significant.

The objective of the current study is to document the occurrence of cosmopolitan *Phoma* spp. in fossil records of Indian Cenozoic and to investigate a possible ectoparasite interaction between the species of fossil fungal form *Phomites* (comparable to modern plant pathogen *Phoma*) and their angiosperm host plants in the monsoon-influenced tropical forests of eastern Himalaya during Siwalik sedimentation (middle Miocene to lower Pleistocene). Furthermore, reconstruction of palaeoecology during deposition of sediments (Mio-Pleistocene) has also been attempted.

18.2 Geological Settings of the Eastern Himalaya

The Siwalik sediments are characterized by alternating sandstone and mudstone facies, with the finer sediments often containing abundant angiosperm fossils (Bora and Shukla 2005). From west to east along their length, the Siwaliks have been divided into seven sectors: Jammu, Himachal Pradesh, Uttar Pradesh, Nepal, Darjeeling, Bhutan and Arunachal Pradesh (Ranga Rao et al. 1981). The eastern Himalayan Siwalik sectors, namely, Darjeeling, Arunachal Pradesh and Bhutan, are the focus of the work presented here. In the Darjeeling area, generalized lithostratigraphy of Siwalik sediments is based on Ganguly and Rao (1970), Acharya (1994) and Taral et al. (2017); in Bhutan, it is based on Coutand et al. (2016) and Khan et al. (2019), while that in Arunachal Pradesh has been derived from

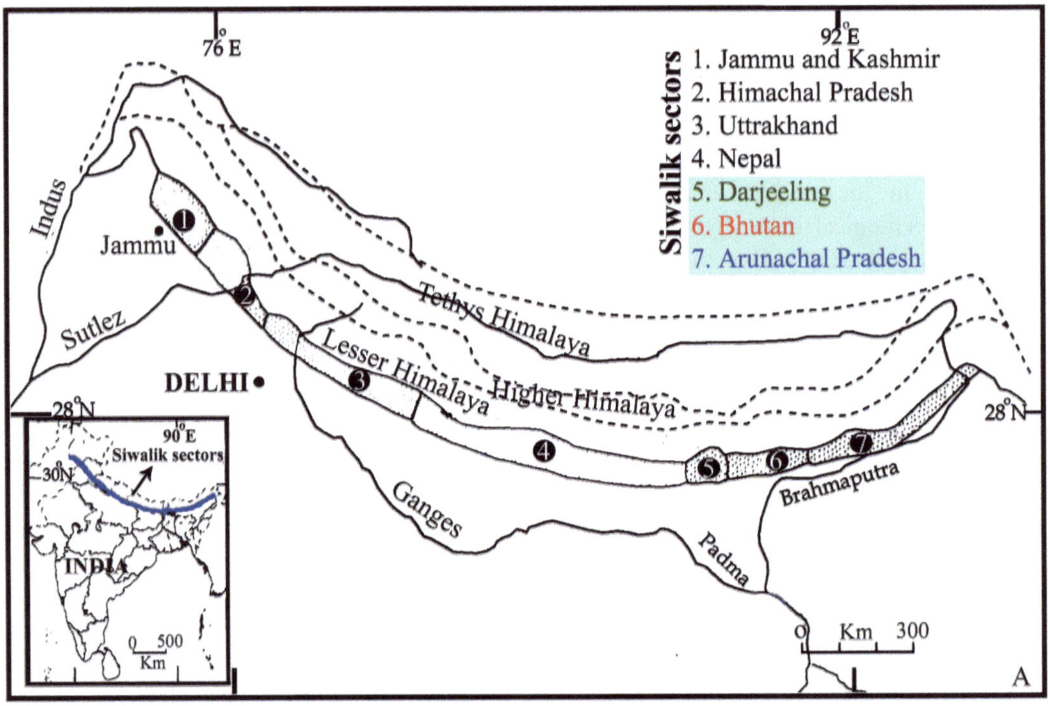

Fig. 18.1 Maps showing the seven sectors of Siwalik belt (Modified after Khan et al. 2019); Darjeeling, Arunachal Pradesh and Bhutan Siwalik sectors are the focus of the work presented here

Prakesh and Singh (2000), Singh (2007) and Khan et al. (2014b). The ages of the sedimentary succession in different Siwalik sectors of eastern Himalaya have been quantified by means of magnetostratigraphy (Chirouze et al. 2012; Coutand et al. 2016). Magnetostratigraphic correlations indicate that the Bhutan Siwalik group was deposited during the latest Miocene and the Pleistocene, between ~7 Ma and ~1 Ma, and that the boundary between lower and middle Siwaliks can be dated to ~6 Ma, the middle to upper Siwaliks to 3.8 Ma and the top of the section to ~1 Ma (Coutand et al. 2016). In Arunachal Pradesh, the Siwalik formation was deposited between 13 and 2.5 Ma. The transition between the lower and middle Siwaliks is dated at about 10.5 Ma and the middle to upper Siwalik transition at 2.6 Ma (Chirouze et al. 2012). The lower part of Siwalik sediments is characterized by fine-grained grey sandstones with clay stones and shale; the middle Siwalik is characterized by medium to fine-grained sandstone, greenish-yellow siltstone and grey silty shale; and the upper part of Siwalik sedimentation is characterized by loosely packed, pebbly, very coarse to fine-grained grey sandstones and shale.

18.3 Systematics *of Phoma*

Phoma, a polyphyletic coelomycetous genus of fungal organisms, belongs to the phylum *Ascomycota*, class *Dothideomycetes*, order *Pleosporales* and family *Didymellaceae* (Bennet et al. 2018). The family *Didymellaceae* accommodates three distinct genera, namely, *Ascochyta*, *Didymella* and *Phoma*, as well as several related *Phoma*-like genera (Chen et al. 2015). *Phoma* is morphologically difficult to distinguish, and species from this taxon has in the past been linked to *Didymella* sexual morphs (Chen et al. 2015). Systematics of the modern genus *Phoma* is based mainly on cultural characteristics and molecular studies (Hawksworth and Cole 2004; Diederich et al. 2007; Rai et al. 2014; Chen et al. 2015). More than 220 species within this genus are currently recognized, but the actual number of species is probably much higher. Identification of species of *Phoma* is difficult for taxonomists due to the asexual nature of most species, the high morphological variability in vivo and the vague generic circumscription (Aveskamp et al. 2008). In recent years, the anamorphic genus *Phoma* has been subdivided into nine sections, namely, *Heterospora*, *Macrospora*, *Paraphoma*, *Peyronellaea*, *Phoma*, *Phyllostictoides*, *Pilosa*, *Plenodomus* and *Sclerophomella*, based on morphological characters such as the formation of conidia (asexual spores), pycnidia (asexual fruiting bodies) and chlamydospores (enlarged, thick-walled vegetative cells), with their type species and associated teleomorph genera suggesting the polyphyly of the genus (Gruyter et al. 2012). The extensive work on the taxonomy of *Phoma* has been done by Boerema et al. (2004) using culturing techniques and morphological data. This has led to the publication of an extensive handbook, the "*Phoma Identification Manual*" (Boerema et al. 2004) providing keys and descriptions of more than 200 taxa at specific and infraspecific level (Gruyter et al. 2012).

18.4 Earlier Records of *Phoma* or *Phoma*-Like fossils

Previously, there were some records of fossil fungi resembling modern-day *Phoma* (Fritel 1910; Chitaley and Patil 1972; Singhai 1974; LePage et al. 1994; Watanabe et al. 1999). *Phoma* or *Phoma*-like fossils were published under the fossil genera, i.e. *Phomites* (Fritel 1910), *Palaeophoma* (Singhai 1974) and *Archephoma* (Watanabe et al. 1999). Fritel (1910) named the fossil analogue of *Phoma* as *Phomites* Fritel. *Phomites myricae* Fritel is known from the fossil leaves of Myricaceae from the Palaeocene of Paris, France (Fritel 1910), and *Phomites ebenoxyloni* Chitaley and Patil from the silicified wood of *Ebenoxylon mohgaoensis* Chitaley and Patil (comparable to modern family Ebenaceae) from the Deccan intertrappean beds of Mohgaonkalan, Madhya Pradesh, India (Chitaley and Patil 1972). Fritel (1910) described *Phomites* as an imperfect fruiting structure, while Chitaley and Patil (1972) described it as a silicified pycnidium with conidiospores. The *Phomites* recovered from the Palaeocene of Paris (Fritel 1910) represents pycnidia without conidia developing on a fungal colony. Fossil-species *Palaeophoma intertrappea* Singhai having various forms of aseptate, hyaline and lunate conidia is known from the Deccan intertrappean beds (late Cretaceous) of Mohgaonkalan, Madhya Pradesh (Singhai 1974). LePage et al. (1994) identified fossil *Phoma* in the middle Eocene permineralized chert from Canada. The Canadian specimen has single-septate conidia, but conidiogenesis is not known. Another fossil-species *Archephoma cycadeoidellae* Watanabe et al. is known from a permineralized bisexual cone of a cycadeoid, *Cycadeoidella japonica* Ogura, from the Late Cretaceous sediments of Japan (Watanabe et al. 1999). *Archephoma* can be differentiated from *Phomites* by the presence of conidia. Mature pycnidia of *A. cycadeoidellae* are similar to those of modern *Phoma* except for undergoing conidiogenesis.

18.5 *Phomites* from Siwalik Sediments of Eastern Himalaya

The fossil-genus *Phomites* Fritel is characterized by well-preserved dark brown, globose, pycnidial fruit body with distinct ostiolar opening and interwoven network of hyphae forming pseudoparenchymatous tissue and some scattered threadlike conidiospores. Pycnidium is partially sunken in leaf cuticle, single or in groups, and ostiole is surrounded by a collar of thick-walled cells. As *Phomites* is a legitimate name, here we describe all recovered epiphyllous fungal morphs from Siwalik sediments of eastern Himalaya having affinity with modern plant pathogen *Phoma* under the legitimate genus *Phomites*. In the present study, we report a large number of well-preserved fossil fungal species of *Phomites* from diverse compressed angiospermic leaf remains of *Dipterocarpus* Gaertn., *Shorea* Roxb., (Dipterocarpaceae), *Amherstia* Wall., *Albizia* Durazz., (Fabaceae), *Dysoxylum* Blume (Meliaceae), *Chonemorpha* G. Don (Apocynaceae), *Actinodaphne* Nees, *Litsea* Lam. (Lauraceae), *Anthocephalus*, *Mitragyna* Korth. (Rubiaceae), *Ficus* L. (Moraceae), *Miliusa* Lesch. ex, A. Dc. (Annonaceae), *Litsea* Lam. (Lauraceae) *Terminalia* L. (Combretaceae), *Macaranga* Thouars (Euphorbiaceae) and Poaceae Barnhart recovered from the Siwalik sediments (middle Miocene to early Pleistocene) of the eastern Himalaya (Fig. 18.2). The recovered fossil fungal species of *Phomites* from Siwalik sediments of eastern Himalaya and their respective hosts are listed in Table 18.1.

18.6 Palaeoecological Significance

The in situ occurrence of fruiting bodies, i.e. pycnidia of *Phomites*, on the cuticular surface of fossilized leaf cuticular fragments of different carbonized angiosperm leaf remains recovered from the Siwalik sediments of eastern Himalaya provides unambiguous evidence that this fungal taxon has existed in the Siwalik forests of eastern Himalaya since middle Miocene. Fossil epiphyllous fungi can serve as environmental indicators, because they generally reflect a humid environment (Dilcher 1965; Phadtare 1989; Taylor and Osborn 1996; Phipps and Rember 2004; Shi et al. 2010). So, the occurrence of *Phomites* in appreciable numbers indicates a warm and humid tropical climate during Siwalik sedimentation (Mio-Pleistocene) in the ancient forests of eastern Himalaya. This is also in consistency with the earlier evidence of other thermophilic epiphyllous fungal morphs, namely, *Asterina*, *Meliola*, *Vizella*, etc., recovered from the same locality (Mitra and Banerjee 2000; Mitra et al. 2002; Das et al. 2007; Mandal et al. 2009, 2011; Khan et al. 2015; Vishnu (née Mandal) et al. 2017; Khan et al. 2018a, b; Bera et al. 2019). So, the present and earlier fossil records of epiphyllous fungal morphs collectively suggest that eastern Himalaya experienced a warm and humid tropical climate associated with high rate of precipitation during the time of deposition.

The above assumption is also in conformity with our earlier published qualitative and quantitative (CLAMP, Climate Leaf Analysis Multivariate

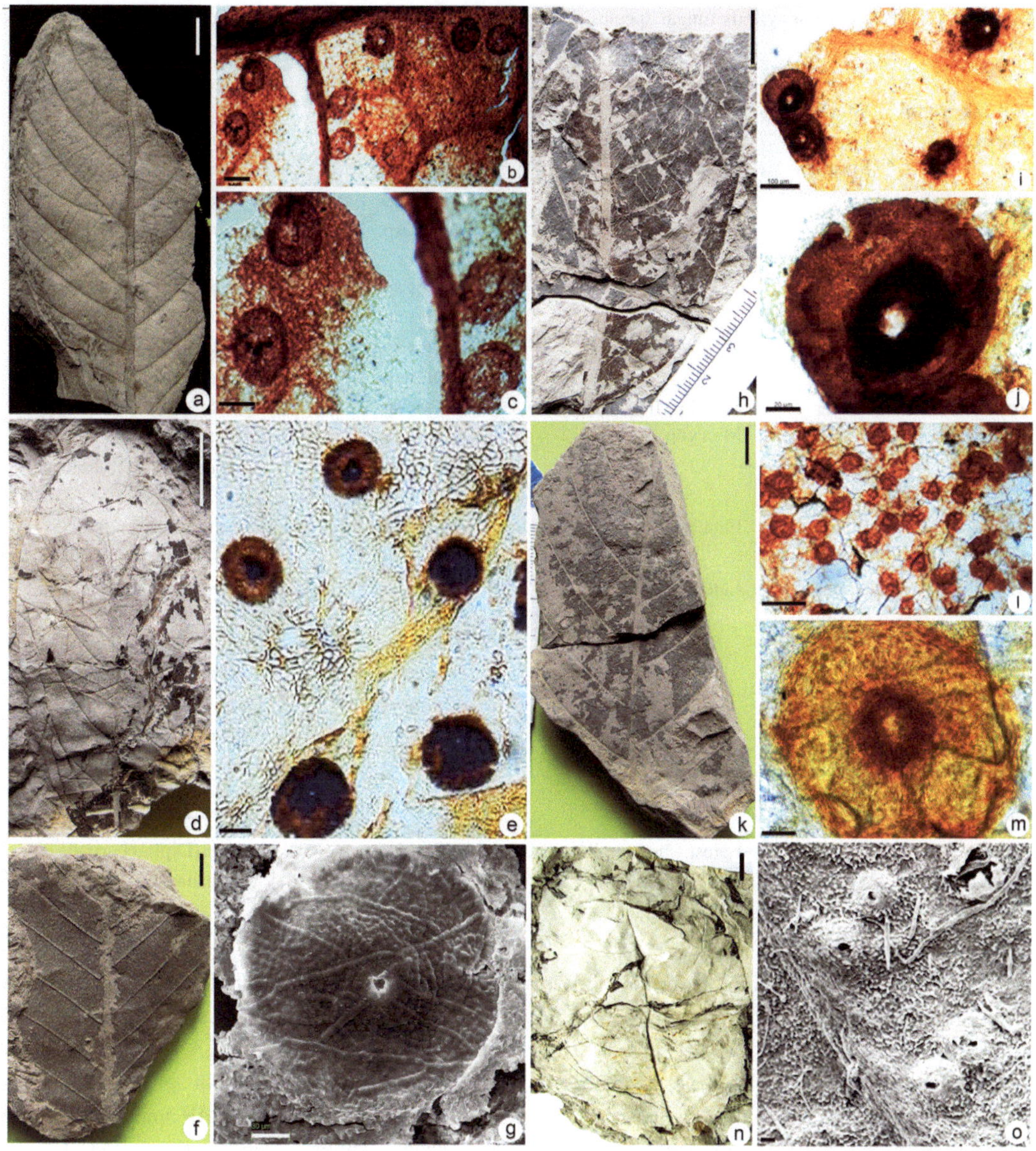

Fig. 18.2 Light and SEM photographs of *Phomites* recovered from the leaf cuticle of angiosperm fossil leaf remains. (**a**) Fossil leaf of *Litsea* (scale bar =1 cm); (**b, c**) scattered pycnidia of *Phomites* on the adaxial surface of *Litsea* (scale bar = 20 μm); (**d**) fossil leaf of *Dysoxylum* (scale bar =1 cm); (**e**) scattered dark, brown pycnidia of *Phomites siwalicus* Vishnu, Khan et Bera on the adaxial surface of *Dysoxylum* (scale bar = 20 μm); (**f**) fossil leaf of *Dipterocarpus* (scale bar =1 cm); (**g**) SEM photograph of pycnidium of *Phomites neogenicus* Vishnu, Khan et Bera recovered from the leaf cuticle of *Dipterocarpus* (scale bar = 30 μm); (**h**) fossil leaf of *Amherstia* (scale bar =1 cm); (**i**) scattered dark, brown pycnidia of *Phomites* having a distinct ostiole recovered from the leaf cuticle of *Amherstia* (scale bar = 100 μm); (**j**) enlarged view of single mature pycnidium showing a central ostiole and thicker cells in collar (scale bar = 20 μm); (**k**) fossil leaf of *Shorea* (scale bar =1 cm); (**l**) scattered pycnidia of *Phomites* recovered from the leaf cuticle of *Shorea* (scale bar = 100 μm); (**m**) enlarged view of single mature pycnidium showing an ostiole and a collar (scale bar = 20 μm); (**n**) fossil leaf of *Chonemorpha* (scale bar =1 cm); (**o**) SEM photograph of pycnidia of *Phomites* on the adaxial surface of *Chonemorpha* (scale bar = 20 μm)

Table 18.1 The fossil epiphyllous fungal species of *Phomites* from the Siwalik sediments of eastern Himalaya and their hosts

Siwalik sector	Strata	Age	Fungal species of *Phomites*	Host
Darjeeling	Lower Siwalik	Middle Miocene-upper Miocene)	*Phomites* sp.	*Terminalia* (Combretaceae)
			Phomites sp.	*Miliusa* (Annonaceae)
			Phomites sp.	*Anthocephalus* (Rubiaceae)
			Phomites sp.	*Dipterocarpus* (Dipterocarpaceae)
			Phomites sp.	*Shorea* (Dipterocarpaceae)
			Phomites sp.	*Ficus* (Moraceae)
			Phomites sp.	*Mitragyna* (Rubiaceae)
			Phomites sp.	*Albizia* (Fabaceae)
Arunachal Pradesh	Lower Siwalik	Middle Miocene-upper Miocene	*Phomites* sp.	Moraceae
			P. neogenicus	Poaceae
			Phomites sp.	*Anthocephalus* (Rubiaceae)
			Phomites sp.	*Litsea* (Lauraceae)
	Middle Siwalik	Pliocene	*Phomites* sp.	*Persea* (Lauraceae)
			Phomites sp.	*Shorea* (Dipterocarpaceae)
	Upper Siwalik	Late Pliocene-early Pleistocene	*P. neogenicus*	*Dipterocarpus* (Dipterocarpaceae)
			P. siwalicus	*Dysoxylum* (Meliaceae)
			P. arunachalensis	*Amherstia* (Fabaceae)
			P. macarangensi	*Macaranga* (Euphorbiaceae)
			Phomites sp.	*Albizia* (Fabaceae)
			Phomites sp.	*Actinodaphne* (Lauraceae)
			Phomites sp.	*Chonemorpha* (Apocynaceae)
Bhutan	Middle Siwalik	Latest Miocene-Pliocene	*P. neogenicus*	*Shorea* (Dipterocarpaceae)
			Phomites sp.	Anacardiaceae

Program) climate data obtained from the study of different megafossil plant remains (Khan et al. 2011, 2014a, b, 2015, 2016, 2017a, b, 2018c, 2019; Khan and Bera 2014). Several records of mega plant fossils such as leaves, woods and fruit remains from the Siwalik sediments of eastern Himalaya confirm this interpretation (Khan et al. 2011, 2014a, b, 2015, 2016, 2017a, b, 2018c, 2019; Khan and Bera 2014). The present-day distribution of modern equivalent taxa and the foliar physiognomic characters or structural features of the fossil leaves such as tip, base, size, margin and venation pattern clearly suggest that tropical evergreen forests were present in eastern Himalaya under a tropical warm, humid condition during the Siwalik sedimentation.

The CLAMP results of the fossil leaf assemblages recovered from the Siwalik sediments of eastern Himalaya support the interpretation of climatic conditions in the area (Khan et al. 2014b, 2019). We used the foliar physiognomic characters of the fossil leaves to run CLAMP. Therefore, the physiognomic characters are in agreement with the CLAMP analysis. The CLAMP analysis measures a mean annual temperature (MAT) of 24.7 °C ± 2.8 °C, a cold month mean temperature (CMMT) of 19.1 °C ± 4 °C, a warm month mean temperature (WMMT) of 28 °C ± 3.3 °C, a growing season precipitation (GSP) of 197.2 ± 91.6 cm and a relative humidity (RH) of 80.6 ± 10.7% (Khan et al. 2014b, 2019). This climate provided a great opportunity to a wide range of leaf-inhibiting fungi (mostly pathogenic) including *Phomites* to establish themselves as an important part of the ancient Siwalik forest ecosystem. This diversification of leaf-inhabiting fungi (both parasite and pathogen) points towards the eventual establishment of pathogenic fungi in a large number of present-day hosts.

18.7 Host-*Phomites* Interaction in the Siwalik Forest

The dynamic interaction of fungi with their host is an important parameter in the modern-day forest ecosystem (Price 1977). Fungi are generally host-specific, but sometimes a fungus may have a wide range of hosts. During palaeomycological investigation, an assessment of the past host-fungal interaction is usually made on the basis of recovered fossil fungal taxa and their respective hosts. The factors controlling host-parasite interaction and its change through time are believed to be either inherent (e.g. changes in host defence, fungal pathogenicity, genetic variability, changes in chemical nature) or environmental (e.g. climatic and physical changes, i.e. pH, temperature and humidity) (Price 1977; Caroll and Wicklow 1992). Some fossil fungi are sometimes known to develop parasitic relationship with hosts which they do not seem to prefer in present-day forest ecosystem.

The fungal fruit body remains of *Phomites* reported here represent the significant evidence of fungi being associated with diverse angiosperm leaves recovered from the eastern Himalayan Siwalik foreland basin. The scarcity of documented evidence of this epiphyllous fungal form from the Indian Cenozoic makes this study an important one. We report here epiphyllous fruit bodies of *Phomites* from angiosperm fossil leaf remains belonging to 11 families, namely, Dipterocarpaceae, Meliaceae, Lauraceae, Fabaceae, Anacardiaceae, Combretaceae, Apocynaceae, Annonaceae, Moraceae, Rubiaceae, Euphorbiaceae and Poaceae (Table 18.1). So, it can be said that *Phomites* had established itself in eastern Himalaya as a common leaf infecting fungal morph since middle Miocene. At present, extant *Phoma* is known to be associated with the members of all the above-mentioned families (Boerema 1976; Singh et al. 2004; Thaung 2008; Gruyter et al. 2009; Pharamat 2009; Zhang et al. 2015; Vishnu (née Mandal) et al. 2017). Thus, an almost similar host choice of *Phomites* since middle Miocene suggests a general host-ectoparasite interaction in the Siwalik ancient monsoon-influenced tropical evergreen forests of eastern Himalaya. Generally, assessment of inherent characters is rather difficult for the study of fossil host-parasite association (Caroll and Wicklow 1992). From this point of view, changing environment and chemically active principles in the host plants may be two key reasons for deviation in choice of host.

Acknowledgements M. B., S. B. and K. A. gratefully acknowledge the UGC-CAS VII Department of Botany, University of Calcutta, for providing infrastructural facilities to accomplish this work. M. K. acknowledges the Department of Botany, Sidho-Kanho-Birsha University, for necessary facilities.

References

Acharya S.K., 1994. The Cenozoic foreland basin and tectonics of the Eastern Sub-Himalaya: problem and prospects. *Himalayan Geol.* 15, 3–21.

Aveskamp M.M., Gruyter J. D.E., Crous P.W., 2008. Biology and recent developments in the systematics of *Phoma*, a complex genus of major quarantine significance. *Fungal Divers.* 31, 1–18.

Babič M.N., Gunde-Cimerman N., Vargha M., Tischner Z., Magyar D., Veríssimo C., Sabino R., Viegas C., Meyer W., Brandão J., 2017. Fungal Contaminants in Drinking Water Regulation? A Tale of Ecology, Exposure, Purification and Clinical Relevance. *Int. J. Environ. Res. Public Health* 14, 636

Bennet A., Ponder M.M., Garcia-Diaz J., 2018. *Phoma infections*: classification, potential food sources, and their clinical impact. Microorganisms 6, 58. https://doi.org/10.3390/microorganisms6030058

Bera, M., Khan, M.A., Bera, S., 2018. Two new species of *Phomites Fritel* from the phyllosphere of Siwalik. *J. Mycopathol. Res.* 56(1), 11–14.

Bera, M., Khan, M.A., Bera, S., 2019. A new foliicolous melioloid fungus from the Pliocene of eastern Himalaya. *Mycol. Progress* 18, 921–931.

Boerema G.H., 1976. The *Phoma* species studied in culture by Dr. R. W. G. Dennis. *Trans. Br. Mycol. Soc.* 67, 289-319.

Boerema G.H., Gruyter J.D.E., Noordeloos M.E., Hamers M.E.C., 2004. *Phoma* identification manual. Differentiation of specific and infra-specific taxa in culture. CABI Publishing, Wallingford, UK.

Bora D.S., Shukla U.K., 2005. Petrofacies implication for the Lower Siwalik Foreland Basin evolution, Kumaun Himalaya, India. Special Publication, J. Palaeontol. Soc. India 2, 163–179.

Caroll G.C., Wicklow D.T., 1992. The Fungal community: its organization and role in the ecosystem, Second edition. Mycology Series, Vol. 9.

Chen Q., Jiang J.R., Zhang G.Z., et al. (2015) Resolving the *Phoma* enigma. *Stud Mycol* 82:137–217.

Chirouze F., Dupont-Nivet G., Huyghe P., van der Beek P., Chakraborti T., Bernet M., Erens V., 2012. Magnetostratigraphy of the Neogene Siwalik Group in the far eastern Himalaya: Kameng section, Arunachal Pradesh, India, *J. Asian Earth Sc.* 44, 117-135.

Chitaley S.D., Patil G.V., 1972. Ebenaceous fossil wood infected with deuteromycetaceous fungus from the Deccan Intertrappean Beds of India. *Botanique* 3(2), 99-106.

Coutand, I., Barrier, L., Govin, G., Grujit, D., Hoorn, C., Dupont Nivet, G., Najman, Y., 2016. Late Miocene-Pleistocene evolution of India-Eurasia convergence partitioning between the Bhutan Himalaya and the Shillong Plateau: new evidences from foreland basin deposits along the Dungsam Chu section, eastern Bhutan. *Tectonics* 35, 2963–2994.

Das P., Khan M.A., De B., Samajpati N., Bera S., 2007. Evidence of exoparasitic relationship between *Asterina* (Asterinaceae) and *Chonemorpha* (Apocynaceae) from the Upper Siwalik (Kimin Formation) sediments of Arunachal subhimalaya, India. *J Mycopathol Res.* 45(2), 225–230

Diederich P., Kocourková J., Etayo J., Zhurbenko M., 2007. The lichenicolous *Phoma* species (Coelomycetes) on *Cladonia*. *Lichenology* 39 (2), 153-163.

Dilcher D.L., 1965. Epiphyllous fungi from Eocene deposits in western Tennessee, U.S.A. *Palaeontogra. Abt. B.* 116, 1–49.

Fitt B.D.L., Brun H., Barbetti M.J., Rimmer S.R., 2006. World-wide importance of *Phoma* stem canker (*Leptosphaeria maculans* and *L. biglo-bosa*) on oilseed rape (*Brassica napus*). *Eur. Journal Plant Pathol.*114, 3-15.

Fritel P.H., 1910. Étude sur les vegetaux fossils de l'étage sparnacien du Bassin de Paris; Mémoires de la Société géologique de France. *Paláeontologie* 40, 5-37.

Ganguly, S., Rao, D.P., 1970. Stratigraphy and structure of the Tertiary foothills of Eastern Himalaya, Darjeeling District. West Bengal. *Quart. J. Geol. Min. Metall. Soc. India* 42, 185–195.

Gruyter J.D.E., Aveskamp M.M., Woudenberg J.H.C., Verkley G.J.M., Groenewald J.Z., Crous P.W., 2009. Molecular phylogeny of *Phoma* and allied anamorph genera: Towards a reclassification of the *Phoma* complex. Mycol. Res. 113, 508-519.

Gruyter J.D.E., Woudenberg J.H.C., Aveskamp M.M., Verkley G.J.M., Groenewald J.Z., Crous P.W., 2012. Redisposition of *Phoma*-like anamorphs in *Pleosporales*. *Stud. Mycol.* 75, 1-36.

Hawksworth D.L., Cole M.S., 2004. *Phoma fuliginosa* sp. nov., from *Caloplaca trachyphylla* in Nebraska, with a key to the known lichenous species. *Lichenology* 36, 7-13.

Joshi A., Tewari R., Mehrotra R.C., Chakraborty P.P., De A., 2003. Plant remains from the Upper Siwalik sediments of West Kameng District, Arunachal Pradesh. *J. Geol. Soc. Ind.* 61, 319-324.

Khan M.A., Bera, S., 2014. New lauraceous species from the Siwalik forest of Arunachal Pradesh, eastern Himalaya, and their palaeoclimatic and palaeogeographic implications. *Turk. J. Bot.* 38, 453-464.

Khan M.A., Ghosh R., Bera S., Spicer RA, Spicer TEV, 2011. Floral diversity during Plio-Pleistocene Siwalik sedimentation (Kimin Formation) in Arunachal Pradesh, India, and its palaeoclimatic significance. *Palaeobio. Palaeoenv.* 91, 237-255.

Khan M.A., Spicer T.E.V., Spicer R.A., Bera S., 2014a. Occurrence of *Gynocardia odorata* Robert Brown (Achariaceae, formerly Flacourtiaceae) from the Plio-Pleistocene sediments of Arunachal Pradesh, northeast India and its palaeoclimatic and phytogeographic significance. *Rev. Palaeobot. Palynol.* 211, 1-9.

Khan M.A., Spicer R.A., Bera S., Ghosh R., Yang J., Spicer T.E.V., Guo S., Su T., Jacques F., Grote P.J., 2014b. Miocene to Pleistocene floras and climate of the eastern Himalayan Siwaliks, and new palaeoelevation estimates for the Namling-Oiyug Basin, Tibet, *Glob. Planet. Change* 113, 1-10.

Khan M.A., Bera S., Ghosh R., Spicer R.A., Spicer T.E.V., 2015. Leaf cuticular morphology of some angiosperm taxa from the Siwalik sediments (middle Miocene to lower Pleistocene) of Arunachal Pradesh, eastern Himalaya: Systematic and palaeoclimatic implications. *Rev. Palaeobot. Palynol.* 214, 9-26.

Khan M., Spicer R.A., Spicer T.E.V., Bera S., 2016. Occurrence of *Shorea* Roxburgh ex C. F. Gaertner (Dipterocarpaceae) in the Neogene Siwalik forests of eastern Himalaya and its biogeography during the Cenozoic of Southeast Asia. *Rev. Palaeobot. Palynol.* 233, 236-254.

Khan M., Spicer R.A., Spicer T.E.V., Bera S., 2017a. Evidence for diversification of *Calophyllum* L. (Calophyllaceae) in the Neogene Siwalik forests of eastern Himalaya. *Plant Syst. Evol.* 303, 371-386.

Khan M., Bera M., Spicer R.A., Spicer T.E.V., Bera S., 2017b. First occurrence of mastixioid (Cornaceae) fossil in India and its biogeographic implications. *Rev. Palaeobot. Palynol.* 247, 83-96.

Khan M.A., Bera M., Bera S., 2018a. *Vizellopsidites siwalika*, a new fossil epiphyllous fungus from the Plio-Pleistocene of Arunachal Pradesh, eastern Himalaya. *Nova Hedwigia* 107(3–4), 543–555

Khan M.A., Bera M., Spicer R.A., Spicer T.E.V., Bera S., 2018b. Evidence of simultaneous occurrence of tylosis formation and fungal interaction in a late Cenozoic angiosperm from the eastern Himalaya. *Rev Palaeobot Palynol* 259,171–184

Khan, M.A., Bera, M., Spicer, R.A., Spicer, T.E.V., Bera, S., 2018c. Floral diversity and environment during the middle Siwalik sedimentation (Pliocene) in the Arunachal sub-Himalaya. *Palaeobio. Palaeoenv.* https://doi.org/10.1007/s12549-018-0351-2.

Khan, M.A., Bera, M., Spicer, R.A., Spicer, T.E.V., Bera, S., 2019. Palaeoclimatic estimates for a latest Miocene-Pliocene flora from the Siwalik Group of Bhutan: Evidence for the development of the South Asian

Monsoon in the eastern Himalaya. *Palaeogeogra. Palaeoclimatol. Palaeoecol* 514, 326–335.

Kocourkova J., 2000. Lichenicolous fungi of the Czech Republic (The first commented checklist). *Acta Musei Nationalis Pragae, Series B, Historia Naturalis* 55, 59-169.

LePage B.A., Currah R.S., Stockey R.A., 1994. The fossil fungi of the Princeton chert. *Int J Plant Sci* 155:828–836.

Mandal, A, Samajpati, N., Bera, S., 2009. In situ occurrence of epiphyllous fungus *Phomites* Fritel from the lower Siwalik sediments of Darjeeling foot hills. *J. Bot. Soc. Bengal* 63, 37–40.

Mandal A., Samajpati N., Bera S., 2011. A new species of *Meliolinites* (fossil Meliolales) from the Neogene sediments of sub-Himalayan West Bengal, India. *Nova Hedwigia* 92(3-4), 435-440.

Mitra S., Banerjee M., 2000. On the occurrence of epiphyllous Deuteromycetous fossil fungi *Palaeocercospora siwalikensis* gen. et. sp. nov. and *Palaeocolletotrichum graminioides* gen. et. sp. nov. from Neogene sediments of Darjeeling foot hills, Eastern Himalaya. *J. Mycopathol. Res.* 37 (2), 7-11.

Mitra S., Bera S., Banerjee M., 2002. On a new epiphyllous fungus *Palaeoasterina siwalika* gen. et. sp. nov. from the Siwalik (Middle Miocene) sediments of Darjeeling foothills, India with remarks on environment. *Phytomorphology* 52 (4), 285-292.

Phadtare N.R., 1989. Palaeoecologic significance of some fungi from the Miocene of Tanakpur (U.P.) India. *Rev. Palaeobot. Palynol.* 59 (1–4), 127–131.

Pharamat T., 2009. Bioactive compounds-producing endophytic fungi from kra thum na *Mitragyna javanica* Koord and Val. Leaves. Ph. D. Thesis, Chulalongkorn University.

Phipps C.J., Rember W.C., 2004. Epiphyllous fungi from the Miocene of Clarkia, Idaho: reproductive structures. *Rev. Palaeobot. Palynol.* 129, 67-79.

Anand Prakesh, Singh, T. 2000. Nature, composition, rank (maturation) and depositional environment of Siwalik coals from Arunachal Himalaya. *Himalayan Geol.* 21 (1&2), 17–29.

Price, P.W. 1977. General concepts on the evolutionary biology of parasites. *Evolution* 31: 405-420.

Rai M.K., Tiwari V.V., Irinyi L., Kövics G.J., 2014. Advances in taxonomy of genus *Phoma*: Polyphyletic nature and role of phenotypic traits and molecular systematics. *Indian J Microbiol* 54, 123–128.

Ranga Rao, A., Khan, K.N., Venkatachala B.S., Sastri V.V., 1981. Neogene/Quaternary boundary and the Siwalik, in: Sastri, et al. (Eds.), Proc. Field Conference on Neogene-Quaternary boundary, India (1979).

Shi G.L., Zhou Z.Y., Xie Z.M., 2010. A new *Cephalotaxus* and associated epiphyllous fungi from the Oligocene of Guangxi, South China. *Rev. Palaeobot. Palynol.* 161 (3–4), 179–195.

Singh, T., 2007. Geology of Itanagar capital complex, Arunachal Himalaya, with special reference to neotectonics. *J. Geol. Soc. India* 70, 339–352.

Singh A.N., Borah T.R., Sharma G.S., Borah N.J., 2004. Nursery Diseases of *Dipterocarpus retusus. The Indian Forest* 130, 742-748.

Singhai L.C., 1974. Fossil fungi from the Deccan Intertrappean Beds of Madhya Pradesh, India. *J. Biol. Sci.* 17, 92-102.

Sullivan R.F., White Jr. J.F., 2000. *Phoma glomerata* as a mycoparasite of powdery mildew. *Appl. Environ. Microbiol.* 66, 425-427.

Taral, S., Kar, N., Chakraborty, T., 2017. Wave-generated structures in the Siwalik rocks of Tista valley, eastern Himalaya: implication for regional paleogeography. Curr. Sci. 113.

Taylor T.N., Osborn J.M., 1996. The importance of fungi in shaping the paleoecosystem. *Rev. Palaeobot. Palynol.* 90 (3–4), 249–262.

Thaung M.M., 2008. Biodiversity survey of Coelomycetes in Burma. *Austra. Mycol. Soc.* 27 (2).

Vishnu (née Mandal) A., Khan M.A., Bera M., Dilcher D.L., Bera S., 2017. Fossil Asterinaceae in the phyllosphere of the eastern Himalayan Neogene Siwalik forest and their palaeoecological significance. *Bot. J. Linean Soc.* 185(2), 147–167.

Vishnu, A., Khan, M.A., Bera, M., Achrya, K., Dilcher, D.L., Bera, S., 2019. Occurrence of Phoma Sacc. in the phyllosphere of Neogene Siwalik forest of Arunachal sub-Himalaya and its palaeoecological implications. Fungal Biology 123, 18-28.

Watanabe K., Nishida H., Kobayashi T., 1999. Cretaceous deuteromycetes on a cycadeoidalean bisexual cone. *Int. J. Plant Sci.* 160 (2), 435–443.

Zhang K.E., Shivas R.G., Cai L., 2015. Synopsis of *Phyllosticta* in China. *Mycology* 6 (1), 50–75.

Zhang T., Wang N.F., Zhang Y.Q., Liu H.Y., Yu L.Y., 2016. Diversity and distribution of aquatic fungal communities in the Ny-Ålesund Region, Svalbard (high Arctic): aquatic fungi in the Arctic. *Microb. Ecol.* 71, 543–554.

Index

M. Rai et al. (eds.), *Phoma: Diversity, Taxonomy, Bioactivities, and Nanotechnology*,
https://doi.org/10.1007/978-3-030-81218-8

MIX
Papier aus verantwortungsvollen Quellen
Paper from responsible sources
FSC® C105338

If you have any concerns about our products,
you can contact us on
ProductSafety@springernature.com

In case Publisher is established outside the EU,
the EU authorized representative is:
Springer Nature Customer Service Center GmbH
Europaplatz 3, 69115 Heidelberg, Germany

Printed by Libri Plureos GmbH
in Hamburg, Germany